Scientific Seminar on Antarctic Ocean Variability and its Influence on Marine Living Resources, Particularly Krill
Paris, 2–6 June 1987

Sponsors

Intergovernmental Oceanographic Commission (IOC)

Commission for the Conservation of
Antarctic Marine Living Resources (CCAMLR)

Supported by

Scientific Committee on Antarctic Research (SCAR)
Scientific Committee on Oceanic Research (SCOR)

Organizer

Steering Group for the Seminar

Dietrich Sahrhage (Ed.)

Antarctic Ocean and Resources Variability

With 232 Figures

Springer-Verlag
Berlin Heidelberg New York
London Paris Tokyo

Professor Dr. Dietrich Sahrhage
Institut für Seefischerei der
Bundesforschungsanstalt für Fischerei
Palmaille 9
2000 Hamburg 50, FRG

ISBN-13: 978-3-642-73726-8 e-ISBN-13: 978-3-642-73724-4
DOI: 10.1007/978-3-642-73724-4

Library of Congress Cataloging-in-Publication Data. Scientific Seminar on Antarctic Ocean Variability and its Influence on Marine Living Resources, Particularly Krill (1987 : Paris, France) Antarctic Ocean and resources variability / Scientific Seminar on Antarctic Ocean Variability and its Influence on Marine Living Resources, Particularly Krill, Paris, 2–6 June 1987 ; D. Sahrhage (ed.). p. cm. "Sponsors: Commission for the Conservation of Antarctic Marine Living Resources (CCAMLR), Intergovernmental Oceanographic Commission (IOC). Includes Index. ISBN-13: 978-3-642-73726-8 1. Marine ecology–Antarctic Ocean–Congresses. 2. Marine resources–Antarctic Ocean–Congresses. 3. Ocean circulation–Antarctic Ocean–Congresses. 4. Krill–Antarctic Ocean–Congresses. I. Sahrhage, D. II. Commission for the Conservation of Antarctic Marine Living Resources. III. Intergovernmental Oceanographic Commission. IV. Title. QH95.58.S35 1987. 574.5'2636'09167–dc 19 88-23481

Softcover reprint of the hardcover 1st edition 1988

2131/3130-543210 – Printed on acid-free paper

Preface

During the intensification of research activities in the Southern Ocean in recent years, both at national levels and through international cooperation in such projects as BIOMASS with its FIBEX and SIBEX phases, the need was felt increasingly for closer collaboration between biologists, meteorologists, and oceanographers in the study of the interaction between the atmospheric forces, the water masses, and the living resources. Better knowledge in this regard is not only of scientific interest but also of practical importance, especially for the management of the resources and the protection of the Antarctic marine ecosystem.

As a follow-up on a recommendation by the IOC Program Group for the Southern Oceans made in March 1983, the Intergovernmental Oceanographic Commission organized a meeting of experts on oceanography related to the dynamics of the Antarctic ecosystems. During this meeting, held in Kiel, Federal Republic of Germany, in May 1984, biologists and oceanographers involved in BIOMASS activities met with the oceanographers of SCOR Working Group 74 to discuss ways and means for additional physical and chemical observations in the oceanographic research within BIOMASS. It was the time when large fluctuations in the distribution of krill with subsequent detrimental effects on predator species dependent on krill had just been observed, and the question arose whether this was possibly the result of changes in the Antarctic water circulation. The need for further interdisciplinary research in this field was identified, and out of this the idea developed to hold a Scientific Seminar on Antarctic Ocean Variability and its Influence on Marine Living Resources, particularly Krill.

The Commission for the Conservation of Antarctic Marine Living Resources (CCAMLR), especially its Scientific Committee, has much interest in studies on this subject, particularly at the stage when its ecosystem monitoring program is being developed. Consequently, CCAMLR and IOC agreed to sponsor jointly the Seminar and also to provide financial support for publishing the Seminar contributions in this book. Support for the Seminar was also obtained from SCAR and SCOR.

The Seminar was held in Paris from 2 to 6 June 1987. About 80 scientists from 18 countries and several international organizations participated. The Seminar provided a forum for the presentation of scientific contributions and for the discussion of results on the following major aspects:

- What is known on the variability in the Antarctic Ocean circulation system?
- What are the possible causes for such variations?
- To what extent and how does this ocean variability influence primary productivity, zooplankton and other organisms?
- To what extent and how does this ocean variability influence krill distribution and abundance?
- What are the effects on krill predators (seals, penguins, fishes, etc.)?

The Seminar was prepared and carried out with the assistance of an international Steering Group with nine prominent scientists as its members. Contributions to the

Seminar, presented in this book, were accepted by the Steering Group and were reviewed by a considerable number of specialists in the various fields.

The editor, who acted also as convenor of the Seminar, wishes to thank all colleagues, members of the Steering Group, session chairmen and rapporteurs, participants in the meeting, reviewers of the manuscripts, and especially the authors of the contributions, for their work.

Hamburg, September 1988 D. Sahrhage

Contents

Part 1 Meso/Large-Scale Variability in the Environment

Part 2 Meso/Large-Scale Variability in the Biota (Related to the Environment)

Part 4 Krill Variability Detected from Predator Studies

List of Contributors

You will find the adresses at the beginning of the respective contribution

Part 1
Meso/Large-Scale Variability in the Environment

A Survey of the Atmospheric Elements at the Ocean's Surface South of 40°S

H. VAN LOON and D.J. SHEA[1]

Summary. This chapter surveys on broad scales in space and time the climatic elements at the surface south of the subtropical ridge. It demonstrates the zonal asymmetry in the temperature distribution, and discusses the implication of the latitudinal changes in the annual temperature range for the interseasonal change in latitudinal temperature gradients. Examples are shown of tracks of cyclones and anticyclones and of the interseasonal changes in position and intensity of the circumpolar trough of low pressure. We stress the importance of the semiannual wave in wind and pressure over the area, and discuss also the interannual variability of the wind in February and August.

1 Introduction

The following description of atmospheric elements covers the ocean between the subtropical ridge and the Antarctic coast. There are few permanent meteorological stations in this area and, apart from the pressure observations, none of the elements observed on these island and coastal stations is representative of the conditions over the open sea. Nearly all of the major shipping lanes lie north of 40°S, with the result that the only area at higher latitudes which is comparatively well covered with ships' observations, but only in summer, is in the latitudes south of 50°S where the whaling fleets operate (Venter 1957). It is therefore limited what one can say about the atmospheric elements and their interannual variability south of 40°S, and the emphasis in the following is necessarily on the most representative of them: sea level pressure and the derived, geostrophic wind.

2 Air Temperature

The temperature south of 40°S decreases in both summer and winter as one goes south (Fig. 1A,B, pp. 4 and 5), but the decrease is not a steady one (van Loon 1966). In the Atlantic and Indian Oceans the temperature difference between 40° and 50°S in summer, where it is largest, is four times greater than the difference between 55° and 65°S, where it is smallest. In the Pacific Ocean there are no such large discontinuities in gradient. The meridional gradients of temperature in middle latitudes are approximately equal in summer and winter, but the gradients steepen over the expanded pack ice in winter, eliminating the weak meridional temperature contrasts found over the open water in this zone in summer.

At its widest, the almost isothermal area spans 20° latitude in summer (50°S to 70°S). Its effect on the frequency distribution of temperature can be seen in Fig. 2, p. 6). The mean temperature over this extensive region, observed on whaling ships, is barely 0.8°C and most of the values lie between −1 and +3°C. Temperatures below freezing in summer occur in almost a third of the observations.

Figure 3A and B (p. 6) show lines of equal deviation of the air temperature from the mean of the latitude. The pattern is similar in February and August; over the Atlantic and Indian Oceans the temperature is below the mean of the latitude, whereas it is above the mean over the South Pacific Ocean. The amplitude of the pattern is somewhat bigger in August than in February with the latitude anomalies exceeding +4°C over the open water in the eastern Pacific.

Table 1 (p. 6) shows the extreme minima observed on three islands. Even as far north as Marion Island one must expect temperatures below freezing in summer. In this season the small area of the island is not likely to lower the minimum much. Negative temperatures occur in all months on the islands in Table 1, except for Macquarie Island in January. Although South Georgia is at the same latitude as Macquarie, its position in a cold current and in winter near the pack ice (in extreme instances within the pack) makes for appreciably lower mean and extreme temperatures.

The annual cycle of the air temperature in our area of interest can be approximated by the first harmonic in a 12-months series. The amplitude of the harmonic (half the range) in Fig. 4A (p. 7) is large in the subtropics and decreases to a minimum at middle latitudes, after which it rises toward Antarctica. Over all of this area the first harmonic dominates the annual cycle, explaining above 90% of

1 National Center for Atmospheric Research, P.O. Box 3000, Boulder, Colorado 80307-3000, USA

Antarctic Ocean and Resources Variability
(ed. by D. Sahrhage)

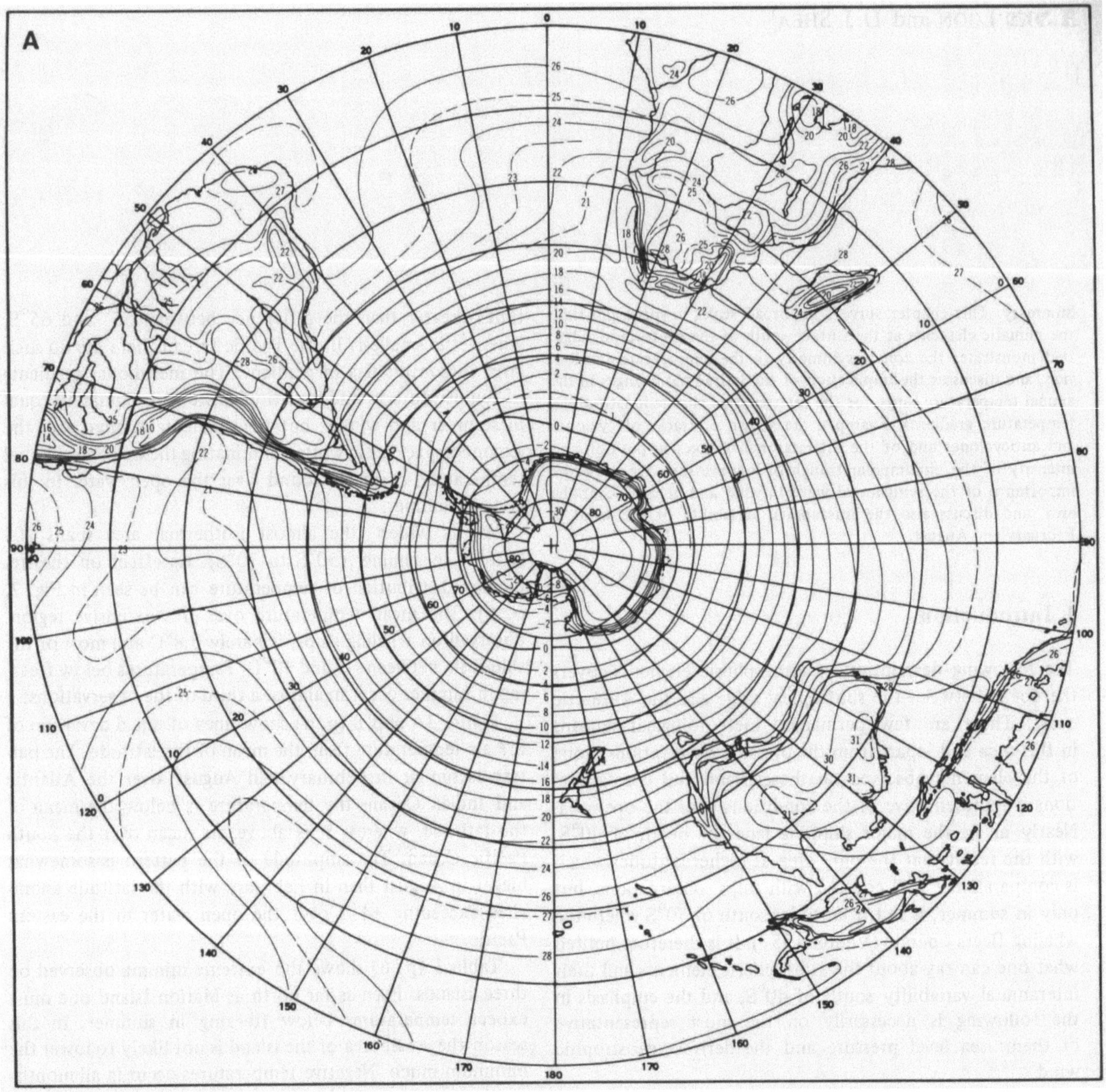

Fig. 1. A Mean air temperature at the surface in January (°C). (van Loon 1972a). **B** Mean air temperature at the surface in July (°C). (van Loon 1972a)

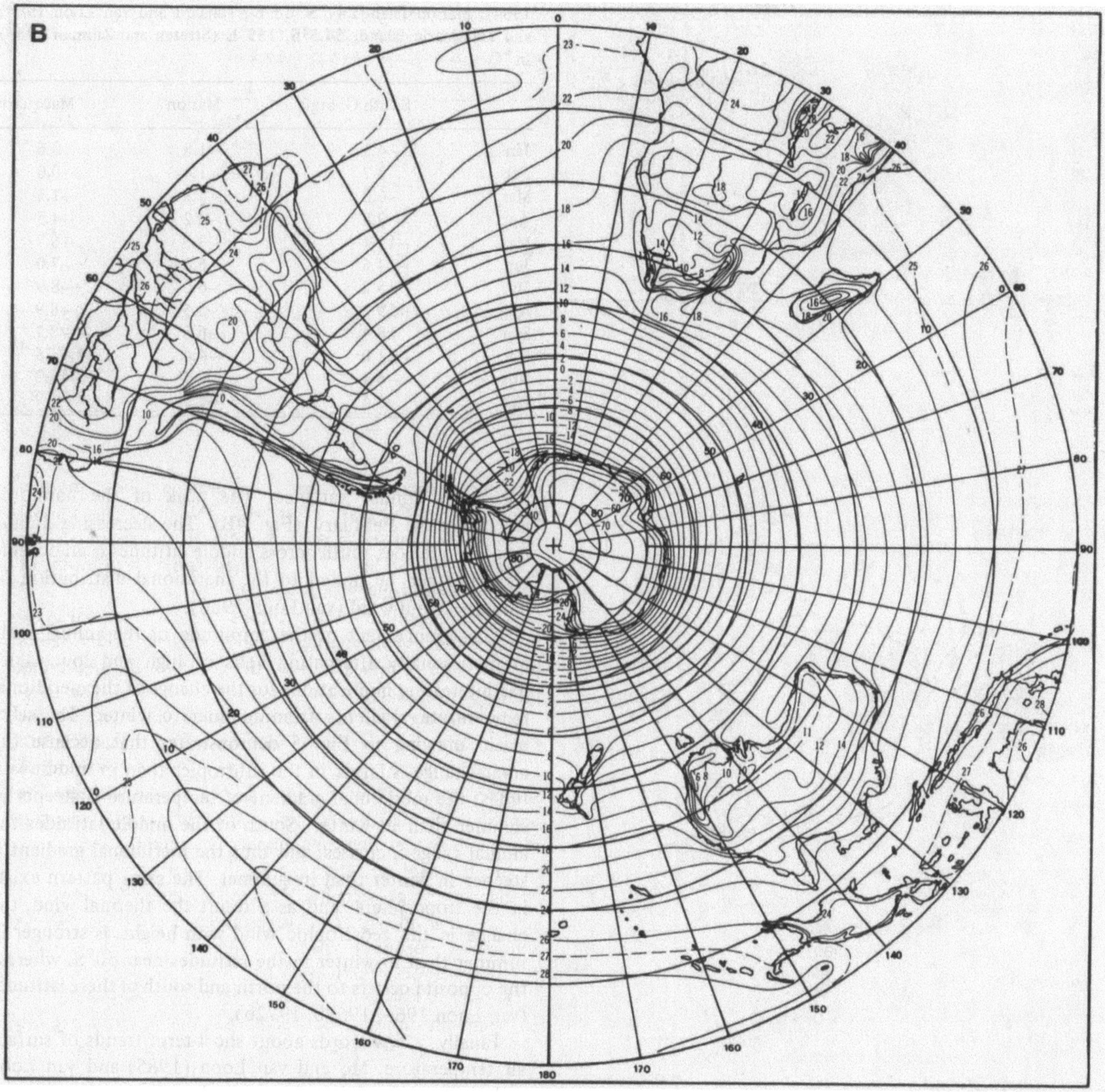

Fig. 1B

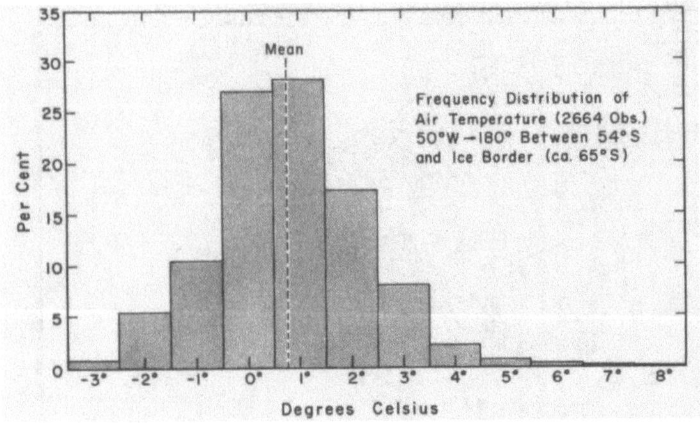

Table 1. Extreme minima at South Georgia, 54°S, 36.5°W (Höflich 1984); Marion Island, 47°S, 38°E (Taljaard and van Loon 1984); and Macquarie Island, 54.5°S, 159°E (Streten and Zillman 1984). In °C

	South Georgia	Marion	Macquarie
Jan	−4.1	−1.8	0.6
Feb	−3.7	−1.4	−0.6
Mar	−6.3	−0.8	−1.1
Apr	−9.8	−2.2	−4.5
May	−11.4	−3.0	−6.7
Jun	−14.6	−6.0	−7.0
Jly	−15.2	−6.0	−8.9
Aug	−19.2	−5.5	−8.9
Sep	−18.4	−6.8	−8.7
Oct	−11.0	−4.4	−4.6
Nov	−6.4	−2.5	−3.3
Dec	−5.4	−1.5	−1.7

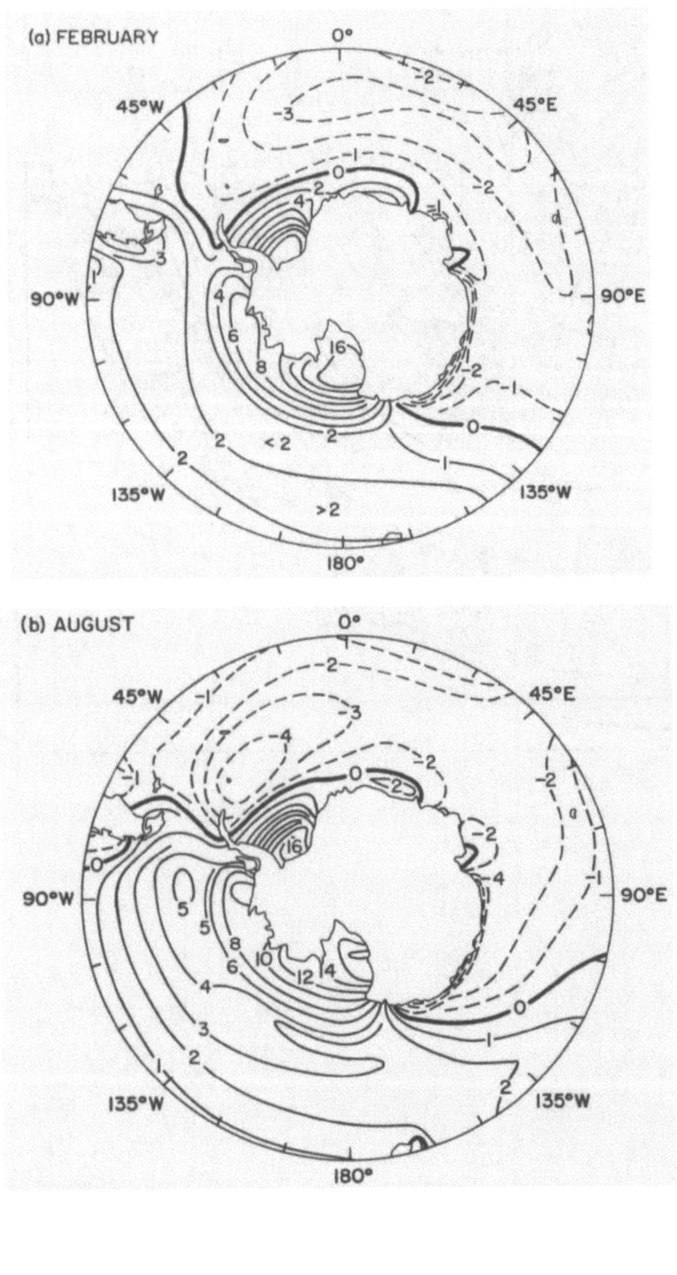

the mean annual variance. The peak of the harmonic is reached in February (Fig. 4B). The decreasing annual range as one goes south across middle latitudes is an oceanic trait and can be linked to the meridional distribution of cloudiness and wind (van Loon 1966).

The arrangement of the amplitude of the annual cycle in meridionally alternating belts of high and low values has interesting implications for the change of the meridional temperature gradients from summer to winter. The schematic drawing in Fig. 5 demonstrates that because the annual range is larger in the subtropics than in middle latitudes, the meridional gradient of temperature is steeper in summer than in winter. South of the middle latitudes the annual range increases, and thus the meridional gradient is steeper in winter than in summer. The same pattern exists in the troposphere, and as a result the thermal wind, the change in the geostrophic wind with height, is stronger in summer than in winter in the latitudes near 50°S, whereas the opposite occurs to the north and south of these latitudes (van Loon 1966, 1967b, 1972b).

Finally, a few words about short-term trends of surface air temperature. Mo and van Loon (1985) and van Loon and Williams (1977) have shown that as a rule there is no uniform trend on the hemisphere and that the trend at a given place often changes sign from one period to another. These circumstances are illustrated by Fig. 6 which shows lines of equal linear trend for an arbitrary period. The temperature trends are on the scale of the long waves, which is also observed on the Northern Hemisphere (van Loon and Williams 1976). Figure 6 shows the danger of using a few stations in one part of the hemisphere to derive the trend at other places.

Fig. 3. a Deviation of the air temperature (°C) in February from the mean of the latitude. **b** Deviation of the air temperature (°C) in August from the mean of the latitude

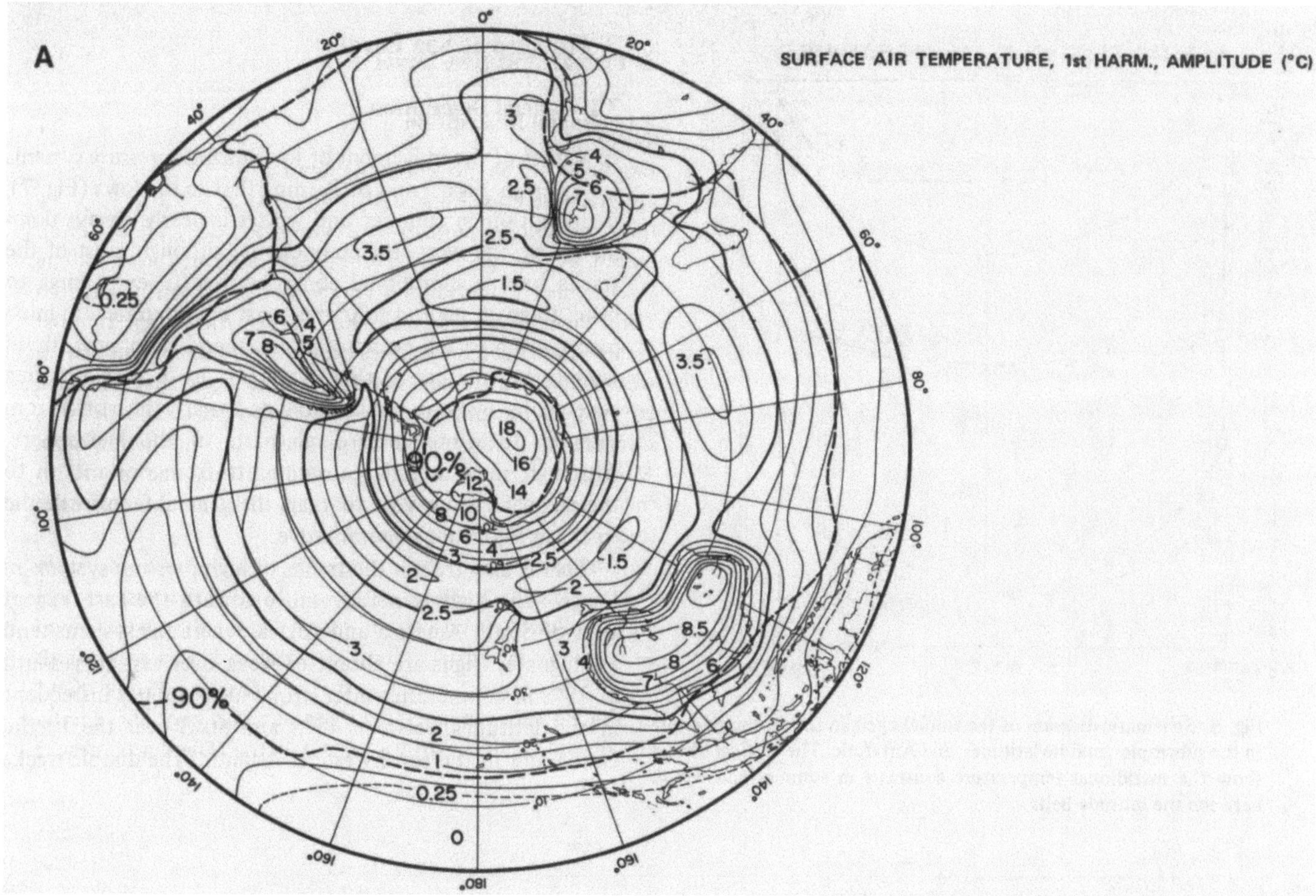

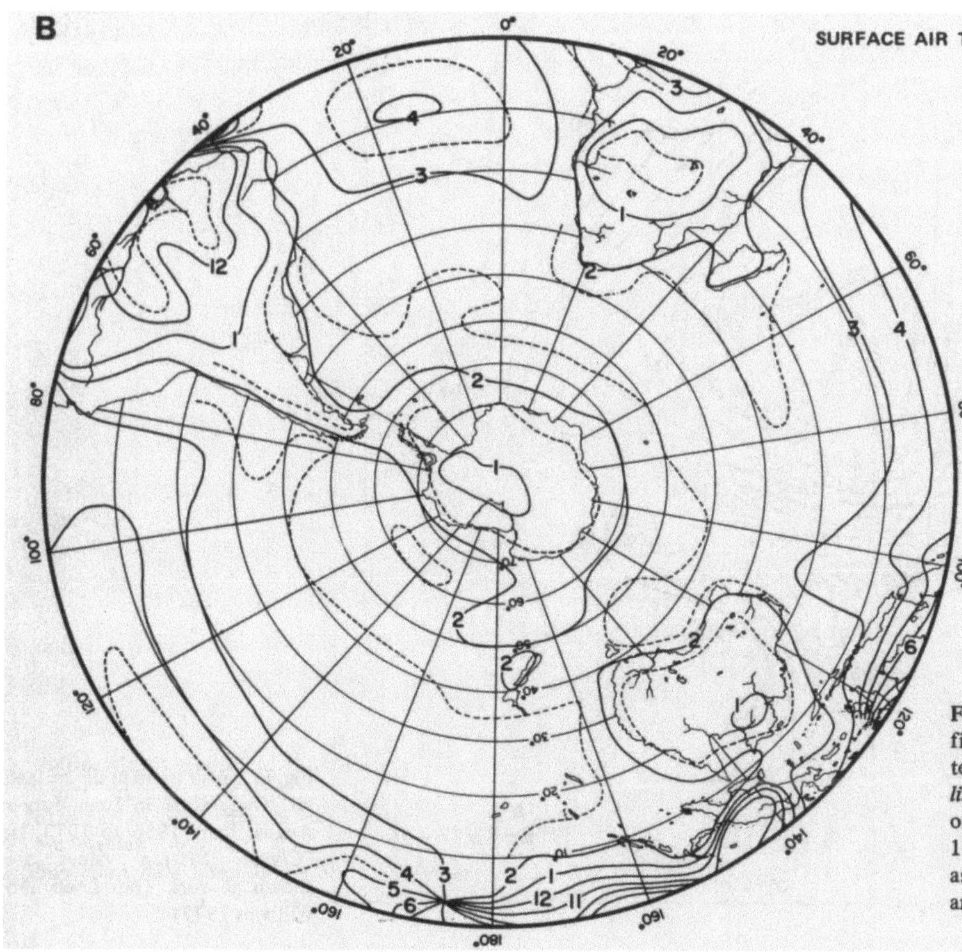

Fig. 4. A Mean amplitude (half range) of the first harmonic in a 12-month series of air temperature (°C). *Between the two dashed lines* the harmonic explains more than 90% of the total annual variance. (van Loon 1972a). B Phase of the first harmonic given as the month of the peak (e.g., 2 = 1 February) of the harmonic. (van Loon 1972a)

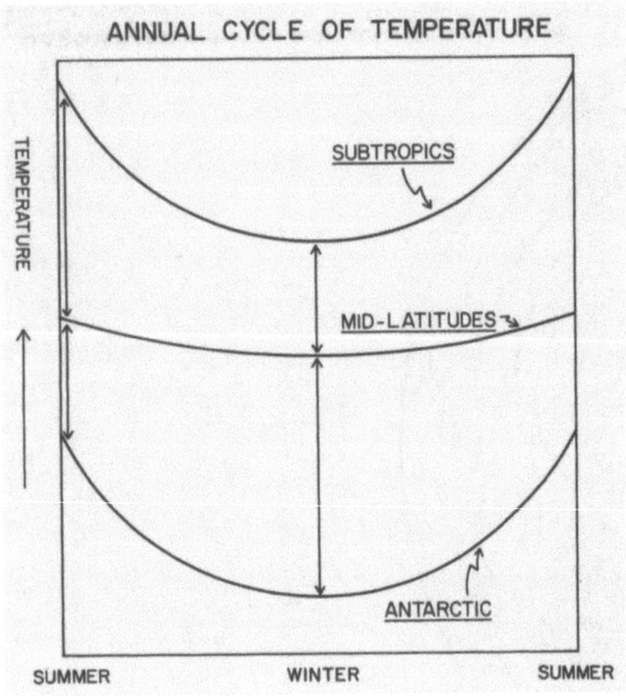

Fig. 5. Schematic diagram of the annual cycle in the air temperature in the subtropics, middle latitudes, and Antarctic. The *vertical arrows* show the meridional temperature contrasts in summer and winter between the latitude belts

3 Pressure at Sea Level

3.1 General Description

Examples of the movement of low and high pressure systems are given in Figs. 7 and 8. Turning first to the lows (Fig. 7), their motion in summer and winter is nearly always dominated by the west-east component, although most of the tracks have a meridional component sufficiently large to bring them to an end near the coast of Antarctica. In most places seven to ten cyclones cross a meridian in a month; in winter the number is slightly larger. Note that the area covered by cyclone tracks expands toward the north from summer to winter over large parts of the hemisphere. Although there is a large variability from one month to the next and from year to year, the general features of the tracks in Fig. 7 are representative.

This is also true of the tracks of high pressure systems in Fig. 8. The highs generally move toward the east, except west of South America and Africa, where the systems tend to stagnate. Highs are absent or weak over vast areas south of 45°S in summer. In winter strong highs are not infrequent in the latitudes between 45°S and 60°S over the Pacific Ocean and in the southwestern Atlantic. The double tracks

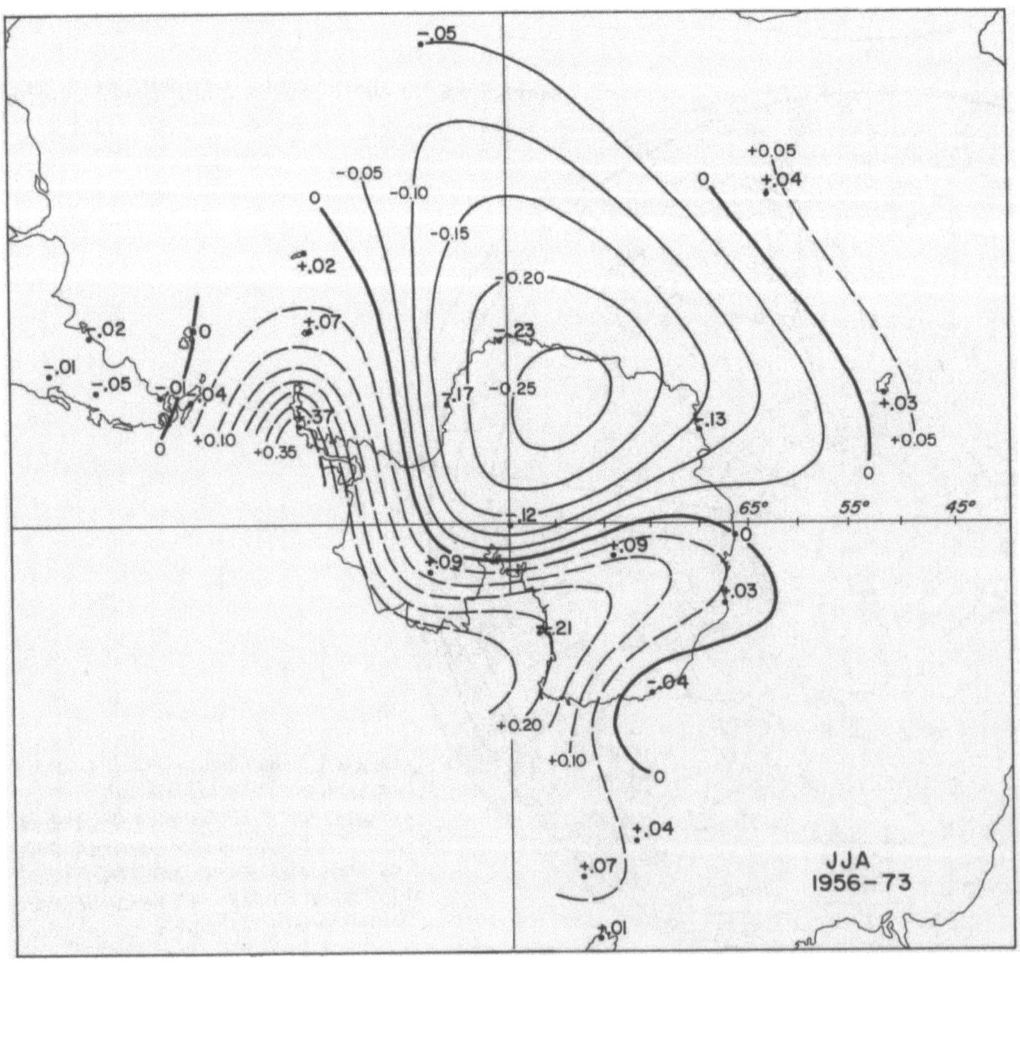

Fig. 6. Linear trend in the surface air temperature in June–July–August from 1956 to 1973. In °C/year. Station values are shown at *dots.* (van Loon and Williams 1977)

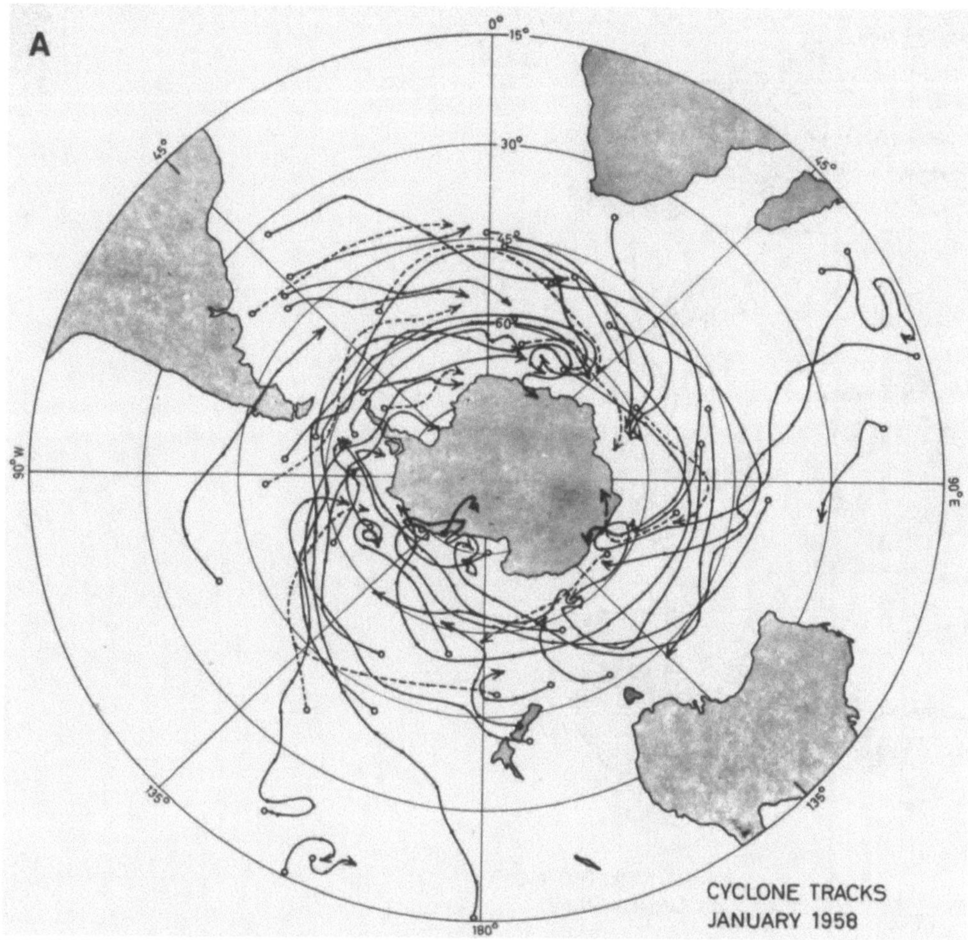

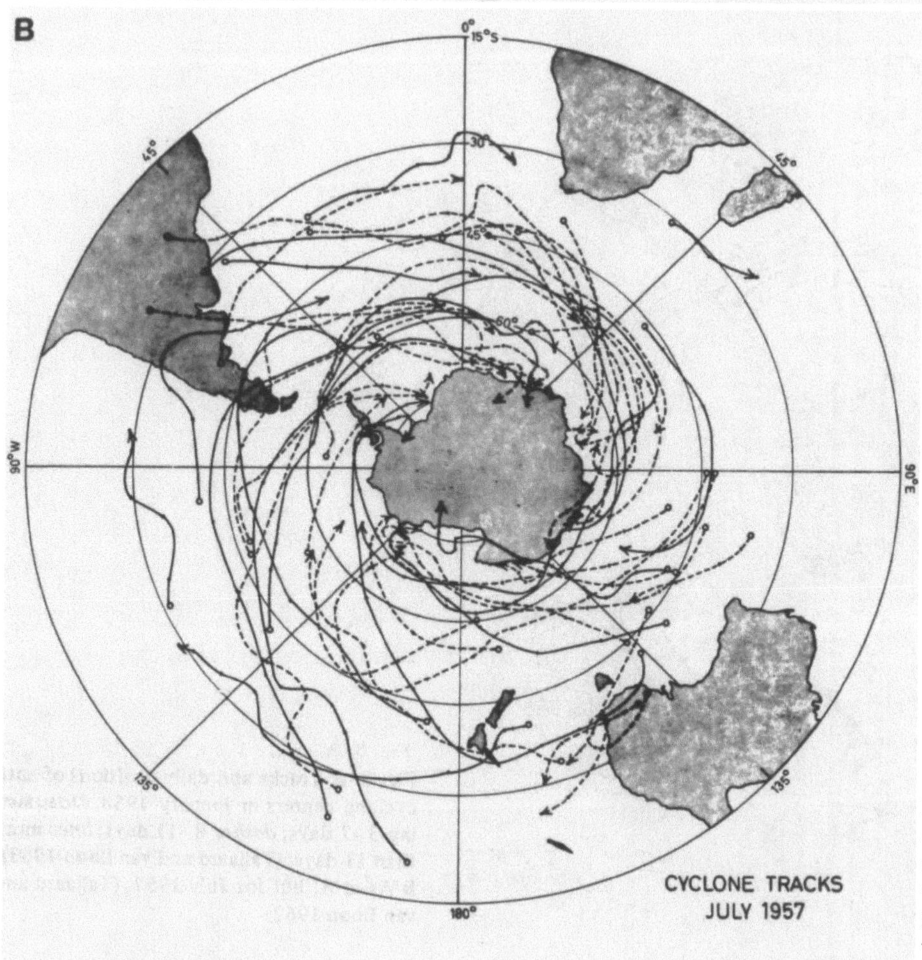

Fig. 7. A Tracks and daily positions of cyclone centers in January 1958. ***Dots*** lasting 4–7 days; ***dashes*** 8–11 days; ***lines*** more than 11 days. (Taljaard and van Loon 1963). **B** As in **A**, but for July 1957. (Taljaard and van Loon 1962)

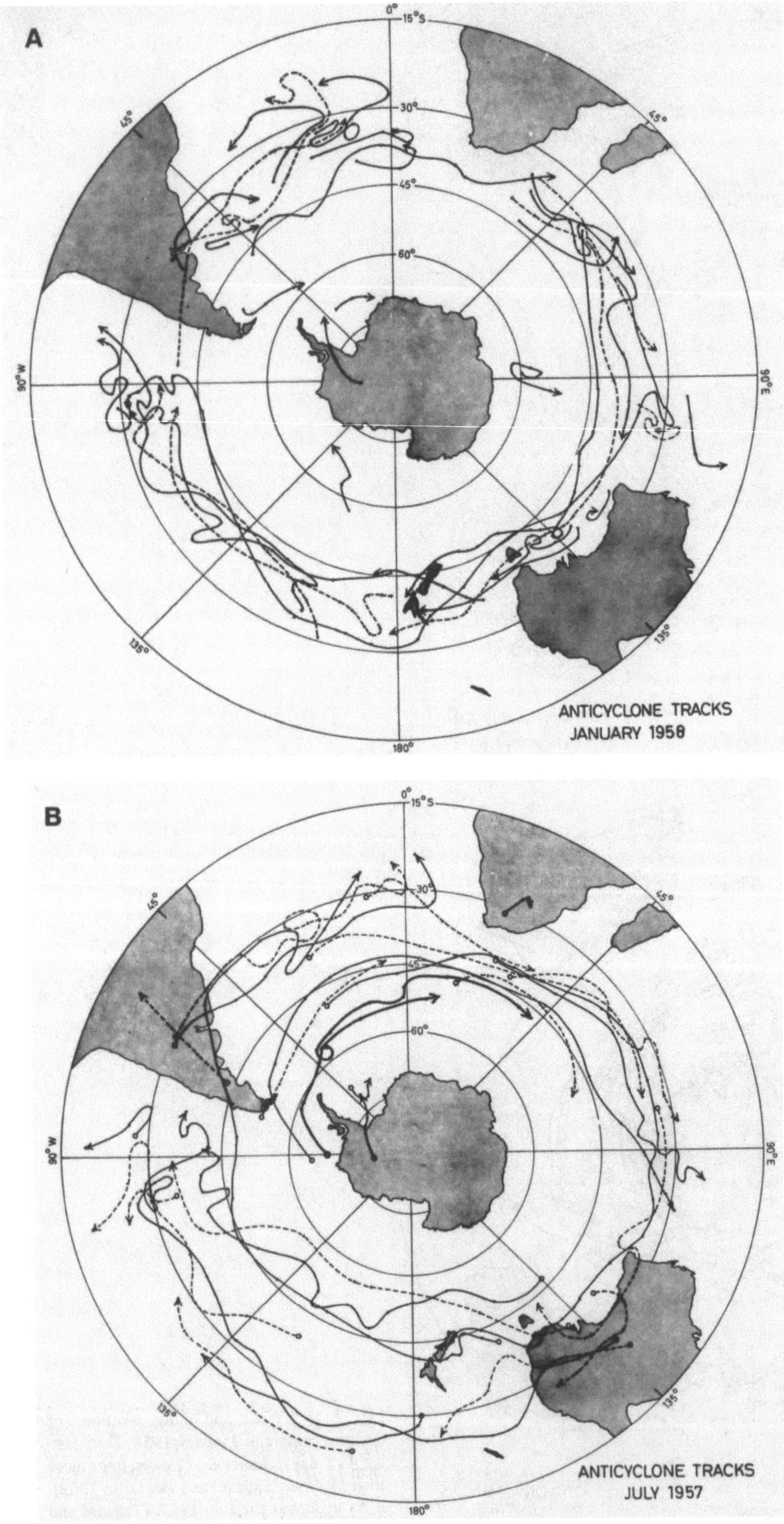

Fig. 8. A Tracks and daily positions of anticyclone centers in January 1958. *Dots* lasting 3–7 days; *dashes* 8–11 days; *lines* more than 11 days. (Taljaard and van Loon 1963). **B** As in **A**, but for July 1957. (Taljaard and van Loon 1962)

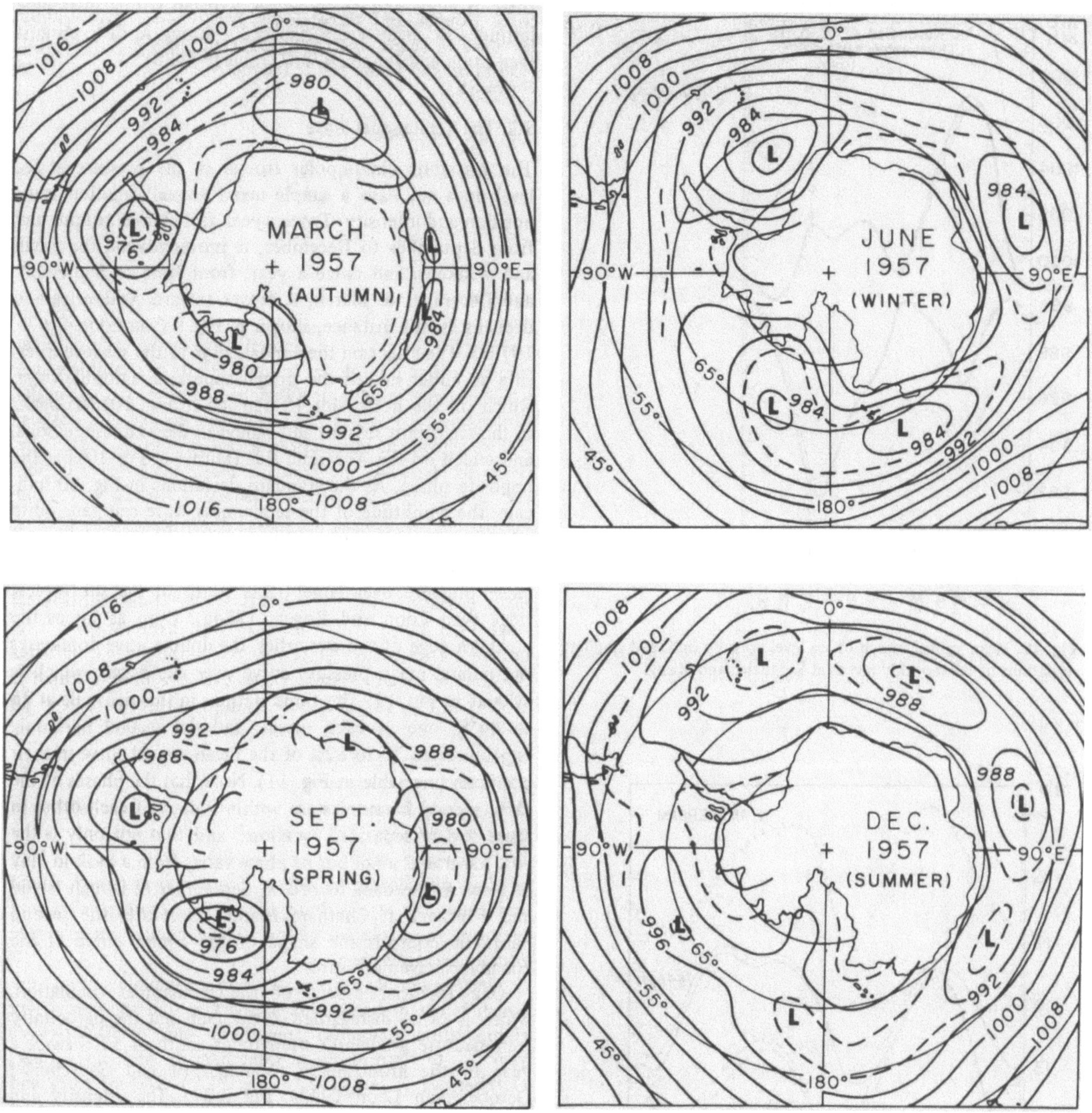

Fig. 9. Mean pressure at sea level (mb) in March, June, September, and December 1957. (van Loon 1967a)

in these two places reflect the tendency for highs to block the westerly flow in these areas (van Loon 1956; Trenberth and Mo 1985; Trenberth 1986; Lejenäs 1984).

Although there is little land between 40°S and the Antarctic coast, there are nevertheless notable deviations of sea level mean pressure from the average of the latitude; that is, there are quasi-stationary waves. Just as in the case of the temperature, the mean pressure at higher latitudes is higher over the Pacific than over the other oceans. This asymmetry is so large that it makes sense to deal with it as the first zonal harmonic wave in the longitudinal pressure distribution. Between 45 and 60°S the amplitude of the first harmonic wave at the 500 mb level (about 5.5 km) is between 60 and 100 m. The phase, which varies little, is such that the crest of the wave lies between 120 and 150°W and the valley on the other side of the hemisphere, 180° away from the crest.

There are other quasi-stationary waves over the ocean in these latitudes than wave number one. On the average they are smaller than the first wave. The third wave, e.g., with

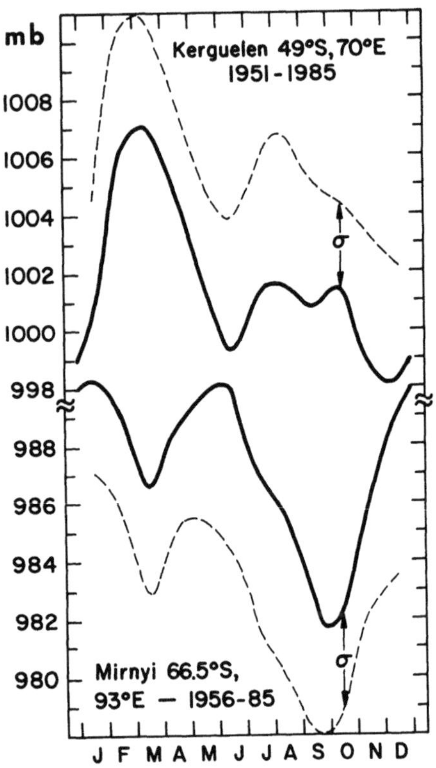

Fig. 10. Mean annual march of sea level pressure (mb) and standard deviations of the monthly means at Kerguelen and Mirnyi

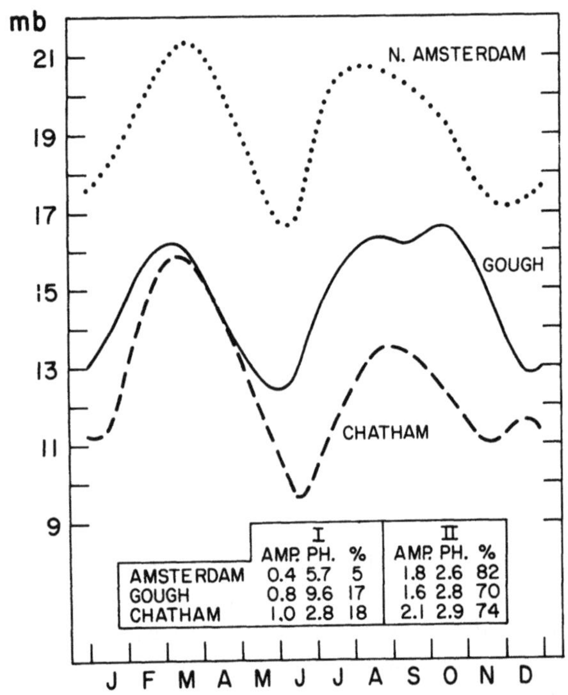

	I			II		
	AMP.	PH.	%	AMP.	PH.	%
AMSTERDAM	0.4	5.7	5	1.8	2.6	82
GOUGH	0.8	9.6	17	1.6	2.8	70
CHATHAM	1.0	2.8	18	2.1	2.9	74

Fig. 11. Mean annual march of sea level pressure (mb) at New Amsterdam, Gough, and Chatham Islands. The table shows the amplitudes (mb) of the first and second harmonics in the mean annual curves; the phase is given as the date of the peak (5.7 is 21 May, etc.); and the percentage is the share of the mean annual variance accounted for by the harmonics

three troughs and three ridges, is often strong in middle latitudes in single years, especially during April to August (van Loon and Jenne 1972; Trenberth 1980).

3.2 The Semiannual Wave

The Antarctic circumpolar trough in the pressure at sea level does not have a simple mean annual variation in its position and intensity. Twice a year, from March to June and from September to December, it moves toward the north and weakens; and twice a year, from June to September and December to March, it moves toward Antarctica and deepens as, for instance, shown in Fig. 9 (van Loon 1967a, 1972c). The effect on the annual curve of the sea level pressure on either side of the trough appears as a double wave, which on the north side (Kerguelen in Fig. 10) has peaks in the transition seasons and valleys in the extreme seasons, and which on the Antarctic side (Mirnyi in Fig. 10) has the opposite phase. As the standard deviations in Fig. 10 indicate, the amplitude of the semiannual wave can vary quite a lot from one year to another; but the phase is stable and the semiannual wave therefore dominates the long term *mean* pressure over large tracts south of the subtropical ridge (van Loon and Rogers 1984a). Even as far as the northern edge of the westerlies, the double wave dominates the annual mean pressure curve *over the ocean,* which is evident in Fig. 11. The three stations in this figure lie at 38 to 44°S, one in each ocean; and the second harmonic explains from 70 to 82% of the mean annual curve in their positions (see table in Fig. 11). Note that the phases of the three second harmonics are within 9 days of each other in these widely separated locations, and that not only is the first harmonic small but its phase varies from a peak in May at New Amsterdam to one in September at Gough Island and February at Chatham Island. Therefore, the second harmonic controls the shape of the annual curve of the zonally averaged pressure.

One essential aspect of the semiannual oscillation, which is often disregarded, is the fact that the baroclinity (temperature gradients) strengthens south of 50°S *twice* a year in the troposphere (March/April and September/October; van Loon 1972a: Fig. 3.37). The intensity and tracks of the cyclones react accordingly with a semiannual oscillation in these latitudes, as reflected in the sub-Antarctic trough (Fig. 9). The pack ice expands and contracts only *once* a year, and the pressure is lowest in the trough and the trough is closest to Antarctica *both* when the ice reaches its maximum *and* its minimum extent. This observation contradicts the idea that the edge of the ice should be especially conducive to cyclogenesis or to determining the movement of the cyclones.

The comparatively small size of the first harmonics in middle and high southern latitudes which one sees in the long-term mean pressure (van Loon and Rogers 1984b) does not mean that this harmonic cannot reach a substantial amplitude in individual years. We illustrate this by means of

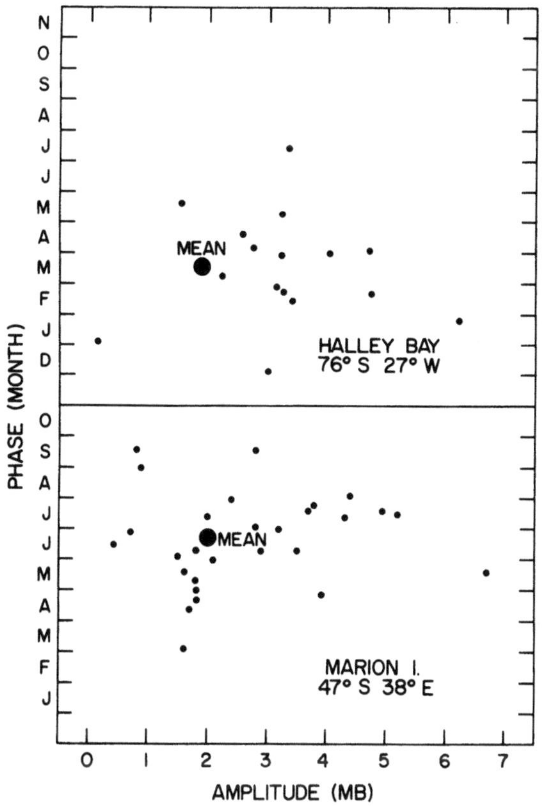

Fig. 12. Amplitudes (mb) of the annual wave in sea level pressure in single years plotted against phase (date of peak) at Halley Bay and Marion Island. The *large dot* is at the amplitude and phase of the long term mean. (van Loon and Rogers 1984b)

Fig. 12, in which the amplitude of the first harmonic in the sea level pressure in single years at an oceanic and an Antarctic station is plotted as a function of phase (date of peak). The large black dot is the first harmonic of the mean pressure for the whole period (28 years at Marion; 16 at Halley Bay). The amplitude in a single year is frequently appreciably larger than the mean, but the phase is extremely variable. The peak can occur in any of 8 months at Halley Bay and of 7 months at Marion Island. The different years thus tend to offset each other and leave the second harmonic with its stable phase to dominate the shape of the mean annual curves (Figs. 10 and 11).

3.3 An Example of a Longer-Period Variation

Besides the variation of the pressure from one season to another, there is of course a large interannual variability (van Loon and Madden 1983; Mo and van Loon 1984, 1985; Trenberth 1981). One phenomenon which influences the pressure distribution irregularly is the Southern Oscillation. Although its influence often is worldwide, its principal domain is the Indian Ocean, Australasia, and the Pacific Ocean. Figure 13A and B shows the average pressure anomalies during two stages in a warm extreme in the Southern Oscillation (van Loon and Shea 1987). That is, the extreme when pressure is high over Australasia and low in the tropical-subtropical South Pacific Ocean, anomalously warm water appears along the equator in the Pacific and on the coast of Peru (El Niño), and heavy rains fall in usually dry areas on the equator and the coast of South America. Both maps show the well-known distribution of above-normal pressure over Australasia and below-normal over the Pacific. The principal difference is in middle and high latitudes. Here the anomalies toward the end of the year (NDJ) show considerable zonal symmetry with above-normal pressure over polar latitudes and below-normal pressure over middle latitudes. This distribution translates into weaker westerlies south of about 45°S over most of the hemisphere. In early winter, MJJ in Fig. 13A, there is no zonal symmetry south of 40°S, where the anomalies are arranged in a pattern of alternating positive and negative sign around most of the circumference, indicating a large south-north component in the mean circulation. In individual warm extremes of the Southern Oscillation there is of course much "noise", which gives the event its specific character. The cold extreme of the Southern Oscillation when the South Pacific high is exceptionally strong also has a recognizable signal at middle and high southern latitudes.

4 Wind

4.1 Intra-Annual Variations

The strength of the wind is proportional to the gradient of pressure, and in the following we therefore discuss the variations of the wind in terms of the pressure gradients in Figs. 14–16. The schematic diagram in Fig. 14 shows the second harmonics in the sea level pressure described in Section 3.1. In Antarctica the peaks are in the extreme seasons; the phase changes abruptly near 60°S, and farther north the peaks are in the transitional seasons. Obviously the annual curve of the pressure gradients between the latitudes where the harmonic is largest in either phase (50°S for the equinox peaks and 65°S for the solstice peaks, see Fig. 10) must contain a second harmonic with peaks in the transitional seasons and valleys in the extreme seasons, as indicated by the dashed lines between 50 and 65°S in Fig. 14. Because the amplitude of the second harmonic decreases north of 50°S while the phase stays the same (Fig. 11) and the mean pressure increases north of 60°S, the annual march of the pressure gradients between 50°S and the latitudes to the north contains a second harmonic with peaks in the extreme seasons and valleys in the transition seasons. This phase distribution is shown by the dashed lines between 40 and 50°S in Fig. 14.

Since the second harmonic in the sea level mean pressure is large and the phase of the first harmonic is quite variable south of the subtropical ridge, the second harmonic in the sea level pressure *differences* between the three belts schematically depicted in Fig. 14 controls the shape of the annual mean curve. We illustrate the difference between

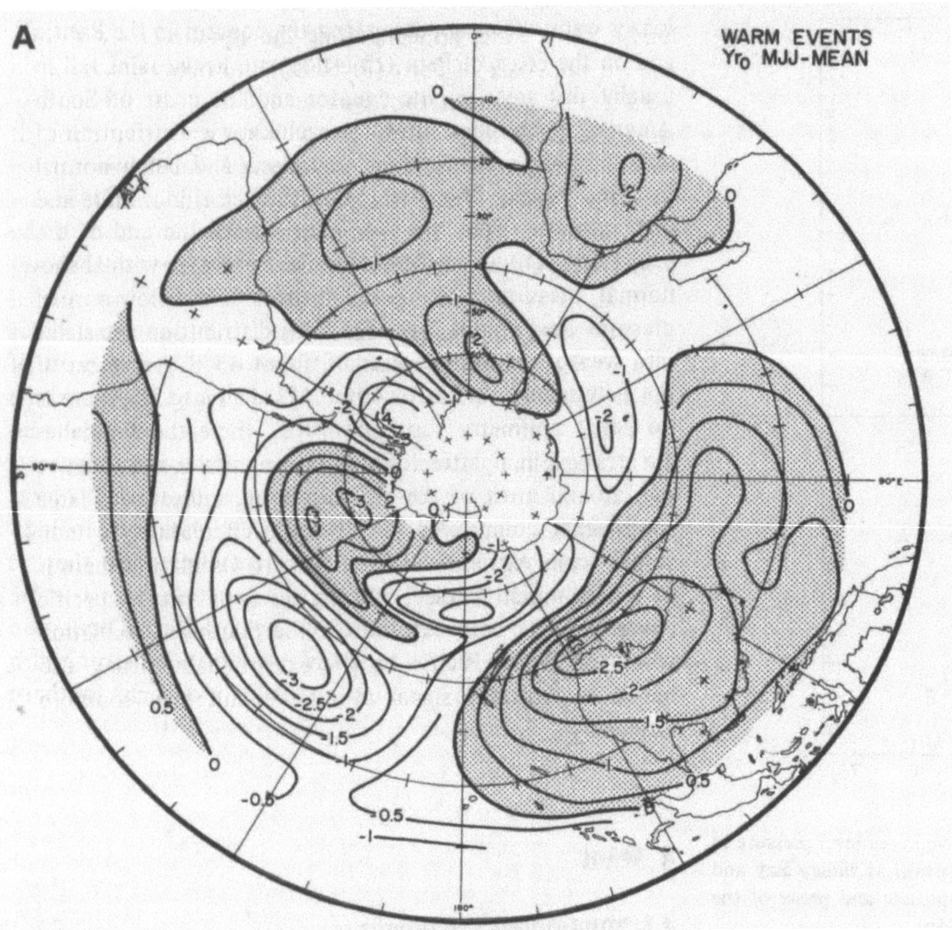

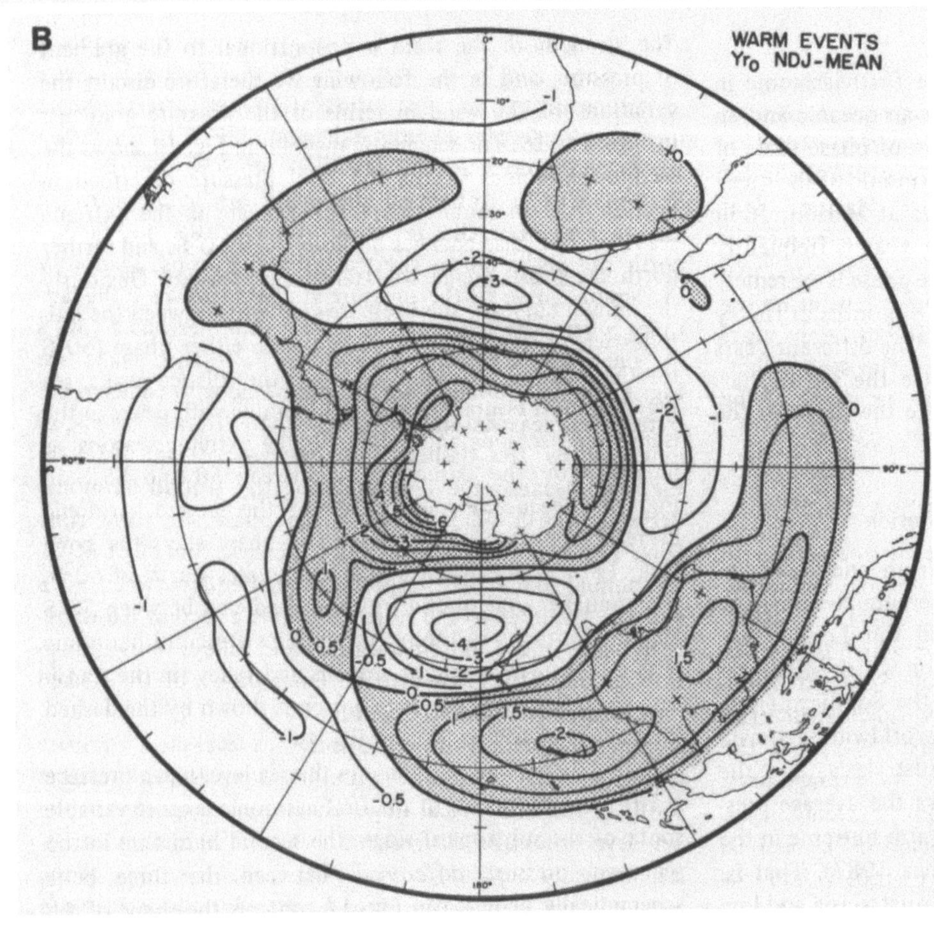

Fig. 13. A The average pressure anomalies (mb) in May–June–July during years of Warm Events in the Southern Oscillation (El Niño events). (van Loon and Shea 1987). **B** The average pressure anomalies (mb) in November–December–January during years of Warm Events in the Southern Oscillation (El Niño events). (van Loon and Shea 1987)

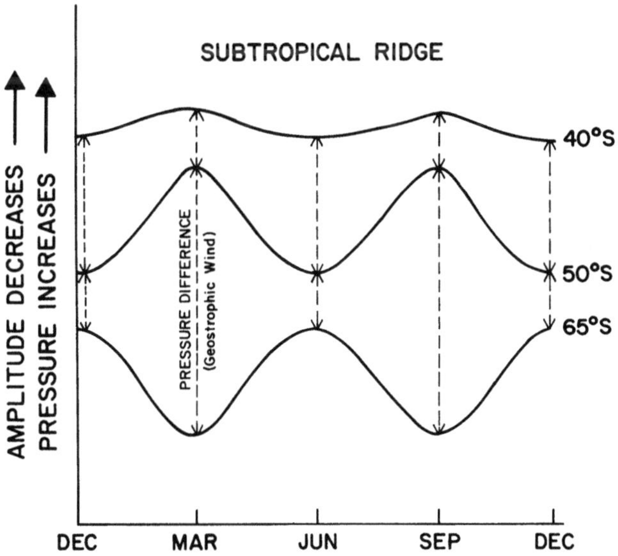

Fig. 14. A schematic drawing of the second harmonics in the sea level pressure in the latitudes near 40° S, 50° S, and 65° S. The *vertical arrows* indicate the pressure differences (geostrophic zonal wind) between the latitude belts in the extreme and transitional seasons

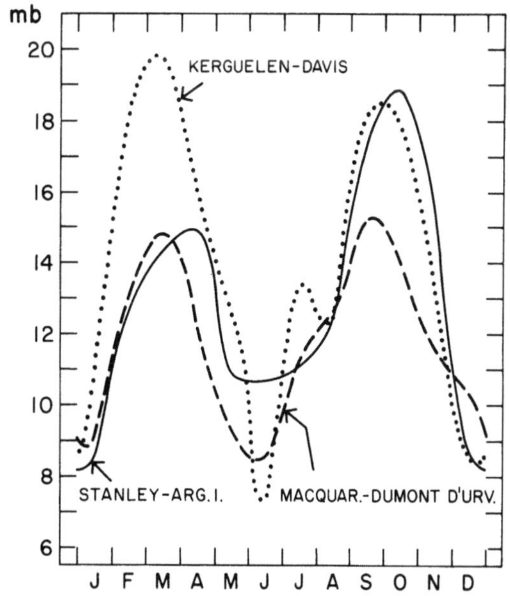

Fig. 15. The annual march of the pressure differences (mb) between stations near 50° S and stations on the Antarctic coast

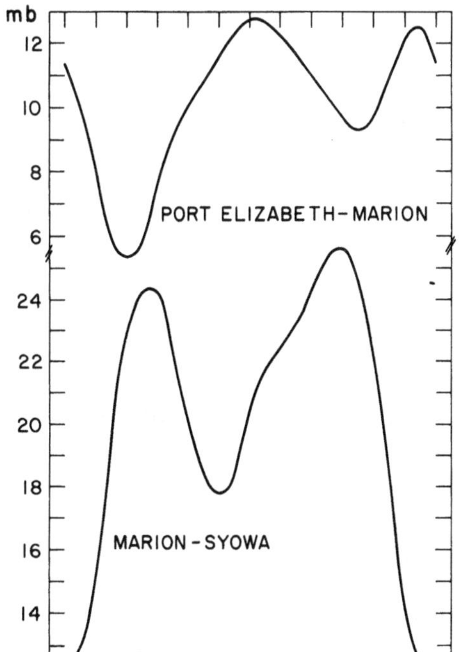

Fig. 16. The annual march of the pressure differences (mb) Port Elizabeth–Marion Island–Syowa

middle and high latitudes by three pairs of stations as widely separated as the data coverage allows (Fig. 15). A double wave with peaks in March–April and September–October dominates the three pressure differences. The zonal geostrophic wind between these latitudes therefore contains a dominant semiannual wave with peaks in these months. The first to describe the equinoctial peaks in the wind over the Antarctic Ocean, i.e., south of the Antarctic Convergence, was W. Schwerdtfeger (e.g., Schwerdtfeger and Prohaska 1956).

The reversal of the second harmonic in the meridional pressure difference (wind) north of 50°S is shown by the difference between Port Elizabeth and Marion Island in Fig. 16. For contrast, the figure contains also the difference between Marion Island and Syowa on the coast of Antarctica.

The distribution of the amplitude of the second harmonic in the mean west-east component of the surface wind can be seen in Fig. 17. There is a maximum along 60°S where the two peaks of the harmonic occur in March/ April and September/October. On the north side of this maximum the amplitude goes toward zero in a circumpolar belt along 50°S. In this belt an abrupt phase reversal takes place and the peaks of the harmonic north of 50°S thus occur in December/January and June/July. Along 60°S the second harmonic in several places explains more than 80% of the mean annual variance. In the area with solstitial peaks its share of the variance is smaller, exceeding 50% in only a few places (van Loon and Rogers 1984a).

In the zone along 50°S where the phase of the semiannual wave reverses and the amplitude goes toward zero,

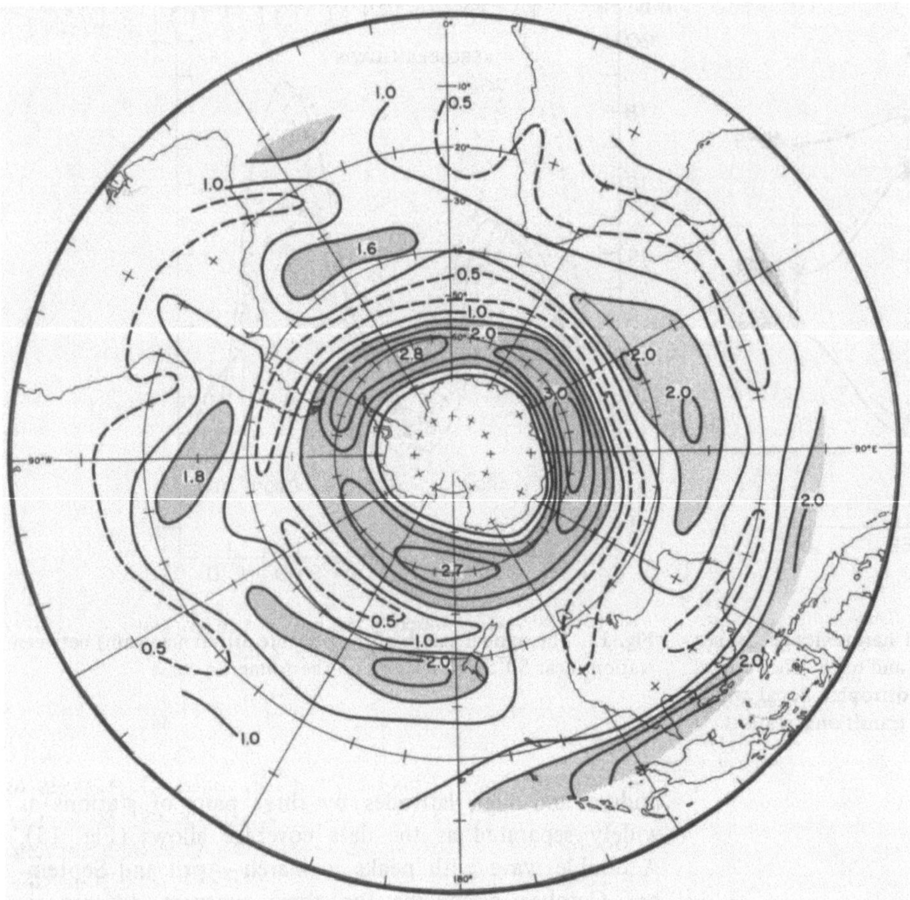

Fig. 17. Amplitude (m/s) of the second harmonic in the mean zonal geostrophic wind. (van Loon and Rogers 1984a)

the wind reaches its peak in summer, although the difference between winter and summer is small. This can be seen in Fig. 18A,B which shows the mean resultant wind speed, averaged over 12 years and based on daily analyses made in the Australian Bureau of Meteorology. The distribution of maxima and minima is similar in February and August, but in the latitudes near 50°S the wind is clearly a little stronger in summer than in winter. The minimum just outside Antarctica follows the axis of the sub-Antarctic trough along which the zonal component of the wind changes from west on the north side to east on the south side.

4.2 Interannual Variations

The interannual variation of the monthly mean u (west–east) and v (south–north) components of the wind is illustrated by the standard deviation of the monthly means in Fig. 19A–D. In both February and August the standard deviation of u is smaller than the monthly mean; in February it is so small in the westerlies north of 60°S that it is improbable that a monthly mean of easterlies will occur anywhere there.

In August the interannual variation is a good deal larger than in February. This is especially so south of Australia and south and east of New Zealand, where the standard deviation amounts to 40% to 80% of the mean. The latter area is one where blocking of the westerly flow is quite common, and a month with easterlies in the mean is not unusual.

The interannual variability of the monthly mean v-component is not larger in August than in February. Since the v-component itself is small, the standard deviation indicates frequent changes in the monthly mean waves from one year to another. This does not mean that the phase of the quasi-stationary waves is variable, but rather that a single monthly mean often contains a large transient component in addition to the stationary component. The longer the period which one includes in an average, the more the stationary component will dominate. Usually a seasonal mean suffices to eliminate the transient parts.

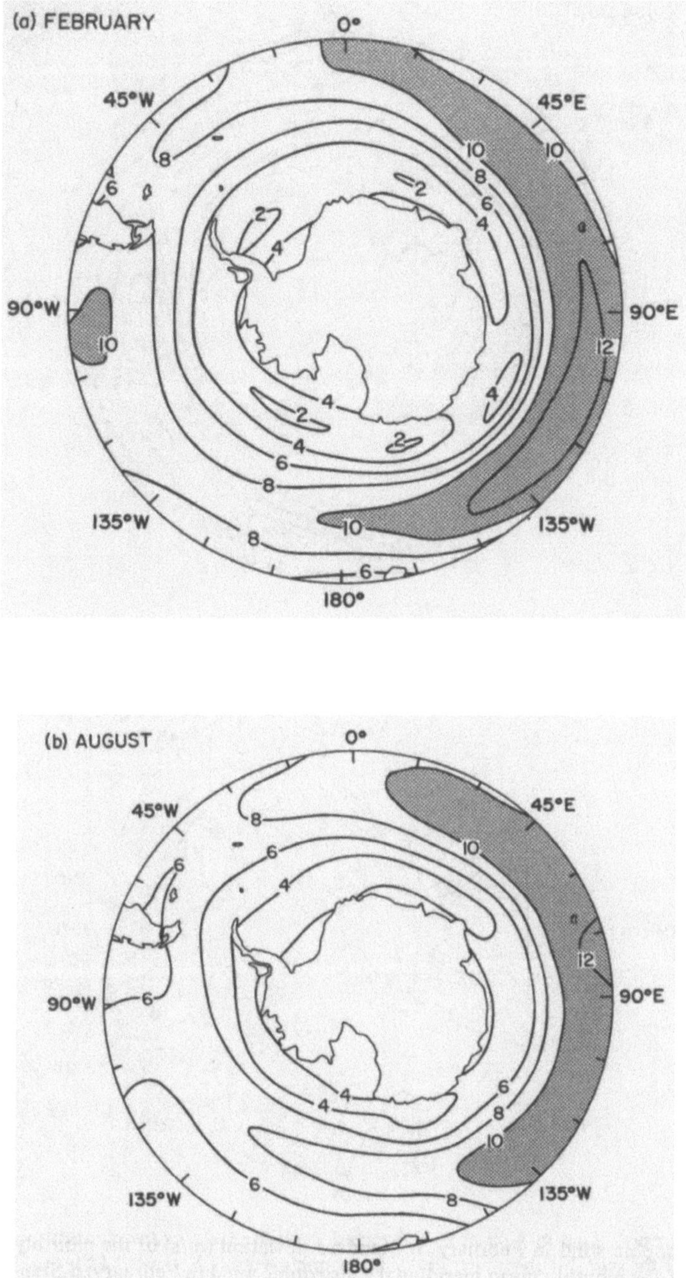

Fig. 18. a Speed (m/s) of the mean resultant geostrophic wind in February. b Speed (m/s) of the mean resultant geostrophic wind in August

5 Clouds and Precipitation

5.1 Average Total Cloudiness

Our knowledge is weakest about precipitation and clouds, and we shall restrict ourselves to a few general statements and refer those who want more information to Chapters 1, 3, and 5 in *Climates of the Oceans* (van Loon 1984) and Chapters 2 and 6 in *Meteorology of the Southern Hemisphere* (van Loon et al. 1972). The heat balance at the surface is also dealt with in these two publications.

Figure 20 shows the average amount of total cloudiness (all cloud types combined) for the southern summer. Over the ocean it is based on observations made aboard ships, and over our area of interest the numbers are fairly reliable owing to the many observations from whalers during the summer. In winter middle and high latitudes are almost devoid of ships and a similar map for July would be only tentative. Observations from satellites confirm the general features of Fig. 20.

The largest amounts of total cloudiness are found in a circumpolar zone at subpolar latitudes where the mean exceeds 80%. The few fixed observation points in the subpolar area report somewhat fewer clouds in winter than in summer, as do most of the Antarctic coastal stations. The eccentricity south of 45°S which exists in temperature, pressure, and other elements is also evident in the total cloudiness whose maximum lies farther south in the Pacific than in the Atlantic and Indian Oceans.

5.2 Precipitation

Several analyses have been made of the precipitation over the globe (for instance, Jäger 1976; Albrecht 1960, 1961). The precipitation over the oceans is of special interest as it involves (along with the evaporation) the largest component of the energy budget of both the atmosphere and the ocean, and of the exchange between them. Direct measurements of oceanic precipitation are few, and although the main patterns are well established, the magnitudes are at best uncertain. There is a precipitation maximum in middle southern latitudes which broadly reflects the patterns of cyclone frequency and movement.

Acknowledgment. Work sponsored by the National Science Foundation.

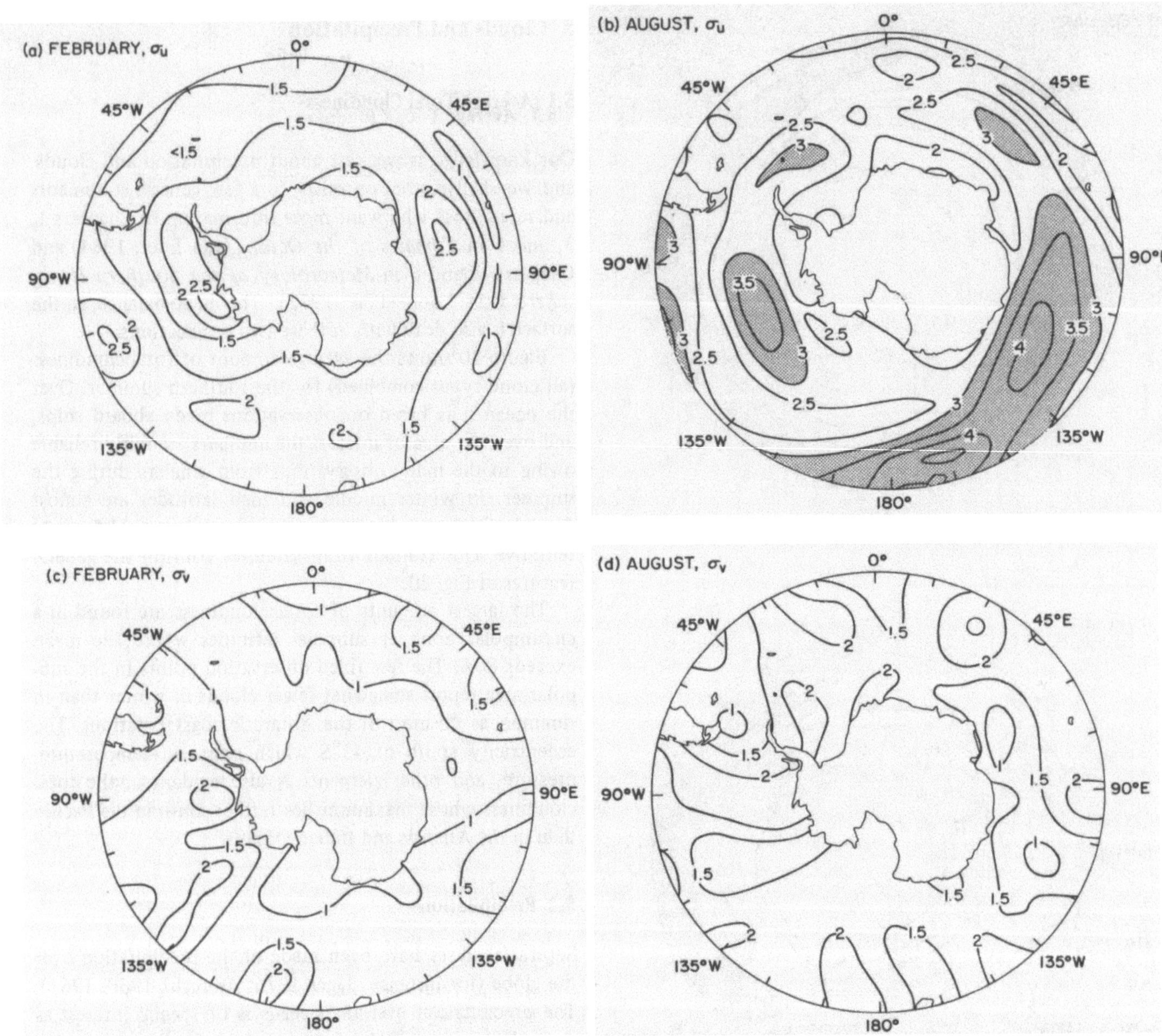

Fig. 19. a Standard deviation (m/s) of the monthly mean zonal geostrophic wind in February. b Standard deviation (m/s) of the monthly mean zonal geostrophic wind in August. c Standard deviation (m/s) of the monthly mean meridional geostrophic wind in February. d Standard deviation (m/s) of the monthly mean meridional geostrophic wind in August

Bibliography

Meinardus W (1911) Deutsche Südpolar-Expedition 1901–1903. Meteorologie, vol III. Reimer, Berlin, pp 43–129

Meinardus W (1929) Die Luftdruckverhältnisse und ihre Wandlungen südlich von 30° südl. Breite. Meteorol Z 46:41–49, 86–96

Meinardus W (1938) Klimakunde der Antarktis. In: Köppen W, Geiger R (eds) Handbuch der Klimatologie, vol IV(U). Bornträger, Berlin, 133 pp

Meinardus W (1940) Die interdiurne Veränderlichkeit der Temperatur auf der südlichen Halbkugel. Meteorol Z 57:165–176, 219–233

Schwerdtfeger W (1962) Meteorologia del Area del Pasaje Drake. Serv Hidrogr Nav, Buenos Aires, 76 pp

Schwerdtfeger W (1970) The climate of the Antarctic. In: Orvig S (ed) World survey of climatology, vol 14. Elsevier, Amsterdam, pp 253–355

van Loon H, Taljaard JJ, Sasamori T, London J, Hoyt DV, Labitzke K, Newton CW (1972) Meteorology of the Southern Hemisphere. In: Newton CW (ed) Meteorol Monogr 13(35):263

van Loon H (ed) (1984) Climates of the oceans. In: World survey of climatology, vol 15. Elsevier, Amsterdam, XVIII + 716 pp

van Rooy MP (ed) (1957) Meteorology of the Antarctic. Weather Bureau, Pretoria, 240 pp

Vowinckel E (1955) Southern Hemisphere weather map analysis: Five-year mean pressures. Notos 4:17–50, 204–216

Vowinckel E (1956) Das Klima des Antarktischen Ozeans: I and II. Arch Meteorol Geophys Bioklim B7:317–369

Vowinckel E, van Loon H (1957) Das Klima des Antarktischen Ozeans: III. Arch Meteorol Geophys Bioklim B8:75–102

Vowinckel E, Oosthuizen CM (1953) Weather types and weather elements over the Antarctic Ocean during the whaling season. Notos 2:157–182

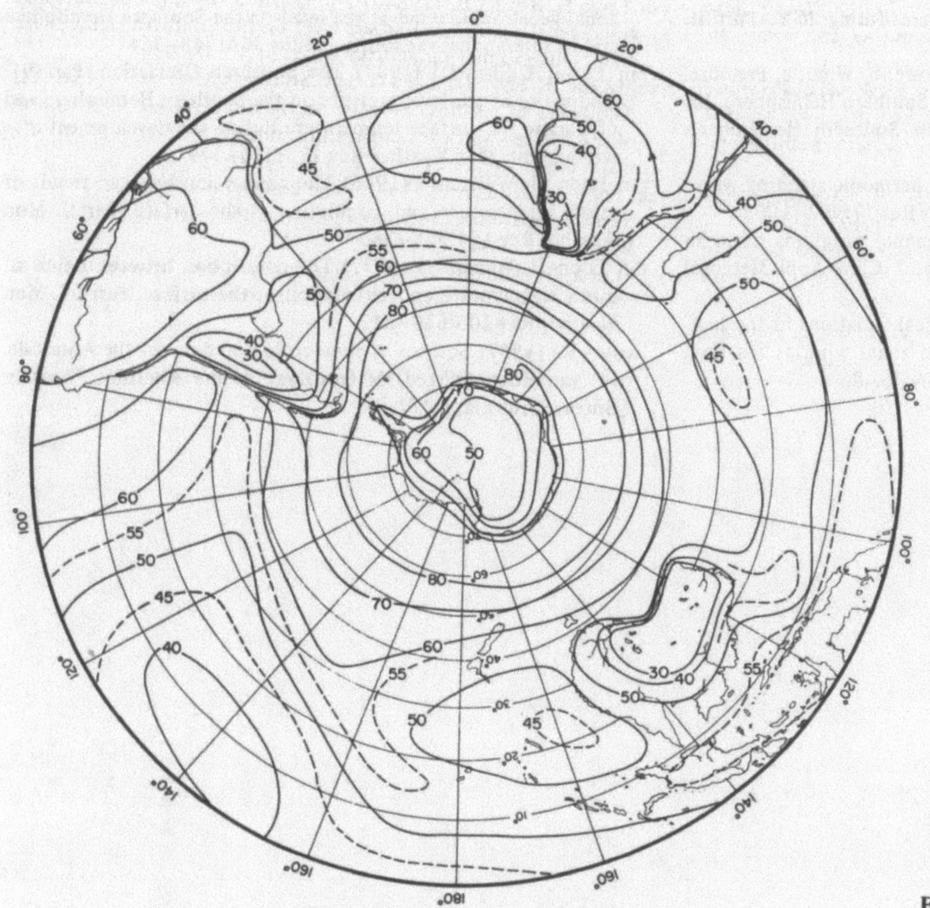

Fig. 20. Average total cloudiness (%) in January. (van Loon 1972d)

References

Albrecht F (1960) Jahreskarten des Wärme- und Wasserhaushaltes der Ozeane. Ber Dtsch Wetterd 9:19

Albrecht F (1961) Der jährliche Gang der Komponenten des Wärme- und Wasserhaushaltes der Ozeane. Ber Dtsch Wetterd 11:24

Höflich O (1984) Climate of the South Atlantic Ocean. In: van Loon H (ed) Climates of the oceans. Elsevier, Amsterdam, pp 1–191

Jäger L (1976) Monatskarten des Niederschlags für die ganze Erde. Beitr Dtsch Wetterd 18(139):38

Lejenäs H (1984) Characteristics of Southern Hemisphere blocking as determined from a time series of observational data. Q J R Meteorol Soc 110:967–979

Mo KC, van Loon H (1984) Some aspects of the interannual variation of the mean monthly sea level pressure on the Southern Hemisphere. J Geophys Res 89:9541–9546

Mo KC, van Loon H (1985) Climatic trends in the Southern Hemisphere. J Clim Appl Meteorol 24:777–789

Schwerdtfeger W, Prohaska F (1956) Der Jahresgang des Luftdrucks auf der Erde und seine halbjährige Komponente. Meteorol Rundsch 9:33–43

Streten NA, Zillman JW (1984) Climate of the South Pacific Ocean. In van Loon H (ed) Climates of the Oceans. Elsevier, Amsterdam, pp 263–429

Taljaard JJ, van Loon H (1962) Cyclogenesis, cyclones and anticyclones in the Southern Hemisphere during the winter and spring of 1957. Notos 11:3–20

Taljaard JJ, van Loon H (1963) Cyclogenesis, cyclones and anticyclones in the Southern Hemisphere during summer 1957–1958. Notos 12:37–50

Taljaard JJ, van Loon H (1984) Climate of the Indian Ocean South of 35°S. In: van Loon H (ed) Climates of the oceans. Elsevier, Amsterdam, pp 505–601

Trenberth KE (1980) Planetary waves at 500 mb in the Southern Hemisphere. Mon Weather Rev 108:1378–1389

Trenberth KE (1981) Interannual variations of the Southern Hemisphere 500 mb flow: Regional characteristics. Mon Weather Rev 109:127–136

Trenberth KE (1986) The signature of a blocking episode on the general circulation in the Southern Hemisphere. J Atmos Sci 43:2061–2069

Trenberth KE, Mo KC (1985) Blocking in the Southern Hemisphere. Mon Weather Rev 113:3–21

van Loon H (1956) Blocking action in the Southern Hemisphere. Notos 5:171–177

van Loon H (1966) On the annual temperature range over the southern oceans. Geogr Rev 56:497–515

van Loon H (1967a) The half-yearly oscillations in middle and high southern letitudes and the coreless winter. J Atmos Sci 24: 472–486 van Loon H (1967b) A climatological study of the atmospheric circulation in the Southern Hemisphere during IGY, Part II. J Appl Meteorol 6:803–815

van Loon H (1972a, b, c, d) a. Temperature; b. Wind; c. Pressure. d. Cloudiness and precipitation in the Southern Hemisphere. In: Newton CW (ed) Meteorology of the Southern Hemisphere. Meteorol Monogr 13(35):25–112

van Loon H, Jenne RL (1972) The zonal harmonic standing waves in the Southern Hemisphere. J Geophys Res 77:992–1003

van Loon H, Madden RA (1983) Interannual variations of mean monthly sea-level pressure in January. J Clim Appl Meteorol 22:687–692

van Loon H, Rogers JC (1984a) Interannual variations in the half-yearly cycle of pressure gradients and zonal wind at sea level on the Southern Hemisphere. Tellus 36A:76–86

van Loon H, Rogers JC (1984b) The yearly wave in the pressure and zonal geostrophic wind at sea level on the Southern Hemisphere and its interannual variability. Tellus 36A:348–354

van Loon H, Shea DJ (1987) The Southern Oscillation. Part VI: Anomalies of sea level pressure on the Southern Hemisphere and of Pacific sea surface temperature during the development of a warm event. Mon Weather Rev 115:370–379

van Loon H, Williams J (1976) The connection between trends of mean temperature and circulation at the surface. Part I. Mon Weather Rev 104:365–380

van Loon H, Williams J (1977) The connection between trends of mean temperature and circulation at the surface. Part IV. Mon Weather Rev 105:636–647

Venter RJ (1957) Sources of meteorological data for the Antarctic. In: van Rooy MP (ed) Meteorology of the Antarctic. Weather Bureau, Pretoria, pp 17–38

Variability of the Atmospheric Circulation over the Drake Passage, Scotia Sea and Weddell Sea

L. KAUFELD[1]

Summary. Grid point values of air pressure from 50 to 70°S and 0 to 70°W are used to show not only monthly and seasonal means of the atmospheric circulation but also its anomalies. A low pressure belt is situated at 65 to 70°S with strong westerly winds on its northern flank; additionally, southerly wind components produce a relatively cold climate in the Weddell Sea. The actual situation may deviate strongly from this pattern, and also individual months can differ considerably. The zonal and meridional flow show more or less irregular fluctuations with statistically significant periods of some months (3.5 and 8 months for the zonal component, and 5 months and 1 year for the meridional components at some longitudes). Periods longer than 1 year are very weak. A long-term trend cannot be detected.

1 Data

Grid point values of air pressure from 50 to 70°S and 0 to 70°W with 5° latitudinal and 10° longitudinal spacing for the period 1957–1985 were extracted from monthly charts of global sea level pressure, which had been analyzed in the Seewetteramt.

Data were stored on magnetic tape, checked for errors and are the basis of further investigations. The quality of charts in the area west of 30°W is better than east of that longitude because there are many more observing stations in the west (on the southern tip of South America, on the Antarctic Peninsula, on the Falkland Islands, South Georgia and the South Orkney Islands) than in the east.

2 Mean Conditions

2.1 General Characteristics

Initially, monthly charts for the whole period of 29 years were constructed. The pressure charts of some neighboring months look similar and could be combined to form seasonal charts. To show the characteristic features of the different seasons in this region, the conventional combination of months (December–February = summer, March–May = autumn, . . .) was better than that of Taljaard (1967), who took December–March for summer, April–May for autumn, June–September for winter and October–November for spring.

Figures 1–4 show the characteristic features of each season. All charts show low pressure in the belt between 65 and 70°S with a center near 30°W. North of this zone there is a pressure gradient to the high pressure belt, which is situated near 30°S. The isobars which extend from west to east indicate a steady and strong zonal westerly flow; near 70°W there is an indication of a meridionalization with a high pressure ridge over South America and the Antarctic Peninsula.

2.2 Seasonal Variations

The intensity and location of the cyclone center varies from month to month. Figure 5 shows that its intensity is strongest in autumn and spring with minimum core pressure in March and October and weak in summer (January) and early winter (June). In most months the center is located near 67.5°S, 27°W; extreme positions are 70°S, 40°W in February and 65°S, 20°W in June.

When the pressure is low in the sub-Antarctic cyclone, it becomes high north of 50°S (see Fig. 6, March, Sept./Oct.) and vice versa (Dec./Jan. and May/June). Therefore the intensity of the westerlies follows a marked seasonal course. The pressure differences between neighboring latitudes are shown in Fig. 7; the full lines stand for the whole area from 70°W to 0° and the dashed lines for the zone from 70°W to 30°W, which has more reliable values. From the pressure differences the mean geostrophic winds have been computed (scales on the right hand side); positive values indicate westerly wind, negative values, e.g., at 65–70°S in summer, easterlies.

All latitudinal belts show half-yearly cycles with maxima in autumn and spring, minima in summer and winter. This half-yearly oscillation has already been demonstrated by Schwerdtfeger and Prohaska (1956), van Loon (1967), Schwerdtfeger (1970), van Loon et al. (1971) and van Loon and Rogers (1984). According to Schwerdtfeger (1960), the equinoctial wind maxima are related to the differential heating between middle and higher latitudes with water and ice surfaces, which is greater in spring and autumn than

1 Deutscher Wetterdienst, Seewetteramt, Bernhard-Nocht-Straße 76, 2000 Hamburg 4, Fed. Rep. of Germany

Antarctic Ocean and Resources Variability
(ed. by D. Sahrhage)

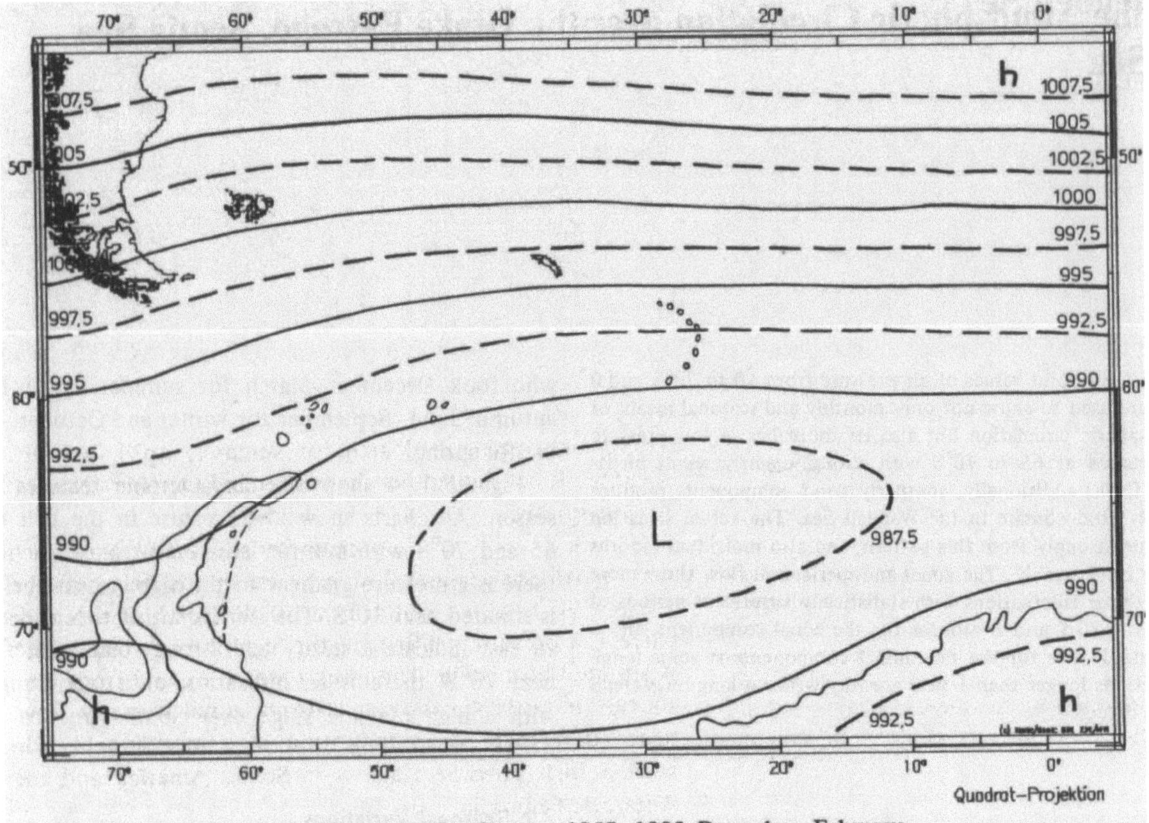

Fig. 1. Mean air pressure at sea level in hPa for the years 1957–1985, December–February

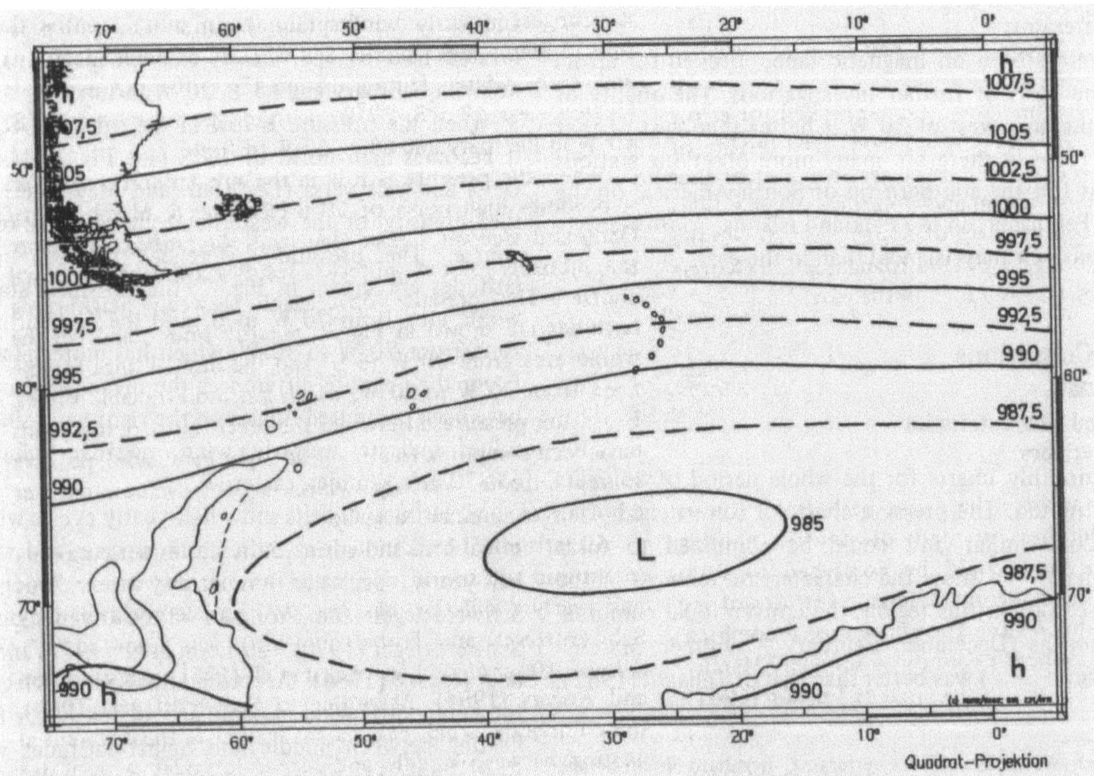

Fig. 2. As Fig. 1, March–May

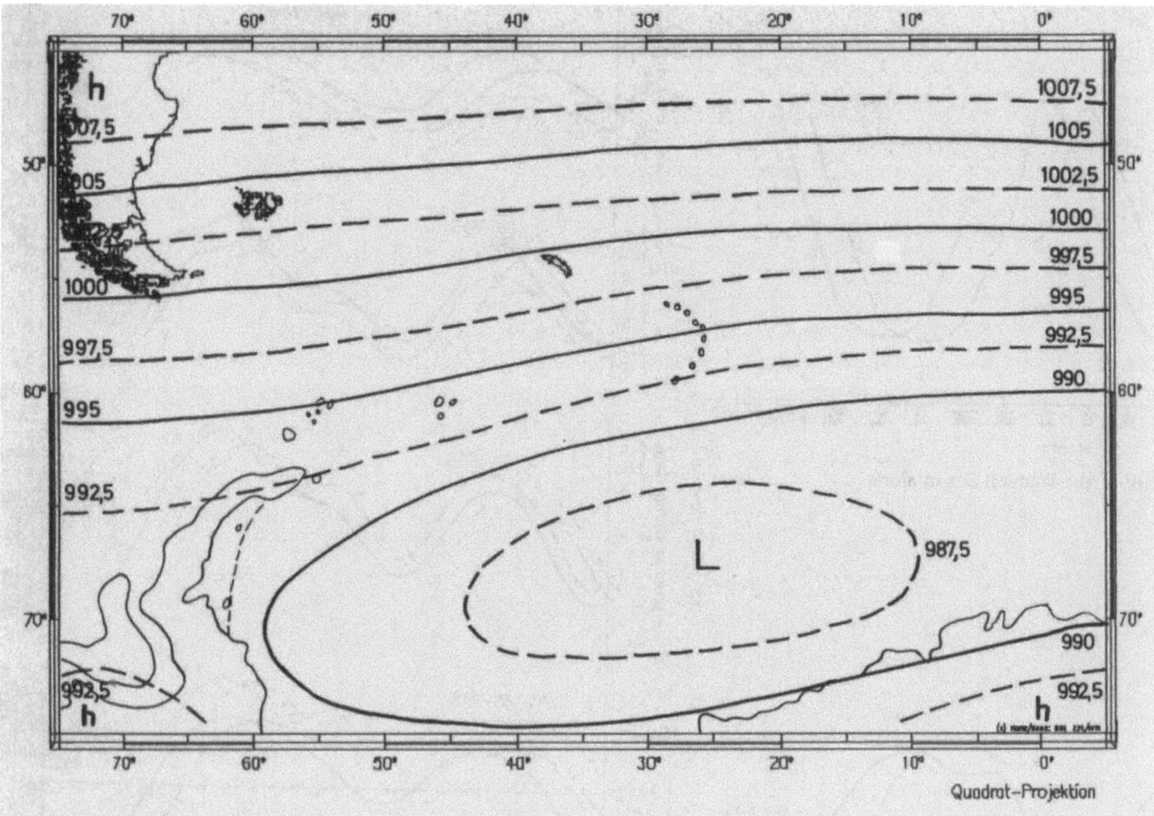

Fig. 3. As Fig. 1, June–August

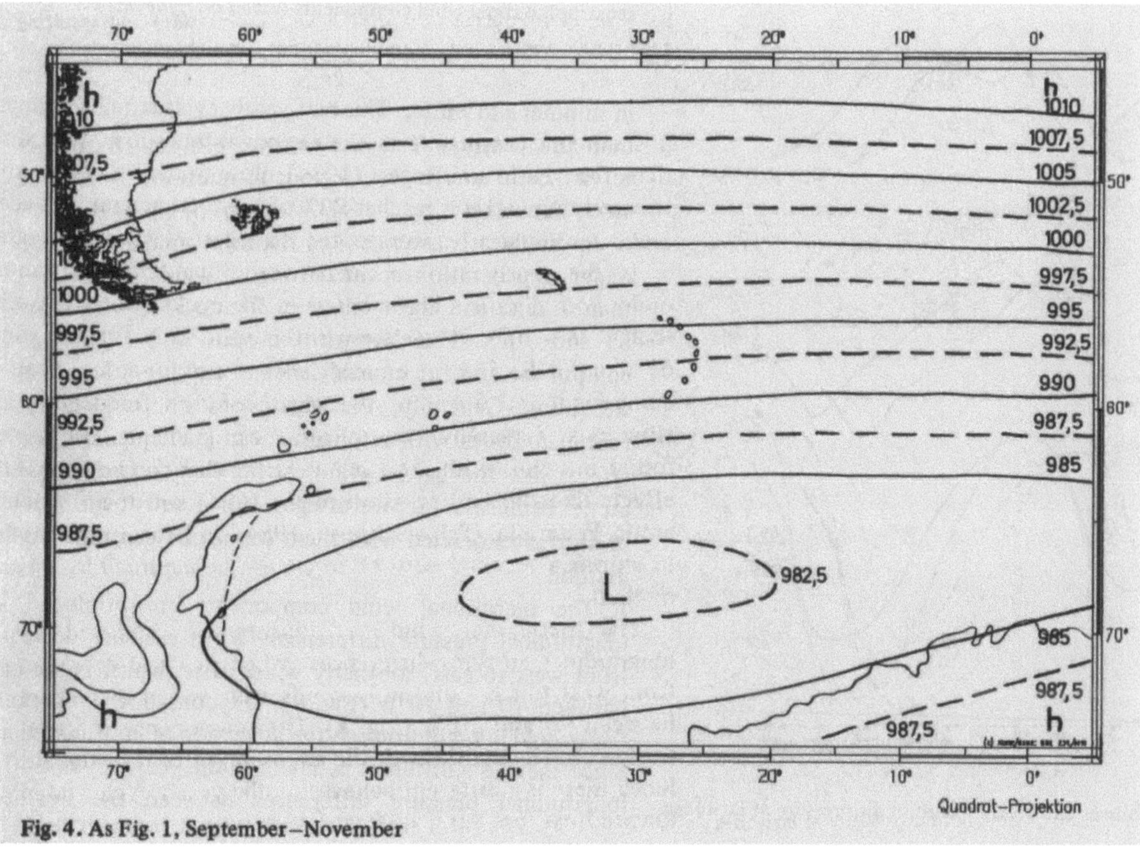

Fig. 4. As Fig. 1, September–November

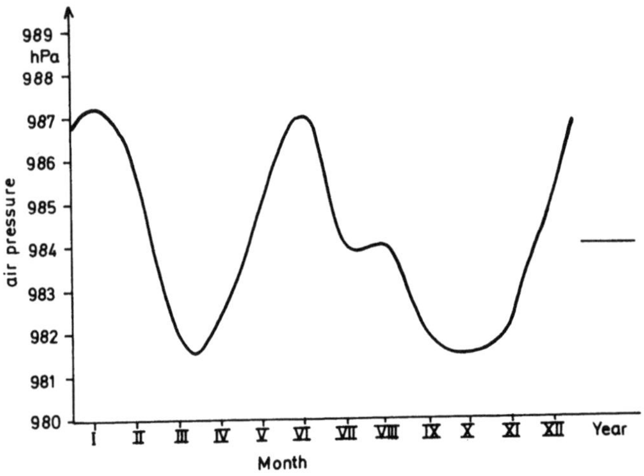

Fig. 5. Core pressure of the Weddell Sea cyclone

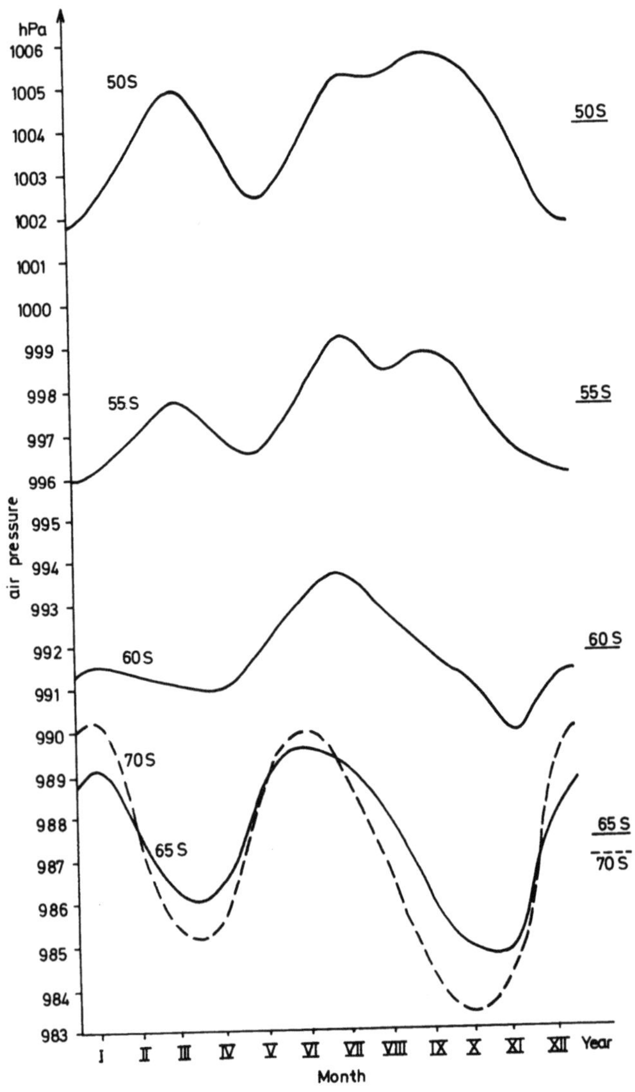

Fig. 6. Seasonal courses and yearly means of air pressure at latitudes

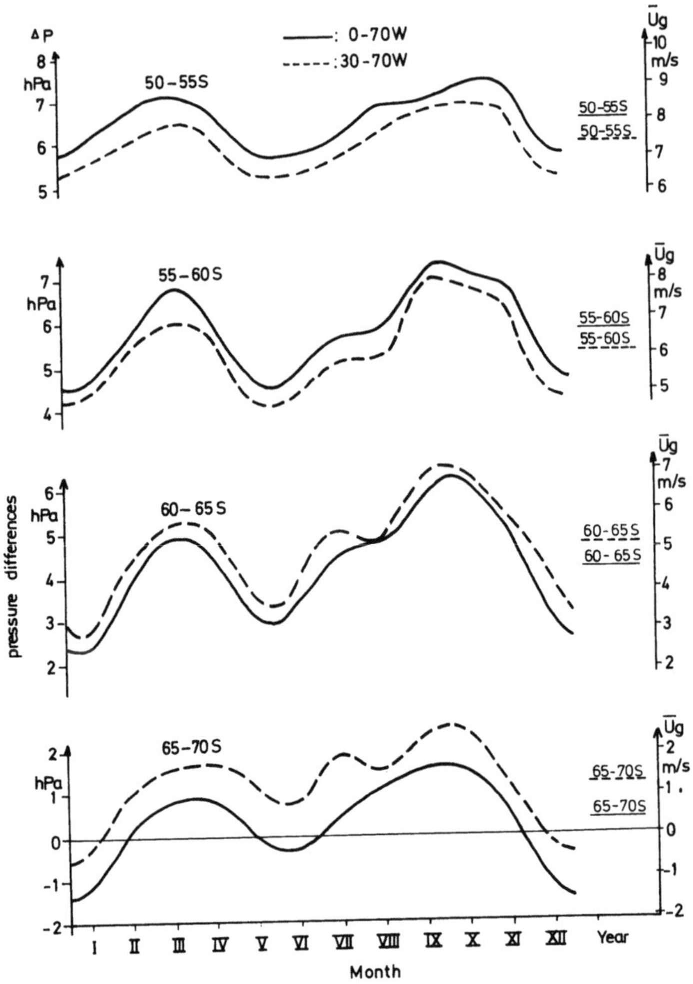

Fig. 7. Seasonal courses and yearly means of pressure differences between neighbouring latitudes (scales on ***left side***) and resulting geostrophic zonal wind components (scales on ***right side***)

in summer and winter. This half-yearly cycle is much stronger than the yearly one in the regions in question. According to van Loon and Rogers (1984), close to the tip of the Antarctic Peninsula it reaches 90% of the total annual variance.

In the yearly average the strongest winds in the area under consideration occur between 50 and 55°S with more than 7 m/s, and the weakest in the cyclonic belt between 65 and 70°S. This statement is valid only for the zonal component and for monthly means, not for actual weather situations. Generally, the greatest storm frequencies are found in zones with strongest mean gradients. Orographic effects like the barrier winds at the east coast of the Antarctic Peninsula (Schwerdtfeger 1984) and the frequency of storms associated with them cannot be captured by this method.

The meridional wind components are produced by longitudinal pressure differences. When pressure decreases from west to east, southerly winds arise, which is the case between 70 and 20°W (Fig. 8). The consequence is a relatively cold climate through the advection of continental air. Since there is a different behavior in the yearly cycle and in longitudinal pressure differences between the northern

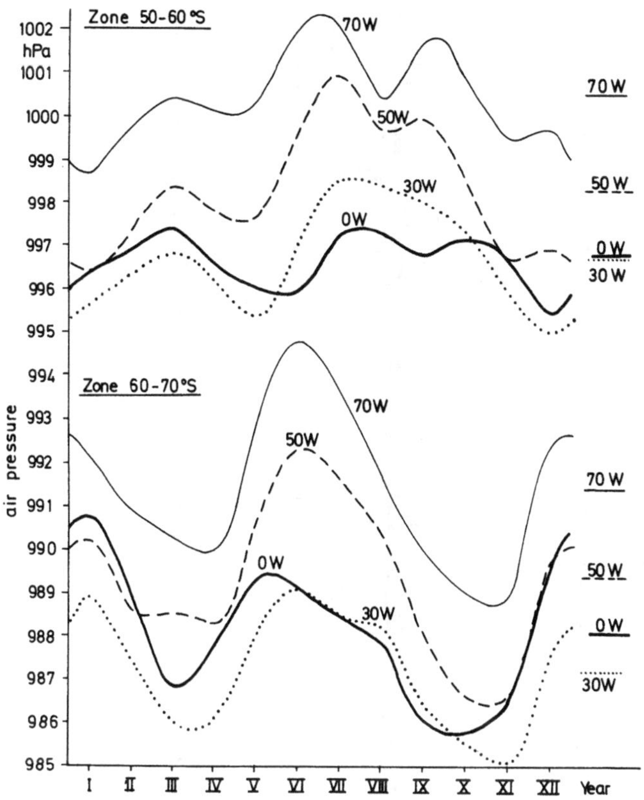

Fig. 8. Seasonal courses and yearly means of air pressure at selected longitudes

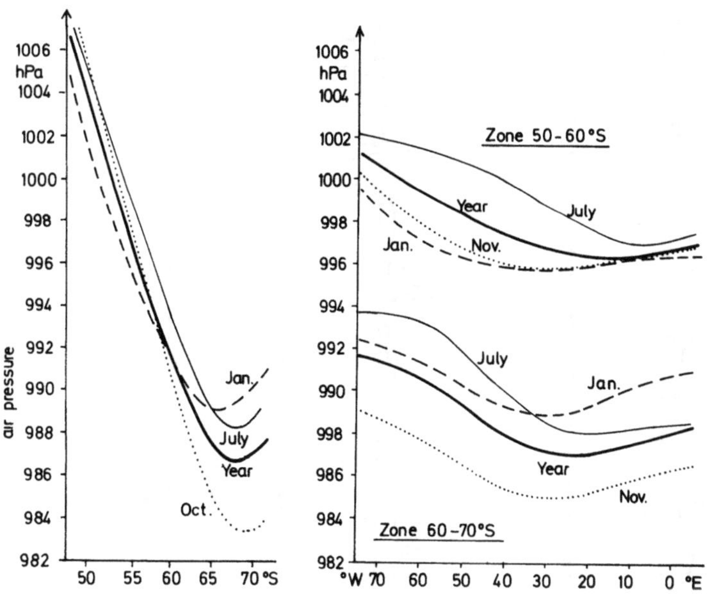

Fig. 9. Latitudinal (*left side*) and longitudinal (*right side*) pressure profiles for selected months and the year

(50–60°S) and the southern (60–70°S) zones, both were separated. In the eastern parts of both zones pressure increases from west to east, but gradients are small.

Figure 9 comprises the results of Figs. 6–8 for the yearly mean and for some selected months. It is shown that the zonal pressure differences (westerly winds) are much stronger than the meridional ones (southerly and northerly components). The low pressure belt moves from 65°S in January to 70°S in October; the steepest gradient (strongest westerly wind) is found in November, weakest zonal winds in January.

In the zone 50–60°S the strongest southerly component occurs from October to February in the westernmost area (70–60°W) with about 2 m/s; in winter (June–September) it occurs in the region 40–30°W. Weak northerly components prevail east of 30°W in summer, east of 10°W in winter. In the southern zone (60–70°S) strongest southerly components of 1.4 to 2.0 m/s are generally found between 60 and 40°W, east of the Antarctic Peninsula. Northerly components reach their highest intensity in summer east of 30°W with 1.1 m/s; in other seasons they are weaker and occur east of 20°W.

3 Variability of the Circulation

3.1 Monthly Latitudinal and Longitudinal Anomalies

Figure 10 shows pressure differences between latitudes (the notation "52.5 S" stands for differences between 50 and 55°S). The unsmoothed curves represent anomalies of individual months from monthly mean values. For this evaluation only the area from 70 to 30°W was taken (more reliable values). Positive values indicate that the actual pressure differences were greater than the averages. This results in stronger westerly winds. Negative values indicate weaker, or even easterly winds if they exceed the monthly average (see Fig. 7). This is the case in July 1958 at 55–60°S with a negative deviation of 7.0 hPa. Since the mean value for that zone in July is 4.3 hPa, this deviation results in a (geostrophically calculated) easterly wind of 3.4 m/s. Other strong negative deviations with easterly winds are found in May 1986 at 57.5 and 62.5°S, in June 1977 at 62.5°S, in December 1978, June 1980, March 1982, and January 1984 at 67.5°S. In the latter case an easterly component of 6.0 m/s is calculated instead of 0.3 m/s in the average. There are also strong positive deviations which result in an increased westerly wind like in March 1957 at 57.5°S with a monthly mean of 14.4 m/s instead of "normal" 7.8 m/s. Other strong positive deviations are found in September 1974 at 52.5°S with 6.4 hPa, in July 1985 at 62.5°S, in September 1970 and July 1973 in the region between 65 and 70°S, where in January 1960 westerly winds reached an average speed of 5.3 m/s instead of "normal" weak easterly winds.

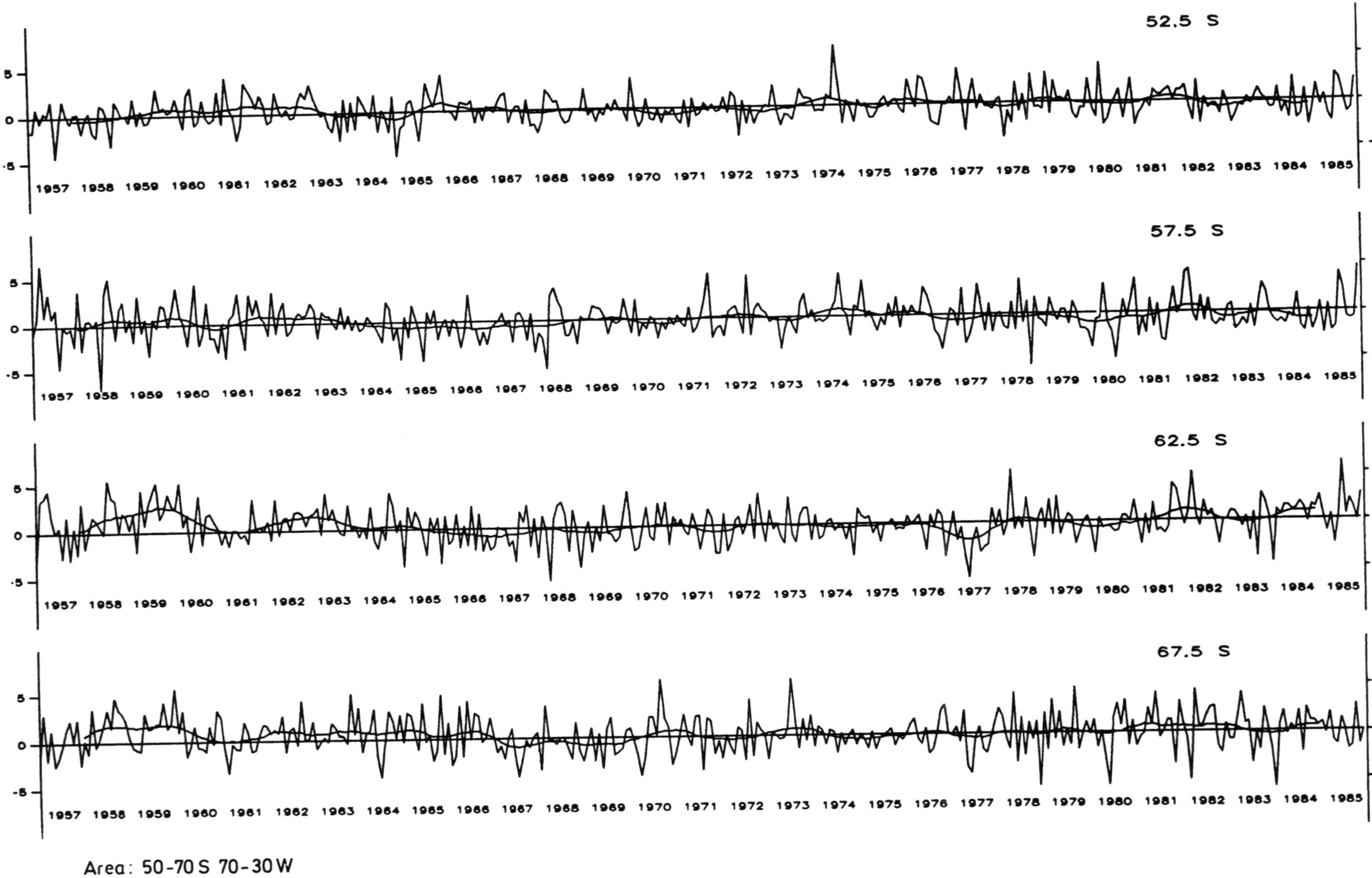

Fig. 10. Anomalies of pressure differences between latitudes in hPa (52.5° S = 50–55° S): deviations of individual months from the long range average (*unsmoothed curves*), low pass filtered values (*smoothed curves*)

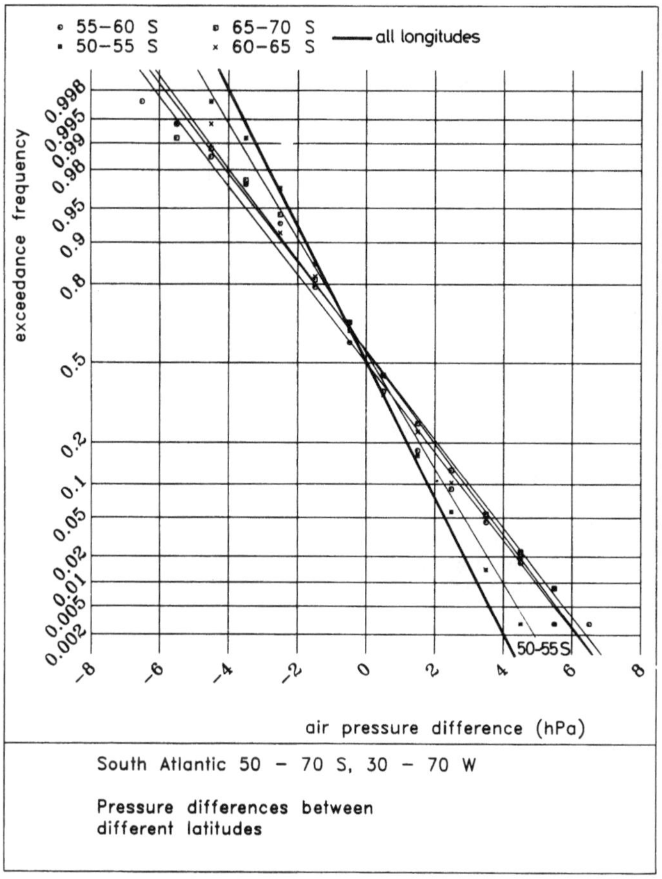

Fig. 11. Gaussian diagram with exceedance frequencies for pressure differences of latitudinal belts (*thin lines*) and for all longitudinal differences (*full line*) between 70° and 30°W

The distribution of deviations for longitudinal means (not shown) looks very similar to that for latitudinal values (Fig. 10). In both zones their amplitudes reach about +5 and –6 hPa. Positive values indicate a stronger southerly flow, negative ones a weakened southerly or even a northerly flow. In the zone 50–60°S the strongest negative deviation is –5.8 hPa in September 1970 at 45°W, that means a northerly component of 6.2 m/s instead of southerly winds with 0.8 m/s in the long time mean. The minimum of zone 60–70°S also amounts to –5.8 hPa and is found in September 1968 at 65°W. Strong southerly winds occurred in October 1963 in both zones at 55 and 45°W. In this month the southerly flow with 6 m/s was nearly as strong as the westerly component (7 m/s). At Orcadas this was one of the coldest Octobers since 1904.

To investigate whether the anomalies are randomly distributed, the anomalies of Fig. 10 were plotted on a Gaussian diagram (Fig. 11). All values lie on straight lines or close to them, which indicates a random distribution. Except for the 50–55°S zone, all latitudinal distributions lie closely together. This means that their anomalies have the same standard deviation. The graph of the northermost zone 50–55°S differs significantly from the other ones and indicates a smaller standard deviation than the other zones. Here the winds are more steady, depressions and fronts move quickly and only a few central lows exist which could appear in monthly charts. The behavior of the anomalies of the meridional air flow, which were described two paragraphs before, was also investigated. Here again only the area between 70 and 30°W was evaluated. In the Gaussian diagram the values of all longitudinal anomalies are also positioned on straight lines which look very similar to those of the latitudes. Since they lie closely together, they were combined to the full line of Fig. 11. The standard deviation of longitudinal anomalies is smaller than that for the latitudinal anomalies.

3.2 Periods

In order to remove oscillations below 20 months, a low pass filter was applied to the monthly deviations of Fig. 10, which results in the smoothed curves, also included in the figure. With some exceptions the deviations of the smoothed curves from the zero line are small and do not exceed ±1 hPa. Only in 1959/60 are the positive anomalies greater than 2 hPa at 62.5°S. The low pass filtered deviations from the longitudinal means show the same pattern: normally they are small and do not exceed ±1 hPa. The greatest positive deviation of 1.5 hPa is found in 1958/59 at 65°W, the greatest negative one (–1.2 hPa) in 1969/70 at the same longitude.

There seem to be periods of 2 or 3 years. Therefore a spectral analysis was performed with the monthly deviations. To obtain sufficient statistically stable spectral estimates, the raw spectrum was smoothed, yielding 12 degrees of freedom. The smoothed spectral estimates are presented in Fig. 12a. Besides some irregular peaks in higher frequency ranges, the power spectra show an increase in variance near a period of 0.7 years at all latitudes. Highest values are in the 55–60°S zone with 6.6 units (hPa^2). The coherency spectrum between 52.5 and 57.5°S and also between 62.5 and 67.5°S (Fig. 12b) indicates correlations with values of >0.80 within this frequency band without pronounced phase differences (Fig. 12c). High spectral power values and great coherences are found for periods of about 140 days at 62.5 and 67.5°S; furthermore, phase differences are negligible. Periods of 1 year are only evident at 65–70°S and of 1.6 years at 52.5°S, but they reveal no significant correlation with the neighboring latitude belts. Furthermore, spectral variance is low for periods of 3 years or more.

The same investigation was performed with the deviations from longitudinal means, divided into the northerly (50–60°S) and the southerly (60–70°S) zone. In most cases the power of the peaks is not as high as that of latitudinal deviations, but in the northerly zone at 65°W and in the southerly zone at 65 and 45°W periods of about 0.45 years (5 months) reach high values of more than 5 hPa^2. These periods are also found at other longitudes of both zones generally with high coherences and small phase differences to neighboring zones.

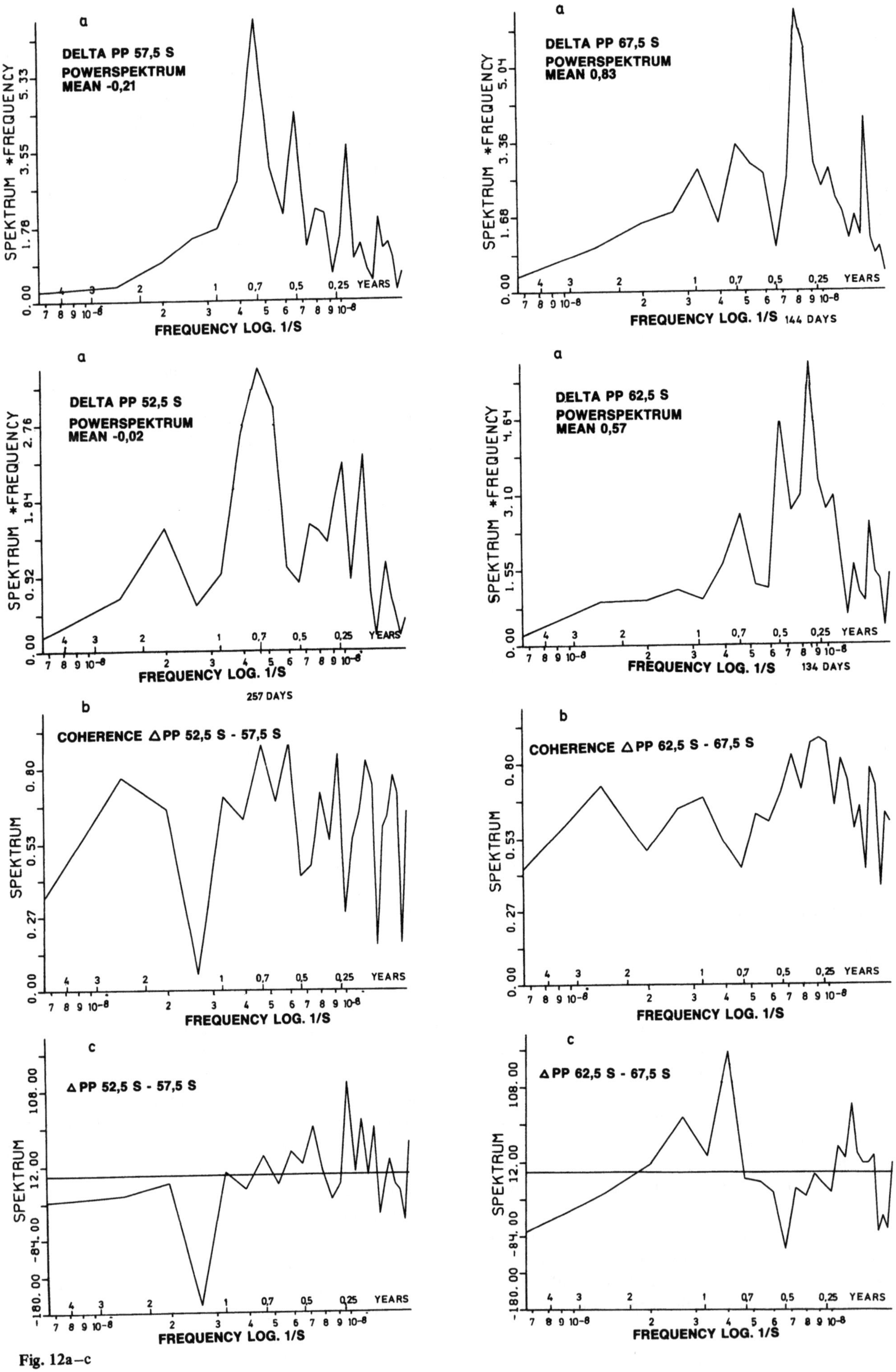
a
DELTA PP 57,5 S
POWERSPEKTRUM
MEAN -0,21
SPEKTRUM *FREQUENCY
0.00
1.78
3.55
5.33
FREQUENCY LOG. 1/S
YEARS
a
DELTA PP 67,5 S
POWERSPEKTRUM
MEAN 0,83
SPEKTRUM *FREQUENCY
0.00
1.68
3.36
5.04
FREQUENCY LOG. 1/S
144 DAYS
YEARS
a
DELTA PP 52,5 S
POWERSPEKTRUM
MEAN -0,02
SPEKTRUM *FREQUENCY
0.00
0.92
1.84
2.76
FREQUENCY LOG. 1/S
257 DAYS
YEARS
a
DELTA PP 62,5 S
POWERSPEKTRUM
MEAN 0,57
SPEKTRUM *FREQUENCY
0.00
1.55
3.10
4.64
FREQUENCY LOG. 1/S
134 DAYS
YEARS
b
COHERENCE ΔPP 52,5 S - 57,5 S
SPEKTRUM
0.00
0.27
0.53
0.80
FREQUENCY LOG. 1/S
YEARS
b
COHERENCE ΔPP 62,5 S - 67,5 S
SPEKTRUM
0.00
0.27
0.53
0.80
FREQUENCY LOG. 1/S
YEARS
c
ΔPP 52,5 S - 57,5 S
SPEKTRUM
-180.00
-84.00
12.00
108.00
FREQUENCY LOG. 1/S
YEARS
c
ΔPP 62,5 S - 67,5 S
SPEKTRUM
-180.00
-84.00
12.00
108.00
FREQUENCY LOG. 1/S
YEARS

Fig. 12a–c

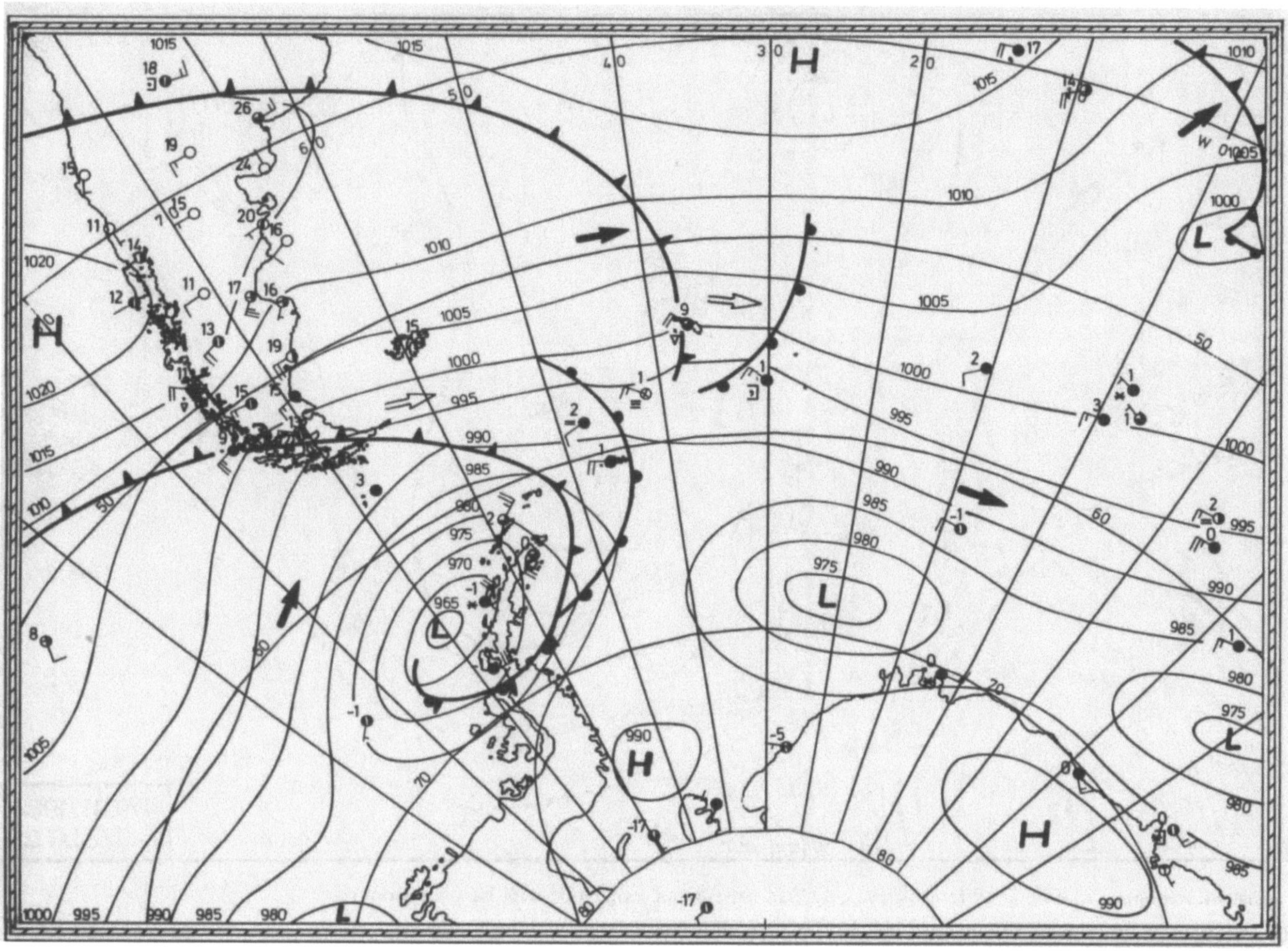

Fig. 13. Synoptic situation of 12 January 1958. 12 UTC: Typical summerly pressure distribution

Longer periods of 0.7 and 1 years generally have lower power values; only at 55°W in both zones do the yearly peaks reach a high intensity with a high coherence to 65°W and little phase lag. A very weak peak with a period of 3.5 years can be found at most longitudes in both zones.

Summarizing these results, it can be said that significant periods shorter than 1 year are observed in the latitudinal and longitudinal deviations. Longer periods (e.g., 3.5 years) have less variance and occur mainly in the longitudinal data.

A long period trend cannot be detected in our 29 years time series. V. Micka (personal communication) has investigated the relatively long series (1904–1950 for air pressure, 1903–1980 for temperature and 1906–1978 for precipitation) of Orcadas (61°S, 45°W) and also did not find any climatic change for any element.

4 Weather Situations

4.1 Synoptic Features

Normally, daily weather situations differ from monthly and seasonal means as presented in Figs. 1–4. Tracks of high and low pressure systems and distributions of anticyclone and cyclone centers per unit area around Antarctica are given, for example, by Taljaard (1967) for the IGY (1957/58); Kirk (1985) presents the figures for the year 1977 from 90°W to 40°E. Seifert (1984) shows multiannual low pressure tracks for the regions of the Drake Passage, Scotia Sea, and Weddell Sea.

There is a great variety of individual patterns. The four weather situations which are described herein are far from being a complete picture, but they demonstrate some characteristic and extreme conditions. Most weather maps originate from the International Geophysical Year 1957/58 (Republic of South Africa, Weather Bureau 1962) with a

Fig. 12a–c. Power spectra of latitudinal monthly deviations of pressure differences of Fig. 10 (lower 4 graphs, **a** units: hPa²), coherence spectra (**b**) and phase difference spectra between adjacent latitudinal belts (in degrees **c**)

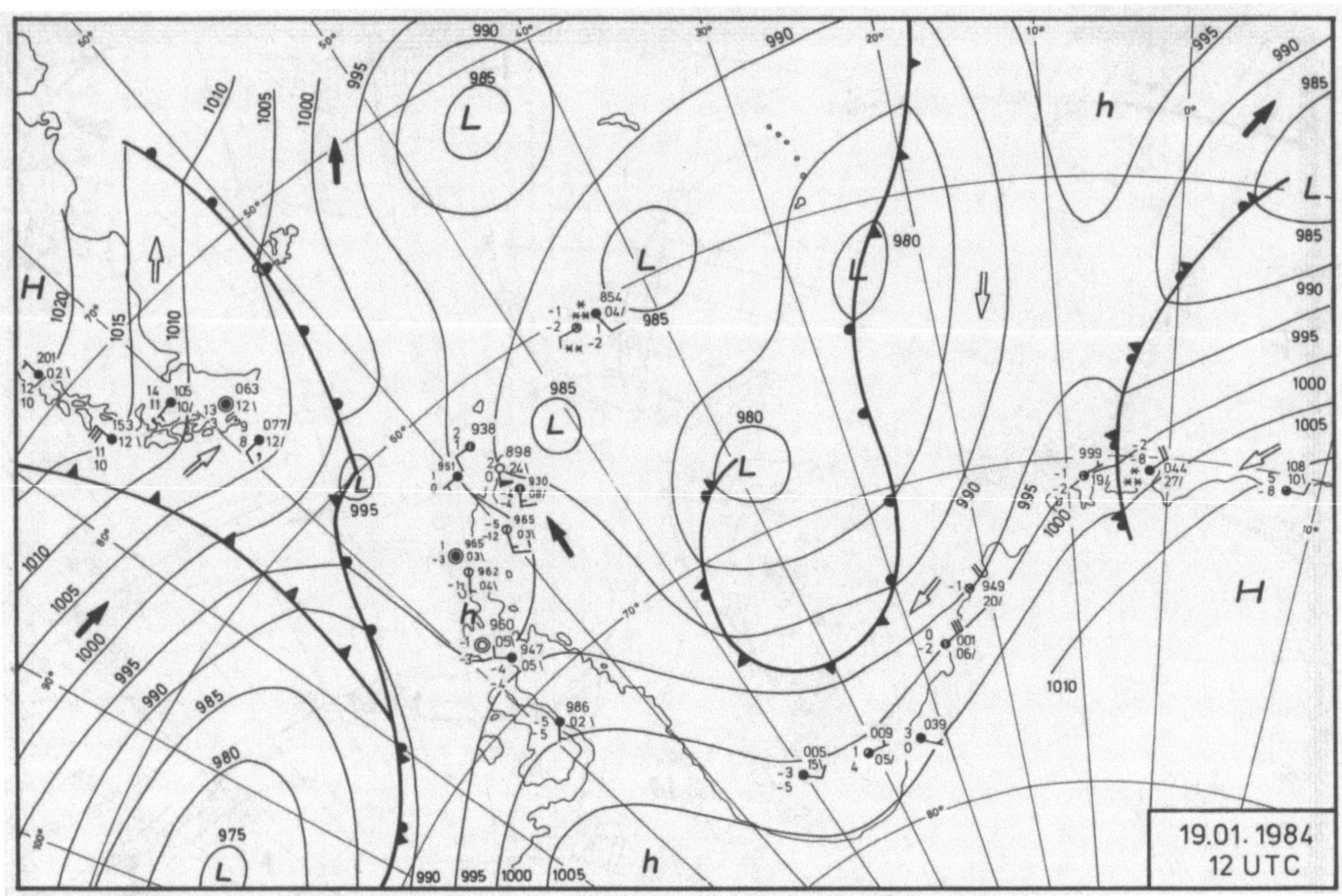

Fig. 14. Synoptic situation of 19 January 1984, 12 UTC: Meridional circulation with high temperatures

maximum of observational coverage. One example (19 Jan. 1984, Fig. 14) is taken during a voyage of the German research vessel *Polarstern*. The wind speeds are plotted in knots.

4.2 "Normal" Conditions: Westerlies North of 65°S, Low Pressure Systems near the Antarctic Peninsula and in the Weddell Sea

On January 12, 1958 (Fig. 13), an old low, which moved slowly eastward, had reached the central part of the Weddell Sea and was in the mean climatological position and slightly more intense than "normal". Another strong cyclone was approaching from the west and had reached the Antarctic Peninsula. Both systems were separated by a weak anticyclonic ridge. North of the low pressure belt there was a strong westerly flow with 55 knots at 500 hPa over the Falklands. Embedded were the fronts which propagated quickly eastward. One day later, the warm sector had already reached the South Sandwich Islands and its cyclone had moved to the region south of the South Orkneys. This situation produces strong contrasting winds along the coasts of the Antarctic Peninsula (barrier winds), moderate to strong westerlies north of 65°S and strong easterlies along the Antarctic coast between 10 and 20°W. Temperatures were near normal; the fronts were associated with drizzle and rain, the cyclones with snow.

4.3 Meridional Circulation

A meridional circulation with extreme high temperatures over the southern part of the Weddell Sea was observed from January 18 to 20, 1984 (Fig. 14). On 19th January at 12 UTC, a low over the Weddell Sea extended far to the north. On its front side a warm air stream reached the Antarctic continent between 15°E and 40°W. At places where foehn effects occur, like at Belgrano II station (78°S, 34°W), positive temperatures were recorded, which nearly reached record values (+4°C on January 18).

In the southerly flow on the east side of the Antarctic Peninsula lower temperatures of –4° to –5°C were observed. The station Esperanza (63.2°S, 56.6°W) experienced a strong barrier wind of 50 knots, which was much stronger than the undisturbed geostrophic wind (see stations south of it). Strong katabatic winds were found near the Antarctic coast at Halley station (75.5°S, 26°W).

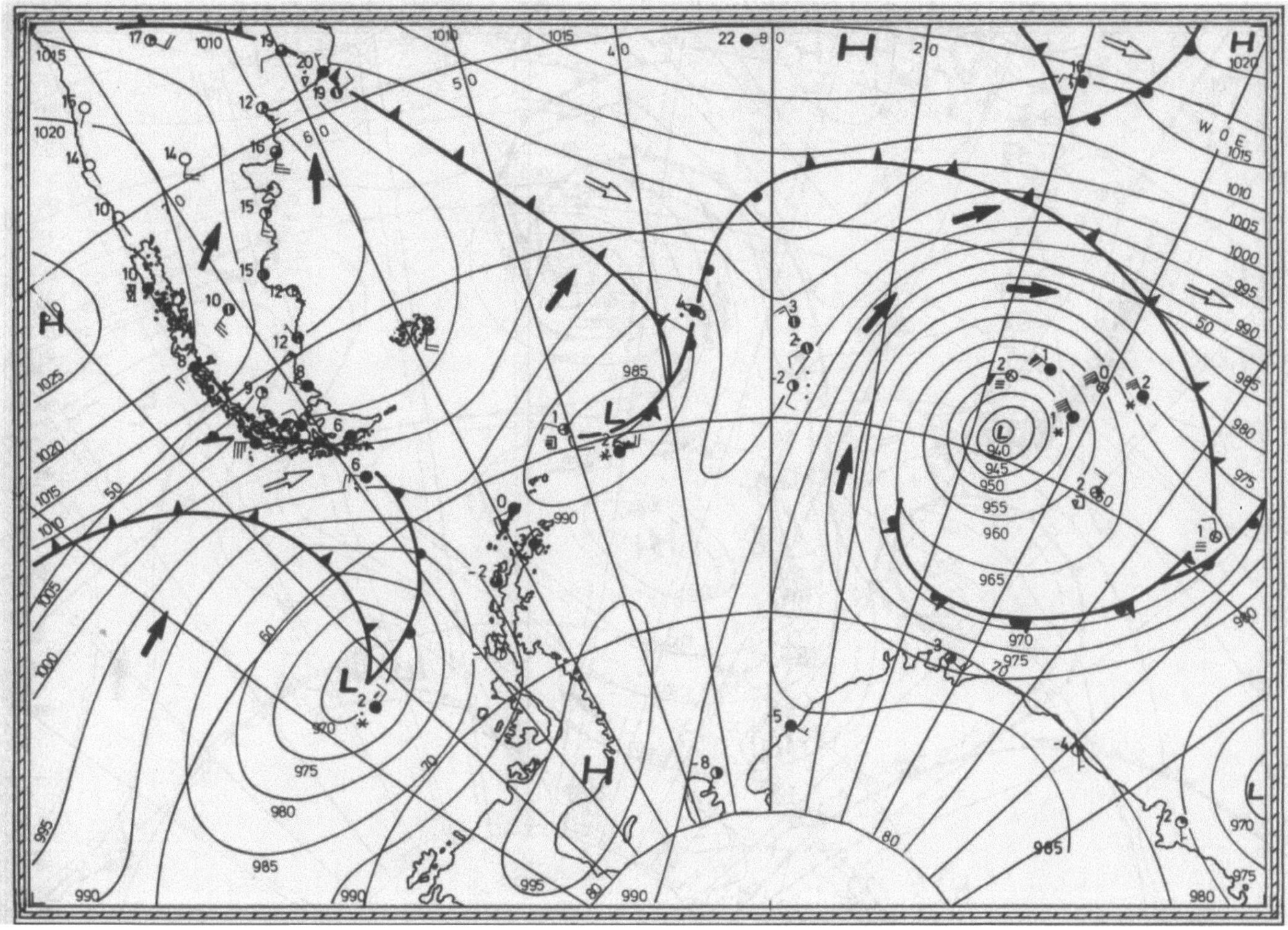

Fig. 15. Synoptic situation of 18 January 1958, 12 UTC: Strong cyclones with hurricance force winds; low temperatures over the Weddell Sea

There are few observations within the low pressure system, but these indicate relatively weak winds (see Orcadas). Low wind speeds were also experienced on the west side of the Antarctic Peninsula, but there a strong increase was expected due to a new strong depression approaching from the west. The conditions of this weather situation are regarded as optimal for the fishery and other activities in most areas.

4.4 Strong Cyclones

Twenty-six years earlier the situation was not so good (Fig. 15). A very strong storm cyclone, which had intensified one day before near the South Sandwich Islands, had reached the position of 59°S, 8°W on January 18, 1958. In its vicinity, hurricane force winds were observed on the previous day and also at the actual time. The Antarctic coast south of the cyclone will have had gales at some places, too, due to the strong gradient.

A weaker depression with moderate winds had reached the South Orkneys and a strong cyclone with stormy winds approached the Antarctic Peninsula and the Drake Passage from the west. In front of this an anticyclonic ridge brought calm but cold weather to the southern Weddell Sea. All temperatures were lower than on 19th January 1984.

This example shows that – unlike the Northern Hemisphere – even in summer very intense cyclones do occur. In fact, they are frequent.

4.5 Northward Displaced Frontal Zone, Anticyclone over the Weddell Sea

Normally, October is the month with the most intense cyclones and very stormy weather over the northern part of the Weddell Sea (see Fig. 4). This was not the case for the end of October 1957 and the first days of November 1957 (Fig. 16). Instead of a low pressure belt, an anticyclone

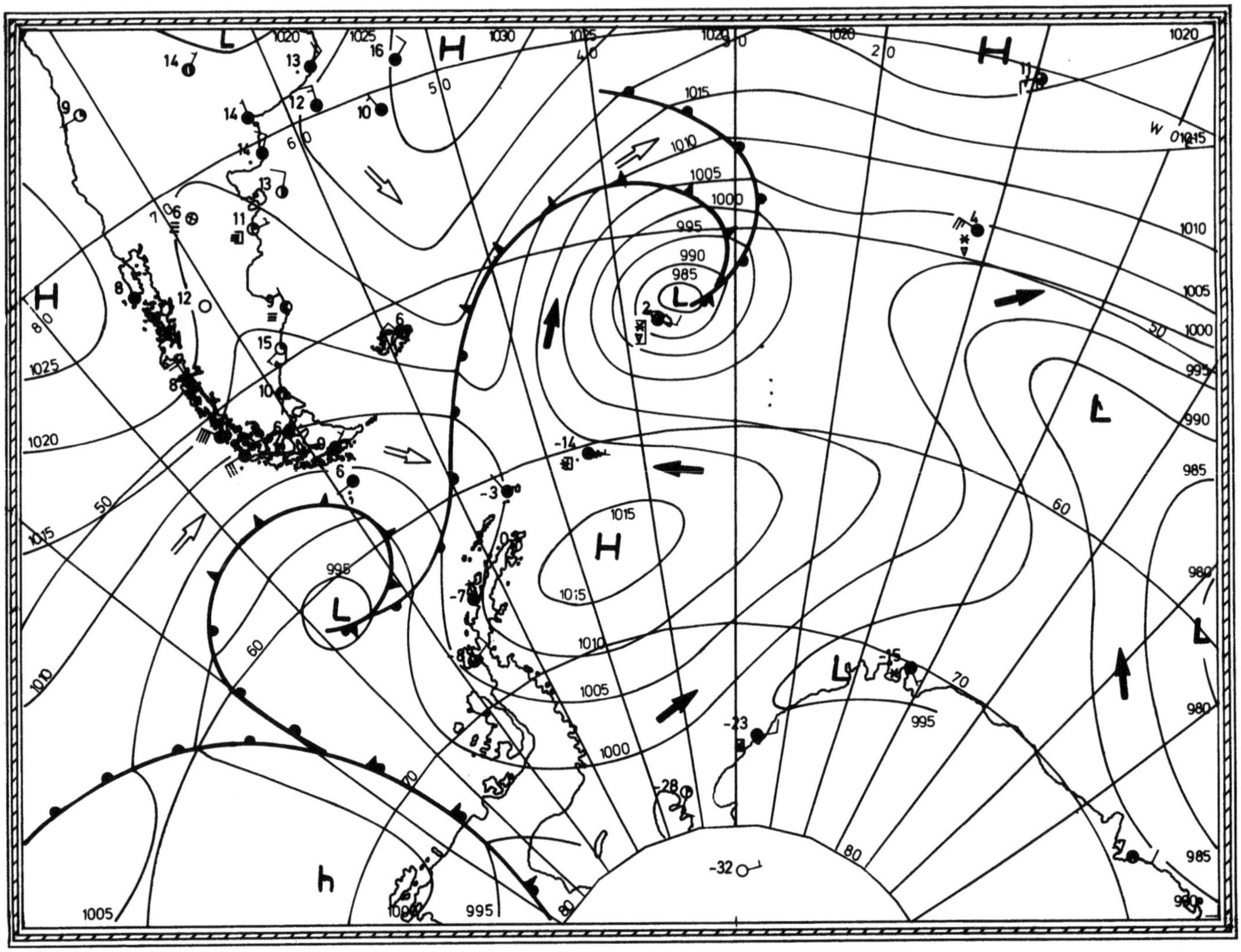

Fig. 16. Synoptic situation of 31 October 1967, 12 UTC: Northward displaced frontal zone; anticyclone over the Weddell Sea

of 1016 hPa was situated east of the Antarctic Peninsula, where it was nearly stationary. The depressions migrated from the Drake Passage northeastward to South Georgia where some of them re-intensified. Therefore, the westerly belt was displaced far northwards over the South Atlantic and there were moderate to weak winds in the Weddell Sea. In the realm of the anticyclone temperatures were lower than normal over the ice-covered sea surface, especially at the South Orkneys (–14°C instead of normal –3°C).

References

Kirk A (1985) Untersuchungen zur Allgemeinen Zirkulation im Weddell See-Gebiet (Antarktis). Mitt Inst Geophys Meteorol, Univ Köln, 106 p

Republic of South Africa, Weather Bureau (1962) International Geophysical Year (1957–58), World Weather Maps Part III, Southern Hemisphere South of 20°S, Sea Level and 500 mb Charts. Weather Bureau, Pretoria

Schwerdtfeger W (1960) The seasonal variation of the strength of the southern circumpolar vortex. Mon Weather Rep 88(6):203–208

Schwerdtfeger W (1970) The climate of the Antarctic, chap 4. In: Orvig S (ed) World survey of climatology, vol XIV. Elsevier, Amsterdam, pp 253–355

Schwerdtfeger W (1984) Weather and climate of the Antarctic. Developments in atmospheric science, vol 15. Elsevier, Amsterdam, 261 p

Schwerdtfeger W, Prohaska F (1956) Der Jahresgang des Luftdrucks auf der Erde und seine halbjährige Komponente. Meteorol Rundsch 9:33–43, 186–187

Seifert W (1984) Probleme der Wettervorhersage in küstennahen Gebieten der Antarktis. Promet 2/3 1984. Deutscher Wetterdienst, Offenbach aM, pp 3–7

Taljaard JJ (1967) Development, distribution and movement of cyclones and anticyclones in the Southern Hemisphere during the IGY. J Appl Meteorol 6:973–987

van Loon H (1967) The half-yearly oscillations in middle and high Southern latitudes and the coreless winter. J Atmos Sci 24(5): 472–486

van Loon H, Rogers JC (1984) Interannual variations in the half-yearly cycle of pressure gradients and zonal wind at sea level on the Southern Hemisphere. Tellus 36A:76–86

van Loon H, Taljaard JJ, Jenne RL, Crutcher HL (1971) Climate of the upper air: Southern Hemisphere, vol II, NAVAIR 50-1C-56 and NCAR TN/STR-57. Boulder, Colorado, 39 p

Some Indications for Environmental and Krill Resources Variability in the Southern Ocean

D. Sahrhage[1]

Summary. Indications for long-term and interannual variations in environmental parameters in the Southern Ocean, particularly the Atlantic sector, are described. Trends in average annual air temperatures show considerable changes and so does the extent of the ice cover from year to year. An attempt is made to identify years with particularly "warm" or "cold" conditions in the Atlantic sector (South Orkney Islands), and to relate this to variations in the distribution of the krill.

1 Introduction

Variation in the environment and corresponding variation in distribution and abundance of living organisms is a phenomenon widely observed in nature. Changes in the abiotic conditions often have wide-ranging consequences for the living resources and the ecosystem. This might be particularly true for regions with extreme conditions, like the Southern Ocean.

Meteorologists and oceanographers tend to use average values of parameters in their investigations to describe the structure and processes in the atmosphere and the ocean. However, it is the interannual change and seasonal abnormality in the weather, ice, and hydrographic conditions which is of greatest influence on the distribution, growth, reproduction, and mortality of the living organisms. For this reason the biologists are mainly interested in the environmental conditions during a specific year or season.

Examples of environmental changes over long periods of time are the fluctuations between glacial and interglacial phases in the Antarctic which can be traced back for more than 150 000 years (Budd and Smith 1981), and climatic changes, for instance those apparent from comparison of cooler conditions on South Georgia during Captain Cook's visit in the 18th century with the present more favorable climate (Everson 1977). Historical information is also provided by the advances and retreats of glaciers in Antarctica.

Some indications of environmental and krill resources variability are described below for much shorter time spans, i.e., interannual changes within the last few decades.

1 Institut für Seefischerei, Bundesforschungsanstalt für Fischerei, Palmaille 9, 2000 Hamburg 50, Fed. Rep. of Germany

2 Environmental Variability

2.1 Variations and Trends in Air Temperature

There are unfortunately few long time series of observations available from the Antarctic which permit thorough analysis. Those existing are from only a few localities, mostly at or close to the Antarctic Peninsula, which has to some extent different climatic characteristics from other parts of the continent.

Based on such long-term monitoring from the Argentine station Orcadas at the South Orkney Islands (Jacka et al. 1984), Fig. 1 shows the trends in the surface air temperature during the period 1908–1986. It is evident that from 1908 until 1949 the 10-year-running means of annual temperature were below the overall long-term average (–4.3°C), with a particularly cold period between 1922 and 1935. On the other hand, temperatures well above the long-term average were measured from 1950 to the present.

This warming trend during the last 30 years has not only been observed in the Scotia Sea area. The mean annual surface temperatures increased also in the other Southern Ocean regions, and this increase was even more pronounced in Antarctica (Jacka et al. 1985; Mo and van Loon 1985). As shown by Wishart (1985), time series of temperatures in New Zealand, Southern Ocean sea surface temperatures, and isotope oxygen profiles from ice cores at the Antarctic coast of the southeastern Indian Ocean suggest that following an extremely cold period around 1830, a warming of about 1.5°C occurred, with a secondary minimum about 1900. Temperatures in New Zealand also indicate a relatively accelerated increase since 1950. However, Limbert (1984) concluded that temperature trends constructed for West Antarctica and the Antarctic as a whole are inconclusive and do not show any significant warming outside the normal range of temperature variability. He pointed out that the trend of increase in temperature ceased between 1970 and 1975. It appears, nevertheless, that the Antarctic Peninsula area warmed markedly also since 1975. Future observations will show the actual trend over recent years.

Antarctic Ocean and Resources Variability
(ed. by D. Sahrhage)

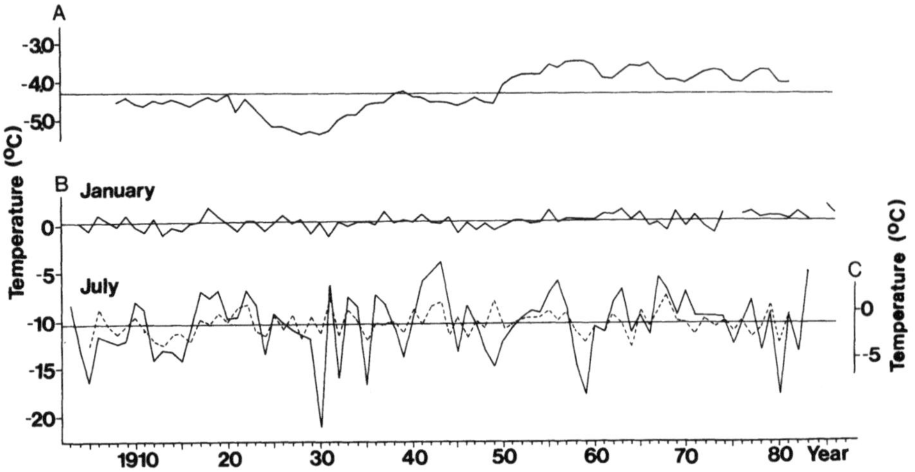

Fig. 1. A Ten-year running means of annual mean surface air temperature at station Orcadas, South Orkney Islands, 1908–1981 (based on data 1903–1986). **B** Average surface air temperatures in January and July at station Orcadas, South Orkney Islands, 1903–1986 (*unbroken line* and *lower left hand* scale). **C** Average surface air temperatures in July at Grytviken, South Georgia, 1905–1981 (*dashed line* and *right hand scale*). *Horizontal straight lines* show means over total period of observations

It seems, however, that the changes in temperature do not follow the same trend in all areas. Schwerdtfeger (1976) found that in the period 1962–1975 the mean annual temperatures showed pronounced positive deviations from the long-term average at stations west of the Antarctic Peninsula, while to its east and northeast, in the Weddell Sea, a decrease of the mean annual and mean winter temperatures was observed. Furthermore, the increasing trend on the western side of the Antarctic Peninsula was faster at Adelaide Island than at the South Orkney Islands, although in earlier years there was a strong positive correlation between the temperature regimes of the stations Argentine Islands and Orcadas. This development was combined with a decrease in the duration of the ice cover in the region of Marguerite Bay. Schwerdtfeger attributed this to a change in the atmospheric circulation pattern with a strengthening of the geostrophic wind component from the northwest. The little information which is available for recent years seems to indicate that the waters of Marguerite Bay are usually open to navigation from late December into February (BAS personal communication).

In contrast to Schwerdtfeger's opinion, Kaufeld (this Vol.) did not detect a long period trend in the atmospheric circulation over the Drake Passage, Scotia Sea, and Weddell Sea during a 29-year time series. Kaufeld refers to similar results by Micka (personal communication) for Orcadas.

In Fig. 1 the two lower curves (scale B) present the average surface air temperatures observed at the Argentine station Orcadas (South Orkney Islands) in January and in July of the years 1903–1986. It appears that in January the average temperatures fluctuate around the long-term average (0.3°C) to a much lesser extent than in July (long-term average −10.5°C) when large changes from one year to the next have been observed. Thus, the July values are better suited to characterize "warm" and "cold" years and seasons. They are taken here as indicators for the general conditions preceding the next summer season. As Limbert (1984) pointed out, the very large winter variations have the greatest effect on the annual average temperature.

The dashed curve (scale C) in Fig. 1 illustrates the mean surface air temperatures in the month of July, observed in Grytviken, South Georgia. It can be seen that the fluctuations around the long-term average (−1.5°C) are much less strong than winter temperatures at Orcadas; however, they follow mostly the same pattern as the curve for Orcadas (a deviation in 1949 is possibly due to a single warm month – the average temperature in August was −1.4°C).

From the curves of mean air temperatures near the South Orkneys in July it appears that the following years could be identified as "warm" or "cold" in the Atlantic sector:

Relatively warm	*Relatively cold*
1917–1919, 1922, 1931, 1941–*1943, 1956,* 1963, 1967, 1968, 1970, *1983, 1985*	1905, 1912–1915, 1924, *1930,* 1932, 1935, 1939, 1945, 1958–1949, 1958–*1959,* 1978, *1980,* 1982

Years with extremes are *in italics.* It is assumed that warm or cold conditions usually continue into the following austral summer season, so that for example *1980* means cold conditions in the beginning of the 1980/81 season, as actually observed. The years listed coincide to a great extent with those identified by Limbert (1974).

On the basis of data from Grytviken, Deacon (1977) characterized the seasons 1927/28 and 1930/31 as cold years and the seasons 1929/30 and 1936/37 as warm years. This is in agreement with the results of Mackintosh (1972), who furthermore showed for the period 1926–1938 that variations of the air temperatures for the middle of each year at the South Orkney Islands were very similar to those of sea temperatures at South Georgia 6 months later.

In this context it has also to be realized that particularly cold years at the Antarctic Peninsula can be particularly warm years at coastal stations outside the Antarctic Peninsula region (like the 1980/81 season), and vice versa (Jacka et al. 1985).

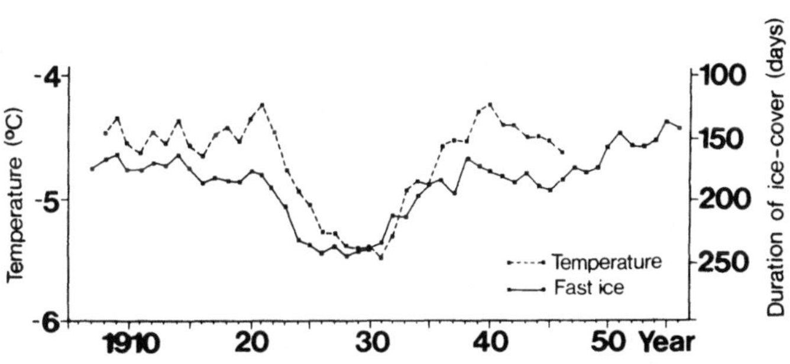

Fig. 2. Ten-year running means of annual temperature and annual duration of fast ice at Laurie Island, South Orkney Islands. (Everson 1977, from Heap 1963)

2.2 Southern Oscillation

One phenomenon which influences the air pressure distribution over the southern hemisphere irregularly is the Southern Oscillation, resulting in the El Niño off Peru (van Loon and Shea 1987 and this Vol.). No long-term trend has been described but it is well known in what years the El Niño was observed (Glantz et al. 1987). From a comparison between El Niño years (1902, 1905, 1911, 1918, 1929, 1939, 1941, 1953, 1957, 1965, 1972, 1976 and 1982) and indications of temperature conditions in the Atlantic sector of the Southern Ocean presented here in Fig. 1 no correlation is apparent. However, Carleton (ms) found that mean monthly sea level pressure indices for higher latitudes of the southwest Atlantic, grouped according to extremes of the Southern Oscillation, show a close systematic relationship between the summer ice anomalies and the antecedent regional atmospheric circulation. Smith et al. (this Vol.) refer to the possible effects on primary productivity. Croxall et al. and Priddle et al. (this Vol.) point out that the seasons 1977/78 and 1983/84, when krill was scarce around South Georgia and land-based predators experienced poor breeding success, both followed years with strong El Niño/ Southern Oscillation (ENSO) events. This will certainly need further investigation. Since the principal domain of the Southern Oscillation is the Indian and the Pacific Oceans, it would be desirable to compare its appearance with long-time data series from these areas of the Southern Ocean, if they exist.

2.3 Variations in the Ice Cover

An important factor which can be of great influence on the living resources is the interannual variability in the ice cover of the Southern Ocean. The long-term trend in the degree of ice coverage at Laurie Island in the Atlantic sector of the Southern Ocean resembles rather closely the variations in the average annual temperatures (Fig. 2). Again, it is evident that the period between 1922 and 1935 was particularly cold. The decrease in the annual duration of fast ice after 1950 is also indicated by the curve. Mackintosh (1972) evaluated available information on variations in ice limits over the periods 1926–1939 and 1950/51. On the basis of deviations from mean latitudes of the ice edge by years and longitudes, he was able to show that in the Southwest Atlantic sector there was an anomalous northward extension of the ice edge during the cold season 1930/31, and a decrease in cover during relatively warm seasons 1929/30, 1932/33, and 1936/37. At the same time he found that there was an inverse relationship between the mean latitude of the ice edge in these waters and in the area east of about 10°W. This means that when the ice extends farther north in the Southwest Atlantic it is usually less extensive in the Southeast Atlantic and Southwest Indian Ocean. This suggests that in some years the flow of cold water from the Weddell Sea, or of the prevailing winds which are generally in the same direction, has a strong northerly component, carrying ice northwards into the Scotia Sea and some way beyond the South Sandwich Islands, while in other years the flow is more to the ENE, leaving the Scotia Sea relatively clear but carrying ice well to the north in the more easterly region. Mackintosh found the suggestion of another point of inflexion about 80–100°E, but not so clear. He concluded that the explanation of variations in the drift of ice is no doubt to be sought in meteorological conditions.

Carleton (ms) found that the substantial annual variations in the rate of ice advance and retreat and the timing of the maximum and minimum extent are linked with large-scale variations in the atmospheric circulation, particularly with cyclonic activity, and with ice advection associated with oceanic circulation. He suggested that the equatorward movement of the sea ice edge may help to promote cyclogenesis over ocean latitudes immediately to the north. However, according to van Loon and Shea (this Vol.), other observations contradict this idea. Greatest interannual variations in ice extent occur in the major Antarctic embayments (Weddell and Ross Seas). Such variations can be rather different at the same time in different sectors (Zwally et al. 1983a).

Certainly strong winds from one direction over longer time can have considerable influence on the ice cover, depending also on the topography of the area. Kaufeld (this Vol.) provides information on periods with particularly strong east and west wind components in the Atlantic sector. For June 1977 and May 1986 he indicates strong east winds, and the result of this weather condition was rather extensive ice coverage in the Bransfield Strait and adjacent waters.

During the period from 1973 to 1977, the mean positions of the ice edge in the Southwest Atlantic sector were found increasingly further south, while during the following years they were shifting back north and remained in the average rather stable during the 1980's (Fig. 3).

In some more detail, the following information is available on the ice conditions in various years or seasons in the Atlantic sector of the Southern Ocean:

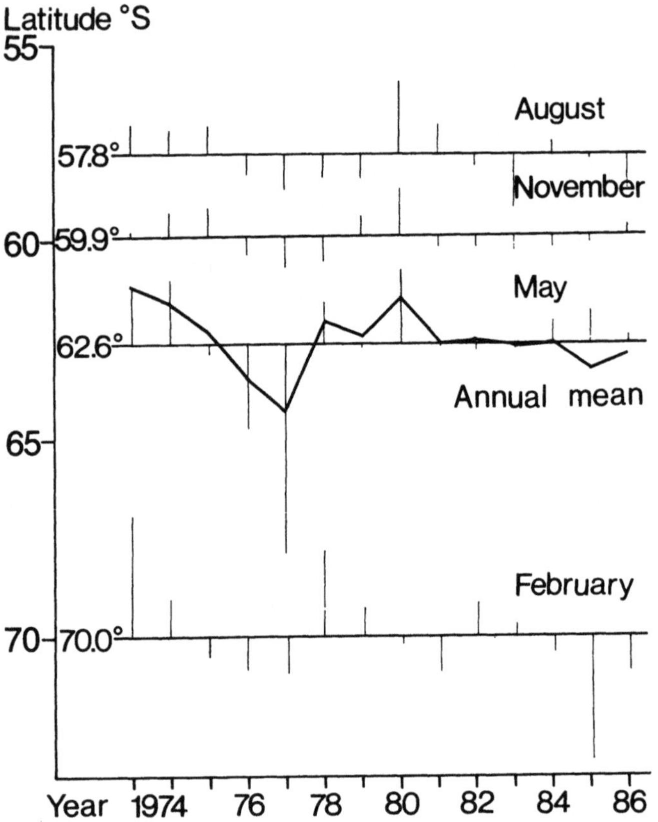

Fig. 3. Mean positions and annual variations of the ice edge in the Southern Ocean between 0° and 60°W at the end of the months February, May, August and November, 1973–1986. Data source: Weekly ice charts of the Navy/NOAA Joint Ice Center, USA. Figure provided by K. Strübing, DHI, Hamburg

1927/28: Deacon (1977) reported the presence of an abnormal number of icebergs near South Georgia.

1966/67: For February/March 1967 Makarov et al. (1970) reported severe hydrometeorological conditions in the South Orkneys area.

1967–73: As pointed out by Zwally et al. (1983b), this was a period with an increase in the areal extent of Antarctic sea ice in approximately the same manner as the decrease during the following years.

1973: According to Carleton (1983) covering the period 1973–77 this was the most severe winter in terms of total area of ice coverage and high ice concentrations, particularly with marked ice increases in the ice extent between about 10°W and 25°E longitude (see also Fig. 3).

1974: Heavy ice winter, particularly during the second half, greatest equatorward extent of sea ice during the 1973–77 period (Carleton 1983).

1976 and 1977: Light ice winters (see Fig. 3).

1977/78: Experience during the German (Fed.Rep.) expedition (November–April) showed heavy ice concentrations in the areas of the Antarctic Peninsula and the southern Scotia Sea, especially before the end of 1977. The situation was much more severe than during a preceding expedition in the 1975/76 season.

1980/81: In winter 1980 the pack ice advanced unusually far north and extended at certain times even as far as north of South Georgia. In November/December 1980 German (Fed. Rep.) research vessels found large ice fields still adrift between Elephant Island and Joinville Island, and in March 1981, during FIBEX, the whole eastern part of Bransfield Strait was covered with ice and icebergs. Similar observations were made by vessels from other countries participating in FIBEX.

1981/82: During the winter of 1981, according to Heywood et al. (1985), satellite observations revealed pack ice extending to the north of South Georgia, indicating a northward shift of the water regimes. Evidence of severe environmental conditions around South Georgia were still present during the 1981/82 summer.

1983/84: During the winter of 1983 the extent and movements of ice in the waters from South Georgia to Elephant Island were average (Heywood et al. 1985). In February/March 1984 American investigations near Elephant Island revealed the presence of warmer than normal surface water (Shulenberger cruise report). Also Chilean observations in January/February 1984, during SIBEX, showed that temperatures and salinities in the upper water layers were in general higher than during FIBEX (Kelly et al. 1985; Rojas 1985).

1986: In May/June the German (Fed. Rep.) research vessel *Polarstern* found that the waters south of Elephant Island and the entire Bransfield Strait right to the eastern and northern entrances into Gerlache Strait were covered with heavy pack ice drifted into these waters from the Weddell Sea during an extended period of easterly winds. Heavy pack ice from the Bellingshausen Sea was also encountered off Adelaide Island.

2.4 Oceanographic and Meteorological Factors

A considerable number of publications are concerned with the relationship between water mass circulation and the distribution of krill and other plankton organisms (e.g., Amos 1984; Lubimova et al. 1985; Witek et al. this Vol.). It seems that krill is often particularly abundant within frontal zones and concentrated in meanders and eddies at boundaries between strong and weak flows of water. However, there are also contradicting observations from the Bransfield Strait, where krill abundance decreased in the boundary zone between waters of Bellingshausen Sea and Weddell Sea origin, while it was higher both north and south of this zone (Siegel 1985). This discrepancy shows that further investigations are required on this important question.

An area of special interest in the Atlantic sector of the Southern Ocean is the Weddell-Scotia Confluence (WSC). Results of USSR investigations in the South Orkney Islands area (summarized by Lubimova et al. 1985), and German (Fed. Rep.) research near Elephant Island (Stein 1986) have shown that there are substantial interannual variations in the position of the WSC and its meanders. From the USSR data it appears that the WSC north of the South Orkneys was found in 1965 close to the islands, while in 1967, 1969, and 1978 it was situated further north, particularly northeast of the islands (Maslennikov and Solyankin this Vol.). There is some indication of stability in the position of the WSC directly north of Coronation Island, and this is one of the main and persistent krill fishing areas. Further to the west, near Elephant Island, changes in the frontal zone are evident but less strong than northeast of the South Orkneys. Again, in an area with relatively stable conditions northwest of Elephant Island, where the isobaths turn into NW direction, there is a seasonally persistent krill-rich area and one of the most important krill fishing grounds around Antarctica (Nast et al. this Vol.). From observations by Nast (1986), it seems that it is not the WSC itself which leads to the aggregation of krill but that the high krill abundance might be the result of the island mass effect with relatively calm waters in the lee of the island. A similar effect could possibly lead also to the high krill abundance north of Coronation Island.

It appears that, dependent on changes in the water circulation pattern as a result of variations in the atmospheric circulation, the WSC location differs rather widely in the waters between the South Orkneys, South Georgia, and the South Sandwich Islands (for details see Everson 1977). Such changes could possibly explain the paucity of krill around South Georgia in certain years (observed in 1977/78 and 1983/84), since krill is not reproducing in these waters, but is carried there from the southwest. Priddle et al. (this Vol.) present a hypothesis for a mechanism which could be instrumental for the aggregation and redistribution of krill in certain areas. In this connection it should be noted that depressions coming from the west first tend to favor a track directed toward the center of the Drake Passage, but then there is a splitting of the zone of maximum frequency into a northern and a southern branch (Fig. 4). Depending on variations in the frequency of depressions going along these tracks, the influence on the location of the WSC may be different with direct consequences for the distribution of krill and other organisms.

Another factor to be taken into account is the variation in the inflow of Weddell Sea water into the southern Bransfield Strait around Joinville Island and through the Antarctic Sound, which can show considerable differences on the seasonal and interannual scales. This water has been observed in the Bransfield Strait to the west as far as the entrance to Gerlache Strait (Heywood 1985). More regular monitoring of the water mass distribution is required.

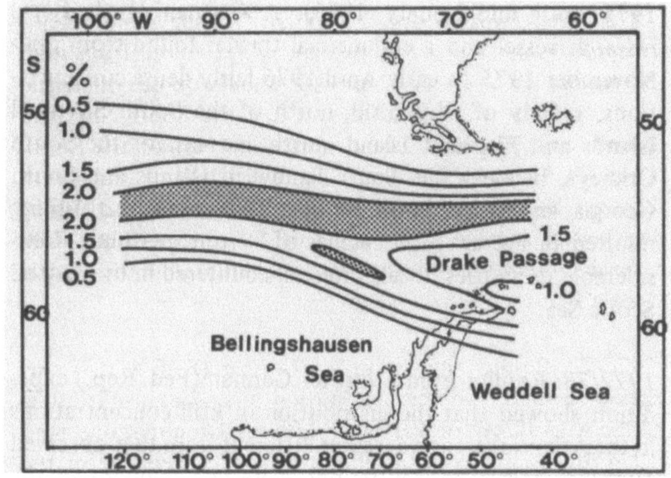

Fig. 4. Frequency distribution (%) of low pressure systems (minima 985 mb and lower) in the Drake Passage region for the period 1901–1904. (Schmitt 1957)

3 Krill Resources Variability

Perhaps the first attempt to deal with annual variation in krill catches was made by A. de C. Baker (in Deacon 1977), to see if there was any obvious correlation between the annual variation in temperature and the abundance of krill. He found that in the particularly cold years 1927/28 and 1930/31 the scientific krill catches of the *Discovery* around South Georgia were considerably higher than the overall mean, and in the warm year 1936/37 considerably lower. Priddle et al. (this Vol.) point out that years when blue whale catches were small around South Georgia can tentatively be identified as seasons with low krill availability in this area. The 1925/26 season, and perhaps the 1929/30 season, could possibly have seen such events.

Further information became available with the intensification of krill research from the mid 1960's onwards, first from the USSR and then more widely, with particularly important input from research activities during the international BIOMASS program with the FIBEX and SIBEX phases:

1966/67: According to results from the USSR, the hydrometeorological conditions in February 1967 were rather severe and the general level of larval development of krill near the South Orkneys was distinctly retarded in comparison with March 1965 and 1969, when it was warmer. No krill larvae were found in February 1967 north, east, and southeast of Elephant Island.

1972/73 and 1973/74: Japanese scientists found that in the southwestern Indian Ocean and Southeast Atlantic krill spawning started earlier in 1973/74 than in 1972/73.

1975/76: Polish investigations indicated that krill spawning in the Antarctic Peninsula area occurred early with a peak already in December 1975 (overall period mid-November

1975 until mid-January 1976). A German (Fed. Rep.) research vessel and a commercial trawler found from mid-November 1975 to early April 1976 fairly dense concentrations, mostly of adult krill, north of the South Shetland Islands and Elephant Island, north and east of the South Orkneys, between the South Sandwich Islands and South Georgia, and east of South Georgia. Semicommercial fishing resulted in average krill catches of 11 tons per hour. Considerable quantities of salps were encountered in the central Scotia Sea.

1977/78: Results from a further German (Fed. Rep.) expedition showed that the distribution of krill concentrations in the Atlantic sector was rather different from that observed 1975/76. Dense concentrations were less frequent in the Antarctic Peninsula region, and most echo traces indicated diffuse krill distribution. Large krill catches of commercial interest were mostly made in the central Scotia Sea but were generally lower than in 1975/76. Spawning of krill lasted from mid January until mid March 1978, with a peak in early February. Off South Georgia, echo traces of small krill concentrations were observed only in early summer but not later during the season. RMT catches from this area contained some krill; however, Mysidacea and *Euphausia triacantha* were much more abundant. Stomach contents of fishes caught at 200 m and deeper consisted largely of krill, indicating that krill was available as food. However, it is not possible to specify at what depths the fish had eaten this krill, and whether krill were possibly abundant deeper than normal. The results coincide with observations by British scientists that there was a dearth of krill swarms near South Georgia which coincided with reduced reproductive success of krill-predating birds and fur seals (Croxall et al. this Vol.).

1978/79: According to Polish observations, massive appearance of krill near South Georgia was delayed until late summer (March). In the Antarctic Peninsula region rather poor reproduction of krill was found mainly in January/February 1979, and larval abundance was fairly low.

1980/81: Heavy ice conditions were observed in the eastern Bransfield Strait during March. Research vessels participating in FIBEX reported very rich catches of krill larvae all along the continental slope from the southern Drake Passage to the waters east of the South Orkneys, indicating a large krill reproduction. It seems that the main period of spawning in this region was relatively short, from mid-December until mid-January, and that spawning had already terminated before February. Siegel (1986) attempted to follow the fate of the brood from the 1980/81 season and also some other year-classes of krill. From the length composition of the krill catches it appears that larvae, born during the 1979/80 season, were rather scarce. It is possible that the 1979/80 year-class suffered from high mortality.

1981/82 and 1982/83: An unusually high proportion of animals born during the 1980/81 season was apparent in the length composition of krill catches.

1983/84: In the Atlantic sector this was an unusual season in several respects. According to Heywood et al. (1985), around South Georgia the mean catch volume of zooplankton in winter 1983 was only about half of what had been observed in summer 1981/82, krill catches with the RMT were 30 times lower, and krill swarms absent. There was some evidence, though, that krill occurred in deeper water layers near the bottom. The paucity of krill led to a serious decrease in breeding success of black-browed albatrosses, macaroni and gentoo penguins, and fur seals at South Georgia in the following summer (Croxall et al. this Vol.). In the Bransfield Strait and Elephant Island area the pack ice had retreated considerably in October, and there was an early onset of the summer season. During German investigations in October/November unusually (for this time of the year) high abundance of krill (age group 1+) was observed. However, research vessels from other nations participating in SIBEX-I during acoustic surveys in December/January found only a rather low biomass of krill. Thus, it is likely that between November and December a change in the distribution of krill occurred. Research during SIBEX-I showed that sea surface temperatures, salinities, and nutrient concentrations were higher than in 1980/81, chlorophyll a concentrations lower. Krill was widely replaced by salps, which were highly abundant (however, not in waters coming from the Weddell Sea) (Witek et al. 1985). But in the main krill fishing areas, particularly northwest of Elephant Island and north of the South Orkneys, substantial krill concentrations were still observed during surveys and also during the fishery (Everson this Vol.). In 1983/84 only small quantities of krill eggs and larvae were found, indicating a very low spawning success. The krill stock consisted mainly of rather small-sized individuals, many of which were born in the previous season. It may be assumed that in the Atlantic sector unusual conditions were created by a shift of the water regimes to the south and the inflow of water from the north. In the Prydz Bay area, in the southern Indian Ocean, there seems to have been also much less krill than during the 1980/81 season. However, observations were less than in the Atlantic sector. In contrast, according to Inagake et al. (1985), a Japanese acoustic survey south of Australia in January/February 1984 showed that the estimated krill stock was about four times larger than that observed in 1980/81 during FIBEX in almost the same area.

1984/85: The failure of the 1983/84 year-class of krill was apparent when in early 1985 extremely few juveniles (1+) were caught in the Bransfield Strait and near Elephant Island. In the former area krill abundance still seemed fairly low, whereas west and northwest of Elephant Island krill concentrations were again observed and fished.

4 Conclusions

From these indications it is apparent that large interannual and seasonal variations in the environmental conditions and the krill resources can be observed.

The results of long time series of measurements, showing the trends in temperature over a period of almost 90 years, prove the great value of such observations. Unfortunately there are still wide areas along the Antarctic coast which are not covered by this monitoring. The establishment of further measuring points in such areas should be encouraged.

It must be realized that measuring stations for logistic reasons are often situated at wind-protected or otherwise favorable positions. Therefore, in certain cases the data obtained may not reflect truly the actual conditions in unprotected neighboring oceanic areas. This deficiency should be kept in mind when establishing further stations. For comparative evaluations of data from various stations it would be beneficial if the method of data collection could be specified.

There is no doubt that changes in the physical environment of the Southern Ocean, with subsequent influences on the living resources, and also variations in the ice cover, are to a great extent forced by atmospheric processes. The interaction between the various factors, however, is so far poorly understood. Besides the few publications which address this problem, there are large meteorological, oceanographic, and biological data sets that could be evaluated jointly by closer collaboration between scientists in these fields.

The variations observed in the krill resources appear to be related mainly to changes in the distribution of krill rather than to changes in the overall abundance. However, there are also indications for good or poor recruitment of krill in certain years.

Acknowledgements. The author wishes to thank Dr. J. Croxall, Dr. I. Everson, Dr. D.W.S. Limbert, Dr. J. Priddle, and Mrs. G. Joan Smith, all from the British Antarctic Survey in Cambridge, for their help and advice. Thanks are due also to Dr. K-H. Kock, Mr. F. Nast, Dr. V. Siegel, and Mr. M. Stein, Hamburg, for useful comments and suggestions.

References

Amos AF (1984) Distribution of Krill (*Euphausia superba*) and the hydrography of the Southern Ocean: Large-scale processes. J Crust Biol 4 (Spec No 1):306–329

Budd WF, Smith IN (1981) The growth and retreat of ice sheets in response to orbital radiation changes. In: Allison I (ed) Sea level, ice and climatic change. IAHS Publ No 131:369–409

Carleton AM (1983) Variations in Antarctic sea ice conditions and relationships with Southern Hemisphere cyclonic activity, winters 1973–77. Arch Meteorol Geophys Bioclimatol Ser B 32: 1–22

Carleton AM (ms) Antarctic sea ice-atmosphere signal of the Southern Oscillation. In: 2nd Int Conf Southern Hemisphere Meteorol, Dec 1–5, 1986, Wellington, NZ Am Meteorol Soc, Boston, pp 431–434

Deacon GER (1977) Seasonal variations in the water temperature and salinity near South Georgia 1925–1937. IOS Rep No 49: 29 pp (mim)

Everson I (1977) The living resources of the Southern Ocean. FAO South Ocean Fish Surv Programme GLO/SO/77/1, 156 pp

Glantz M, Katz R, Krenz M (eds) (1987) The societal impacts associated with the 1982–83 worldwide climate anomalies. Workshop Rep Nat Cent Atmos Res, Boulder, Col USA, 105 pp

Heap JA (1963) Sea ice distribution in the Antarctic between longitudes 7°W and 92°W. Hydrogr Dep London (HD 542)

Heywood RB (1985) Environmental conditions in the Antarctic Peninsula area of the Southern Ocean during the Anglo-German Joint Biological Expedition, February 1982. Meeresforschung 30(4):220–239

Heywood RB, Everson I, Priddle J (1985) The absence of krill from the South Georgia zone, winter 1983. Deep-Sea Res 32(3): 369–378

Inagake D, Matsura N, Kurita Y (1985) Stock and quantitative distribution of the Antarctic krill (*Euphausia superba* Dana) in the Antarctic Ocean south of Australia in January and February 1984. Trans Tokyo Univ Fish 6:139–147

Jacka TH, Christou L, Cook BJ (1984) A data bank of mean monthly and annual surface temperatures for Antarctica, the Southern Ocean and South Pacific Ocean. ANARE Res Not 22

Jacka TH, Christou L, Cook BJ (1985) Updating the sea ice and climate monitoring programs. ANARE Res Not 28:59–62

Kelly R, Blanco JL, Blanco JL, Diaz M (1985) Hydrography of the Bransfield Strait during 1984 southern summer. SIBEX – Phase I. Ser Cient INACH 33:15–45

Limbert DWS (1974) Variations in the mean annual temperature for the Antarctic Peninsula, 1904–1972. Polar Rec 17(108): 303–306

Limbert DWS (1984) West Antarctic temperatures, regional differences, and the nominal length of summer and winter seasons. In: Environment of West Antarctica: Potential CO_2-induced changes. Workshop in Madison, Wisc, 5–7 July 1983. Natl Acad Press, Washington D.C., pp 116–139

Lubimova TG et al. (1985) The ecological pecularities, stocks and role of *E. superba* in the trophic structure of the Antarctic ecosystem. Part II. Select Pap Present Sci Comm CCAMLR 1982, 1984, pp 391–505

Mackintosh NA (1972) Life cycle of Antarctic krill in relation to ice and water conditions. Discovery Rep 36:1–94

Makarov RR, Naumov AG, Shevtsov VV (1970) The biology and the distribution of the Antarctic krill. In: Holdgate MW (ed) Antarctic ecology, vol 1. Academic Press, London New York, pp 173–176

Mo KC, van Loon H (1985) Climatic trends in the Southern Hemisphere. J Clim Appl Meteorol 24(8):777–789

Nast F (1986) Changes in krill abundance and in other zooplankton relative to the Weddell Scotia Confluence around Elephant Island in November 1983, November 1984 and March 1985. Arch Fischereiwiss (Beih 1) 37:73–94

Rojas R (1985) Description of the thermal structure of the Bransfield Strait based on XBT observations. Ser Cient INACH 33: 83–114

Schmitt W (1957) Synoptic meteorology of the Antarctic. In: van Rooy MP (ed) Meteorology of the Antarctic. Weather Bureau, Pretoria, pp 209–231

Schwerdtfeger W (1976) Changes of temperature field and ice conditions in the area of the Antarctic Peninsula. Mon Weather Rev 104 (11):1441–1443

Siegel V (1985) The distribution pattern of krill, *Euphausia superba,* west of the Antarctic Peninsula in February 1982. Meeresforschung 30:292–305

Siegel V (1986) Untersuchungen zur Biologie des antarktischen Krill *Euphausia superba,* im Bereich der Bransfield Straße und angrenzender Gebiete. Mitt Inst Seefisch Hamburg 38:1–244

Stein M (1986) Variability of water masses and currents of the Antarctic Peninsula during SIBEX. Arch Fischereiwiss (Beih 1) 37:25–50

van Loon H, Shea DJ (1987) The Southern Oscillation. Part VI: Anomalies of sea level pressure on the Southern Hemisphere and of Pacific sea surface temperature during the development of a Warm Event. Mon Weather Rev 115:370–379

Wishart ER (1985) Evidence of Southern Hemisphere warming from oxygen isotope records of Antarctic ice. ANARE Res Not 28: 36–44

Witek Z, Kittel W, Czykieta H, Zmijewska MI, Presler E (1985) Macrozooplankton in the southern Drake Passage and in the Bransfield Strait during BIOMASS-SIBEX (December 1983–January 1984). Pol Polar Res 6 (1–2):95–115

Zwally HJ, Comiso JC, Parkinson CL, Campbell WJ, Carsey FD, Gloersen P (1983a) Antarctic sea ice, 1973–1976: Satellite passivemicrowave observations. NASA Sci Technol Branch, SP-459

Zwally HJ, Parkinson CL, Comiso JC (1983b) Variability of Antarctic sea ice and changes in carbon dioxide. Science 220 (4601): 1005–1012

Spatial and Temporal Variability Within the Southern Ocean

A. L. GORDON[1]

Summary. The Southern Ocean is not radially symmetric; many circulation and water mass features vary markedly with longitude. This recommends care in making generalizations of the ocean dynamics and ventilation for the Southern Ocean. The Antarctic Circumpolar Current (ACC) varies in latitude from a position slightly north of 50° S in the Atlantic sector to near 60° S southeast of New Zealand. The ACC displays significant temporal variation, ranging from mesoscale structures of meanders and eddies, to broad regional low frequency variations. South of the ACC are vast areas of more sluggish flow, many of which are organized into cyclonic circulation gyres. The largest is the Weddell Gyre; smaller gyres are located north and east of the Ross Sea, east of the Kerguelan Plateau. Variation in the strength of the western boundary current of the Weddell Gyre may be a primary factor in driving variations in the biology of the region. Wind-induced variations, perhaps coupled to changes in the freshwater balance, would also influence the vertical stability of the water column, the vertical heat flux and the sea ice cover extent. The oceanography along the continental margins is also very dependent on longitude. These variations are related to changes in the geometry of the shelf, the characteristics of the adjacent glacial ice, and on the wind field.

1 Introduction

As we obtain more observations of the physical environment of the Southern Ocean, the better we appreciate the extent of its variability, at a variety of scales. The mean climatic condition exists for only brief periods as the ocean stratification and circulation continuously swings between extremes. Certainly the unique biological community of the Southern Ocean responds to these changes in the habitat. There must be good and bad years for the krill. Our challenge is to understand the causes of environmental variability and the response of krill to these changes. This is a most difficult task. The ocean-atmosphere-ice-coupled system is in itself exceedingly complex. Added to this is the life cycle of krill, which we are only very slowly getting to know.

Southern Ocean processes are responsible for production of most of the water characteristics below the main thermocline of the world ocean. Associated with this is significant poleward heat flux across the Antarctic Circumpolar Current (ACC). This aspect of the Southern Ocean is induced by a combination of regional Ekman upwelling and intense thermohaline or buoyancy forcing by the atmosphere. The cryosphere complicates water mass modification in two ways: the highly spatially and temporally variable sea ice cover strongly influences the coupling of the ocean and atmosphere in regard to momentum, heat, water, and gas exchange; and the ocean interaction with glacial ice influences the characteristics of water masses, and may be a significant factor in glacial ice budgets.

What follows is a brief review of some aspects of the Southern Ocean physical oceanographic environment. Emphasis is placed on the spatial and temporal variability of the water mass and circulation structure. Figures 1 to 4 present the potential temperature and salinity sections along meridians crossing the Southern Ocean. The captions provide some interpretation which supplement the text.

2 The Driving Forces

2.1 The Wind

The wind field is responsible for the Southern Ocean circulation, at least to a first approximation. The westerlies within the circumpolar belt are quite strong, with the maximum westerlies situated in close proximity to the Antarctic Circumpolar Current (ACC). Wind stress within the maximum in excess of 2 dynes/cm^2 occurs over extensive sectors. This is particularly the case in the Indian Ocean sector, where the northern extent of Antarctica's ice cap compresses the atmospheric thermal gradient over the ocean, strengthening the zonal winds. The wind field produces Ekman divergence (upwelling) south of the ACC, convergence (sinking) to the north. This situation, perhaps enhanced by thermohaline forces, is responsible for the very intense shallowing of density surfaces across the ACC, which enables isopycnal "communication" between the colder less saline water near the surface of the ocean south of the ACC with the deep and bottom waters found north of the ACC.

The upwelling poleward of the ACC carries about 45 Sv (1 Sv = 1 × 10^6 m^3/s) into the mixed layer, 2/3 is directed to the north, the rest toward Antarctica, where coastal

1 Lamont-Doherty Geological Observatory of Columbia University, Palisades, New York 10964, USA

Antarctic Ocean and Resources Variability
(ed. by D. Sahrhage)

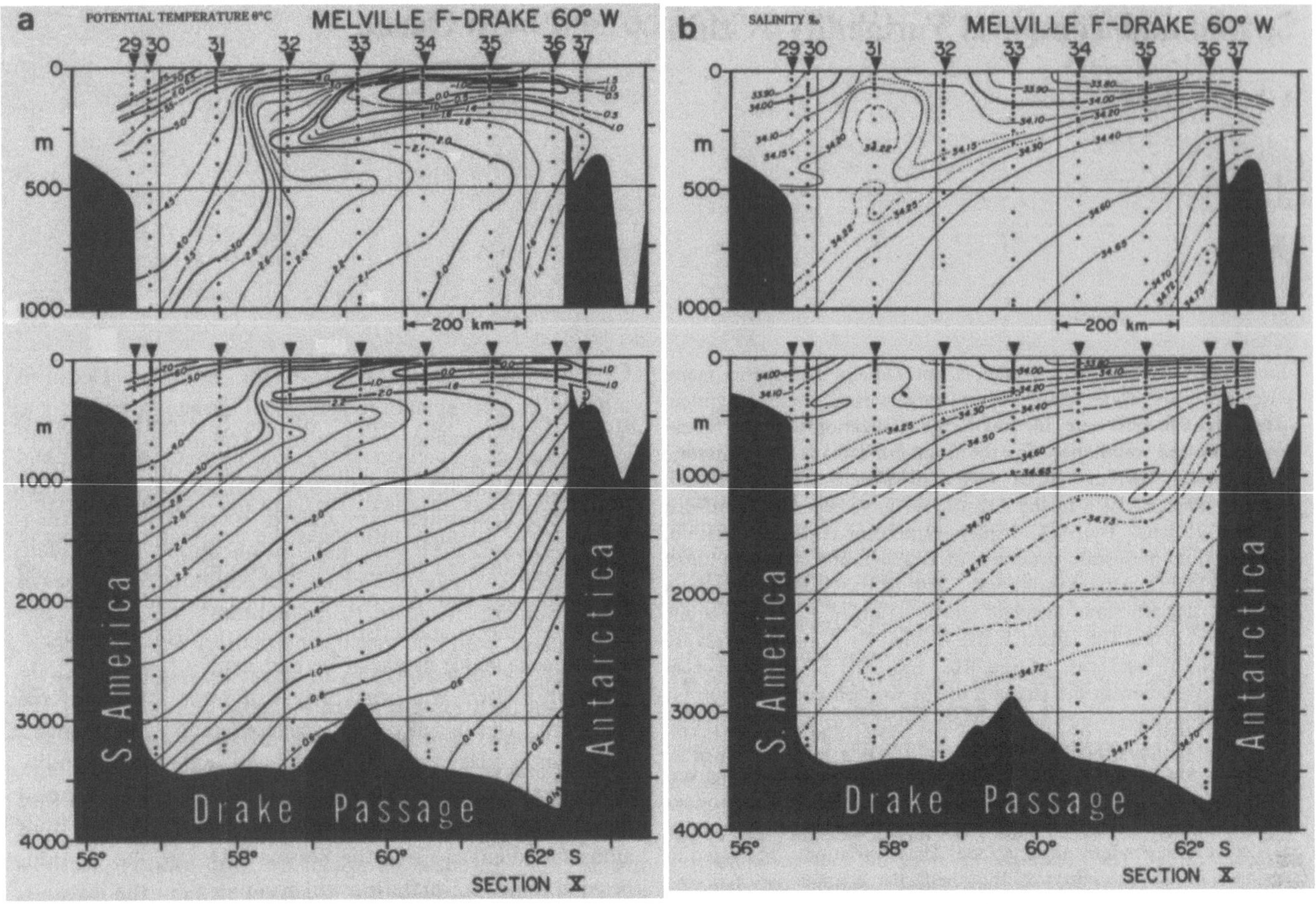

Fig. 1a,b. Potential temperature in °C (a) and salinity (b) across the Drake Passage, from Gordon and Molinelli (1982). The Polar Front (or Antarctic Convergence) falls between stations 31 and 32, near 58° S. Here the temperature minimum, which marks the lower part of the previous winter's mixed layer, begins to deepen and then terminates. North of the Polar Front, a weak salinity minimum extends northward from the temperature minimum layer. This S-MIN marks the initiation of the Antarctic Intermediate Water.

The top of the Circumpolar Deep Water is marked by a temperature maximum near 500 m, south of the Polar Front. The central core of this deep water mass is represented by the salinity maximum running from the bottom at 57° S to 1000 m at 62° 30′S. Below the deep water is Antarctic Bottom Water, which in Drake Passage is not very intense

sinking is expected (Gordon et al. 1977a; Taylor et al. 1978). The residence time within the surface layer poleward of the ACC is approximately only 2 years. There is very little "memory" of thermohaline history on the decadal scale within the surface layer. The wind-induced Ekman upwelling has the effect of shallowing the pycnocline. This effect must be balanced on an annual basis by downward entrainment of surface water. The entrainment is driven by wind-induced mixing and buoyancy removal during the winter period. As discussed in the subpolar gyre section below, this balance is quite critical to the overall heat and freshwater budgets of the mixed layer and to the static stability of the surface layer and seasonal sea ice mass balance.

To the north of the ACC the wind induces a convergent Ekman layer, with sinking of surface water. The maximum northward Ekman transport at the ACC seems inconsistent with the relatively large meridional gradients in surface water properties. This suggests that in the vicinity of the ACC there is strong sinking, which would be associated with formation of Antarctic Intermediate Water (AAIW). North of the ACC deep winter mixed layers may be responsible for the Subantarctic Mode Water (McCartney 1977), which together with AAIW ventilates the lower thermocline and the layer immediately below the thermocline.

Observation of the ACC within the Drake Passage indicates that the ACC transport is closely related to the zonal wind stress (Wearn and Baker 1980). A simple linear Sverdrup model seems to fit the gross latitudinal form of the ACC (Baker 1982). The momentum introduced into the ACC region by the wind field must be dissipated for a steady-state condition. The mechanism to do this is not known. Eddy flux and form drag (mountain drag) are possible dissipators.

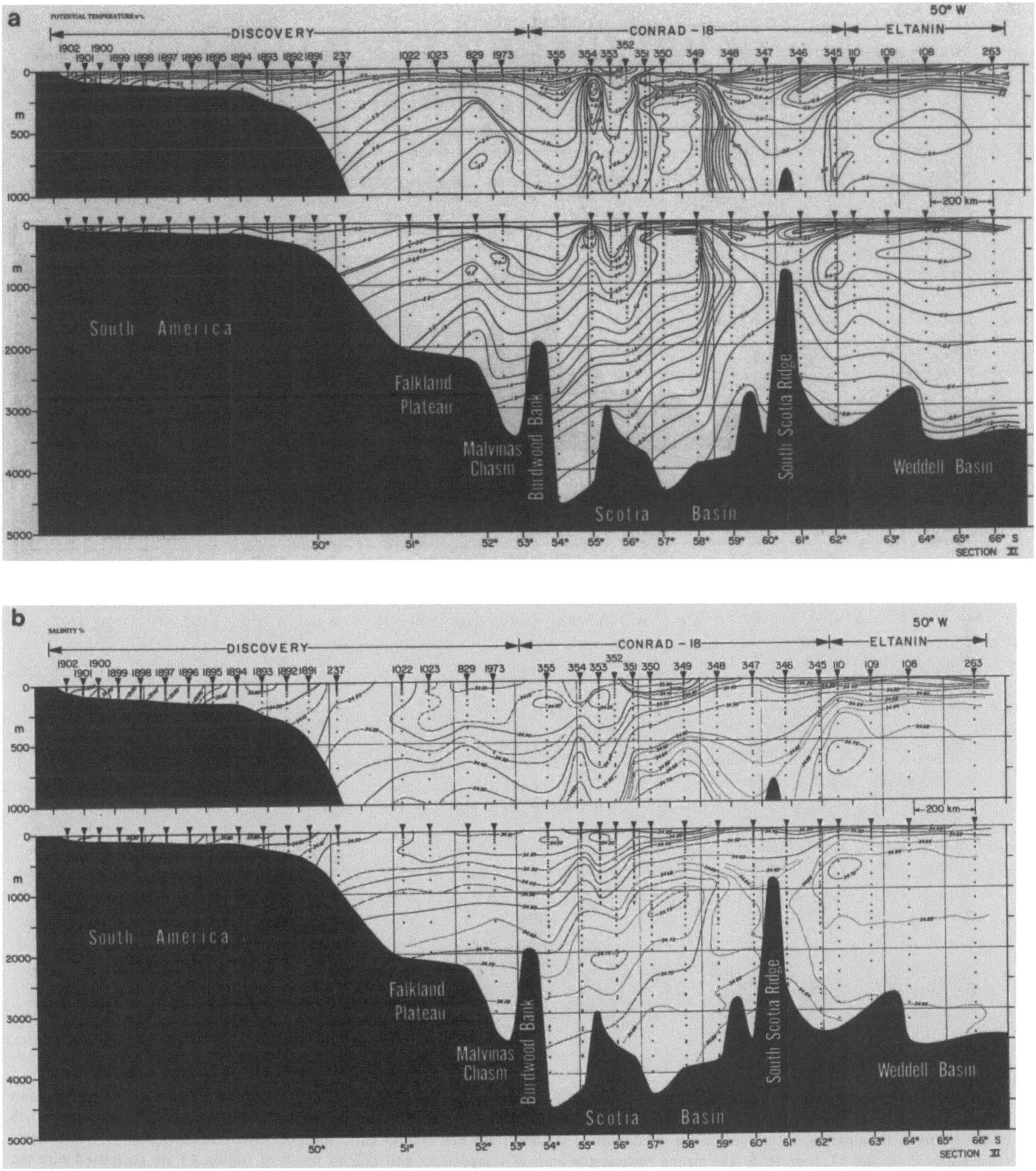

Fig. 2a,b. Potential temperature (**a**) and salinity (**b**) along 50°W, from Gordon and Molinelli (1982). The Polar Front is located just north of station 351 at 57°S, where the temperature minimum layer reaches its northern limit. At station 354 there is a return to the temperature minimum layer within a cold eddy of Antarctic water. Perhaps the warmer surface water at stations 352 and 353 marks a warm eddy of subantarctic water (see Gordon et al. 1977b). These eddies reach quite deep into the water column, as evident by the distortion of the isotherms and isohalines to at least 2000 m.

The strong meridional change of water properties found both south of station 349 (58°S) and the to north of stations 345 (62°S) mark the boundaries of the Weddell-Scotia Confluence. The confluence has colder, fresher deep water than found within the circumpolar water to the north or Weddell Gyre water to the south. The surface layer within the confluence, however, is somewhat warmer and saltier than characteristic of the surrounding region. It is likely that the Weddell-Scotia Confluence is derived from the continental margins of the Antarctic Peninsula and has had a history of strong vertical exchange processes

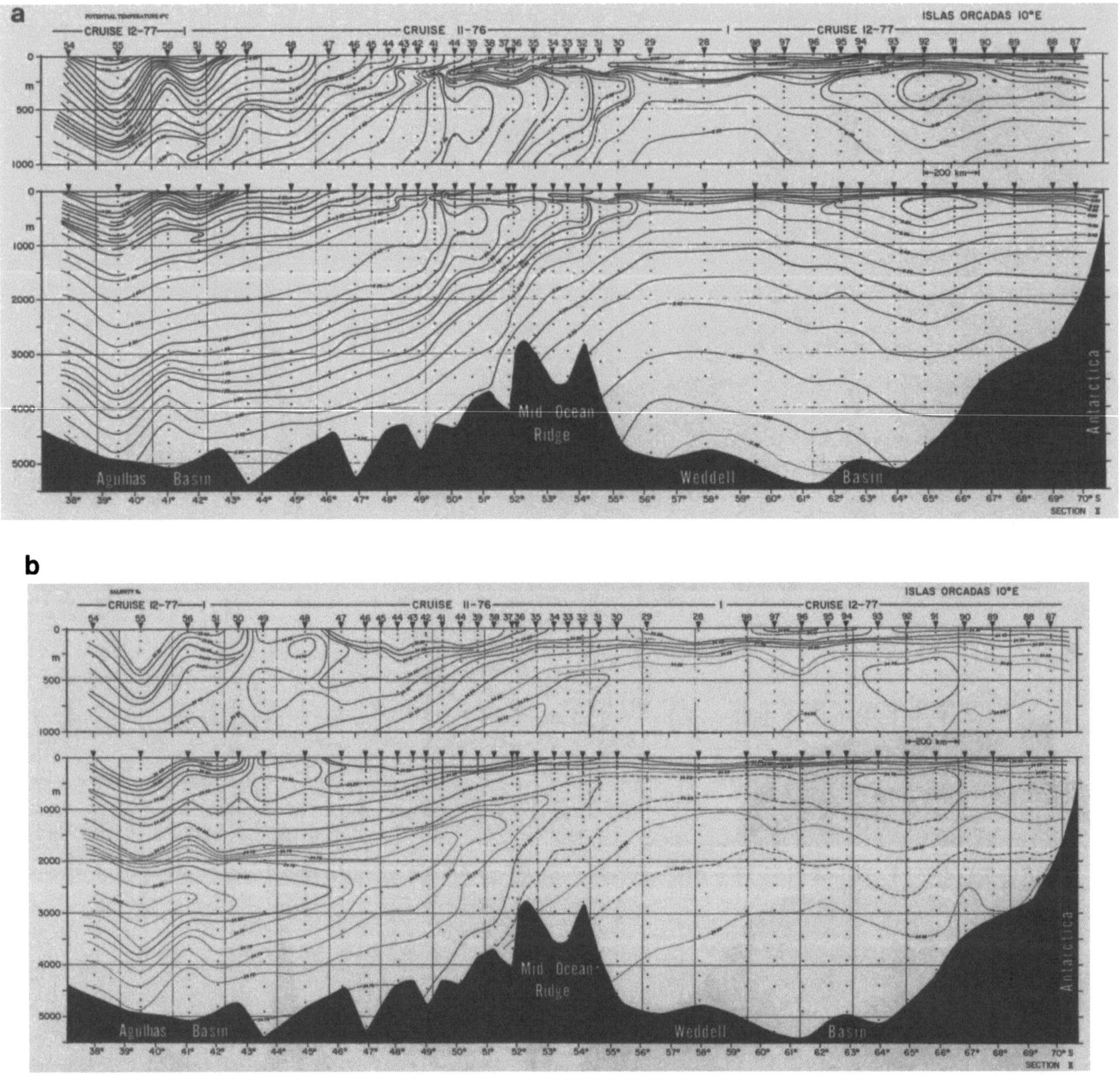

Fig. 3a,b. Potential temperature (a) and salinity (b) along 10°E, from Gordon and Molinelli (1982). The Polar Front falls between stations 41 and 42, near 49°S (the northern most circumpolar limits of the Polar Front). Low salinity surface water extends quite far to the north of the northern termination of the temperature minimum. The subsurface low salinity expression of Antarctic Intermediate Water stems from north of station 47, near 46°S. The strong vertical gradients of temperature and salinity north of station 49 are associated with the Agulhas Retroflection.

At depth the Circumpolar Deep Water salinity maximum is significantly saltier than found in the Drake Passage. This is due to influx of North Atlantic Deep Water within the Atlantic sector. The subsurface temperature maximum south of the polar front, marking the top of the Circumpolar Deep Water, is warmer than 0.9°C to station 30, at 55°S. However, further south it is significantly colder as the northern (eastward flowing) branch of the Weddell Gyre is crossed. At 65°S the return to warm temperature maximum water marks the westward flowing input to the Weddell Gyre.

The bottom water within the eastward flowing branch of the Weddell Gyre is colder than −0.85°C indicating an intense Antarctic Bottom Water flux out of the Weddell Sea

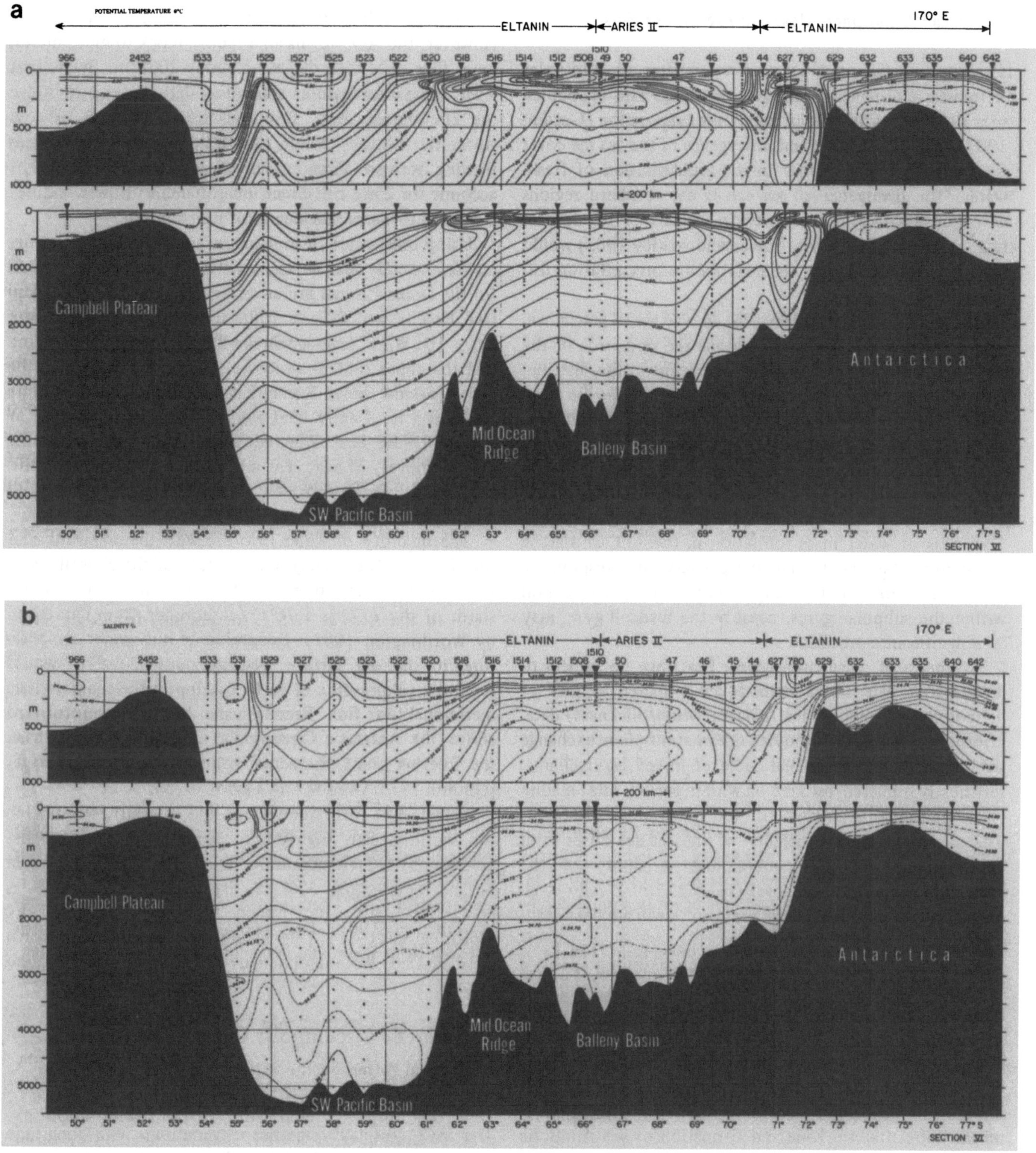

Fig. 4a,b. Potential temperature (**a**) and salinity (**b**) along 170°E, from Gordon and Molinelli (1982). The Polar Front is situated at the northern termination of the temperature minimum layer near 61°S (marking the southernmost limit of the Circumpolar Polar Front). The salinity minimum layer, representing Antarctic Intermediate Water spreads at a depth of 1000 m just south of the Campbell Plateau.

Near 70°S at stations 44 to 46 a low salinity "plume" is seen penetrating to great depth. This feature spreads to the north, forming a deep salinity minimum to the mid-ocean ridge. Below this minimum, the increased salinity to the sea floor can be traced to the Ross Sea. This saline bottom water mass is derived from the salty-cold shelf water of the Ross Sea

2.2 Ocean-Atmosphere Heat and Freshwater Flux

The Southern Ocean is exposed to outflow of cold polar air masses and hence is subject to cooling. This is coupled to a slight excess of precipitation over evaporation (Gordon 1981). The convergence and divergence of the sea ice cover is a very significant factor in the redistribution of freshwater. The divergence of sea ice in the southern regions with convergence in the northern regions of its domain (e.g., in the Weddell Gyre, Hibler and Ackley 1983) would have a larger effect than the evaporation/precipitation difference.

The ocean plays a primary role in wastage of glacial ice, mainly by direct melting of ice along the underside of the shelf ice. The glacial melt water is evident in the distribution of isotopically light oxygen and helium (Jacobs et al. 1985; Schlosser 1986) in the ocean including within Antarctic Bottom Water (AABW). The freshwater input only marginally maintains regional static stability. In some places, where deep water upwelling is more intense and/or there is a divergence of sea ice, deep-reaching convection occurs. The areas of water mass formation in the ACC and along the continental margins are the primary sites responsible for ocean ventilation, though deep-reaching convection within the subpolar gyres, notably the Weddell gyre, may be a significant contributor.

Estimates of ocean-atmosphere heat flux are subject to large errors. The direct methods are hindered by lack of meteorological and sea ice (concentration/thickness) data. The indirect method of studying ocean-atmosphere exchange by inspection of the annual cycle of mixed layer characteristics is hindered by lack of winter period data, mainly below the extensive (20×10^6 km^2) winter sea ice cover. The strong seasonal variability of the sea ice cover with approximately 10–20% interannual variability of the maximum winter sea ice cover (see Fig. 5 of Zwally et al. 1983) further complicates estimates of mean annual ocean-atmosphere energy exchange.

Ocean-atmosphere heat flux has been estimated by a few authors for various sites or latitudes around Antarctica (Gordon and Owens 1987). South of 60°S, which marks the northern limit of the seasonal sea ice cover, heat loss is large (an annual average of approximately 30 W/m^2, Gordon 1981), but highly dependent on ice cover. The heat loss north of the polar front leads to the deep mixed layers of the sub-Antarctic zone and formation of sub-Antarctic mode water.

Heat flux across the ACC and polar front (average 53°S) is 3.1×10^{14} W. The heat flux across 60°S is larger, amounting to 5.4×10^{14} W. This net poleward heat flux must be accomplished by the ocean to maintain a steady state condition. The ocean can accomplish this by three methods: mean flow, eddy flux, and Ekman layer transport. The Ekman flux carries heat to the north (1.5×10^{14} W), so the mean and eddy flux must carry 4.6×10^{14} W to the south. Mean flow is ruled out as being significant in regard to meridional heat flux by deSzoeke and Levine (1981). However, the ACC forms an effective block to mean meridional flux only to a depth of about 2000 m. Below that level, submarine features allow boundary currents to bridge the circumpolar belt. The near-bottom currents of cold Antarctic Bottom Water (AABW) balancing southward-flowing warmer Circumpolar Deep Water (CDW), might account for 25%, but direct measurement of these features is needed to assess this means of heat transfer.

Using time series current meter and thermistor measurements, the eddy heat flux across the ACC has been estimated for the Drake Passage and at a site southeast of New Zealand (see Nowlin and Klinck 1986 for a review); values are: 3.7×10^3 W/m^2 (average Drake Passage) to 30×10^3 W/m^2 (for the upper 2000 m in northern Drake Passage and at the New Zealand site). Extrapolation around the Antarctic leads to a total eddy heat flux value of 3 to 6×10^{14} W poleward heat flux. Thus eddy heat flux seems to be the prime "mover" of heat. The unknown is how representative the Drake Passage and New Zealand areas are of the full circumpolar belt.

The quantity of water involved with the poleward heat flux is large. The average temperature of the abyssal ocean that has isopycnal communication with the surface water south of the ACC is 1.76°C (determined from data given by Worthington 1981). Reduction of this water temperature to the sea surface freezing point, −1.9°C, would require a volume flux of 32 Sv to support the characteristic meridional heat flux. Because water that returns northward across the Antarctic Circumpolar Current is warmer than the freezing point, the actual volume flux would be larger (Gordon 1975; Gordon and Taylor 1975).

Evidence of this large volume flux is clearly observed in the world ocean. Eighty-four percent of the 200 most voluminous T/S (temepraturе/salinity) modes in the world ocean, representing 57% of the world ocean volume is colder than 2°C, determined from data presented by Worthington (1981). They cannot be associated with North Atlantic Deep Water and hence must be Antarctic in origin.

3 Antarctic Circumpolar Current (ACC)

2.1 Spatial Pattern of the ACC

The ACC displays significant variations with longitude (Fig. 5; Gordon et al. 1978). These are related to bottom topography, although the exact nature of this coupling still needs further study, particularly in regard to momentum dissipation.

South of Australia the ACC follows the northern flank of the zonally oriented mid-ocean ridge. The ACC tends to track the ridge crest as it shifts poleward, south of the Tasman Sea, although the ACC displays many waves and eddies in this region, as shown by satellite altimeter and drifter trajectories. The ACC seems to split south of the Campbell

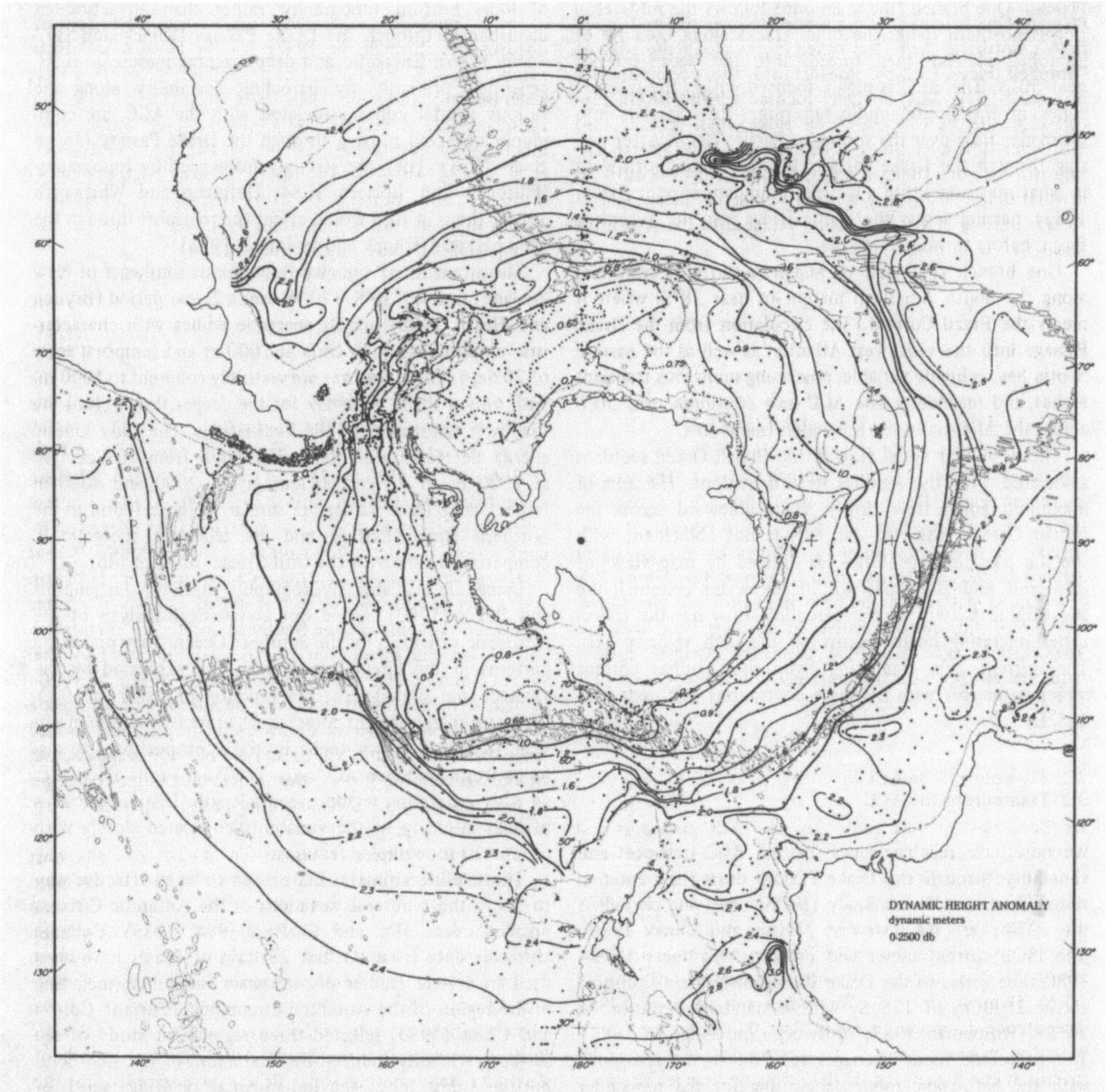

Fig. 5. Dynamic height anomaly of the sea surface relative to 2500 decibars (approximately 2500 m), from Gordon and Molinelli (1982). The lines of equal dynamic height anomaly represent streamlines of geostrophic currents at the sea surface, assuming zero geostrophic current at 2500 db. Since the flow at 2500 db is small when compared to the surface current within the circumpolar belt north of 60°S, the pattern shown for the Antarctic Circumpolar Current is believed to represent the actual mean surface circulation. South of 60°S the barotropic circulation component (flow at 2500 db) is stronger, but the surface circulation pattern is probably still a valid representation.

The cyclonic (clockwise flowing) feature of the southern extreme of the Atlantic sector (south of 55°S) is the Weddell Gyre. The cyclonic feature north of the Ross Sea is the Ross Gyre. The Agulhas Retroflection is the area of closely spaced streamlines immediately south of Africa.

Note: The direction of the geostrophic flow is such that lower values of dynamic height are found to the right of the current. The speed of the current is proportional to the separation of the streamlines.

Plateau. One branch (the main one) follows the mid-ocean ridge's northern flank, the other tracks along the edge of Campbell Plateau, then spreads into the ocean interior near 50°S. The ACC is highly focused within the fracture zones in the Pacific mid-ocean ridge, then spreads into a broader flow over the southeast Pacific Basin. After passing through the Drake Passage, the ACC turns northward in what may constitute a western boundary current (Baker 1982), passing across the Scotia Ridge into the Argentine Basin, before turning to the east.

One branch (Falkland or Malvinas Current) continues along the South American margin to near 38°S, where it meets the Brazil Current. The circulation from the Drake Passage into the southwest Atlantic, as well as the eastern Scotia Sea, is highly variable, possessing numerous transient eddies and meanders. The ACC axis continues near 50°S across the Atlantic to the Kerguelen Island area.

The dominant zonal flow in the Indian Ocean sector is associated with the Agulhas Return Current. The axis of maximum zonal flow slowly shifts poleward across the Indian Ocean. However, the axis is not coincident with specific particle tragectories (as defined by map views of isotherms and isohalines within the water column); the geostrophic pattern of the baroclinic flow has the trajectories migrating north relative to the high velocity axis. The current axis emanating from the Agulhas current structure merges with the more continuous ACC axis near 125°E.

3.2 Transport of the ACC

We now have reliable values for the ACC transport and variability through the Drake Passage during the International Southern Ocean Study (ISOS) observing period of the 1970's (see the review by Nowlin and Klinck 1986). The ISOS current meter and pressure transducer 1979–1980 time series in the Drake Passage indicate a transport above 2500 m of 125 Sv with a standard deviation of 10 Sv (Whitworth 1983; Whitworth and Peterson 1985). The baroclinic mode accounts for 70% of the transport, with the barotropic mode accounting for the remainder mainly at the higher end of the frequency range. The pressure transducer data set (from sensors placed at 500 m depth on both sides of the Drake Passage), which extend to March 1982 (Whitworth and Peterson 1985) agrees with the current meter results to within 25 Sv.

3.3 Variability of the ACC

Significant variability occurs within the ACC at a variety of spatial and temporal scales. It extends throughout the water column (Klinck and Hofmann 1986). However, time lags of 1 to 3 days are observed in the deep levels relative to the 500-m variability, which is attributed to the effects of local bottom topography rather than advection of disturbances through the Drake Passage (Klinck and Hofmann 1986). Energetic and deep-reaching mesoscale rings, generated primarily by baroclinic instability along the various frontal zones associated with the ACC, are commonly observed moving through the Drake Passage (Joyce et al. 1978). They are strongly influenced by topography (Pillsbury and Bottero 1984; Hofmann and Whitworth 1985); they, in turn would affect the transport through the deep passages (Klinck and Hofmann 1986).

Measurement of velocity fluctuations southeast of New Zealand, near 49°30'S, 170°W over a 2-year period (Bryden and Heath 1985) reveals energetic eddies with characteristic amplitudes of 20 cm/s at 1000 m and temporal scale of 20 days. These features are vertically coherent to 5000 m, with only a slight tendency for the deeper flow to lead the shallower expression of the fluctuations. The eddy kinetic energy decreases with increasing depth from 169 cm^2/s^2 at 1000 m to 58 and 38 cm^2/s^2 at 2000 and 4000 m respectively. These values are similar to those found in the northern Drake Passage and are relatively energetic in comparison to those of the Gulf Stream or Kuroshio.

Using all available hydrographic stations, Lutjeharms and Baker (1980) carried out a statistical analysis of the mesoscale variability of the Southern Ocean. They presented patterns in the circulation intensity as expressed by the variance and spatial structure function of the 0 to 1000-decibar dynamic height interval. The intensity of the mesoscale field was shown to be inversely proportional to distance from the mean ACC axis. The Polar Front may act as an eddy-generating region over its length. Distinct patterns of high intensity of the variability correlated closely with prominent topographic features.

The satellite altimeter has proven to be an effective way to study the temporal variations of the Antarctic Circumpolar Current (Fu and Chelton 1984, 1985). Colinear altimeter data from the last 25 days of Seasat have been used in several studies of mesoscale variability, including in the region of the Antarctic Circumpolar Current. Colton and Chase (1983) selected three regions for study of sea surface variability induced by interaction of the ACC with bottom topography: the Indian-Antarctic Ridge south of Australia, representing zonal flow over a zonal ridge; the Macquarie Ridge southwest of New Zealand, representing zonal flow over an isolated bump; and the Indian Mid-Ocean Ridge south of Africa, representing zonal flow over a meridional ridge. Residuals of colinear tracks from the mean profile were calculated to represent variations of sea surface height.

Cheney et al. (1983) presented a global distribution of mesoscale variability of the sea surface. The high variability of the ACC extended nearly continuously around Antarctica. Only in the extreme southeast Pacific and in the central South Atlantic were values of variability less than 5 cm based on 2° gridded values. Values in excess of 6 cm were found over much of the ACC path, decreasing to north and south. The largest variability again seemed to be associated with areas of major topographic relief. Fu and Chelton (1985) used the 3-month set of Seasat crossover differences to study large-scale temporal variability of the Antarctic Circumpolar Current. They showed eastward acceleration of the current (with some deviations from the trend), which appears to be associated with bottom topographic features. Variability was large over the northern boundary of the Weddell Gyre.

Using data from the drifting buoys deployed for the 1979 First Global Atmospheric Research Program Global Experiment (FGGE), Garrett (1981) computed the mean and eddy kinetic energy on a 10° latitude by 10° longitude grid for the Southern Ocean. The results showed a general increase in kinetic energy levels, particularly the mean, associated with the ACC. However, the grid size used was too large to permit much definition. Patterson (1985) generated hourly buoy positions from smoothed trajectories. The drifter trajectories showed highly zonal distribution of surface mean kinetic energy associated with the Antarctic Circumpolar Current. Relatively high mean kinetic energies (500 cm^2/s^2) appear within the ACC near major bathymetric features. The two regions of relatively low kinetic energy (Patterson 1985) are in the extreme southeast Pacific and south of Australia, upstream of major topographic obstructions. Most of the kinetic energy of the surface circulation is in the eddy field. Eddy kinetic energy was calculated as the difference between total and mean kinetic energies for each 5° x 5° box. Unlike the mean kinetic energy distribution, which is zonal, the highest values of eddy kinetic energy are associated with western boundaries. Secondary maxima occur in patches within the zonal flow of the ACC shown in the mean kinetic energy pattern. This suggests that energy emanates from southern sources in the western boundary regimes, as well as from regions with major topographic features of variability. The contribution to the variance by local wind effects may not be large.

Peterson (1985) compared drifter tracks to the extensive 1979 ISOS current meter array. The deep flow, as measured by the current meters, was well reflected in the drifter tracks. Deviations correspond to the wind action on the drifters and on the surface layer of the ocean. The drifters generally moved at 3.4% of the wind speed at an angle of 25° to the left of the wind.

3.4 Circumpolar Frontal Zones

There are various fronts within the Southern Ocean (Deacon 1982). While many are circumpolar, such as the Polar Front (or Antarctic Convergence), they, like the ACC, display much variability in characteristics with longitude. Within the upper water column, characteristics change abruptly across the fronts. The fronts separate stratification zones (Gordon et al. 1977a,b) with the stronger baroclinic currents associated with the front. The water masses associated with the fronts and zones of the Drake Passage are discussed by Gordon et al. (1977a,b) and Sievers and Nowlin (1984). The ACC is composed of two current cores separated by a zone, the Polar Front Zone, in which the near-surface characteristics are intermediate between those of the Antarctic Zone south of the current and the Subantarctic Zone to the north. The fronts are the Polar Front and the Subantarctic Front. They are probably circumpolar though significant spatial variations occur (Emery 1977; Hofmann 1985). Widths of the fronts are 50 km or less at the Drake Passage (Nowlin and Clifford 1982). At the continental margin is the shelf-slope front and marking the northern limit of the Weddell Gyre is the Weddell-Scotia Confluence. These are discussed below.

The geostrophic shear at the fronts, with reference to measured subsurface currents or to deep reference levels, indicates maximum geostrophic surface speeds at the ACC cores of 30–45 cm/s at Drake Passage. With a very deep pressure surface as the reference, closely spaced stations along the Greenwich Meridian yield similar surface speeds. The geostrophic shear associated with these fronts extends deep into the water column to the bottom in Drake Passage (Nowlin et al. 1977).

3.5 Eddies and Rings

A large amount of mesoscale variability within the Antarctic Circumpolar Current is associated with its zonation. This eddy variability is due to a combination of factors. The current cores migrate laterally (Nowlin et al. 1977), by as much as 100 km in 10 days. Meanders, or waves, form and propagate along the Antarctic Circumpolar Current fronts. The first synoptic observations of their zonal propagation was from satellite and moored current meters (Legeckis 1977; Sciremammano 1979). Subsequent observations by larger arrays of moored instruments, satellites, and surface drifters have documented the commonplace existence of such meanders.

These meanders sometimes develop into closed current rings. The first such observation of formation was reported by Joyce and Patterson (1977); many more observations of rings have followed. Cold and warm core rings have both been observed. They appear to form from both the Subantarctic Front and the Polar Front. Bryden (1983), summarizing the characteristics of eddy variability observed in the Southern Ocean, stated that such variability is found everywhere observations have been made: eddies are 30–100 km wide, their surface velocities are typically 30 or more cm/s, and eddies are vertically coherent from surface to bottom. Numerous current rings and meanders have been reported at Drake Passage. Heat, salt, and mechanical and potential energies of various rings have been estimated and reported (Joyce and Patterson 1977; Peterson et al. 1982; Pillsbury and Bottero 1984).

Cold core rings have been observed forming from the Polar Front on two occasions (Joyce and Patterson 1977; Joyce et al. 1981; Peterson et al. 1982). Counting such occurrences or the ratio of warm core to cold core rings is difficult. From time series at moored arrays, Pillsbury and Bottero (1984) interpreted the records from a mid passage location as showing that five cyclonic rings and one anticyclonic ring passed from June 1975 through January 1976. The cyclonic ring has the properties of the continental water and is most likely derived from the continental water boundary; the anticyclonic (warm core) ring is believed to have separated from the Polar Front. The rings had diameters of 30–130 km and extended at least 2500 m vertically.

4 North of the ACC

The dominant water mass formed near the polar front is Antarctic Intermediate Water (AAIW; Molinelli 1981). Its low salinity characteristic is clearly seen spreading at the base of the thermocline of the Southern Hemisphere, and into the Northern Hemisphere within the Atlantic. During the past decade, traditional theories of AAIW formation were refined and a new interpretation advanced (McCartney 1977). Sub-Antarctic waters (Sub-Antarctic Mode Water) are characterized by a low vertical stability (pycnostad) and are warmest in the western South Atlantic and coolest in the southeast Pacific. McCartney (1977, 1982) found that southeast Pacific Sub-Antarctic Mode Water undergoes additional cooling in the Drake Passage and western Scotia Sea, and passes into the Atlantic, where it can be identified as Antarctic Intermediate Water. However, Piola and Georgi (1982) calculated that the sea-air heat exchanges are insufficient to convert southeast Pacific Sub-Antarctic Mode Water into South Atlantic Antarctic Intermediate Water. Although a subantarctic origin appears unlikely to account for all types of Antarctic Intermediate Water found in the world's oceans (Piola and Georgi 1982), Sub-Antarctic Mode Water appears to contribute volumetrically significant qualities to the subantarctic and subtropical thermohaline structure (Georgi 1979).

5 South of the ACC

5.1 The Deep Ocean

The deep ocean regime south of the Antarctic Circumpolar Current is influenced by the positive wind stress curl, which in combination with general upwelling induces poleward Sverdrup transport. Everson and Veronis (1975) showed that south of the maximum westerlies, with the Antarctic Peninsula serving as both the eastern and western boundaries, the total Sverdrup transport amounts to 300 Sv. This requires a mean poleward baroclinic velocity of only 0.4 cm/s.

Hydrographic data indicate very weak baroclinic flow south of the ACC; characteristic velocities are 1 cm/s. Values of 5–10 cm/s are associated with boundary currents over the continental slope and the western boundary currents of the three cyclonic flowing subpolar gyres. Much of the circulation is with the large cyclonic flowing gyres. These can be considered to be subpolar gyres, southern hemisphere counterparts to the Greenland gyre and Bering Sea gyre.

The largest and best-defined of these is the Weddell Gyre, extending east of the Antarctic Peninsula to 20–30°E and from Antarctica near 70°S to 55–60°S (Deacon 1979; Gordon et al. 1981; Comiso and Gordon 1987). The others are the Ross Gyre, north and east of the Ross Sea (Tchernia and Jeannin 1983) and a poorly defined gyre east of the Kerguelen Plateau (Deacon 1937; Tchernia and Jeannin 1983; Rodman and Gordon 1982).

The Weddell Gyre is likely to be the source of much of the abyssal ocean characteristics (Reid et al. 1977; Gordon 1982), and thus is of prime concern in regard to open ocean water mass conversion.

5.2 Weddell Gyre Inflow

The inflow to the Weddell Gyre is accomplaished as a branch of ACC water which enters the Gyre near 20°E. This relatively warm salty inflow extends to the Maud Rise region, where some of its characteristics apparently pass westward into the central region of the Weddell Gyre, as eddies. However, the bulk of the warm salty water that continues into the western hemisphere enters into the Weddell Sea (the southwest extreme of the Weddell Gyre) as a well-defined current over the continental slope (Gordon and Huber 1984).

The core of relatively warm and saline deep water over the continental slope in the Weddell Sea is associated with strong baroclinic westward flow. This is likely to be the primary source of oceanic heat and salinity for the western portion of the Weddell Gyre and may feed the production of AABW. Surface flow of the slope current is about 10 cm/s, relative to the sea floor, in general agreement with iceberg drift rates, with characteristic volume transport around 10 Sv (these calculations are based on Islas Orcadas cruise 12 near 10°W; also see Carmack and Foster 1975a,b).

Within the slope current the pycnocline capping the deep water is strongly tilted down to the south, where the winter mixed layer can reach thicknesses of several hundred meters. It remains to be determined whether this is a baroclinic adjustment or the result of intense water mass conversion in response to frequent coastal polynyas.

The slope current enters the Weddell Sea west of Cape Norvegia, where it splits into continental slope and the coastline branches (Gill 1973; Carmack and Foster 1975a; Tchernia and Jeannin 1983). Because of the slope current's great depth, significant shear is likely to be associated with this branching.

The water column ejected into the open ocean at the Weddell-Scotia Confluence no longer contains the deep-water characteristics of the slope current. Apparently the "warm" water drawn into the Southern Weddell Sea by the slope current is completely altered during its transit of the Weddell Gyre and boundary current, making it a primary contributor to a major water mass conversion event.

5.3 Weddell Gyre Outflow

The Weddell Gyre is a large, wind-driven subpolar Sverdrup gyre, with a transport of 70 to 90 Sv (Carmack and Foster 1975a; Gordon et al. 1981). The more or less uniform flow from the surface to the sea floor requires only low speeds, within the northward flowing western boundary current. The average speed of the surface current, 8 cm/s, measured by drifting ships and buoys, if extended to the sea floor can easily account for the large transport. The low degree of baroclinicity of the Weddell water column indicates that such an extrapolation is reasonable.

The outer "edge" of the Weddell western boundary current, which follows the continental margins, is highly modified by thermohaline forces. It is this component that feeds the Weddell-Scotia Confluence. The seaward segment of the current appears to be part of the more slowly ventilated central hub of the Weddell Gyre, part of the Weddell cold regime, discussed by Gordon and Huber (1984). The relative magnitudes of these two components is not resolved with the present data set. The order of magnitude estimate (10 Sv) of the slope current flowing into the Weddell Sea, discussed above, suggests that the outer "edge" of the boundary current would be a small part of the total transport, perhaps amounting to 15% of the total. However, since this "outflow" water is highly altered from the "inflow" state, it may represent a significant flux of newly formed water type, namely Antarctic Bottom Water.

The outflow from the western boundary current of the Weddell Gyre displays variability in deep and bottom water characteristics (Foster and Middleton 1979). In 1976 the bottom water was somewhat colder, saltier, higher in oxygen, and lower in silicate than was the case in 1975. There appears to be variability in the mixing ratios of the ingredients of AABW, or perhaps changes in the contributions from specific sites of formation.

Foster and Middleton (1979) report on a year-long current meter record 50 m above the sea floor in the western region of the Weddell Gyre, though somewhat east of the main segment of the western boundary current. The mean northward flow of 1.3 cm/s was superimposed on active variability. The variability, which appears to be a product of passage of small mesoscale features, is more vigorous during the summer.

5.4 Weddell-Scotia Confluence

Along its northern boundary, the Weddell Gyre merges with circumpolar waters entering the Atlantic via the Drake Passage, creating a zone of low stability, with accompanying intense mesoscale activity. This feature is referred to as the Weddell-Scotia Confluence (see Fig. 2a.b; Gordon 1967; Gordon et al. 1977b; Deacon and Moorey 1975; Patterson and Sievers 1980). It displays intense eddy activity, which increases downstream from the initial contact of the two circulation regimes (Foster and Middleton 1984). The Weddell-Scotia Confluence varies in width from 10's to 100 km scale, presumingly a product of mesoscale features. However, there does seem to be a quasi-stationary form which is related to the topographic expression of the South Scotia Ridge. The fronts marking the northern and southern edge of the Confluence have been called the Scotia Front and the Weddell Front, respectively (Gordon et al. 1977b). The sea ice edge along this front was investigated with the Shuttle Imaging Radar-B (SIR-B) by Carsey et al. (1986). Comiso and Sullivan (1986) compared passive microwave data from satellite with field observations along the ice edge of the region. The sea ice edge also displays irregular patterns responsive to the mesoscale activity. At the Confluence the deep water is anomalously cold and the surface water is relatively salty, both characteristics suggesting the action of vertical mixing. The Confluence region, including the Bransfield Strait along the northern tip of the Antarctic Peninsula, is subject to convective processes and high biological activity (Deacon and Moorey 1975; Gordon and Nowlin 1978; Patterson and Sievers 1980). The deep water temperatures along the northern rim of the Weddell Gyre at the Greenwich Meridian are relatively cold (below 0.4°C), suggesting that Weddell-Scotia Confluence characteristics extend eastward to at least the Greenwich Meridian (Gordon and Huber 1984).

5.5 Static Stability Within the Weddell Gyre

Within the Weddell Gyre, and to some extent throughout the region south of the ACC, the pycnocline is weak; amounting to only 0.2 Sigma-Θ (or kg/m^3). The warmer deep water introduced to the upper surface layer by entrainment processes in the general Ekman-induced upwelling of the pycnocline maintains a near steady state. In view of the marginal stability of the pycnocline of the Weddell Gyre, slight changes in the magnitude of the entrainment and upwelling rates could have a major impact on the thermohaline stratification.

Observations in the Weddell Gyre indicate that about 30 m of deep water enter the surface layer each year (Gordon et al. 1984; Gordon and Huber 1984). They show that the winter mixed layer below the sea ice along the Greenwich Meridian near 60°S is slightly above freezing (average +0.035°C) and is undersaturated (86% of full saturation) in oxygen. This condition is attributed to entrainment of deep water into the surface mixed layer, brought about by buoyancy flux at the sea surface and perhaps more importantly, mechanical stirring due to relative ice movements. The incorporation of deep water heat and salt appears to be a significant factor in the heat and freshwater balance, and to have an important impact on the sea ice budget. An increase in the upward movement of the warm salty water without a corresponding increase in freshwater from the atmosphere or sea ice convergence would lead to weakening of the pycnocline and result in convection.

During the 1986 Winter Weddell Sea Project aboard the *Polarstern* cruise ANT V/2, entrainment of deep water into the winter mixed layer was observed to increase with latitude along the Greenwich Meridian. Entrainment of deep water near Maud Rise (65°S) is perhaps twice that at 60°S.

The marginal static stability of the Weddell water column makes the Weddell Gyre susceptible to deep convective processes (Killworth 1979; Martinson et al. 1981; see Killworth 1983 for a review), such as occurred during the Weddell Polynya episode in the austral winters of 1974, 1975, and 1976 (Carsey 1980). Associated with the Weddell Plynya was massive convective overturning. This cooled the deep water by up to 0.5°C to a depth of 2500 m relative to the pre-polynya state (Gordon 1982).

Shorter-lived but recurring polynyas within the subpolar zone have been observed by satellite near 65°S, 5°E (the Maud Rise Polynya), and near 67°S, 45°E (the Cosmonaut Polynya; Comiso and Gordon 1987). Convective processes at several sites may be intermittent but frequent enough to have a climatic and perhaps biological impact.

5.6 Continental Margins

The water over the continental shelf is exposed to the harshest form of the Antarctic atmosphere, as very cold dry air flows off the continent. Strong winter winds often remove the insulating cover of sea ice adjacent to the coast (Zwally et al. 1985). These coastal polynyas become potential sea ice factories, in which massive amounts of sea ice can form and be quickly transported northward. This in turn induces extreme thermohaline alterations of shelf water, increasing the shelf water salinity by as much as 0.3 ppt. The shelf water contacts the fronts and bases of the glacial ice shelves, further modifying the water masses (Weiss et al. 1979; Jacobs et al. 1979, 1985).

Shelf water is colder than the deep-ocean water column, but displays a wide range of salinity, hence density. At several sites the shelf water density is high enough, even in summer, to allow deep-reaching convection if it escapes from the shelf. Deep-reaching convection, not necessarily to the sea floor (Carmack and Killworth 1978), is often suggested by hydrographic data from the vicinity of the continental margin.

Environmental conditions specifically responsible for the wide range of shelf water characteristics (Gordon 1973), its mixing with deep water, and the formation of Antarctic Bottom Water are not known. Killworth (1983) reviewed observations and argued that five ingredients appear to be involved in production of deep water near oceanic boundaries; a reservoir in which to form dense water, a source of dense water within the reservoir, a "reason" for the dense water to leave the reservoir, the involvement of more than one water mass in dense water formation, and a combination of densities, geography, and dynamics that do actually permit the dense water to sink. Regional influences in the process of recent interest include glacial ice and the atmospheric and coastal polynyas.

One of the strongest fronts of the Southern Ocean separates the shelf and deep water (Jacobs 1986). Across this front must pass the supply of new shelf water, potential bottom water, and large volumes of sea ice and glacial ice. Recipes for mixing at the shelf-slope front have been proposed from summer observations of the integrated effects of water mass conversion, but the actual exchange processes are subject to speculation.

The coldest, freshest, and probably the most Antarctic Bottom Water is formed in the southwest corner of the Weddell Gyre (Carmack 1982). Saltier AABW water forms in the Ross Sea (see Fig. 4a,b), and there is evidence of bottom water formation at many other sites around Antarctica (see Gordon 1973; Smith et al. 1984; Jacobs et al. 1985).

Postulated processes and sources contributing to Weddell bottom water formation have included cabbeling, haline convection by evaporation or freezing in open leads and polynyas, cooling under the Filchner Ice Shelf, Ekman-layer effects, sinking along frontal zones, derivation from deeper oceanic areas, overflow of dense water from the Bransfield Strait, and double-diffusive convection. Estimates of the formation rate of bottom water in the Weddell Sea range from 1 to 5 Sv (Foster and Carmack 1976). Estimates of circumpolar production rates of Antarctic Bottom Water are in excess of 13 Sv (Jacobs et al. 1985). The Antarctic Bottom Water's importance to the world ocean warrants more information on its flux and production processes.

Bottom water production in the Weddell Sea region is discussed by Foldvik et al. (1985a,b) using hydrographic and year-long current meter data. They obtained a complete hydrographic section along the barriers of the Filchner and Ronne Ice Shelves. Ice shelf water is observed leaving the Filchner Depression, forming a slope plume with a transport of 1 Sv producing a total of 2 Sv of new Antarctic Bottom Water. The current meter array at the shelf break displays strong tidal currents, dominated by the solar diurnal component although this component weakens during the winter, perhaps in response to reduced thermohaline stratification.

Middleton et al. (1982) discuss low frequency variability measured by a number of year-long current meter moorings along the shelf and slope of the Weddell Sea. They find evidence of wind-induced shelf waves with periods of 3 to 60 days. The higher frequency waves are anticyclonic while the lower range is cyclonic. Topographic trapping of the wave energy is suspected.

Currents and temperature were measured from January to August 1978 and from February 1983 through January 1984 along the Ross Ice Shelf Barrier by Pillsbury and Jacobs (1985). These data show that the warm water feature apparent in summer hydrographic data between 170 and 175°W flows into the ocean volume below the floating ice shelf. Typical velocities are 5 to 9 cm/s. A mooring within the colder water to the west reveals flow away from the ice shelf, in a pattern consistent with circulation schemes inferred from water mass considerations. Higher kinetic energy levels are evident during the winter, as are generally lower average temperatures, though warm events are common during the winter.

The overall water mass structure and resident times in the Ross Sea are discussed by Jacobs et al. (1985), using hydrography and stable isotopes of oxygen and hydrogen. Warm slope water spreads onto the shelf where it is modified by the atmosphere and interaction with the glacial ice, to form various shelf water masses. The slope water provides the primary heat source for melting of glacial ice. A best estimate for shelf water residence time is 6 years with a total circumpolar bottom water production rate of 13 Sv, accompanying 0.4 m/year of basal ice shelf melting and sea ice production of 1.9 m/year.

6 Polynyas

6.1 Sensible Heat Polynyas

During the mid-1970's there was a large ice-free region in the area of Maud Rise, now referred to as the Weddell Polynya (Carsey 1980). It was maintained by deep ocean convection, and hence is referred to as a sensible heat polynya. The convective overturning left the ocean down to 3000 m much colder and fresher than the situation before the polynya (Gordon 1982). Time series studies of deep water temperature near Maud Rise by Gordon (1982) suggest a previous occurrence of a Weddell Polynya in 1961. The initiation and termination of the Weddell Polynya is not fully understood, but it is likely that an increase in salinity of the surface water (due to ice formation and divergence or by increased deep water upward flux) weakened the pycnocline, allowing convection. The termination is likely to be a result of convergence of sea ice which on melting would attenuate the convection (Martinson et al. 1981).

Shorter-lived recurring polynyas have now been observed in microwave satellite data in the Maud Rise region and at the Cosmonaut Sea area (near 65°S and 45°S; Comiso and Gordon 1987). Both the Maud Rise region and the Cosmonaut Sea are located along the eastern margins of the Weddell Gyre (Fig. 5). Maud Rise is near the western "apex" of an anticyclonic bend in the relative geostrophic steamlines. Cosmonaut Sea is positioned at the eastern terminus of a zonally elongated cyclonic trough in the sea level over the southern extreme of the deep Weddell Basin. The oceanographic environment at both sites is remarkably similar, in that both areas are associated with isolated patches of a relatively shallow gynocline (Comiso and Gordon 1987). These recurring polynyas may have a significant climatic influence, since they form for a few weeks nearly every year.

The difference between a polynya that endures an entire winter versus one that lasts for only a few weeks, may be one of size (Comiso and Gordon 1987): the larger the size of the convective region, the less likely that sufficient freshwater can be injected to shut down the convection.

6.2 Latent Heat Polynyas

Along the coast of Antarctica, there are frequent ice-free belts. These are believed to be due to the wind, which removes the ice as fast as it forms. These are referred to as latent heat polynyas. The coastal polynyas can produce great quantities of sea ice and induce significant thermohaline modification of the shelf water (Zwally et al. 1985). It is likely that these features play an important role in the formation of dense water which contributes to AABW formation.

A recurring polynya in Terra Nova Bay, just north of the Drygalski Ice Tongue, also appears to be a latent heat polynya (Kurtz and Bromwich 1985). It may play a role in supplying salt to the high salinity shelf water of the Ross Sea.

7 Concluding Remarks

Southern Ocean circulation, the ice cover, and the water mass modification (including vertical and meridional flux of heat and feshwater) are all intricately linked; the wind-driven features must also be considered in terms of their effect on the thermohaline structure and circulation. The biology must also be linked into these environmental features. Since the physical situation does vary at a variety of scales, both spatial and temporal, similar changes in the biology field are expected.

In order to effectively study the complex variability of the environment and the response of krill to these changes, there must be developed a carefully planned process-oriented field approach coupled with a modeling effort. These "experiments" should be based on testing of conceptual understanding or hypothesis and designed to answer specific research questions.

Acknowledgements. The author's Southern Ocean Research over the last two decades has been supported by the Division of Polar Programs of the National Science Foundation. The presently active Grant (DPP 85-02386) is in support of the research associated with the Polarstern Winter Weddell Sea Project 1986. L-DGO contribution No. 4335.

References

Baker DJ Jr (1982) A note on Sverdrup balance in the Southern Ocean. J Mar Res (Suppl) 40:21–26

Bryden HL (1979) Poleward heat flux and conversion of available potential energy in Drake Passage. J Mar Res 37:1–22

Bryden HL (1983) The Southern Ocean. In Robinson R (ed) Eddies in marine science. Springer, Berlin Heidelberg New York, pp 265–277

Bryden HL, Heath RA (1985) Energetic eddies at the northern edge of the Antarctic Circumpolar Current in the Southwest Pacigic. Progr Oceanogr 14:65–87

Carmack EC (1982) Water characteristics of the Southern Ocean south of the Polar Front. In: Angel M (ed) A voyage of discovery. Pergamon, Oxford New York, pp 15–37

Carmack EC, Foster T (1975a) On the flow of water out of the Weddell Sea. Deep-Sea Res 22(11):711–724

Carmack EC, Foster TD (1975b) Circulation and distribution of oceanographic properties near the Filchner Ice Shelf. Deep-Sea Res 22(2):77–90

Carmack EC, Killworth P (1978) Formation and interleaving of abyssal water masses off Wilkes Land, Antarctica. Deep-Sea Res 25:357–369

Carsey FD (1980) Microwave observations of the Weddell Polynya. Mon Weather Rev 108:2032–2044

Carsey FD, Holt B, Martin S, McNutt L, Rothrock DA, Squire VA, Weeks WF (1986) Weddell-Scotia Sea marginal ice zone observations from space, Oct 1984. J Geophys Res 91(C3):3920–3924

Cheney RE, Marsh JG, Beckley BD (1983) Global mesoscale variability from collinear tracks of SEASAT altimeter data. J Geophys Res 88(C7):4331–4538

Colton MT, Chase RRP (1983) Interaction of the Antarctic Circumpolar Current with bottom topography: An investigation using satellite altimetry. J Geophys Res 88(3):1825–1843

Comiso J, Gordon AL (1987) Recurring polynyas over the Cosmonaut Sea and the Maud Rise. J Geophys Res 92(C3):2819–2833

Comiso J, Sullivan C (1986) Satellite microwave and in situ observations of the Weddell Sea ice cover and its marginal ice zone. J Geophys Res 91(C8):9663–9681

Deacon GER (1937) The hydrology of the Southern Ocean. Discovery Rep 15:3–152

Deacon GER (1979) The Weddell Gyre. Deep-Sea Res 26:981–998

Deacon GER (1982) Physical and biological zonation in the Southern Ocean. Deep-Sea Res 29(1A):1–15

Deacon GER, Moorey JA (1975) The boundary region between currents from the Weddell Sea and Drake Passage. Deep-Sea Res 22(4):265–268

deSzoeke R, Levine M (1981) Th advective flux of heat by mean geostrophic motions in the Southern Ocean. Deep-Sea Res 28A(10):1057–1085

Emery W (1977) Antarctic Polar Frontal Zone from Australia to the Drake Passage. J Phys Oceanogr 7(6):811–822

Everson AJ, Veronis G (1975) Continuous representation of wind stress and wind stress curl over the world ocean. J Marine Res 33:131–144

Foldvik A, Gammelsrod T, Torresen T (1985a) Circulation and water masses on the southern Weddell Sea shelf. In: Oceanology of the Antarctic continental shelf. Antarct Res Ser vol 43. Am Geophys Un, Washington, pp 21–34

Foster TD (1984) The marine environment. In: Laws RM (ed) Antarctic ecology, vol 2. Academic Press, London, pp 345–371

Foster TD, Carmack EC (1976) Frontal zone mixing and Antarctic bottom water formation in the southern Weddell Sea. Deep-Sea Res 23:301–317

Foster TD, Middleton JH (1979) Variability in the bottom water of the Weddell Sea. Deep-Sea Res 26:743–762

Foster TD, Middleton JH (1980) Bottom water formation in the Weddell Sea. Deep-Sea Res 27:367–382

Foster TD, Middleton JH (1984) The oceanographic structure of the eastern Scotia Sea-I. Physical Oceanography. Deep-Sea Res 31(5):529–550

Fu L, Chelton DB (1984) Temporal variability of the Antarctic Circumpolar Current observed from satellite altimetry. Science 226:343–346

Fu L, Chelton DB (1985) Observing large-scale temporal variations of ocean currents by satellite altimetry: with application to the Antarctic Circumpolar Current. J Geophys Res 90(C3):4721–4739

Garrett JF (1981) Oceanographic features revealed by the FGGE drifting buoy array. In: Gower JFR (ed) Oceanography from space. Plenum, New York, pp 61–69

Georgi DT (1979) Modal properties of Antarctic intermediate water in the Southeast Pacific and the South Atlantic. J Phys Oceanogr 9:456–468

Gill AE (1973) Circulation and bottom water production in the Weddell Sea. Deep-Sea Res 20:111–140

Gordon AL (1967) Structure of Antarctic waters between 20° W and 170° W, Folio 6. Antarct Map Folio Ser, Am Geogr Soc

Gordon AL (1973) Varieties and variability of Antarctic bottom water. In: Colloq Int CNRS No 215. Proc Formation des Eaux Oceaniques Profondes, Paris, pp 33–47

Gordon AL (1975) General ocean circulation, "Numerical models of ocean circulation", Symp, Durham NH, 17–20 Oct, 1972, NAS Publ, pp 39–53

Gordon AL (1981) Seasonality of Southern Ocean sea ice. J Geophys Res 86:4193–4197

Gordon AL (1982) Weddell Deep Water variability. J Mar Res (Suppl) 40:199–217

Gordon AL (1983) Polar oceanography. Rev Geophys Space Phys 21(5):1124–1131

Gordon AL, Baker T (1982) Southern Ocean Atlas: Objective contouring and grid point data set. Columbia Univ Press, New York

Gordon AL, Huber BA (1984) Thermohaline stratification below the Southern Ocean sea ice. J Geophys Res 89(C1):641–648

Gordon AL, Molinelli EM (1982) The Southern Ocean Atlas: Thermohaline and chemical distributions and the Atlas data set. Columbia Univ Press, New York

Gordon AL, Nowlin WD (1978) The basin waters of the Bransfield Strait. J Phys Oceanogr 8(2):258–264

Gordon AL, Owens WB (1987) Polar oceans. Rev Geophys 25(2): 227–233

Gordon AL, Taylor HW (1975) Heat and salt balance within the cold waters of the world ocean. In: Numerical models of ocean circulation. Nat Acad Sci, Washington DC, pp 54–56

Gordon AL, Taylor HW, Georgie DT (1977a) Antarctic oceanographic zonation. In: Dunbar MJ (ed) Polar oceans. Proc Polar Oceans Conf. Arct Inst North Am, Calgary, Alberta, pp 45–76

Gordon AL, Georgi DT, Taylor HW (1977b) Antarctic Polar Front Zone in the Western Scotia Sea – Summer 1975. J Phys Oceanogr 7(3):309–328

Gordon AL, Molinelli E, Baker T (1978) Large-scale relative dynamic topography of the Southern Ocean. J Geophys Res 83(C6): 3023–3031

Gordon AL, Martinson DG, Taylor HW (1981) The wind-driven circulation in the Weddell-Enderby Basin. Deep-Sea Res 28A (C2):151–163

Gordon AL, Chen CTA, Metcalf WG (1984) Winter mixed layer entrainment of Weddell Deep Water. J Geophys Res 89(C1): 637–640

Hibler WD, III, Ackley SF (1983) Numerical simulation of the Weddell Sea pack ice. J Geophys Res 88(C5):2873–2877

Hofmann EE (1985) The large scale horizontal structure of the Antarctic Circumpolar Current from FGGE drifters. J Geophys Res 90:7087–7097

Hofmann EE, Whitworth T III (1985) A synoptic description of the flow at Drake Passage from year-long measurements. J Geophys Res 90:7177–7187

Jacobs SS (1986) Injecting ice-shelf water and air into the deep Antarctic Oceans. Nature (London) 321(6067):196–197

Jacobs SS, Gordon AL, Ardai JL (1979) Circulation and melting beneath the Ross Ice Shelf. Science 203:439–443

Jacobs SS, Fairbanks R, Horibe Y (1985) Origin and evolution of water masses near the Antarctic continental margin: Evidence from $H_2{}^{18}O/H_2{}^{16}O$ ratios in seawater. In: Oceanology of the Antarctic Continental Shelf, Antarct Res Ser vol 43. Am Geophys Un, Washington, pp 59–85

Joyce TM, Patterson SL (1977) Cyclonic ring formation at the polar front in the Drake Passage. Nature (London) 265(5590): 131–133

Joyce TM, Patterson SL, Millard RC Jr (1981) Anatomy of a cyclonic ring in the Drake Passage. Deep-Sea Res 28A(11): 1265–1287

Joyce TM, Zenk W, Toole JM (1978) The anatomy of the Antarctic polar front in the Drake Passage. J Geophys Res 83(C12): 6093–6113

Killworth PD (1979) On "chimney" formations in the oceans. J Phys Oceanogr 9:531–554

Killworth PD (1983) Deep convection in the world ocean. Rev Geophys Space Phys 21(1):1–26

Klinck JM (1985) EOF Analysis of Central Drake Passage currents from Drake 79. J Phys Oceanogr 15:288–298

Klinck JM, Hofmann EE (1986) Deep-flow variability at Drake Passage. J Phys Oceanogr 16(7):1281–1292

Kurtz DD, Bromwich D (1985) A recurring, atmospherically forced polynya in Terra Nova Bay. In: Oceanology of the Antarctic Continental Shelf, Antarct Res Ser vol 43. Am Geophys Un, Washington, pp 177–202

Legeckis R (1977) Oceanic polar front in the Drake Passage – satellite observations during 1976. Deep-Sea Res 24:701–704

Lutjeharms JRE, Baker DJ Jr (1980) A statistical analysis of the meso-scale dynamics of the Southern Ocean. Deep-Sea Res 27(2A):145–159

Martinson DG, Killworth PD, Gordon AL (1981) A convective model for the Weddell Polynya. J Phys Oceanogr 11:466–488

McCartney MS (1977) Subantarctic mode water. In: A voyage of discovery. Angel M (ed) George Deacon Forth Anniv Vol. Deep-Sea Res Suppl 103–119

McCartney MS (1982) The subtropical recirculation of mode waters. J Mar Res Suppl 40:427–464

Middleton JH, Foster TD, Foldvik A (1982) Low frequency current and continental shelf waves in the southern Weddell Sea. J Phys Oceanogr 12(7):618–634

Molinelli EJ (1981) The Antarctic influence on Antarctic intermediate water. J Mar Res 39(2):267–293

Nowlin WD Jr, Clifford M (1982) The kinematic and thermohaline zonation of the Antarctic Circumpolar Current at Drake Passage. J Mar Res Suppl 40:481–507

Nowlin WD, Klinck JM (1986) The physics of the Antarctic Circumpolar Current. Rev Geophys Space Phys 24(3):469–491

Nowlin WD Jr, Whitworth T III, Pillsbury RD (1977) Structure and transport of the Antarctic Circumpolar Current at Drake Passage from short-term measurements. J Phys Oceanogr 7(6): 788–802

Nowlin WD Jr, Pillsbury RD, Bottero J (1981) Observations of kinetic energy levels in the Antarctic Circumpolar Current at Drake Passage. Deep-Sea Res 28(1):1–17

Patterson SL (1985) Surface circulation and kinetic energy distributions in the Southern Hemisphere oceans from FGGE drifting buoys. J Phys Oceanogr 15(7):865–884

Patterson SL, Sievers HA (1979/80) Mesoscale thermal structure of the Polar Front Zone in Drake Passage during the austral summer of 1976. Sov Cient Inst Antarct Chileno 25/26:49–112

Patterson SL, Sievers HA (1980) The Weddell-Scotia Confluence. J Phys Oceanogr 10(10):1584–1610

Peterson RG (1985) Drifter trajectories through a current meter array at Drake Passage. J Geophys Res 90(C3):4883–4893

Peterson RG, Nowlin WD Jr, Whitworth T III (1982) Generation and evolution of a cyclonic ring at Drake Passage in early 1979. J Phys Oceanogr 12(7):712–719

Pillsbury RD, Bottero JS (1984) Observations of current rings in the Antarctic Zone at Drake Passage. J Mar Es 42:853–874

Pillsbury RD, Jacobs SS (1985) Preliminary observations from long-term current meter moorings near the Ross Ice Shelf, Antarctica. In: Oceanology of the Antarctic Continental Shelf. Antarct Res Ser, vol 43. Am Geophys Un, Washington, pp 87–108

Pillsbury RD, Whitworth T III, Nowlin WD Jr, Sciremammano F Jr (1979) Currents and temperatures as observed in Drake Passage during 1975. J Phys Oceanogr 9:469–482

Piola AR, Georgi DT (1982) Circumpolar properties of Antarctic intermediate water and Subantarctic mode water. Deep-Sea Res 29(6A):687–711

Reid JL, Nowlin WD, Patzert WC (1977) On the characteristics and circulation of the southwestern Atlantic Ocean. J Phys Oceanogr 7:62–91

Rodman MR, Gordon AL (1982) Southern Ocean Bottom Water of the Australian-New Zealand Sector. J Geophys Res 87(C8): 5771–5778

Schlosser P (1986) Helium: a new tracer in Antarctic oceanography. Nature (London) 321(6067):195–196

Sciremammano F Jr (1979) Observations of Antarctic Polar Front motions in a deep water expression. J Phys Oceanogr 9(1): 221–226

Sciremammano F Jr (1980) The nature of the poleward heat flux due to low-frequency current fluctuations in Drake Passage. J Phys Oceanogr 10:843–852

Sciremammano F Jr, Pillsbury RD, Nowlin WD Jr, Whitworth T III (1980) Spatial scales of temperature and flow in Drake Passage. J Geophys Res 85(C7):4015–4028

Sievers HA, Nowlin WD (1984) The stratification and water masses in Drake Passage. J Geophys Res 83:10489–10514

Smith NR, Zhao Qian D, Kerry K, Wright S (1984) Water masses and circulation in the region of Prydz Bay, Antarctica. Deep-Sea Res 31(9):1121–1147

Taylor H, Gordon AL, Molinelli E (1978) Climatic characteristics of the Antarctic polar front zone. J Geophys Res 83(C9):4572–4578

Tchernia P, Jeannin PF (1983) Quelques aspects de la circulation oceanique Antarctique révélés par l'observation de la dérivé d'icebergs (1972–1983). Centre Nat d'Etudes Spatiales. Mus Nat Hist Nat, Paris, 92 p

Wearn RB Jr, Baker DJ Jr (1980) Bottom pressure measurements across the Antarctic Circumpolar Current and their relation to the wind. Deep-Sea Res 27(11A):875–888

Weiss RF, Ostlund HG, Craig H (1979) Geochemical studies of the Weddell Sea. Deep-Sea Res 26:1093–1120

Whitworth T III (1980) Zonation and geostrophic flow of the Antarctic Circumpolar Current at Drake Passage. Deep-Sea Res 27(7):497–507

Whitworth T III (1983) Monitoring the transport of the Antarctic Circumpolar Current at Drake Passage. J Phys Oceanogr 13: 2045–2057

Whitworth T III, Peterson RG (1985) The volume transport of the Antarctic Circumpolar Current from three-year bottom pressure measurements. J Phys Oceanogr 15(6):810–816

Whitworth T III, Nowlin WD Jr, Worley SJ (1982) The net transport of the Antarctic Circumpolar Current through Drake Passage. J Phys Oceanogr 12(9):960–970

Worthington LV (1981) The water masses of the world ocean: Some results of a fine-scale census. In: Warren BA, Wunsch C (eds) Evolution of physical oceanography. Scientific surveys in honor of Henry Stommel. MIT Press, Cambridge, U.S., pp 42–69

Zwally J, Comiso J, Parkinson C, Campbell F, Carsey F, Gloersen P (1983) Antarctic sea ice, 1973–1976: satellite passive microwave observations. NASA SP-459, Washington DC, p 206

Zwally J, Comiso J, Gordon AL (1985) Antarctic offshore leads and polynyas and oceanographic effects. In: Oceanology of the Antarctic Continental Shelf. Antarct Res Ser, vol 43. Am Geophys Un, Washington, pp 203–226

Upper Ocean Characteristics in Drake Passage and Adjoining Areas of the Southern Ocean, 39°W–95°W

H. A. SIEVERS[1] and W. D. NOWLIN Jr.[2]

Summary. A new description is presented of the hydrographic structure of the upper 1000 m in Drake Passage and adjoining Southern Ocean areas between 39° and 95°W. Based on 949 oceanographic stations (mostly summer observations) by various ships between 1927 and 1980, the characteristics show remarkable consistency, enabling us to produce meaningful horizontal distributions of water characteristics and relative geostrophic flow. Frontal positions and distinct water mass zones are easily recognized and mapped at several depths. Isopleths are constricted where the eastward-flowing Antarctic Circumpolar Current (ACC) enters the Drake Passage and continues over shallower topography to the east. This forces a northward intensification of streamlines within the passage and a deflection to the north once the lateral constriction is passed. Wave-like patterns are associated with topographic features in the Scotia Sea. The largest gradients of near-surface characteristics are found next to the shelves of South America and the Antarctic Peninsula; but the gradients at deeper levels are largest in the Sub-Antarctic and Polar Fronts of the ACC. Within the Bransfield Strait, local processes form characteristics different from those in the surrounding seas, and their eastward extension may contribute to the Weddell-Scotia Confluence.

1 Introduction

Many of the descriptive publications treat the entire Southern Ocean, e.g., Deacon (1937), Mackintosh (1946), Ishino (1963), Gordon and Goldberg (1970), Gordon and Molinelli (1982), and Clifford (1983). By contrast, there are relatively few *detailed* regional descriptions. However, many studies of the Drake Passage region have been performed and published, making it the best-known region of the Southern Ocean. During the past decade or so, the Drake Passage and adjoining oceanic areas (the western Scotia Sea and the Bransfield Strait) have captured most of the research efforts in the Southern Ocean. But even so there are few classical regional descriptions.

Our objective here is to present a new description based on hydrographic data from various levels (20, 30, 200, 500, and 1000 m) of an upper ocean region (50–70°S and 39–95°W) including Drake Passage. Although the main features are known (fronts, zonation, water masses, currents, transport, and coarse horizontal distributions of the water characteristics) it seemed to us that new detailed horizontal charts might portray better than any published previously the distribution of water characteristics in this region. Of course, the distributions are subject to many variations (seasonal, mesoscale, and interannual) but we believe that the isopleths should show as nearly as possible the average distributions of the water characteristics in this region.

2 Data and Methods

The data used in the preparation of the characteristic distributions at 20, 30, 200, 500, and 1000 m in Drake Passage and adjoining oceanic areas came from two sources: (1) hydrographic stations from the set presented by Gordon and Molinelli (1982) in the Southern Ocean Atlas and (2) data collected during the International Southern Ocean Studies (ISOS) expeditions but not included in the atlas data set. For the data sources, selection, and quality control, refer to Gordon and Molinelli's introduction to their Southern Ocean Atlas.

The hydrographic stations are unevenly distributed in time and space and include observations made by various ships from different countries between 1927 and 1980. In most of the study region all available hydrographic stations were included irrespective of the season or the year in which they were made; but a selection had to be made in Drake Passage, because the number of stations between Cape Horn and the South Shetland Islands is very large. The criterion followed was to choose preferentially the more recent stations which include dissolved oxygen and nutrient measurements and the quality of which was confirmed by their use in published studies.

A total of 1009 hydrographic stations were selected in this manner. The data were carefully checked; values that appeared unrealistic when compared with those of other stations in close proximity were eliminated. The final set of 949 stations (Fig. 1) was used to prepare potential temperature, salinity, potential density, and geopotential anomaly charts at the levels to be examined.

1 Instituto de Oceanología, Universidad de Valparaíso, Viña del Mar, Chile

2 Department of Oceanography, Texas A & M University, College Station, TX 77843, USA

Antarctic Ocean and Resources Variability
(ed. by D. Sahrhage)

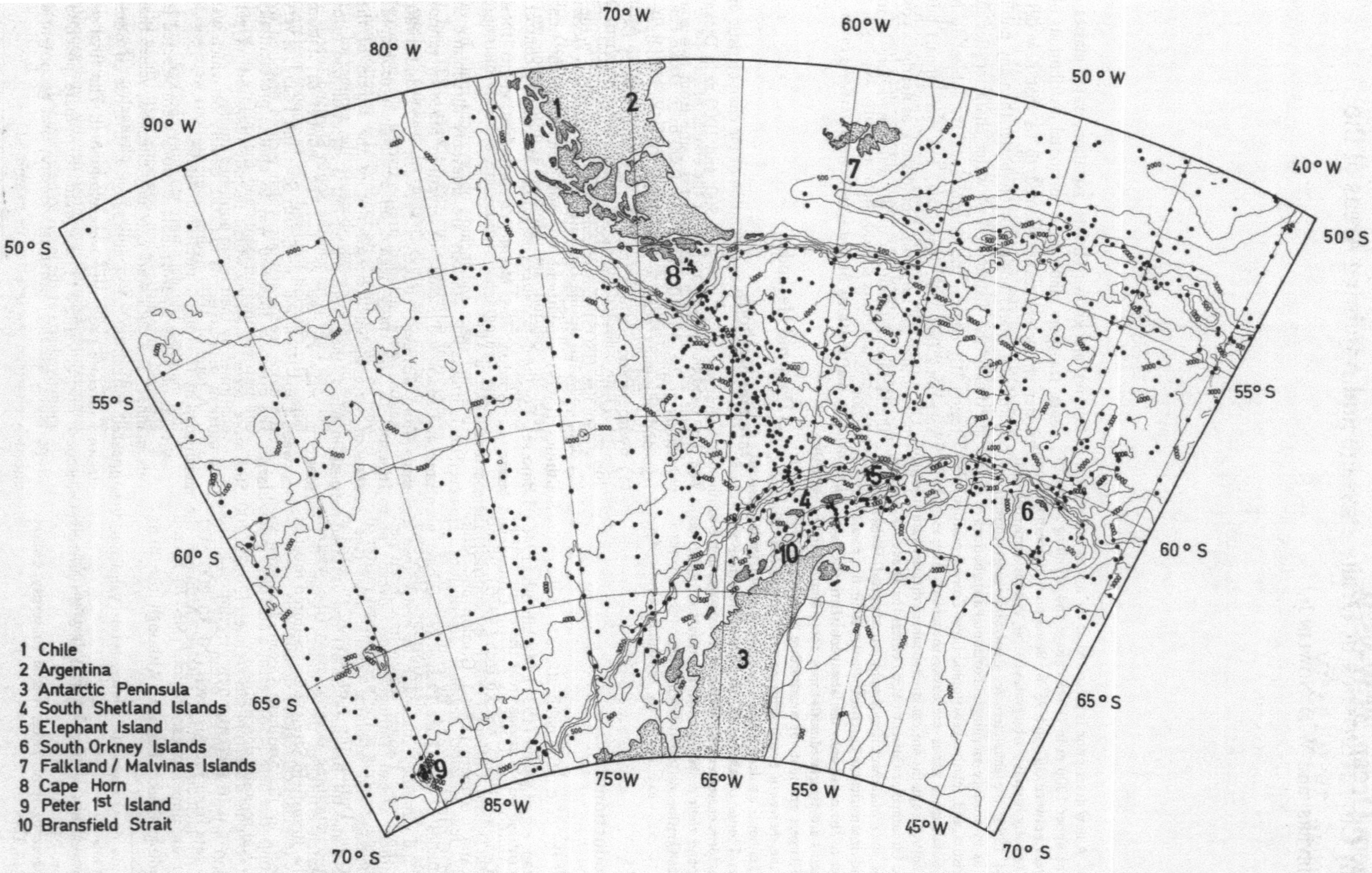

Fig. 1. Bathymetry (contours in m) of the study region and locations of stations from which serial oceanographic data were used

Data were plotted on polar stereographic projections. The bathymetric base chart (Fig. 1) was generated using the DBDB5 5 × 5 minute gridded data set from the National Geophysical Data Center. The 500- and 3000-m isobaths were included on all characteristic distributions to show the major physiographic features. The 20- and 30-m distributions were constructed using only data between January and May because December temperatures were generally colder than other summer data. Distributions in these months seem self-consistent and are taken to represent the summer season. All fields were contoured by hand. In some cases this was difficult and subjective, but in general the data showed remarkable coherence. Each chart was contoured independently to avoid bias. Only after all were finished did we compare them. Surprisingly, only minor adjustments in the isopleths were needed to obtain good agreement between the various water characteristics at each level and among the vertical shears. All distributions were constructed, but those that did not convey additional information are not presented here.

3 Regional Hydrographic Review

3.1 Zonation

The Southern Ocean circulation is dominated by the continuous eastward-flowing Antarctic Circumpolar Current (ACC), which is a deep-reaching current driven by strong, prevailing westerly winds. Waters with diverse properties are entrained into the ACC along its path and are subsequently blended into a more or less zonally uniform flow.

The idea of zonation in the ACC began with Meinardus (1923). He observed a surface temperature gradient separating Antarctic from Sub-Antarctic waters, which was long called the Antarctic Convergence and is now known as the Antarctic Polar Front (PF). Deacon (1937) determined its position based on data from Discovery cruises in 1932–33 and showed that the PF was continuous around Antarctica. Mackintosh (1946) observed considerable latitudinal variations in the position of the PF.

Additional fronts have been identified from more recent studies. Gordon et al. (1974) identified four water mass zones south of Australia and named them, from north to south: Sub-Antarctic, Complex, Antarctic, and Shelf zones. Each one has distinct temperature and salinity characteristics and is separated from the next by a horizontal transition in properties (a front).

The connection between the zonation observed south of Australia and that east of Drake Passage was provided by Emery (1977). He showed the Antarctic Polar Frontal Zone (PFZ), found between Antarctic and Sub-Antartic Surface Water, to be continuous from south of Australia to east of Drake Passage and proposed the name Sub-Antarctic Front (SAF) for the northern boundary of this zone.

Allanson et al. (1981) located the fronts south of Africa based on their temperature sections following closely the definitions used by Sievers and Emery (1978). This may be considered to infer that these fronts are also continuous between Drake Passage and Africa.

Nowlin and Clifford (1982) examined zonation of the ACC south of Australia and Africa in addition to that at Drake Passage and found analogous fronts with similar properties to exist at all three locations. Clifford (1983) demonstrated the circumpolar nature of ACC zonation. She examined data from the Southern Ocean north to the Sub-Tropical Front (STF) or Convergence and found the STF, Sub-Antarctic Front (SAF), and Polar Front (PF) to exist in all seasons in circumpolar extent. The Continental Water Boundary (CWB), first described by Sievers and Emery (1978) for Drake Passage, separates the Antarctic Zone (AAZ) from the continental waters at that location. Its zonal continuity has not been established.

The Weddell-Scotia Confluence (WSC), also called the Weddell-Scotia Front by Deacon (1982), is the boundary zone between the Weddell and Scotia Seas (Gordon 1967). Its anomalous properties have led several workers to suggest that it is not merely a mixture of the adjacent water masses, but includes water masses derived from another source, as in the Bransfield Strait or by local processes (Patterson and Sievers 1980). The position of the WSC closely follows the South Scotia Ridge, indicating that interactions with bottom topography are important to the formation and maintenance of this relatively homogeneous zone (Patterson and Sievers 1980).

Nowlin et al. (1977) described four zones in Drake Passage, differentiating them by the position each occupies in temperature-salinity space. They also showed that the Circumpolar Deep Water (CDW) does not rise uniformly toward the south, but that depth changes occur in steps separating the zones which appear in the vertical density sections as narrow regions of increased isopycnal slope. These fronts, or regions of large horizontal density gradients, are bands of stronger eastward geostrophic speed imbedded in slower eastward flow (Nowlin et al. 1977; Baker et al. 1977). The shear was shown to continue through the entire depth of water. Whitworth (1980) proposed a nomenclature for the fronts and zones which is generally accepted. From north to south they are, with characteristic widths: Sub-Antarctic Zone (SAZ), 200 km; Sub-Antarctic Front (SAF), 50 km; Polar Frontal Zone (PFZ), 200 km; Polar Front (PF), 60 km; Antarctic Zone (AAZ), 300 km; Continental Water Boundary (CWB), 40 km; and Continental Zone (CZ), 100 km. These widths may vary considerably in time and space.

The surface waters differ significantly across fronts. However, Nowlin and Clifford (1982) found that the vertical profiles from the same side of a specific front are remarkably uniform from year to year. After allowing for depth differences in isopycnals across the fronts (of 500 to 1000 m), they detected only few systematic differences between the deep and bottom waters of the different water mass zones.

Meanders are known to form on these fronts, and from time to time these meanders become contorted to the extent that a front closes on itself and an isolated current ring is formed (Joyce and Patterson 1977; Joyce et al. 1981). Peterson et al. (1982) showed that near the fronts the necessary conditions for both baroclinic and barotropic instability are met. Thus, if warm core rings can form to the south of each front and cold core rings to the north, six different ring types are possible. All six possible ring types have been observed (Joyce and Patterson 1977; Joyce et al. 1981; Peterson et al. 1982; Pillsbury and Bottero 1984; Hofmann and Whitworth 1985; Gordon et al. 1977; Nowlin and Zenk 1988).

3.2 Water Masses

The cores of water masses may be identified by either maximum or minimum values (extrema) of temperature, salinity, oxygen, and nutrients. Conservative characteristics, such as potential temperature and salinity, are modified only by diffusion and mixing as they sink and spread. Oxygen and nutrients, as nonconservative properties, also have their source values altered by biological processes or through interaction with bottom sediments. Nonetheless, extrema in the nonconservative properties persist over great distances and are very useful as water mass tracers.

Based on consideration of vertical stability maxima, core layers, and the horizontal zonation, plus the accumulated published knowledge, Sievers and Nowlin (1984) identified and described in detail the seven major water masses as found at Drake Passage. A summary of this material follows, beginning with the uppermost water mass.

1. Sub-Antarctic Surface Water (SASW) forms a thin surface layer less than 150 m thick north of the PF. It has relatively high temperatures (>7°C in summer), salinity of about 34.1 at its base, a weak oxygen maximum, and very low nutrient concentrations.

2. Antarctic Surface Water (AASW) forms a layer about 250 m thick and is located between the PF and the Antarctic coast. Its most prominent characteristic is very low temperatures (below −1°C). Summer heating at the surface induces a very strong subsurface temperature minimum referred to as "winter water" by Mosby (1934). The surface salinity is low. The dissolved oxygen content is very high (>8 ml/l), as are also the nutrient concentrations.

3. Sub-Antarctic Mode Water (SAMW) is found as a relatively homogeneous and low stability layer up to 700 m thick which lies mainly in the SAZ, beneath the SASW. Weak maxima in salinity (>34.2) and oxygen distinguish this water mass.

4. Antarctic Intermediate Water (AAIW) is found beneath SASW and SAMW north of the PF. Its main characteristics are a subsurface salinity minimum and a relative oxygen maximum. It is derived in part from the AASW, as is evidenced by the partial extension of the AASW temperature minimum into the AAIW and occasional near coincidence in characteristic distributions including density. Both AAIW and AASW seem separated from the adjacent water masses by the same stability stata.

5. Circumpolar Deep Water (CDW) fills most of Drake Passage in a layer over 3000 m thick. Based on distinct characteristics it is often divided into Upper (UCDW) and Lower (LCDW) Circumpolar Deep Water. The UCDW core is best characterized by an oxygen minimum, maxima in phosphate and nitrate, and a potential temperature maximum south of the PF. The main characteristic of the LCDW is its salinity maximum (average value 34.73), but it is also distinguished by minima in phosphate and nitrate.

6. Southeast Pacific Deep Water (SPDW) is found south of the PF in the Drake Passage where it is characterized by a maximum in silicate (>135 μmol/l). Potential temperature and salinity decrease, but oxygen and nutrient values increase with depth through this water mass.

7. Weddell Sea Deep Water (WSDW) is the densest water mass that reaches Drake Passage. This cold, fresh, and oxygenated water is found at the bottom near the South Shetland Islands, probably its westernmost extent.

3.3 Circulation

Clowes (1933) found a considerable eastward flow through the Drake Passage which he pictured as continuous from the surface to very great depths. This flow of the Antarctic Circumpolar Current (ACC) is conceded to be driven principally by wind, although the coupling and relative importance of wind and thermohaline driving has not been adequately investigated (Nowlin and Klinck 1986).

Gordon (1967) presented relative dynamic topographies for the Southern Ocean between 20 and 170°W. His surface and 1000-db topographies relative to 2500 db were very similar spatially. Across the southeastern Pacific Ocean the ACC exhibits a more or less uniform eastward flow. Where it approaches Drake Passage and in the Scotia Sea the circulation in the southern region is wavelike (Gordon 1967) and speeds are slower than in the northern region. The current turns northward in the Scotia Sea and most of it enters the Atlantic Ocean west of South Georgia, responding to the bottom topography.

Based on geostrophic shears referenced to directly measured subsurface currents or to deep reference levels, Nowlin et al. (1977) and Nowlin and Clifford (1982) reported high velocity cores at the fronts in Drake Passage with eastward surface speeds of 25–35 cm/s. These cores are narrow bands (∿50 km) embedded in slower and wider zones. The currents are coherent throughout the water column at Drake Passage with speeds above 5–10 cm/s at 2500 m.

The volume transport of the ACC through Drake Passage has been well studied. As summarized by Nowlin and Klinck (1986), the best estimate of net transport above 2500 m has a mean of 125 x 10^6 m^3/s with a standard deviation of 10 x 10^6 m^3/s and an estimated error of not more than 10%. Higher frequency fluctuations occur in the barotropic mode, but 70% of the net is in the baroclinic mode. This instantaneous transport can vary from the mean by as much as 20%, with most of the variation occurring in the barotropic mode. Eastward transport below 2500 m has been estimated as approximately 10 x 10^6 m^3/s by Whitworth (1983), so that the best estimate of mean volume transport is 135 x 10^6 m^3/s.

4 Water Characteristics at Selected Levels

4.1 The Near Surface Levels (20 and 30 m)

Because many hydrographic stations lack surface or very near-surface data, we chose, following Gordon and Goldberg (1970) and Patterson and Sievers (1980), the 20-m level to study the surface distributions of most water characteristics. The oxygen and nutrient distributions were plotted also for the 30-m level. Considering that both levels (20 and 30 m) are in the mixed layer, distributions should be complementary, and were.

The bottom topography generally shallows from about 5000 m in the southeast Pacific Ocean to 3500–4000 m in Drake Passage and continues to shallow east of the passage (Fig. 1). The surface isotherms (Fig. 2) are constricted as the eastward-flowing Antarctic Circumpolar Current (ACC) enters Drake Passage. The Scotia Sea is almost closed by the Scotia Ridge with an average depth of only 1600 m. This forces a northward intensification of streamlines within the passage and a deflection to the north once the lateral constriction is passed. There are wave-like patterns in the isotherms associated with some of these topographic features. Although most isopleths are continuous through Drake Passage, they are contorted and patchy due to the effects of local dynamics and other variability, which are more pronounced in this near-surface layer than in the deeper ones.

Summer surface temperatures in the region range from 11°C, near the west coast of Chile, to below –1.5°C, near Antarctica. The range in Drake Passage is between 0 and 8°C. Surface temperatures in the Bransfield Strait are comparable to values outside except colder waters of origin over the Antarctic Peninsula continental shelf (Clowes 1934; Patterson and Sievers 1980; Sievers 1982). A tongue of subzero water extends eastward from the Bransfield Strait past the South Orkney Islands, marking the near surface expression of the WSC (Patterson and Sievers 1980).

The surface salinity distribution (Fig. 3) appears more contorted and patchy than for temperature. There are two relative maxima: one in northern Drake Passage; the other in the Bransfield Strait with eastward extension. The first can be considered as an induced maximum in this region due to the low salinity off the Chilean coast. This fresher water results from estuarine outflow and high precipitation which modifies the Sub-Antarctic Surface Water. It is advected into the northern region of Drake Passage, as can be seen from the values and orientation of the isohalines (Deacon 1933 and 1937; Gordon and Goldberg 1970; Silva and Neshyba 1979/80).

Sievers and Nowlin (1984: Fig. 17) show meridional distributions of surface characteristics and their relationships to fronts at Drake Passage. The relative maximum in salinity (>34.1) in the northern passage distinguishes the waters of the PFZ, (transition between the Antarctic and Sub-Antarctic Surface Waters). The fresher water of the Antarctic Zone stretches around Antarctica as can be seen in Gordon and Goldberg's (1970) Antarctic Map Folio and in Gordon and Molinelli's (1982) Southern Ocean Atlas. In Fig. 3 the lowest values (<33.1) are found near the Antarctic Continent in the Bellingshausen Sea, as a consequence of precipitation and of summer melting of sea and glacial ice.

At Drake Passage salinity increases south of the CWB and in the Bransfield Strait. There is a strong surface gradient in salinity at the southwest end of the Bransfield, as observed by Clowes (1934) and described recently by Sievers (1982), which is a consequence of the encounter of the saltier Bransfield Strait with the fresher Gerlache Strait water (Clowes 1934; Sievers 1982; Kelly et al. 1985). A high salinity tongue of water (>34.2) extends eastward from the Bransfield Strait and northern tip of the Antarctic Peninsula. Patterson and Sievers (1980) traced this near surface WSC water (S >34.0) east to the South Sandwich Island arc, where it turns northward.

The surface density field for summer (not shown) closely reflects the large salinity gradients in the south. Potential density increases southward, from values less than 26.0 kg/m^3 (units omitted hereafter) in the vicinity of the Chilean coast and east of the Falkland Islands. Another region with low surface density is the southern Bellingshausen Sea, where low temperature is unable to compensate for the very low surface salinity (<33.1). Density increases close to Antarctica; there the salinity is somewhat higher and the temperature is near the freezing point even in summer. The maximum density values (27.6) are found east of Bransfield Strait and between the Scotia and Weddell Seas marking the WSC.

Near-surface dissolved oxygen content is very high, as expected due to low temperatures, especially in the southern region (> 8.0 ml/l). The lowest values are off the Chilean coast and in the vicinity of the Falkland Islands (<7 ml/l).

Silicate is almost absent in surface waters in the north of our region, especially off the Chilean coast (Fig. 4). Values increase gradually southward to the PF which is marked by a gradient, as observed by Patterson and Sievers (1979/80).

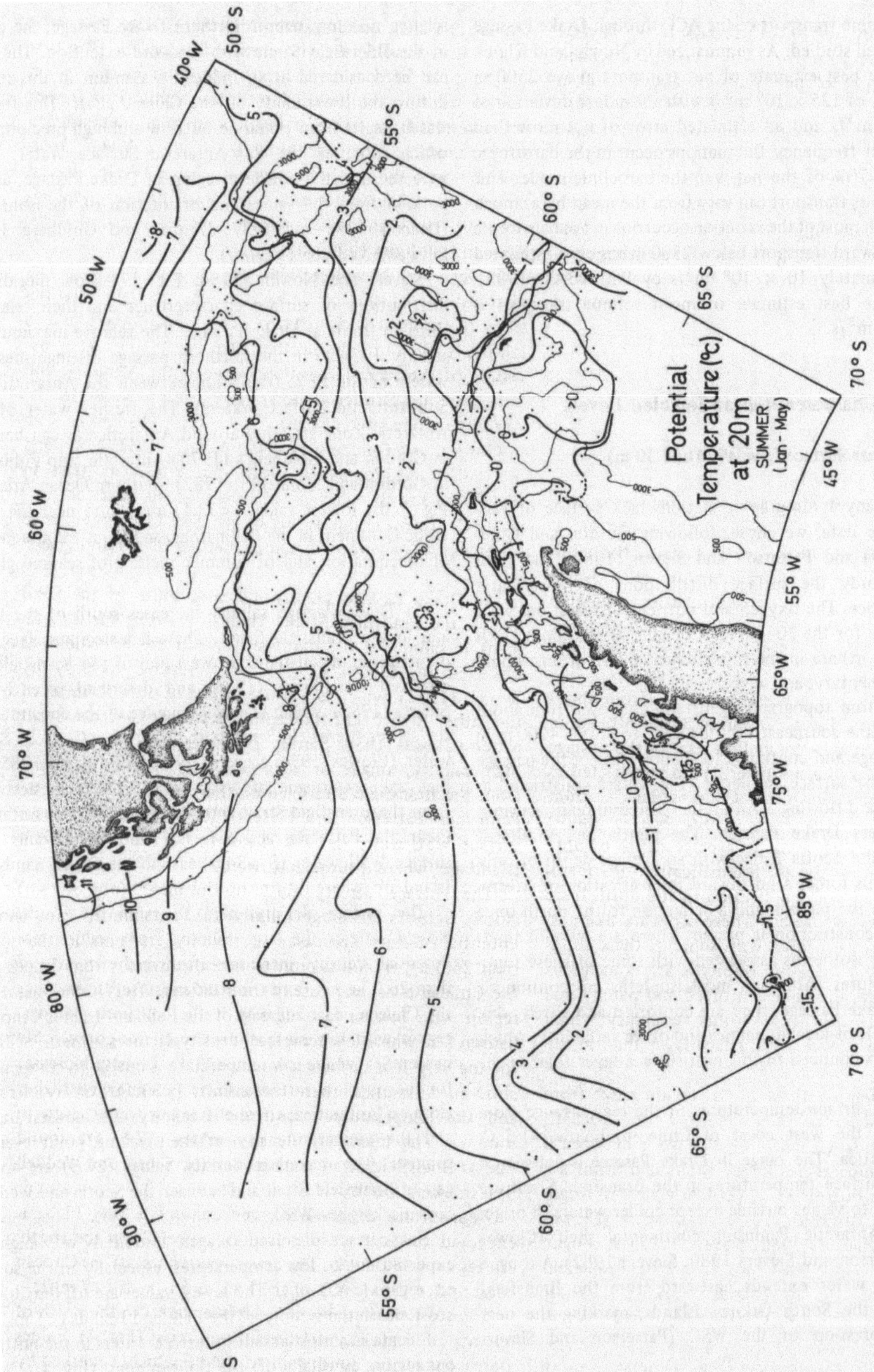

Fig. 2. Potential temperature (°C) at 20 m constructed from January–May observations

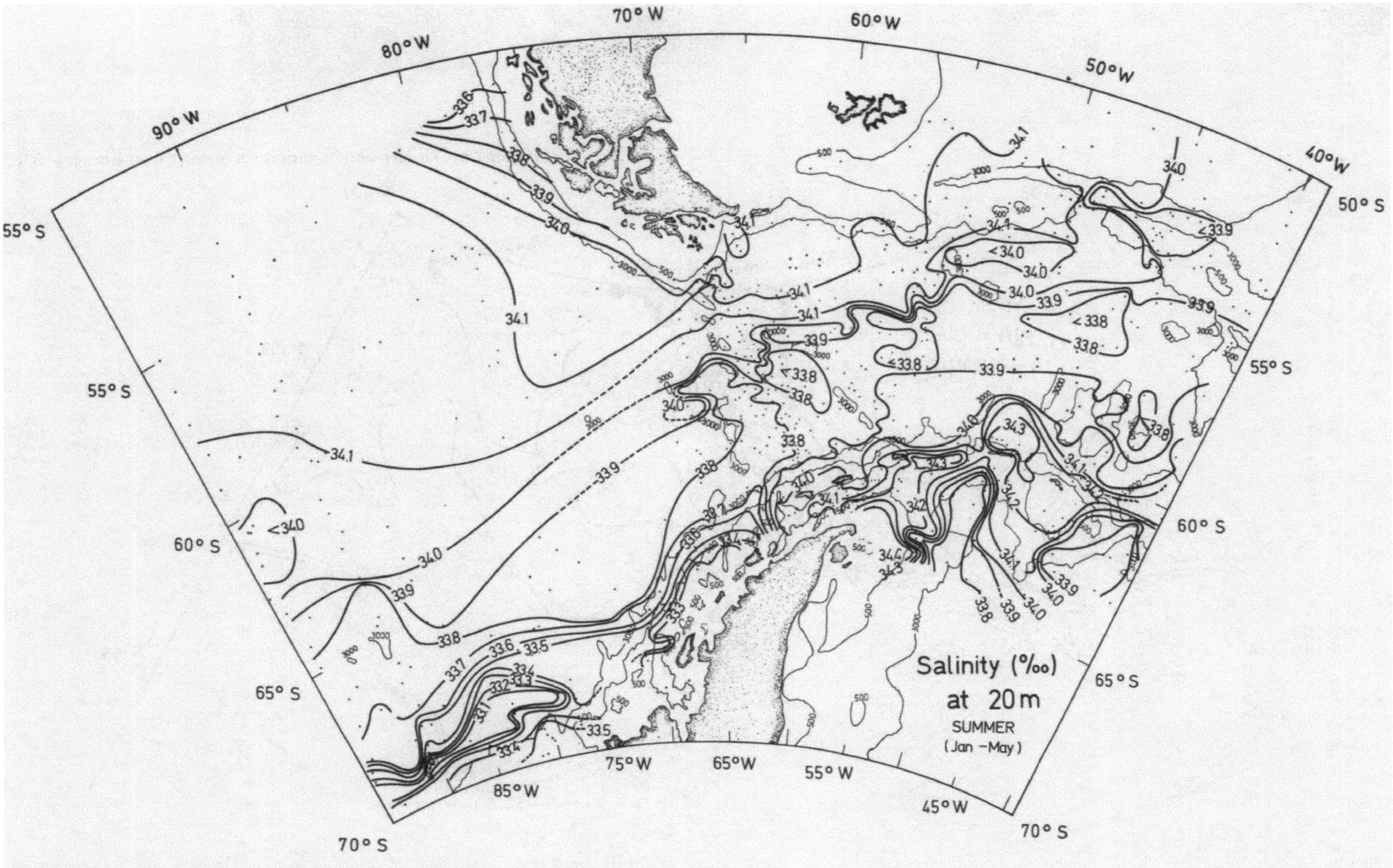

Fig. 3. Salinity at 20 m constructed from January–May observations

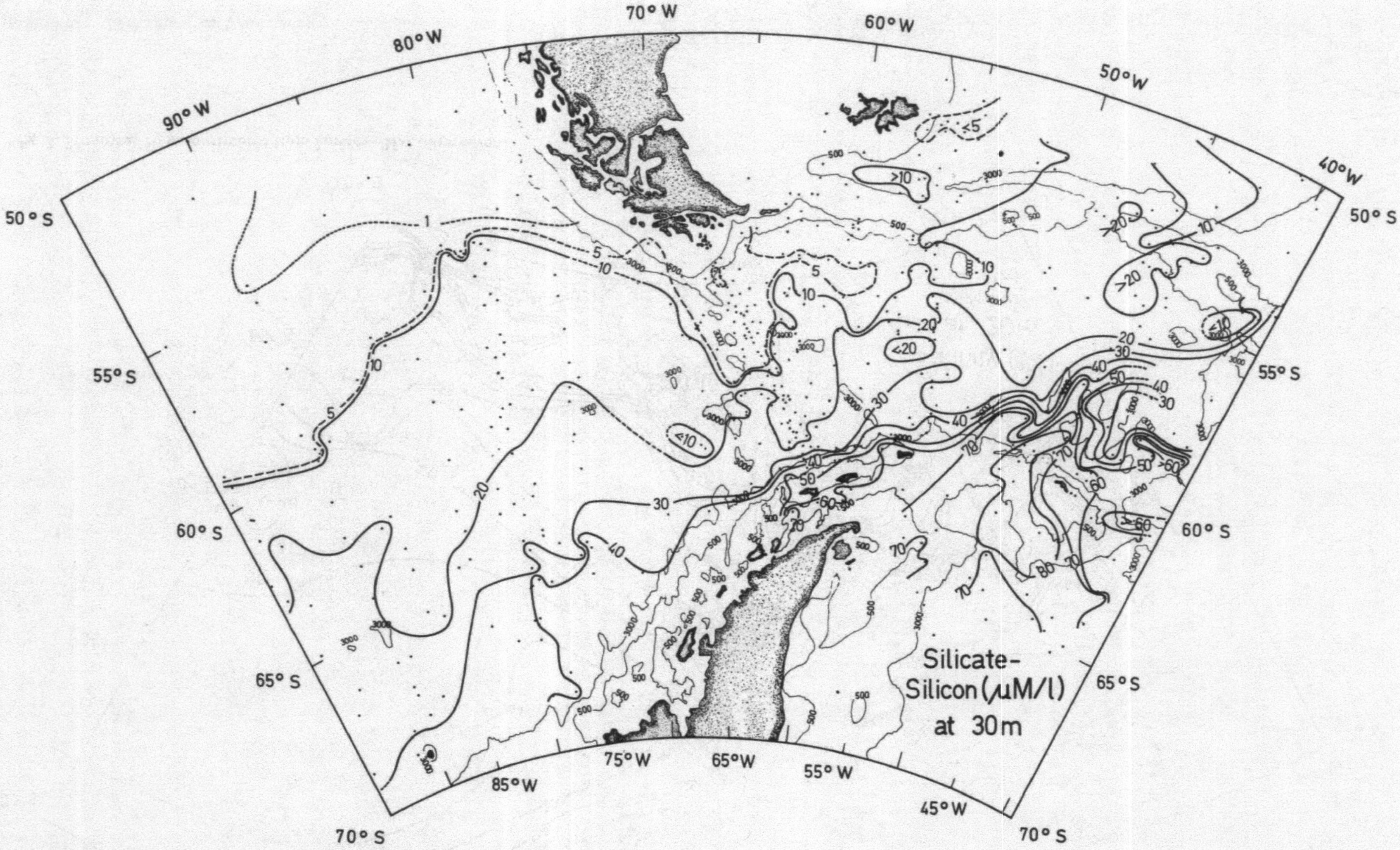

Fig. 4. Silicate (µmol/l) at 30 m constructed from January–May observations

Increases are from 5 μmol/l to 10 or 20 μmol/l in some areas. We cannot explain the wide northern arc of the isopleths in the southeastern Pacific, unless we consider silicate regeneration in spring and early summer (Clowes 1938). A second gradient occurs at the CWB (Sievers and Nowlin 1984) and extends along the WSC (Gordon et al. 1977). Surface silicate values in the Bransfield Strait are high; recent observations have detected values greater than 80 μmol/l in its eastern region (Stein 1978/79) and 90 μmol/l in its western region, north of Trinity Island (Silva 1985). Concentrations decline somewhat into the Weddell Sea.

Based on these distributions of characteristics, it is difficult to define good positions for the SAF and PF because the variability and dynamics of the surface layer mask the details. Also, there is a large gap in summer data in the western study area. We have attempted to locate these fronts based on the historical locations and variability determined by Whitworth (1980) for Drake Passage, and on the distribution and values of the surface water characteristics at the fronts for the center of the passage from Sievers and Nowlin (1984).

We may indicate the positions of major fronts in relation to summer surface characteristics. The SAF lies near the 6°C isotherm in Drake Passage but between 6 and 7°C to the west. The SAF is near the northern 34.1 isohaline for the central region of Drake Passage itself. East of Drake Passage the SAF meanders and turns north, merging into the Falkland Current, and is not marked by distinct isopleths.

The position of the PF is between the 4–5°C surface isotherms in Drake Passage. To the east it is found at slightly greater temperatures – between the 5 and 6°C in the northern Scotia Sea where the front crosses the North Scotia Ridge near 48°W. Judging from the salinity gradient in Drake Passage, the position of the PF is near the 34.0 and 34.1 isohalines in the northern Scotia Sea.

The position of the CWB, paralleling the Antarctic Peninsula in Drake Passage, lies in temperature between the 1 and 2°C isotherms, in salinity between 33.9 and 34.0, and in density between 27.1 and 27.2. The position of the WSC is also clear; it extends east from the Bransfield Strait to beyond the South Orkney Islands with temperatures lower than 0°C, salinity greater than 34.2, and density greater than 27.4.

4.2 Characteristics at 200 m

The most conspicuous feature in the potential temperature distribution at 200 m is the zonal minimum (<1.0°C) seen extending from the Pacific to about 63°W in Drake Passage (Fig. 5). Low temperatures reappear in the upper right corner of the chart. The potential temperature minimum, a consequence of AASW sinking at the PF region, can be seen extending over most of the Pacific sector of the Southern Ocean around 63°S (Gordon and Molinelli 1982) with patches of temperatures even below 0°C between 140 and 130°W. Otherwise, the temperature distribution at this level is quite similar to that at 20 m, but with a range between >5°C in the southeast Pacific to below –1°C in the far south.

A minimum in salinity (<34.1) extends along the central sector of the region (Fig. 6). It overlaps the temperature minimum, except that in the southeast Pacific it is found farther north and it extends farther northeast (to about 55°W) than the temperature minimum. The salinity minimum is seen again in the northeast corner of our region. A much wider gap in this salinity minimum (90 to 30°W) appears in Ostapoff's (1962) description of the low salinity circumpolar belt. He interpreted this low salinity band at 200 m as a result of sinking of relatively fresh AASW at the PF. It is seen clearly in the Southern Ocean Atlas (Gordon and Molinelli 1982).

Salinity at 200 m is low along the coast of Chile resulting in a relative high (>34.2) between there and the salinity minimum band. Salinity increases to the south in a significant salt gradient to values of over 34.5 and then decreases in the vicinity of the Antarctic coast. The high salinity water may have its origin in the Ross Sea, as was suggested by Ostapoff (1962), but most probably it is UCDW which has reached this depth as it ascends to the south. Lower salinities (<34.3 in some places) close to the coast in the Bellingshausen Sea probably result from glacial ice melting. Jacobs et al. (1979) mention a study of 407 Antarctic icebergs by Weeks and Mellor in 1978, who calculated a *mean* subsurface draft of 182 m. Because melting occurs also at the bottom of icebergs, and is not restricted to the summer season but is a continuous process, this appears to be a plausible source of cold, fresh water.

The salinity in Bransfield Strait is much the same as the surrounding waters. The lowest values (<34.4) are found around the South Shetland Islands (Fig. 6; and Sievers 1982). The WSC is not marked strongly in the 200-m salinity distribution as in the near-surface levels; instead, a gradual transition occurs to the higher Weddell Sea values (>34.6).

As with temperature and salinity, dissolved oxygen at 200 m (Fig. 7) has a maximum (>7.0 ml/l) as a zonal tongue from the southeast Pacific into central Drake Passage. This extremum has a broader north-south extent than the temperature and salinity minima, covering the combined area of the other two. It extends farther east than the temperature minimum but not as far as the salinity minimum. The maximum reappears in patches of values greater than 7.0 ml/l east of South America. There is a broad oxygenated (>6.5 ml/l) belt around Antarctica (the Southern Ocean Atlas, Gordon and Molinelli 1982) which generally coincides with the circumpolar belts of temperature and salinity minima. Along the Antarctic Peninsula and in the WSC is found another relative maximum in oxygen with values >6.0 ml/l. Between these maxima, and to the southeast in the Weddell Sea, oxygen concentrations are near or less than 5.0 ml/l.

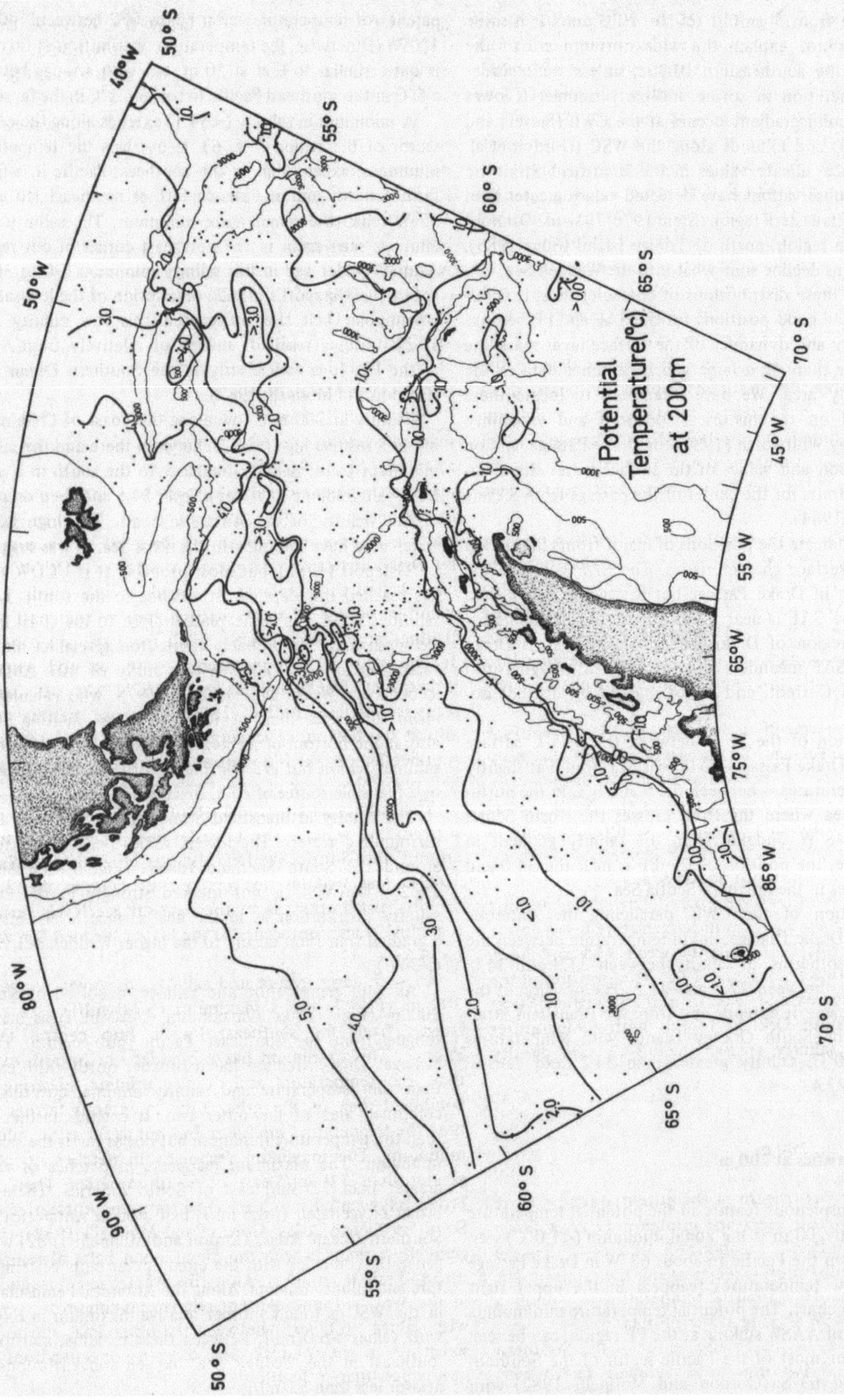

Fig. 5. Potential temperature (°C) at 200 m

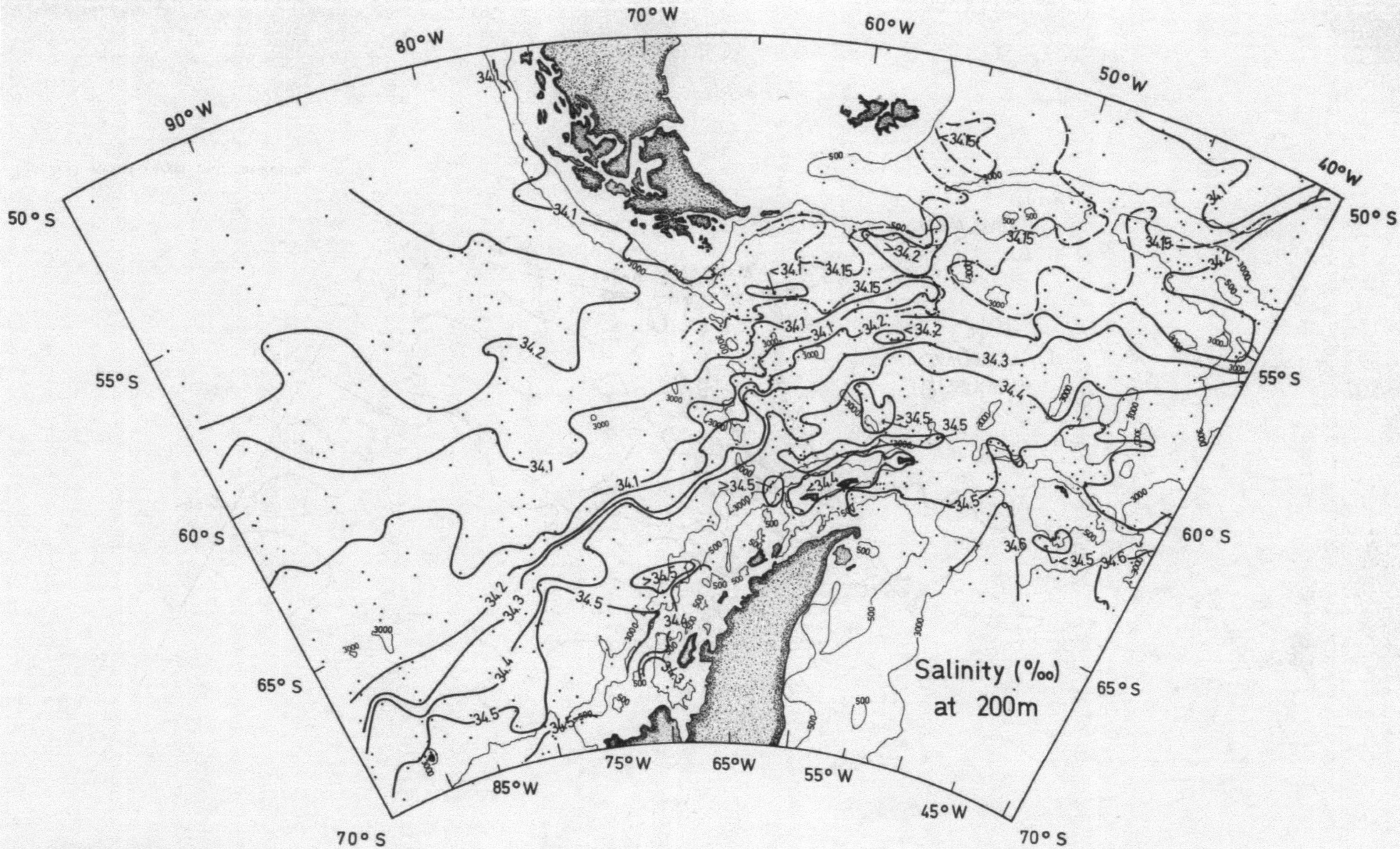

Fig. 6. Salinity at 200 m

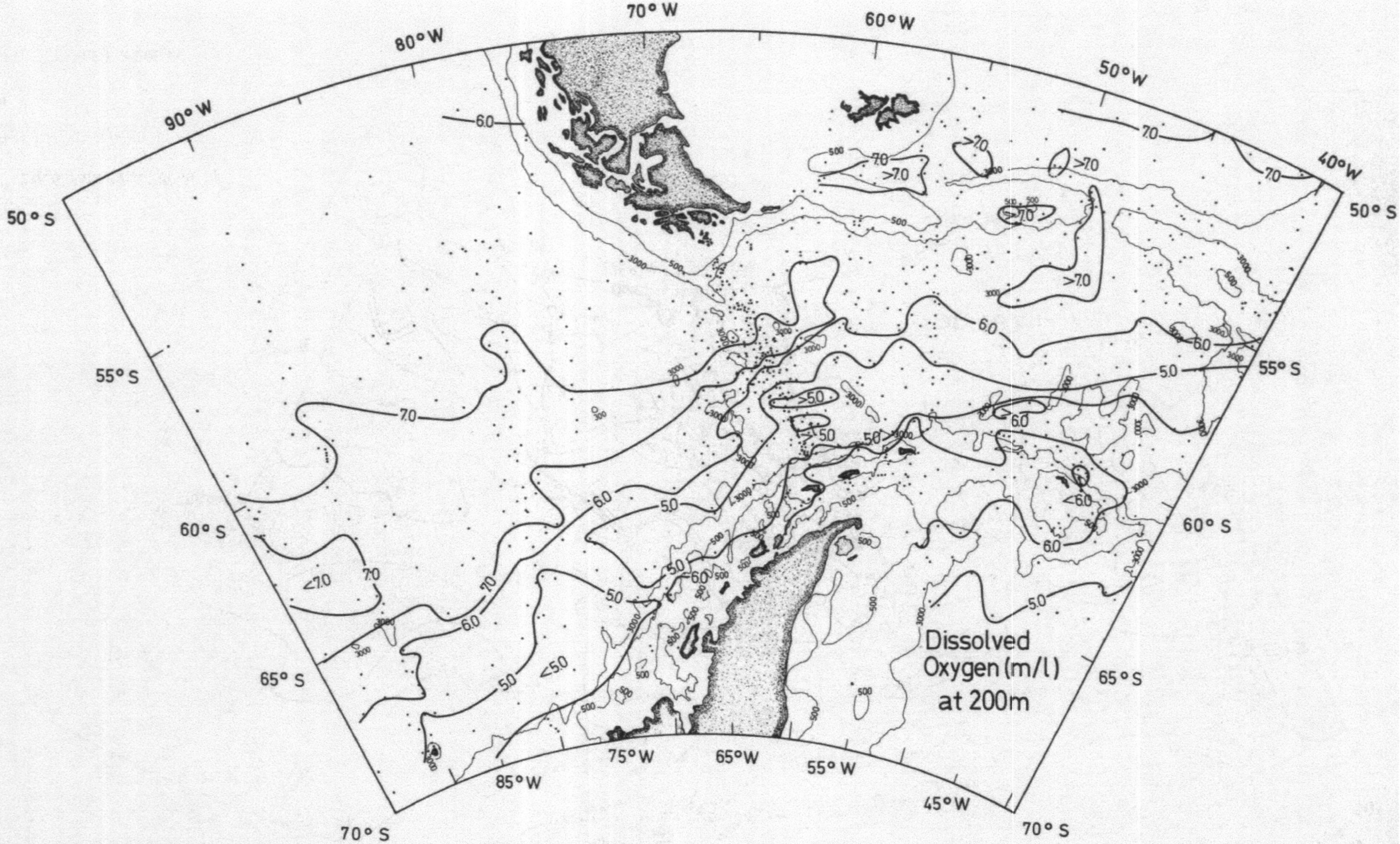

Fig. 7. Dissolved oxygen (ml/l) at 200 m

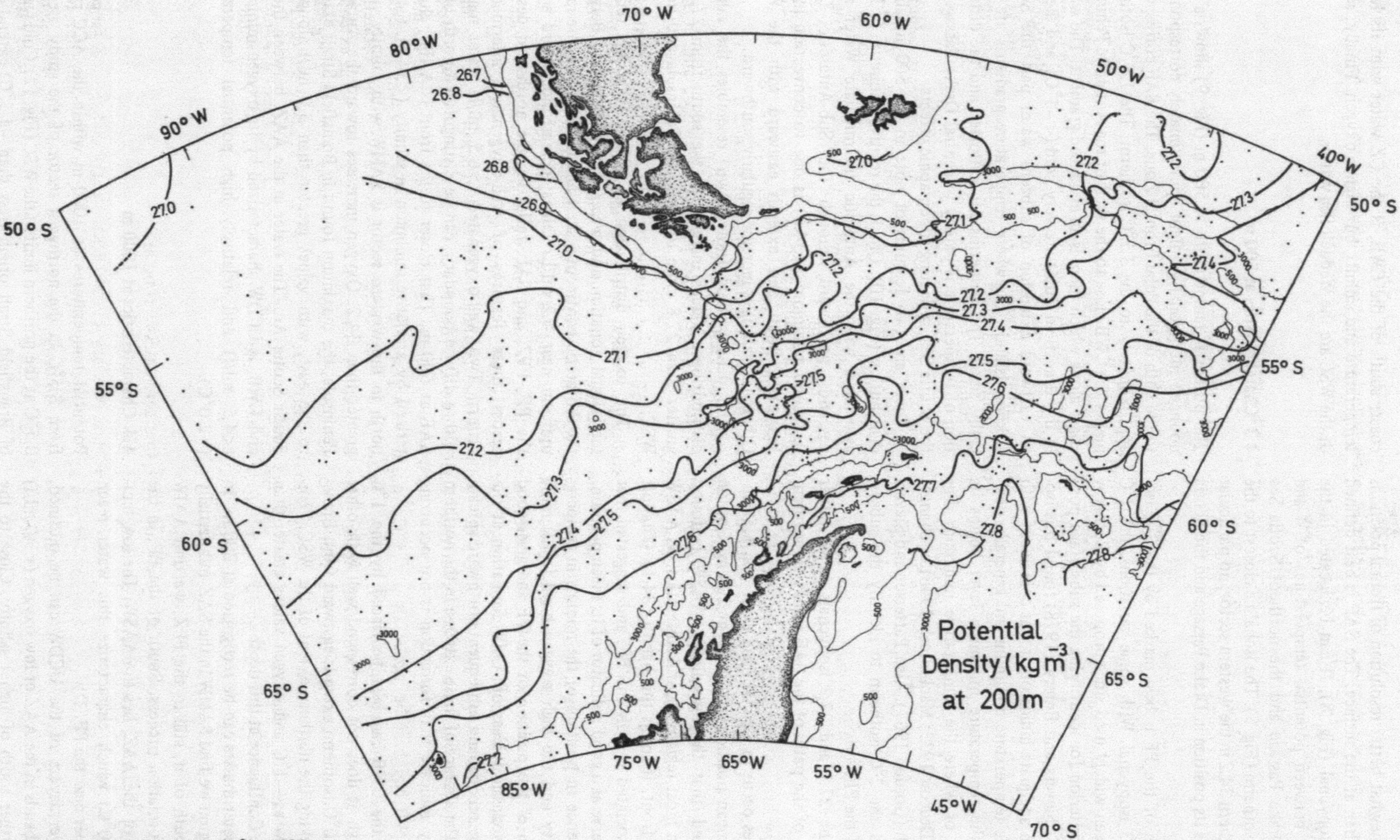

Fig. 8. Potential density (kg/m³) at 200 m

There is somewhat better resolution of frontal positions at this depth than at the surface. The SAF is best defined by the 27.1 isopycnal (Fig. 8). It can be located in the western sector between potential temperatures of 4 and 5°C, and in Drake Passage and the northern Scotia Sea along the 4°C isotherm (Fig. 5). The isohaline closest to the SAF is the southern 34.2 in the western sector; no particular isohaline reveals its position in Drake Passage and the Scotia Sea.

The position of the PF is best marked by the 2°C isotherm and 27.3 isopycnal. With some variations, the 2°C isotherm has been widely used since proposed by Botnikov (1963) as the criterion for identifying the subsurface position of the PF. Sievers and Emery (1978) found this isotherm to be an adequate indicator; it was coincident with the limit of the temperature minimum and generally positioned where the temperature minimum layer begins its rapid descent, or where the temperature minimum dips below 200 m (Deacon 1933; Mackintosh 1946) or changes sharply in depth (Gordon 1967). Also, Patterson and Sievers (1979/80) used the 2°C isotherm to identify the subsurface position of the PF.

Note that the 27.1 and 27.3 isopycnals define rather well separation of the paths of the SAF and PF in northern Scotia Sea.

Low salinities occur in a zonal band at or near Mackintosh's (1946) mean position of the PF at 200 m. Gordon (1967) indicated that the low-salinity AASW is found above 200 m in most regions around Antarctica. As AASW sinks near the front, it forms this band (<34.1). Ostapoff (1962) also associated it with the PF, and suggested that salinity could serve as a good indicator of its mean position.

The coincidence in position of the minima in temperature and salinity and the high oxygen band (>7.0 ml/l) is an indication of the position of the PF and gives good support to the classical ideas of AAIW formation at the front by sinking, mixing, and subsequent northward spreading of AASW. The absence of these extrema in the northern Scotia Sea may indicate that formation is not occurring there.

At 200 m the CWB can best be traced by the 1°C isotherm, though it does not correspond well with other isopleths. The 0°C isotherm extends eastward from Bransfield Strait marking the northern extent of the WSC. Temperatures as low as −1°C and oxygen values greater than 6 ml/l mark the Confluence at this depth.

Up to six water masses can be recognized at 200 m. In the northern region we find SAMW in the SAZ and partially in the PFZ. South of it, still in the PFZ, we find AAIW mostly in its formation process. South of the PF, in the western sector of the AAZ, lies the AASW. The low temperatures (<1.0°C) which characterize this water reach 200 m and deeper near the PF.

The shallower layers of the UCDW can be outlined, mainly by the bands in the AAZ of low oxygen (<5.0 ml/l) and slightly farther south of higher salinity. Close to the coast, south of the CWB, is the CZ water with its lower temperature and salinity but higher oxygen. Finally, south of the WSC are the Weddell Sea waters.

4.3 Characteristics at 500 m

The potential temperature at 500 m (Fig. 9) shows a pronounced meridional gradient approximately corresponding to the path of the Polar Frontal Zone. The SAF corresponds approximately to the 3.5°C isotherm. The 1.5°C isotherm marks the CWB close to the shelf of the Antarctic Peninsula coast along which there is an enhanced gradient. The waters in the Bransfield Strait are very cold (<−1°C), and there is an eastward extension of subzero water past the South Orkney Islands. Elsewhere temperature gradients are generally small. The isotherms are contorted and the distribution of temperature is patchy in central Drake Passage and the Scotia Sea, reflecting topographic effects.

There are two features of note in the 500-m salinity distribution (Fig. 10). One is the relative minimum (<34.2) associated with the Antarctic Intermediate Water as it descends northward through the Sub-Antarctic Zone. A second minimum (<34.6) is due to convection in the Bransfield Strait and extends eastward with the WSC. Otherwise, salinities increase regularly southward.

The density field (not shown) resembles the salinity distributions. Values increase to the south from 27.1 to greater than 27.8 close to Antarctic Peninsula and in the WSC.

The oxygen distribution at 500 m (Fig. 11) reflects meridional zonation and water mass conversions. Between the Antarctic continental shelf and the PF values are rather uniform near 4.2 ml/l. Values increase northward across the PF, PFZ, and SAF, reflecting the northward descent across these features of well-oxygenated intermediate waters. Two well-oxygenated (>6.5 ml/l) water masses, with a diffuse boundary, can be distinguished north of the SAF at 500 m. That closer to the front is AAIW, characterized by a relative minimum in salinity (<34.2). Farther north in the western sector is SAMW with salinity values greater than 34.2. Oxygen increases toward the Antarctic Peninsula; the maximum found in Bransfield Strait extends to the east, with values greater than 5.0 ml/l along the South Scotia Arc. The water in the AAZ, between the PF and CWB, is UCDW characterized by its oxygen minimum (<4.5 ml/l) and relatively high potential temperature (>2.0°C).

4.4 Characteristics at 1000 m

Potential temperatures at 1000 m within the ACC range from 3.5°C in the northwest sector of the study area to 0.5°C at the northern limit of the WSC (Fig. 12). Cold waters of Bransfield Strait origin (less than −1.5°C) extend east-

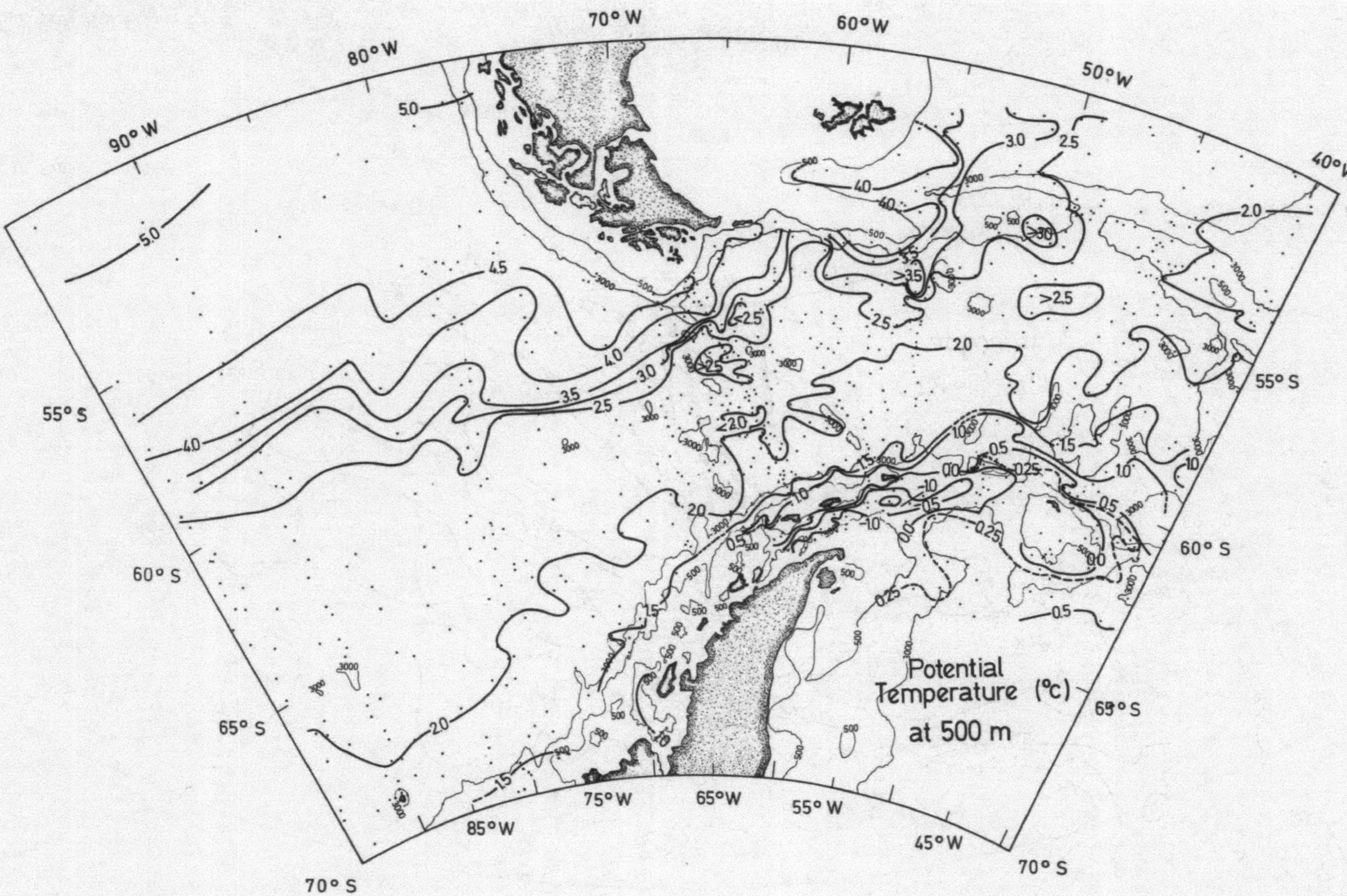

Fig. 9. Potential temperature (°C) at 500 m

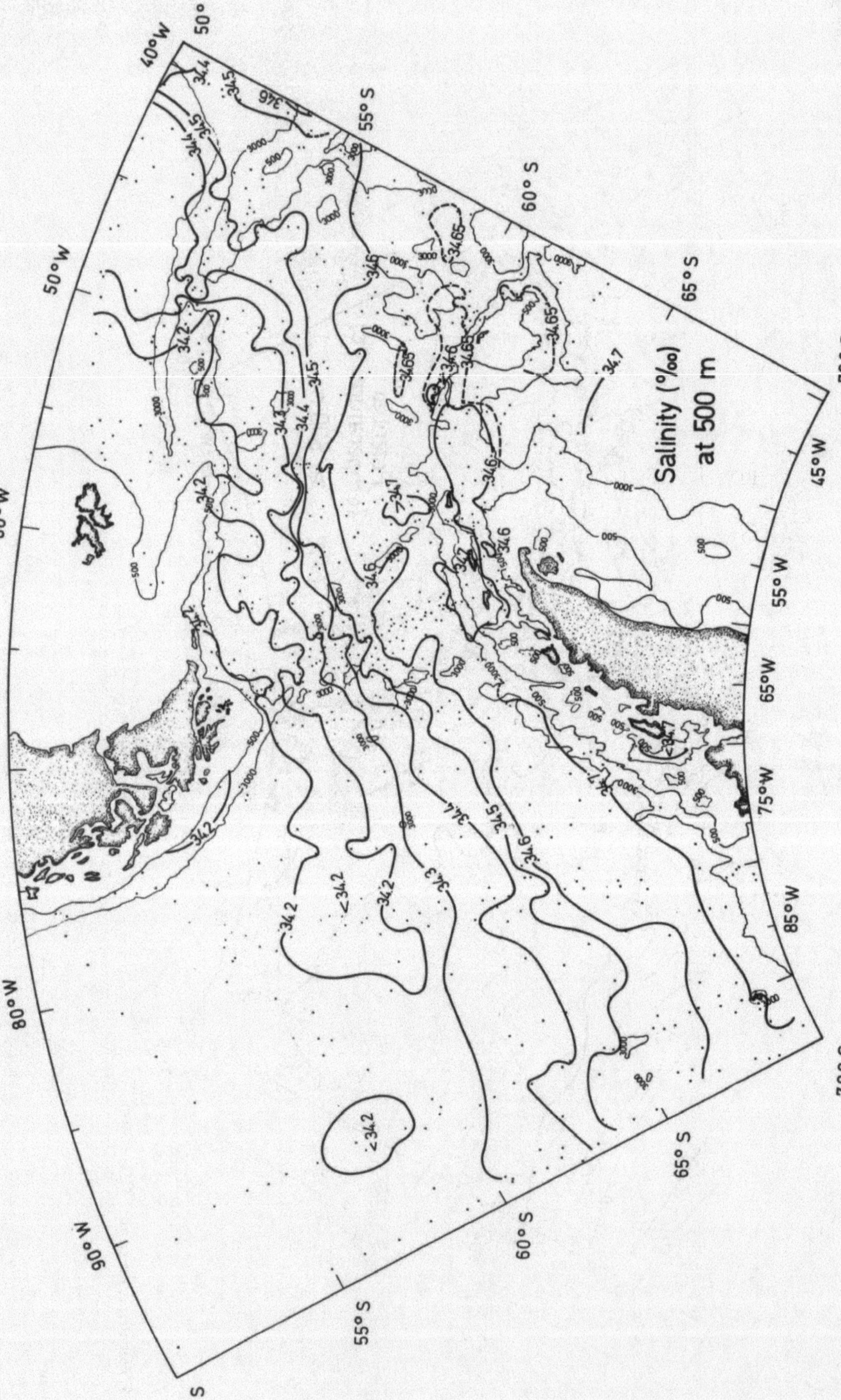

Fig. 10. Salinity at 500 m

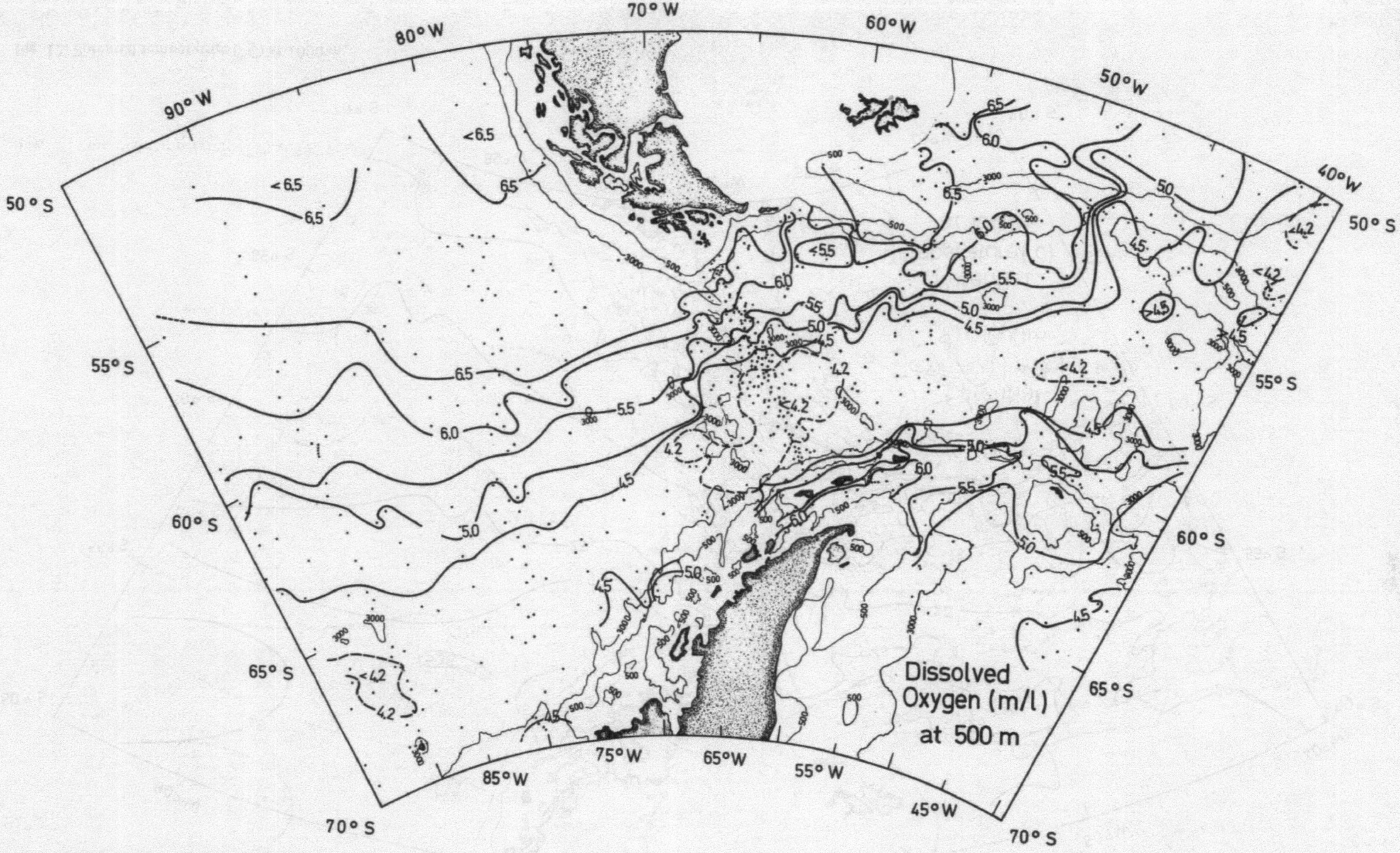

Fig. 11. Dissolved oxygen (ml/l) at 500 m

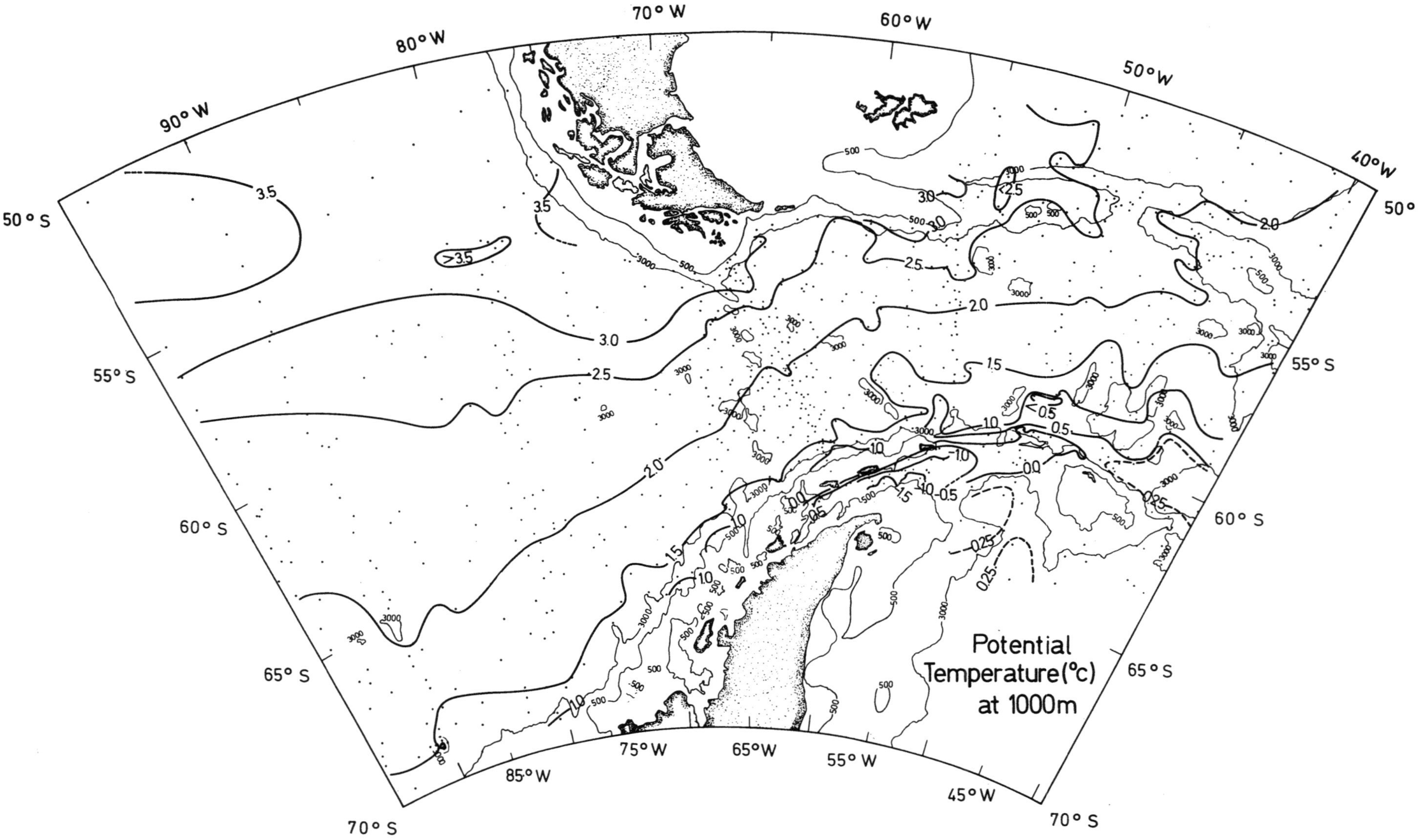

Fig. 12. Potential temperature (°C) at 1000 m

ward, with increasing temperatures, almost to the South Orkney Islands. In the Weddell Sea temperatures are near 0.25°C at this depth.

Three major features can be seen in the salinity field (Fig. 13): values of less than 34.3 at the SAZ in the southeast Pacific (a minimum associated with the AAIW), a broad maximum band with values near 37.73 in the AAZ (the signature of the LCDW), and a new minimum in Bransfield Strait. This last extremum may be traced eastward from the Strait at this depth. Salinity values in Bransfield Strait (<34.6) are similar to those at 500 m. The association of this salinity with low temperature and high oxygen results from local vertical convection.

The oxygen distribution at 1000 m (Fig. 14) evidences a continuous strip of relative minimum values (<4.2 ml/l) extending through the region from SW to NE marking the UCDW at this depth. The highest oxygen concentrations are found south of the Falkland Islands and in the Bransfield Strait and WSC. The high oxygen in the north is confined to the proximity of the wide continental shelf. Another region of high oxygen lies west of South America.

Isopleths of distinct characteristics generally parallel one another at 1000 m. (A good example is the coincidence of positions of the 2.5°C, 34.5, and 27.5 kg/m^3 isopleths). This enables us easily to distinguish horizontal distribution of water masses at 1000 m. The water mass north of the SAF at the depths is AAIW and is characterized primarily by its low salinity (>34.30) and high oxygen (>5.0 ml/l). AAIW is seen to extend east to near the Falkland Islands, where it turns north. Circumpolar Deep Water is found between the SAF and the CWB at 1000 m. The UCDW can be distinguished in the north of our region by an oxygen minimum (<4.2 ml/l). Farther south the salinity maximum (>34.73) is indicative of LCDW found at this depth due to the southward ascent of isopycnals and water masses.

The water characteristics within Bransfield Strait, with sill depth of approximately 1100 m, differ greatly from the surrounding water. Cold water (less than −1.0°C and in certain areas close to the freezing point), low salinity (<34.6), high density (>27.83) and oxygen (>6.8 ml/l) clearly distinguish the Strait. These characteristics, attenuating somewhat to the east, can be seen within the WSC.

5 Circulation

It is important, when studying the water circulation in this region, to keep in mind that Drake Passage is the narrowest constriction through which the Antarctic Circumpolar Current (ACC) flows, that it is relatively shallow compared to the southeast Pacific Basin, and that it is a region of rugged bottom topography.

The dynamic topographies for 0 and 500 db referenced to 1000 db are presented in Figs. 15 and 16. The geostrophic streamlines pictured at these two levels and at 200 m (not shown) are very similar. They are constricted where the eastward flowing ACC enters Drake Passage and the northern streamlines deflect to the north once this lateral constriction is passed. The southern streamlines maintain an easterly direction. There are wavelike patterns associated with some topographic features specially in the Scotia Sea. The northern sector of Drake Passage shows larger gradients, indicating that the upper level shear is stronger there than to the south. The effect of the Shackleton Fracture Zone extending northwest from Elephant Island shows clearly in the streamlines, as it did also in the temperature and density fields at 1000 m.

The contour intervals of these charts (0.10 and 0.05 m^2/s^2) and their nonsynoptic nature make it difficult to distinguish the high velocity cores associated with the fronts, as observed by many workers (cf., Nowlin and Klinck 1986). Based on frontal positions derived from characteristics' distributions, some dynamic height contours can be associated with specific fronts. The surface streamlines which best correspond to positions are: 0.90 for the SAF; 0.70, and between 0.70 and 0.80 in the Scotia Sea, for the PF; and 0.40 for the CWB. At 500 db the values are 0.40 to 0.45 for the SAF; 0.35 for the PF, and 0.15 (mainly in Drake Passage) for the CWB.

Ishino (1963), in searching for an appropriate level of no motion for dynamic topographies in the Southern Ocean, prepared shears relative to various depths. He concluded that the resulting current patterns show little differences. Gordon (1967) found great similarity in the geostrophic velocity fields at the surface and 1000 db referenced to 2500 db. Adding the similarity in our geostrophic shears in these upper water levels referenced to 1000 db, it appears that the circulation patterns between surface and mid-depth are very similar in the Drake Passage region. This vertical coherence is also supported by direct current measurements within Drake Passage, as noted by several workers.

It is interesting to compare these geostrophic flow representations with drifter trajectories, as described, for example, by Patterson (1985) or Hofmann (1985). Peterson (1985) examined in detail the trajectories of six buoys through Drake Passage. The buoys were shown to be fairly efficient water followers. The trajectories of these buoys are similar to the streamlines of Fig. 15. Interesting is the trajectory of Peterson's buoy 54604 which follows very closely the 0.90 m^2/s^2 isopleth (the streamline which defines the SAF) in northern Drake Passage. Even more notable is the trajectory of buoy 54642 which stayed for more than 3 months within the Antarctic Zone and followed quite closely the 0.50 m^2/s^2 isopleth, including the meander in the center of the southern Drake Passage. This agreement may be only accidental, but is worth mention. Peterson (1985) speculated that the trajectory and long residence time of this drifter in that region was in response to surface winds, but did not reach any definite conclusion.

Because of our choice of a deep (1000 m) reference level, the westward shelf and slope currents along the South

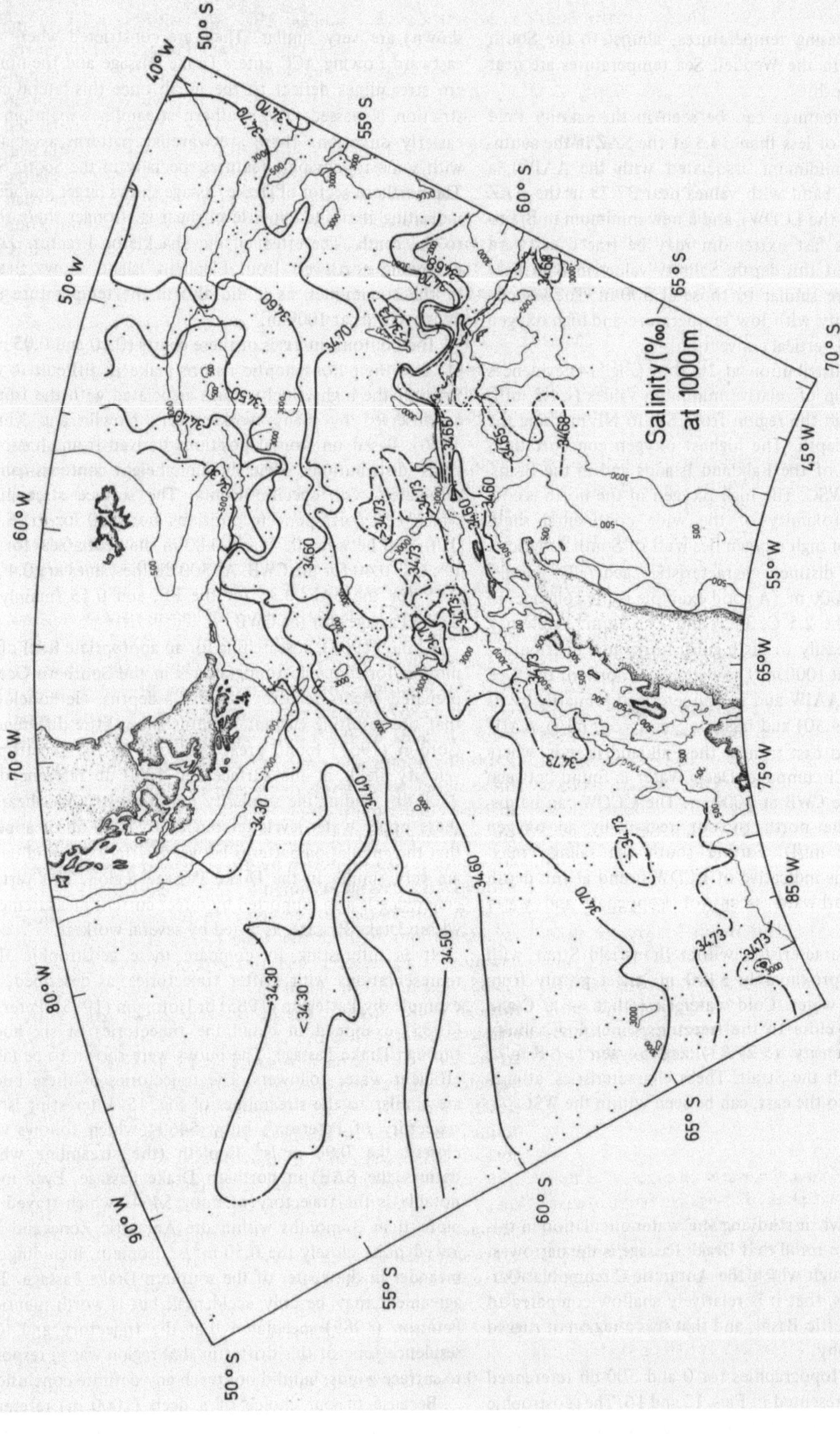

Fig. 13. Salinity at 1000 m

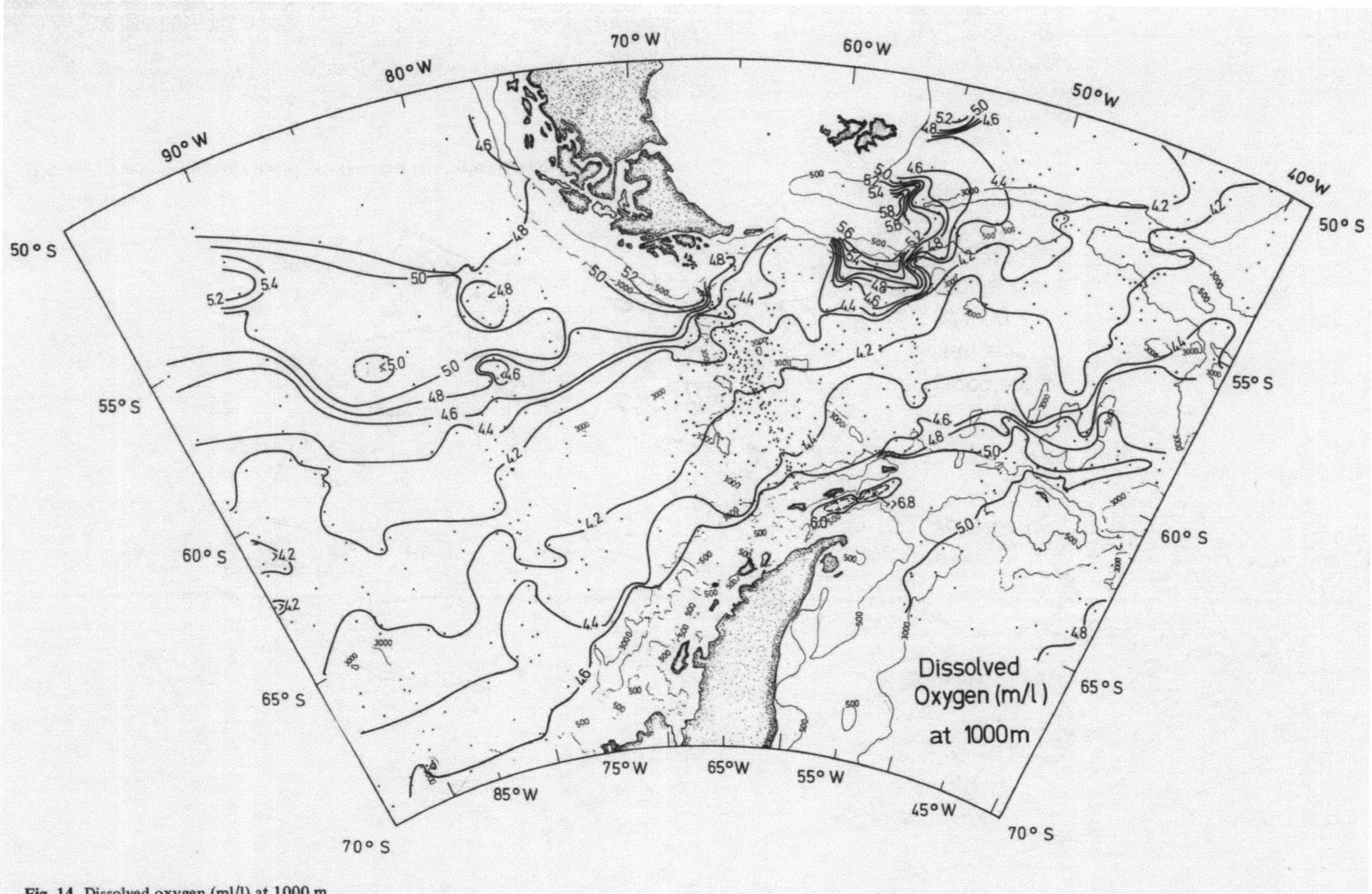

Fig. 14. Dissolved oxygen (ml/l) at 1000 m

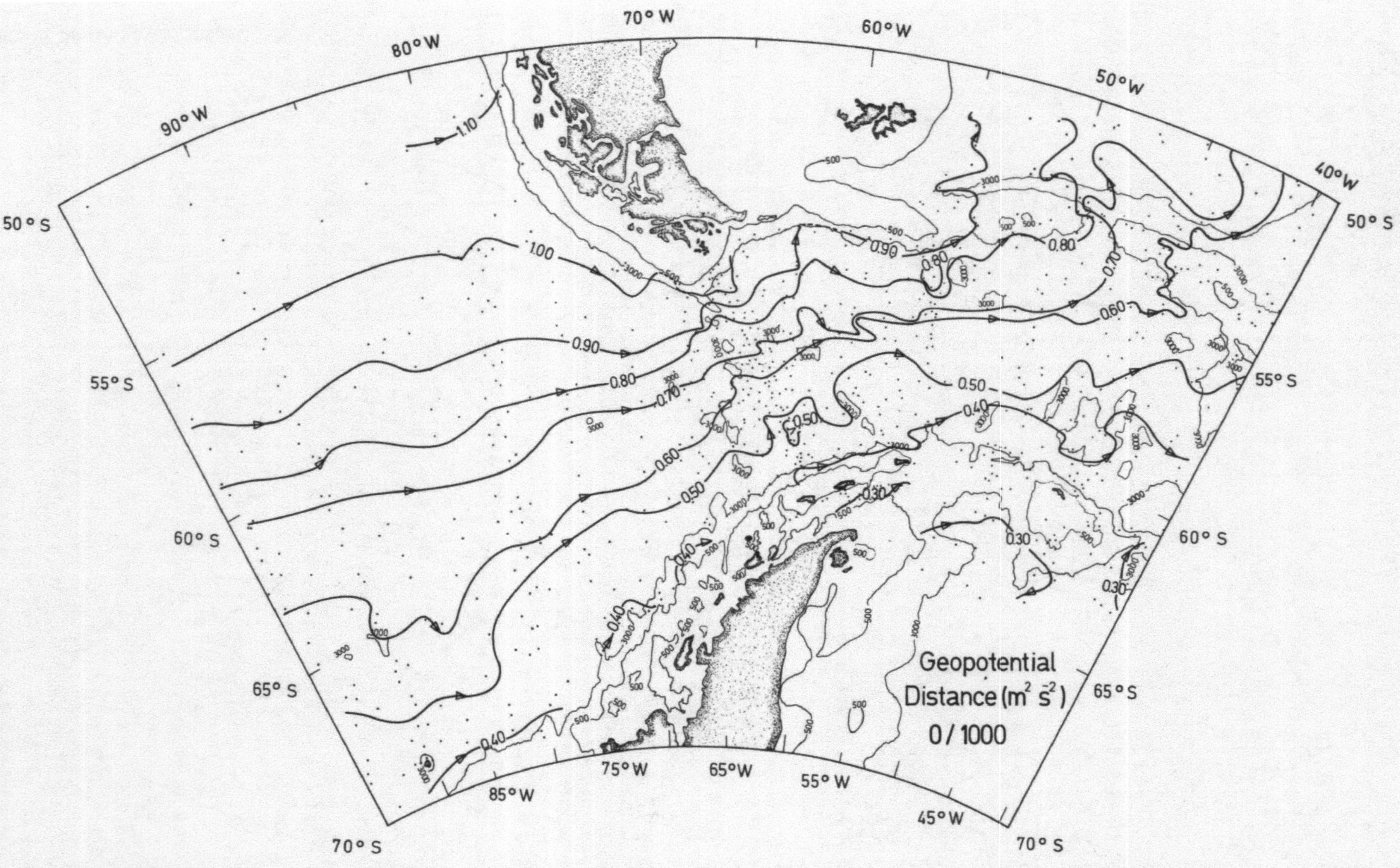

Fig. 15. Geopotential anomaly (m^2/s^2) of sea surface relative to 1000 db

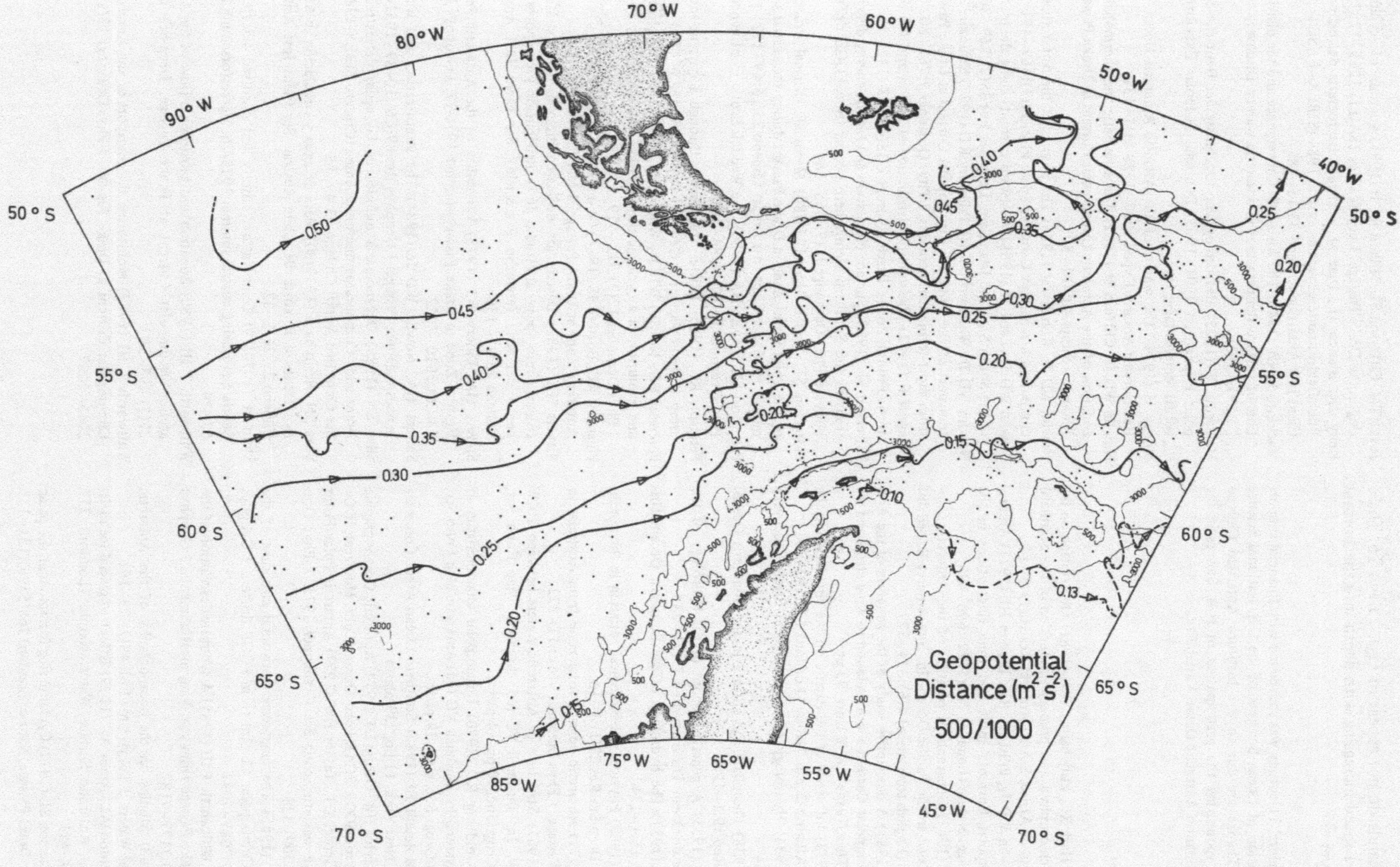

Fig. 16. Geopotential anomaly (m^2/s^2) of 500-db surface relative to 1000 db

Sandwich Islands do not appear in Figs. 15 and 16. These currents have speed increasing with depth into the bottom boundary layer.

Acknowledgments. This study was completed with financial support from the Division of Ocean Sciences of the U.S. National Science Foundation. Thanks are due to the Instituto Antártico Chileno (INACH) for sponsoring the participation of H.A. Sievers in the Scientific Seminar on Antarctic Ocean Variability.

References

Allanson BR, Hart RC, Lutjeharms JRE (1981) Observation on the nutrients, chlorophyll and primary production of the Southern Ocean South of Africa. S Afr J Antarct Res 10/11:3–14

Baker DJ, Nowlin WD Jr, Pillsbury RD, Bryden HL (1977) Antarctic Circumpolar Current: space and time fluctuations in the Drake Passage. Nature (London) 268(5622):696–699

Botnikov VN (1963) Geographical position of the Antarctic Convergence Zone in the Pacific Ocean. In: Sov Antarct Exped Inf Bull No 41 (English translation) 4(6):324–327

Clifford MA (1983) A descriptive study of the zonation of the Antarctic Circumpolar Current and its relation to wind stress and ice cover. MS Thes, Texas A & M Univ, 93 pp

Clowes AJ (1933) Influence of the Pacific on the circulation in the southwest Atlantic Ocean. Nature (London) 131:189–191

Clowes AJ (1934) Hydrology of the Bransfield Strait. Discovery Rep 9:1–64

Clowes AJ (1938) Phosphate and silicate in the Southern Ocean. Discovery Rep 19:1–120

Deacon GER (1933) A general account of the hydrology of the South Atlantic Ocean. Discovery Rep 7:171–238

Deacon GER (1937) The hydrology of the Southern Ocean. Discovery Rep 15:1–124

Deacon GER (1982) Physical and biological zonation in the Southern Ocean. Deep-Sea Res 29(1):1–15

Emery WJ (1977) Antarctic polar frontal zone from Australia to the Drake Passage. J Phys Oceanogr 7(6):811–822

Gordon AL (1967) Structure of Antarctic waters between 20°W and 170°W. In: Bushnell VC (ed) Antarctic Map Folio Ser, No 6. Am Geogr Soc, 10 pp, 14 plates

Gordon AL, Goldberg RD (1970) Circumpolar characteristics of Antarctic waters. In: Bushnell VC (ed) Antarctic Map Folio Ser, No 13. Am Geogr Soc, 6 pp, 19 plates

Gordon AL, Molinelli EJ (1982) Southern Ocean Atlas. Columbia Univ Press, New York, 11 pp, 233 plates

Gordon AL, Taylor HW, Georgi DT (1974) Antarctic Oceanographic Zonation. Proc SCOR/SCAR Polar Oceans Conf, Montreal, IOC

Gordon AL, Georgi DT, Taylor HW (1977) Antarctic Polar Front Zone in the western Scotia Sea -- summer 1975. J Phys Oceanogr 7(3):309–328

Hofmann EE (1985) The large-scale horizontal structure of the Antarctic Circumpolar Current from FGGE drifters. J Geophys Res 90(C4):7087–7097

Hofmann EE, Whitworth T III (1985) A synoptic description of the flow at Drake Passage from year-long measurements. J Geophys Res 90(C4):7177–7187

Ishino M (1963) Studies on the oceanography of the Antarctic Circumpolar Waters. J Tokyo Univ Fish 49(2):73–181

Jacobs SS, Gordon AL, Amos AF (1979) Effect of glacial ice melting on the Antarctic Surface Water. Nature (London) 277 (5696):469–471

Joyce TM, Patterson SL (1977) Cyclonic ring formation at the polar front in the Drake Passage. Nature (London) 265(5590):131–133

Joyce TM, Patterson SL, Millard RC Jr (1981) Anatomy of a cyclonic ring in the Drake Passage. Deep-Sea Res 28A(11):1265–1287

Kelly JR, Blanco JL, Diaz M V (1985) Hidrografía del estrecho Bransfield durante el verano austral 1984 (SIBEX–FASE I). Ser Cient Inst Antártico Chileno 33:15–48

Mackintosh NA (1946) The Antarctic Convergence and the distribution of surface temperatures in antarctic waters. Discovery Rep 23:177–212

Meinardus W (1923) Meteorologische Ergebnisse der Deutsche Südpolar Expedition 1901–1903. Deutsche Südpolar Expedition, vol III Meteorologie 1(1)

Mosby H (1934) The waters of the Atlantic Antarctic Ocean. Sci Res Norweg Antarct Exped 1927–1928 11:1–131

Nowlin WD Jr, Clifford M (1982) The kinematic and thermohaline zonation of the Antarctic Circumpolar Current at Drake Passage. J Mar Res Suppl 40:481–507

Nowlin WD Jr, Klinck JM (1986) The physics of the Antarctic Circumpolar Current. Rev Geophys Space Phys 24(3):469–491

Nowlin WD Jr, Zenk W (1988) Westward currents along the margin of the South Shetland Island arc. Deep-Sea Res 35(2):269–301

Nowlin WD Jr, Whitworth T III, Pillsbury RD (1977) Structure and transport of the Antarctic Circumpolar Current at Drake Passage from short-term measurements. J Phys Oceanogr 7:788–802

Ostapoff F (1962) The salinity distribution at 200 m and the Antarctic Frontal Zones. Dtsch Hydrogr Z 15(4):133–142

Patterson SL (1985) Surface circulation and kinetic energy distributions in the Southern Hemisphere oceans from FGGE drifting buoys. J Phys Oceanogr 15(7):865–884

Patterson SL, Sievers HA (1979/80) Mesoscale thermal structure of the Polar Front Zone in Drake Passage during the austral summer of 1976. Ser Cient Inst Antárt Chileno 25/26:49–112

Patterson SL, Sievers HA (1980) The Weddell-Scotia Confluence. J Phys Oceanogr 10(10):1584–1610

Peterson RG (1985) Drifter trajectories through a current meter array at Drake Passage. J Geophys Res 90(C3):4883–4893

Peterson RG, Nowlin WD Jr, Whitworth T III (1982) Generation and evolution of a cyclonic ring at Drake Passage in early 1979. J Phys Oceanogr 12(7):712–719

Pillsbury RD, Bottero JS (1984) Observations of current rings in the Antarctic Zone at Drake Passage. J Mar Res 42:853–874

Sievers HA (1982) Descripción de las condiciones oceanográficas físicas, como apoyo al estudio de la distribución y comportamiento del Krill (estrecho Bransfield). Ser Cient Inst Antárt Chileno 28:87–136

Sievers HA, Emery WJ (1978) Variability of the Antarctic Polar Frontal Zone in Drake Passage-summer 1976-77. J Geophys Res 83(C6):3010–3022

Sievers HA, Nowlin WD Jr (1984) The stratification and water masses in Drake Passage. J Geophys Res 89(C6):10489–10514

Silva SN (1985) Oceanografía química de las aguas del estrecho Bransfield: Compuestos micronutrientes (Crucero SIBEX-1984). Ser Cient Inst Antárt Chileno 33:49–88

Silva SN, Neshyba S (1979/80) Masa de agua y circulación geostrófica frente a la costa de Chile Austral. Ser Cient Inst Antárt Chileno 25/26:5–32

Stein M (1978/79) Temperature, salinity and dissolved silicate in Scotia Sea during austral summer 1975/76. Meeresforschung 27: 67–74

Whitworth T III (1980) Zonation and geostrophic flow of the Antarctic Circumpolar Current at Drake Passage. Deep-Sea Res 27(7):497–507

Whitworth T III (1983) Monitoring the transport of the Antarctic Circumpolar Current at Drake Passage. J Phys Oceanogr 13(11): 2045–2057

Variation of Geostrophic Circulation off the Antarctic Peninsula and in the Southwest Scotia Sea, 1975–1985

M. STEIN[1]

Summary. Based upon the hydrographic data collected by the Federal Republic of Germany during biological surveys with FRV *Walther Herwig* and RV *Polarstern* between 1975 and 1985, examples of geostrophic circulation patterns for the region of the Antarctic Peninsula and the adjacent Scotia Sea are discussed. The dynamic topography elucidates the nature of a stationary meander in the vicinity of Elephant Island (South Shetland Archipelago) as well as in the western Bransfield Strait. The data indicate the southwestward flow of water west of the Antarctic Peninsula.

1 Introduction

The region studied within the scope of the present paper comprises the waters northwest of the Antarctic Peninsula, the Bransfield Strait as well as its western and eastern approaches, and the southern Drake Passage. Additionally, the general geostrophic circulation of the Scotia Sea as obtained during 1975/76 will be treated. The scheme of surface water circulation in the area of the southern Drake Passage and around the South Shetland Islands Archipelago was first described by Clowes (1934). According to his results, and Gordon and Nowlin (1978), the area between the islands and the Antarctic Peninsula, the Bransfield Strait, is influenced by waters from the Bellingshausen Sea and the Weddell Sea. These water masses penetrate into the Bransfield Strait area from the open boundaries of the Strait (Fig. 1). From the west, Bellingshausen Sea water enters between Low, Smith, and Snow Islands. Parts of this current flow north and south of Deception Island, where a characteristic bend to the southeast is observed. In the vicinity of Trinity Island the current loops to the northeast and joins the branches at the northern edge of the Bransfield Strait, forming a strong flow along the southern coast of the South Shetland Islands. Along the shelf off the Antarctic Peninsula Weddell Sea water influences the southern parts of the Strait.

The kinematic structure of the Antarctic Circumpolar Current (ACC) between the tips of South America and the South Shetland Islands is given by Nowlin et al. (1977). Based on moored current meter measurements and geostrophic computations, they report current bands of high eastward speeds separated by regions of lower speed or even westward flow in the central Drake Passage. Off the north coast of the South Shetland Islands they derive an eastward flow which amounts to about 10 cm/s. Nowlin and Zenk (this Vol.), however, derived a westward flow over the shelf slope increasing with depth.

At the eastern edge of the South Shetland Archipelago a steady flow of about 12 cm/s into 65°TC (true course) was observed by means of moored current meters between the mid of November 1977 and the end of January 1978 (Stein 1979). From six year-long moorings deployed during February 1975 in Drake Passage, Pillsbury et al. (1979) found that the mean flow directions are consistent with those from geopotential anomaly charts. The mean deep flow (2667 m) off the South Shetland Islands was almost directly to the east (east component 5.5 cm/s, north component 0.4 cm/s). Whitworth et al. (1982) report on a counterflow near the South Shetland Islands, which, according to their direct current measurements and calculated baroclinic shears, is persistent throughout the year-long meter records. The west component within the upper 1000 m of the water column amounts to about 8 cm/s. Similar results for the slope region were obtained by Nowlin and Clifford (1982) with characteristic mean speeds of 8.6 cm/s.

Current observations by Wittstock and Zenk (1983) from the southern Drake Passage approximately 110 km north of King George Island (3625 m depth) reveal a current scenario with a mean flow directed 58° TC, vector mean 6.8 cm/s at 439 m depth, at 1489 m depth 142°TC, 18.7 cm/s, and 222°TC, 20.9 cm/s at a depth of 3590 m. The persistent deep flow confirms the observations performed by Whitworth et al. (1982) 220 km towards the west off Livingston Island. From a small-scale study in the eastern Bransfield Strait Schröder (1983) obtains for the baroclinic part of the flow 67°TC, 7.9 cm/s.

Whereas the coverage of the Drake Passage with moored current meters is fairly good due to FDRAKE (First Dynamic Response and Kinematic experiment) 75, 76, 77 and 78 and DRAKE-79, it is less good or nonexistent in the remainder of the area under consideration.

1 Institut für Seefischerei, Bundesforschungsanstalt für Fischerei, Palmaille 9, 2000 Hamburg 50, Fed. Rep. of Germany

Antarctic Ocean and Resources Variability
(ed. by D. Sahrhage)

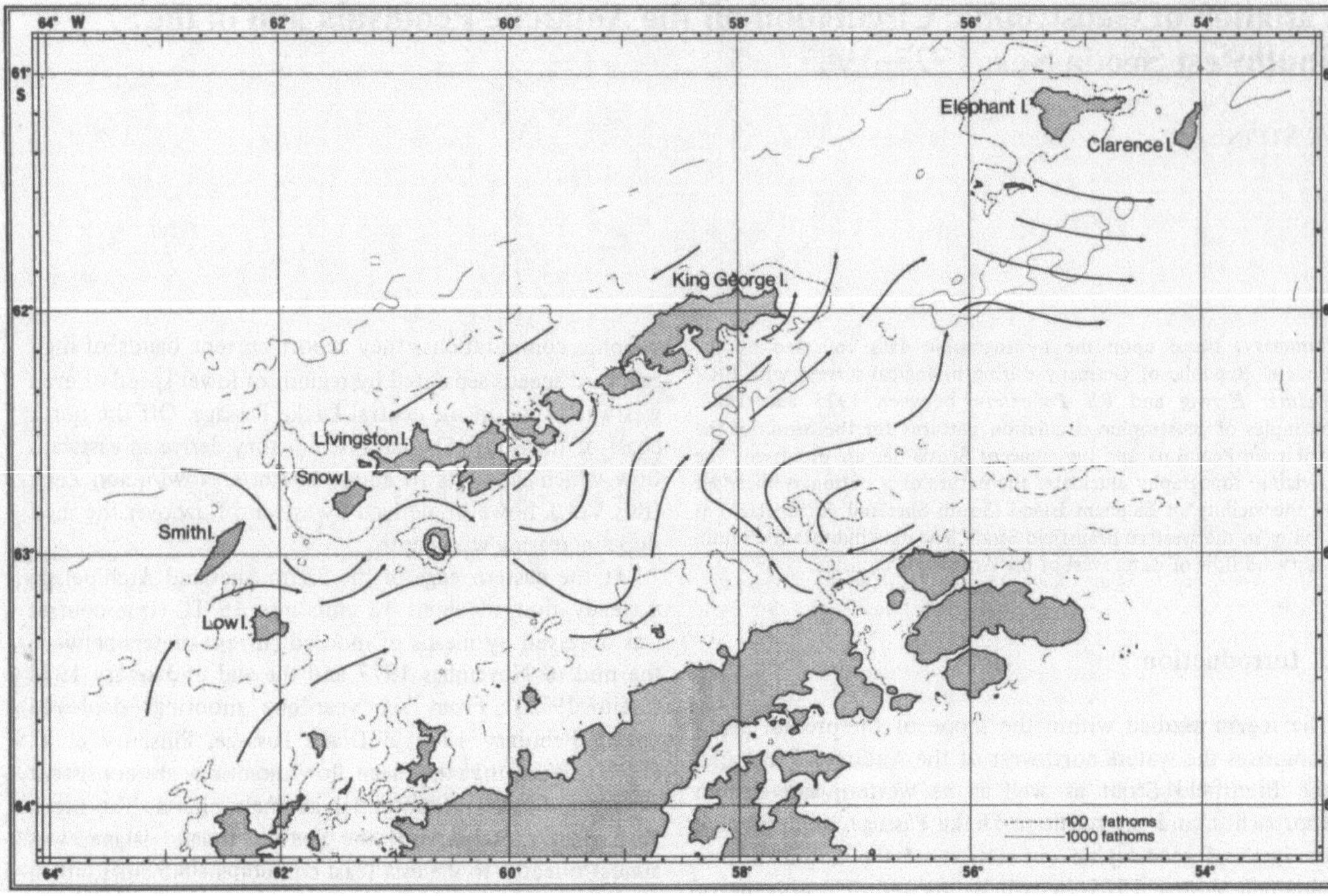

Fig. 1. Mean surface circulation in the Bransfield Strait according to Clowes (1934)

2 Material and Methods

From 1975 onwards, the Federal Republic of Germany conducted a series of biological surveys to the area of the Scotia Sea and off the Antarctic Peninsula. Whereas the first cruise to the area in 1975/76 (Sahrhage et al. 1978) was a large-scale approach, the 1977/78 cruise (Hempel et al. 1979) aimed at observations on the meso-scale.

The 1981, 1983, and 1985 expeditions under the auspices of BIOMASS (Biological Investigations of Marine Antarctic Systems and Stocks) considered the eastern approaches of the Bransfield Strait, the Bransfield Strait and the region north of the South Shetlands, as well as off the Antarctic Peninsula (Anonymous 1982, 1986). Due to the priorities of the program, krill and fish, a compromise had to be found between biologists and oceanographers. This differed from year to year with increasing understanding for the oceanographic demands from the biologists' side. Thus, the increasing depth of profiles, obtained from Nansen bottle sampling during 1975/76 within the upper 500 m to CTD-profiles from 1977 onwards which covered the entire water column in the Elephant Island Box (Fig. 3), reflects year-long discussions between oceanographers and biologists. With the exception of the FIBEX data set which covers only the upper 600 m, the remainder of the CTD data set enables a qualitative analysis of the hydrographic environment.

All data (pressure, temperature, salinity profiles) were treated in the way described by Stein and Rakusa-Suszczewski (1983). The resulting plots display the geopotential anomaly at the sea surface in dyn m relative to the 3 MPa, 5 MPa, 10 MPa, 20 MPa and 30 MPa surface. The direction of the geostrophic current flow is indicated by arrows. From the above-cited literature it is evident that the shear flow represents only parts of the total flow as measured by current meters. Low-pass filtered 65-day time series of moored current meter records indicate a steady flow of about 12 cm/s, 65°TC at the shelf break south off Elephant Island (Stein 1979), and 6.8 cm/s, 58°TC north of King George Island (Wittstock and Zenk 1983). These figures indicate the order of magnitude of the mean currents within the upper 500 m of the water column off the South Shetland Islands. Although there are some constraints with regard to the applicability of the geostrophic method, especially in shallower waters, this method provides us with the general circulation patterns in a very quick way. These patterns show the mean propagation of water masses although mixing across the streamlines might occur. As shown below, some results of this method are to a large extent consistent with the flow pattern obtained by direct current measurements.

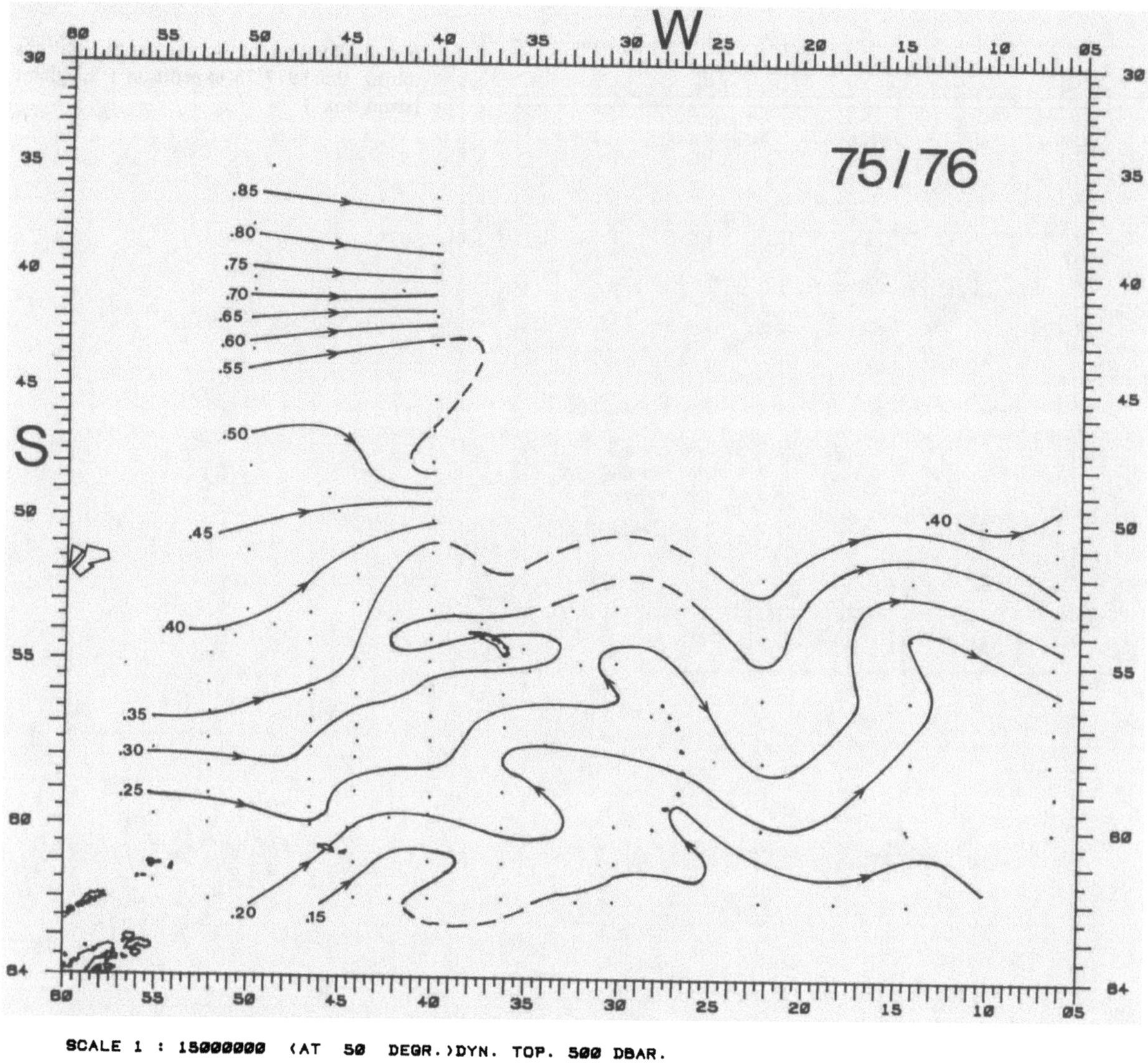

Fig. 2. Geopotential anomaly of the sea surface relative to the 5 MPa surface; November, 1975 to May, 1976

3 General Geostrophic Circulation in the Scotia Sea During the 1975/76 Survey

Figure 2 displays the geostrophic flow in the Scotia Sea and east of the South Sandwich Islands from 50 to 62°S, 55 to 5°W, referenced to the 500 decibar (5 MPa) pressure surface. The geostrophic computation is based on Nansen bottle data collected throughout the entire cruise from November, 1975 to May, 1976. The picture reflects quite well the general circulation pattern as given by Gordon and Molinelli (1982) in the Southern Ocean Atlas on plate 231 (dynamic height anomaly 0/1000 db). The bending of the contour lines 25 dyn cm and 30 dyn cm in the vicinity of the South Sandwich Archipelago and off South Georgia probably reflects topographic influence on the current flow. Similar characteristic curvatures are found between 10 and 15°W west of the Mid-Atlantic Ridge. At the western boundary, the 35 dyn cm contour marks the approximate location of the Antarctic Polar Front, whereas further north between 40° and 45°S the Subtropical Front emerges from the picture. It should, however, be noted that these fronts do not follow positions of these contours as one proceeds into central and eastern regime of figure. The Antarctic Polar Front, for example, turns northward after passing through Drake Passage.

4 Near-Surface Geostrophic Shear off Elephant Island 1977/78

During all three cruise legs a meso-scale sized box of biological/oceanographic stations was performed around Elephant Island (Hempel et al. 1979). The interstation distance along the north/south transects amounted to 15 nautical miles (Fig. 3). Based on this grid the BIOMASS station grid was implemented during the 1983 SIBEX field phase.

4.1 Cruise Leg 1, October 31, 1977 to December 20, 1977

The dynamic isobaths relative to the 3 MPa level indicate a wavy pattern around Elephant Island (Fig. 4). Due to the pack ice situation the southeastern portion of the box

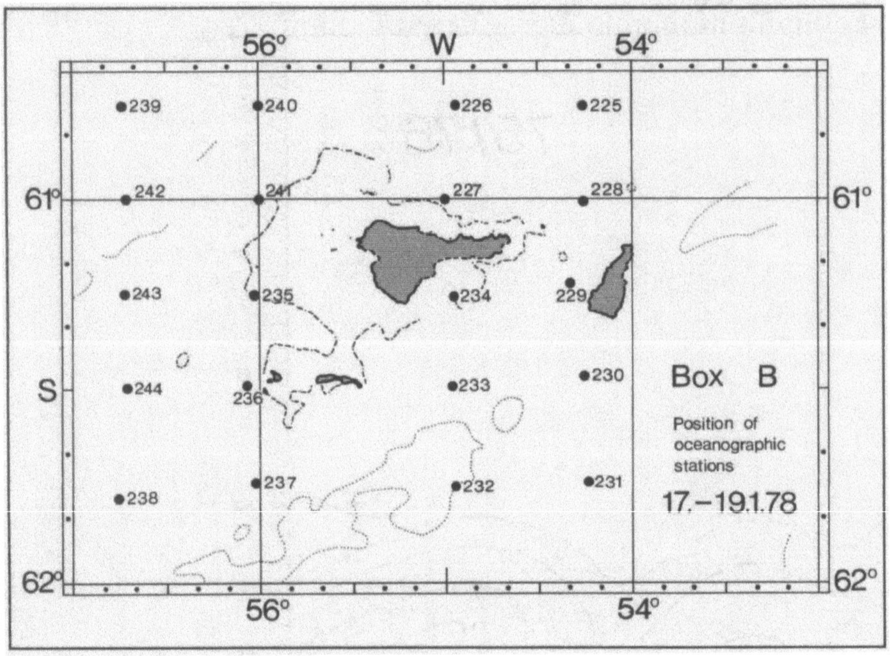

Fig. 3. Position of oceanographic stations during the 1977/78 expedition ("Elephant Island Box")

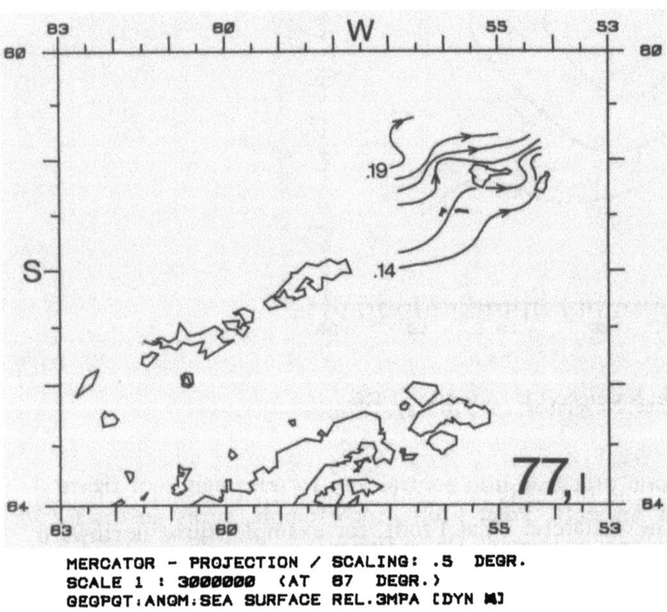

Fig. 4. Geopotential anomaly of the sea surface relative to the 3 MPa surface; leg 1: October 31, 1977 to December 20, 1977

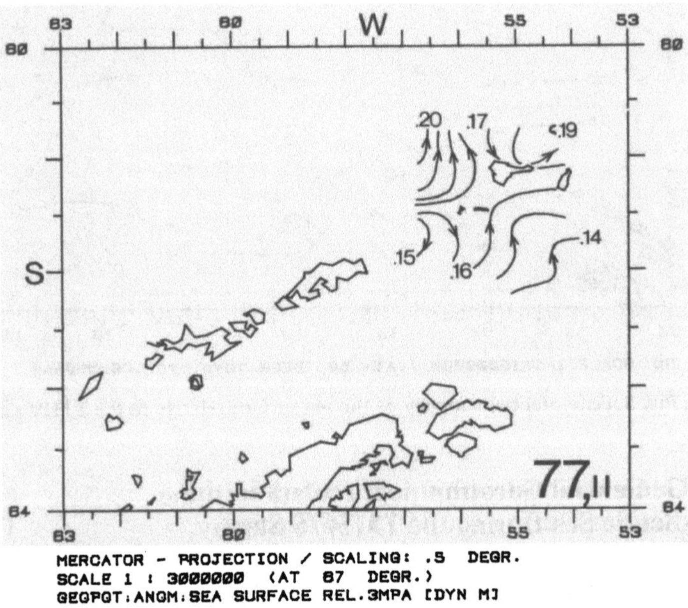

Fig. 5. Geopotential anomaly of the sea surface relative to the 3 MPa surface; leg 2: December 28, 1977 to February 17, 1978

could not be performed during this cruise leg. Of special interest is the curvature of the dynamic topography northwest off Elephant Island and north of Clarence Island.

4.2 Cruise Leg 2, December 28, 1977 to February 17, 1978

Of completely different structure is the geostrophic circulation pattern during the summer season (Figs. 5, 6). Relative to the 3 MPa surface the box is divided into cells which seem to indicate a divergent flow of water masses or eddying. The dynamic heights with reference to 5 MPa (Fig. 6) reflect the strong flow of water from the southwest with distinct gradients of stream lines, a flow of water from the Bransfield Strait, and a meandering north of Clarence Island.

4.3 Cruise Leg 3, February 23, 1978 to April 10, 1978

Similarly to the summer situation (Fig. 5), the autumn data suggest a division into four cells (Fig. 7, 3 MPa level). Of quite different shape, however, is the geopotential anomaly of the sea surface relative to the 5 MPa surface (Fig. 8). This points out the fact that the near-surface shears may not represent the general geostrophic flow

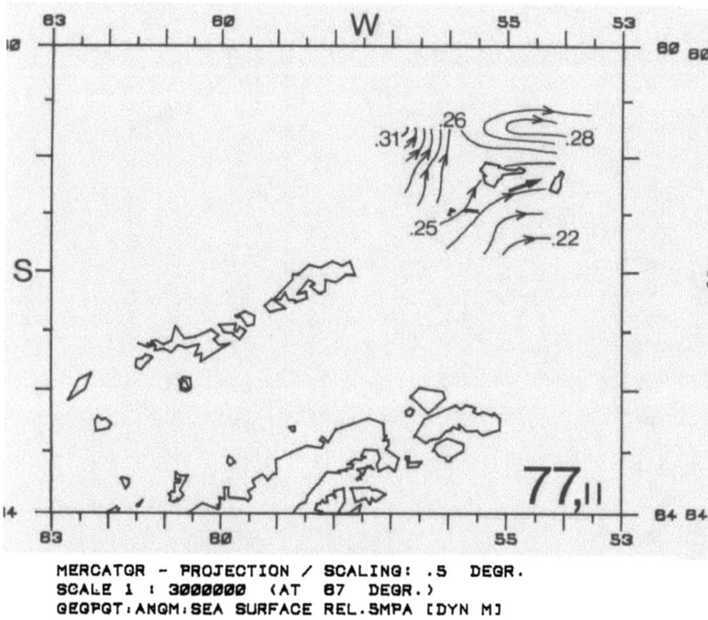

Fig. 6. Geopotential anomaly of the sea surface relative to the 5 MPa surface; leg 2: December 28, 1977 to February 17, 1978. The *arrow* denotes the mean current direction according to Stein (1979)

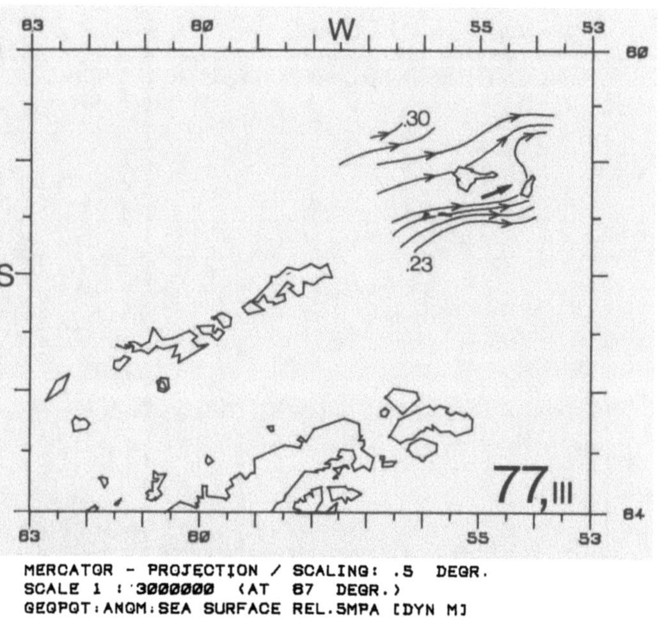

Fig. 8. Geopotential anomaly of the sea surface relative to the 5 MPa surface; leg 3: February 23, 1978 to April 10, 1978. The *arrow* denotes the mean current direction according to Stein (1979)

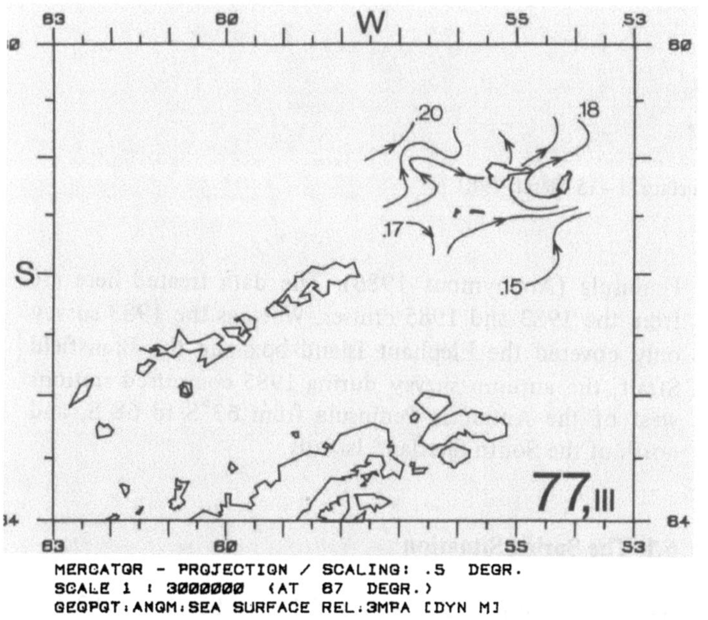

Fig. 7. Geopotential anomaly of the sea surface relative to the 3 MPa surface; leg 3: February 23, 1978 to April 10, 1978

within the Elephant Island Box. Based upon the 500 db pressure level a northeasterly flow dominates throughout the entire box with slow motion in the middle and northwestern parts, but swift currents south of Elephant Island. The meandering north of Clarence Island again clearly emerges from the autumn data. For comparison the mean current vector as obtained at the mooring site south off Elephant Island (61°18.2′S, 54°55.2′W) at 250 m depth is inserted in Figs. 6 and 8.

5 Upper Ocean Geostrophic Shear in the Southern Drake Passage, the Bransfield Strait, and off Elephant Island During FIBEX

This section reflects some of the results published by Stein and Rakusa-Suszczewski (1983) based on the Polish/West German FIBEX (First International BIOMASS Experiment) oceanographic data set. This data set was treated with the oceanographic data handling system of the Institut für Seefischerei, Hamburg, as described by the above-mentioned authors.

5.1 Bransfield Strait

The Bransfield Strait data originate from the first half of March, 1981. The characteristic flow pattern in the western parts of the Strait is well documented in Fig. 9. As outlined in Fig. 1, which is the general circulation pattern as described by Clowes (1934), the summer 1981 situation is consistent with the results based on the observations of the old *Discovery*. The inflowing Bellingshausen Sea water after having passed north or south Deception Island bends to the southeast, reaches Trinity Island and turns back to the northwest before it joins the current along the south coast of the South Shetland Islands. The influence of the Weddell Sea water on the circulation of the eastern and southern parts of the Strait clearly emerges from Fig. 9. Along the shore of the Antarctic Peninsula a southwesterly flow was observed, indicating the spreading of Weddell Sea water to Trinity Island, where it meets the Bellingshausen Sea water.

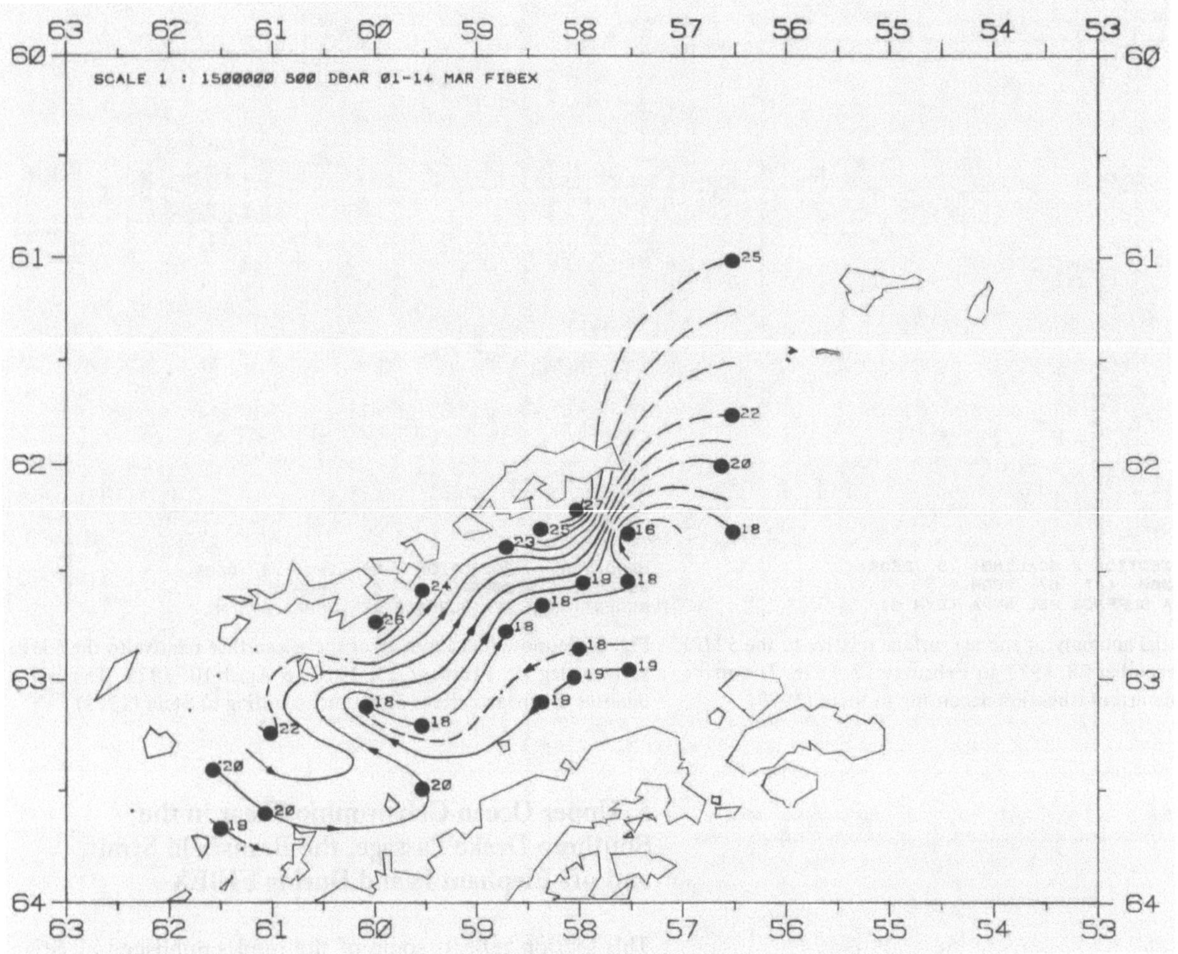

Fig. 9. Geopotential anomaly of the sea surface relative to the 5 MPa surface; 1–15 March 1981

5.2 Geostrophic Circulation off Elephant Island

Northeast of Elephant Island, the detailed structure of a Weddell-Scotia Confluence meander is visible (Fig. 10). Whereas the current border between Pacific and Weddell Sea water masses is very broad around and north of Elephant Island, the stream-lines are concentrating along their further way to the northeast. In the vicinity of 50°W the current border makes a characteristic bend to the southeast. As mentioned by Stein (1981), this meander corresponds with the strong indentation of Weddell Sea water and Scotia Sea water. Due to the topography of the southern Scotian Ridge, the gap between the Elephant Island plateau and the South Orkney plateau is one of the major outlets for Weddell Sea water to flow under the deflecting influence of the Coriolis force into the Scotia Sea.

6 Upper Ocean Geostrophic Shear Pattern off the Antarctic Peninsula During SIBEX

From October, 1983 to April, 1985 the Federal Republic of Germany conducted three cruises with RV *Polarstern* and FRV *Walther Herwig* to the area off the Antarctic Peninsula (Anonymous 1986). The data treated here are from the 1983 and 1985 cruises. Whereas the 1983 survey only covered the Elephant Island box and the Bransfield Strait, the autumn survey during 1985 comprised stations west of the Antarctic Peninsula from 63°S to 68°S, and north of the South Shetland Islands.

6.1 The Spring Situation

Between October 22, 1983 and November 10, 1983 RV *Polarstern* surveyed the area of the Bransfield Strait and off Elephant Island. From the CTD/Rosette profiles the dynamic heights were calculated in the above manner. Figure 11 displays the geopotential anomaly of the sea surface relative to the 3, 5, 10, 20, and 30 MPa surface. The numbers to the right and left of the circulation pattern refer to the minimum and maximum value of dynamic heights in dyn m. The distance of the dynamic isobaths is 0.01 dyn m. A general view of the circulation in the Bransfield Strait and around Elephant Island is displayed in Fig. 11a. As in the FIBEX period (BIOMASS 1983), the major current flow in the Bransfield Strait is to the northeast. A counter current of Weddell Sea water known to flow southwestwards along

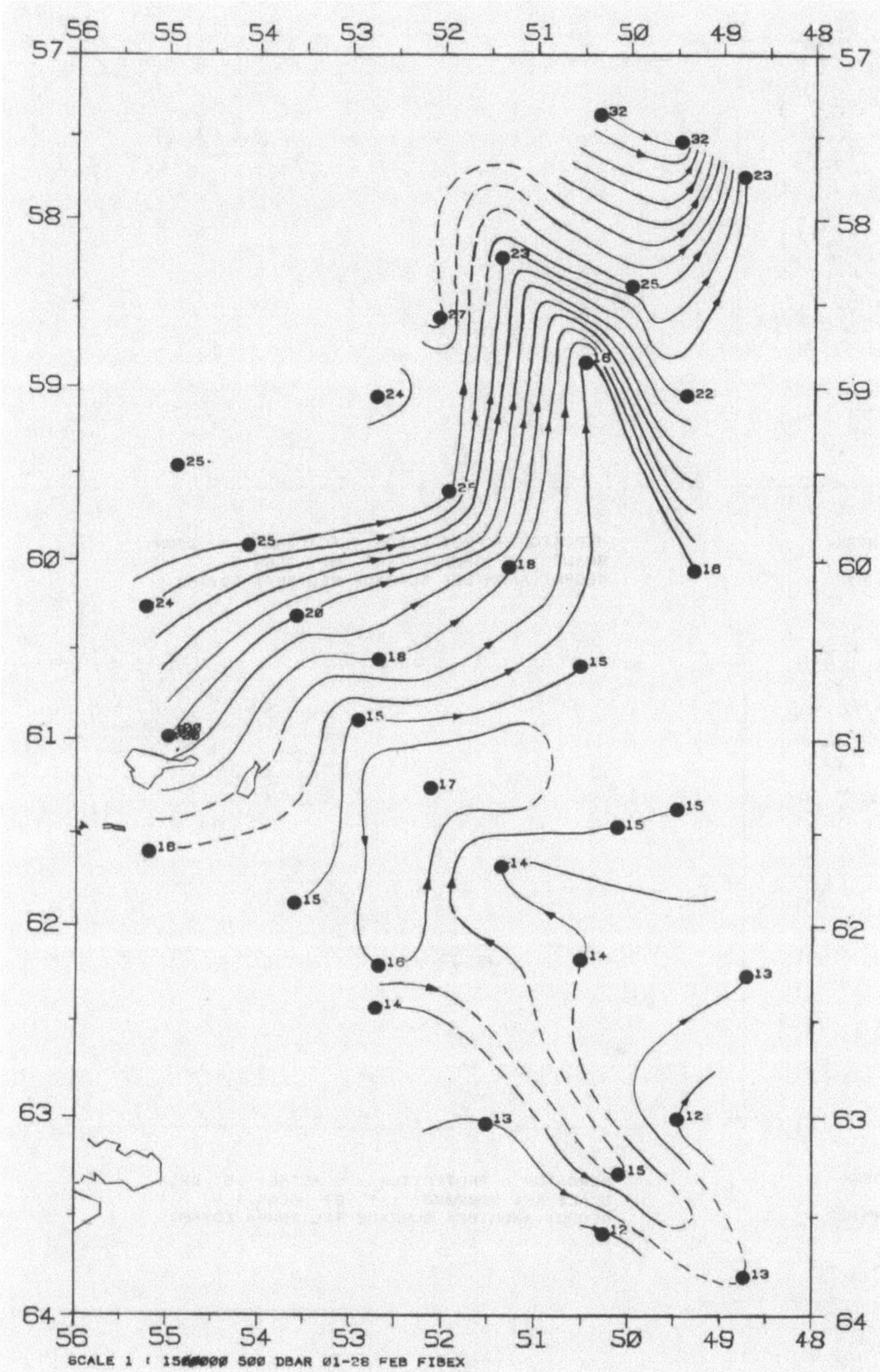

Fig. 10. Geopotential anomaly of the sea surface relative to the 5 MPa surface; FIBEX: January–February 1981

the Antarctic Peninsula is not shown by the geopotential anomaly relative to the 3 MPa surface. The area around Gibbs Island, about 61°30′S, 56°W is a region of slow geostrophic motion. The gradients of dynamic heights relative to the 3 MPa and 5 MPa surfaces, are weak (Fig. 11a,b). The circulation pattern in the northern part of the Elephant Island box reveals a swift current approaching the Elephant Island Plateau from the southwest (Fig. 11d). After meeting with the shelf slope the current diverges, to flow in part to the northeast and in part to the northwest. Calculation of the geostrophic currents relative to the 20 MPa surface yield surface currents ranging from 14 cm/s to 25 cm/s along the southern border of the circulation pattern. Areas of slow geostrophic motion like the northeastern and northwestern corner of the box are characterized by currents ranging from 2 cm/s to 11 cm/s relative to the 20 MPa surface. The areas of strong gradients reflect the approximate location of the Weddell-Scotia Confluence. Figure 11a–e indicates that the shear is shown to continue well into the water column with the same

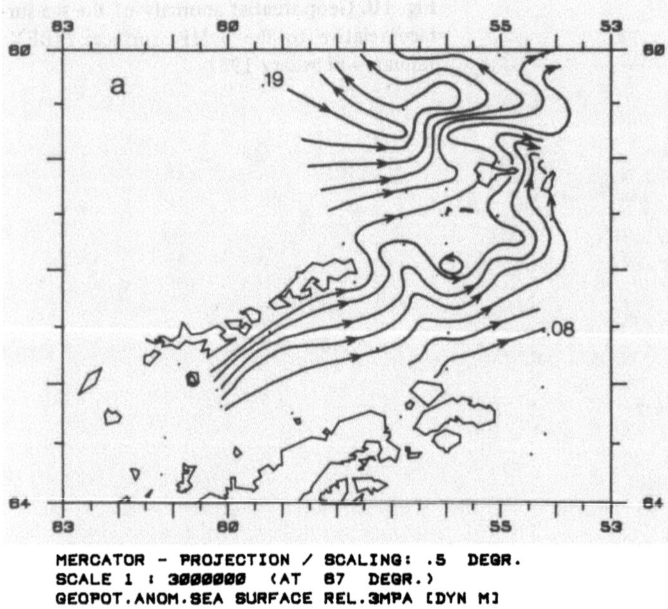

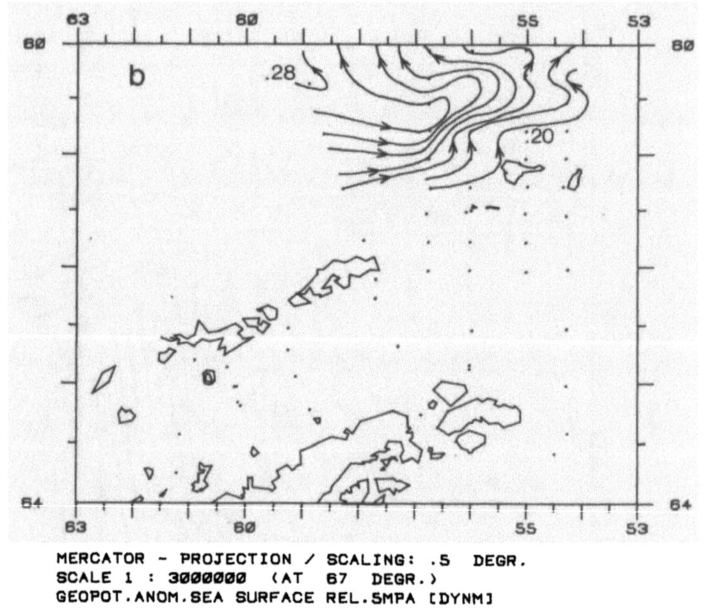

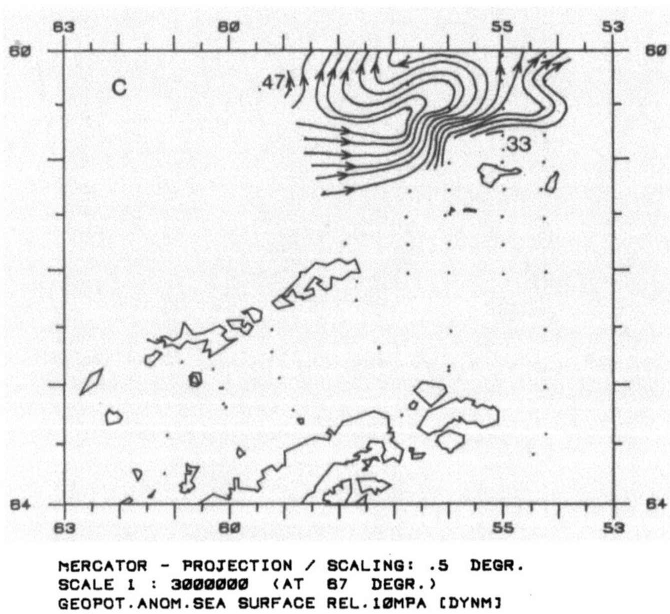

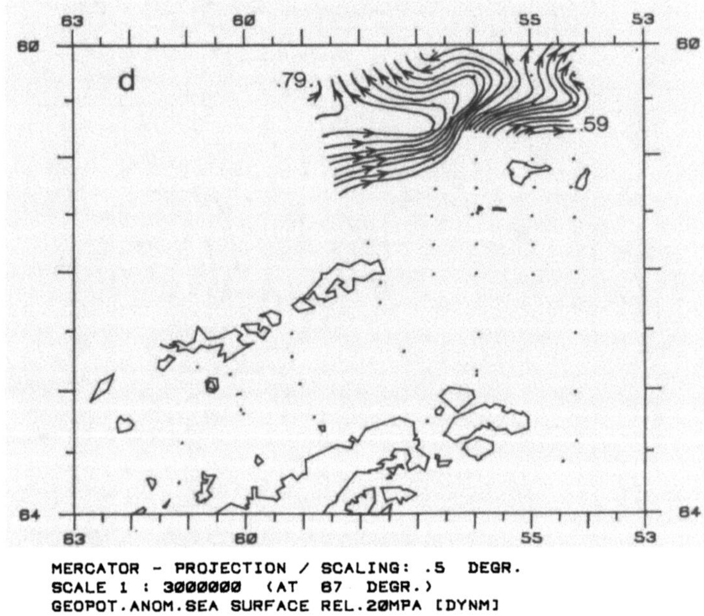

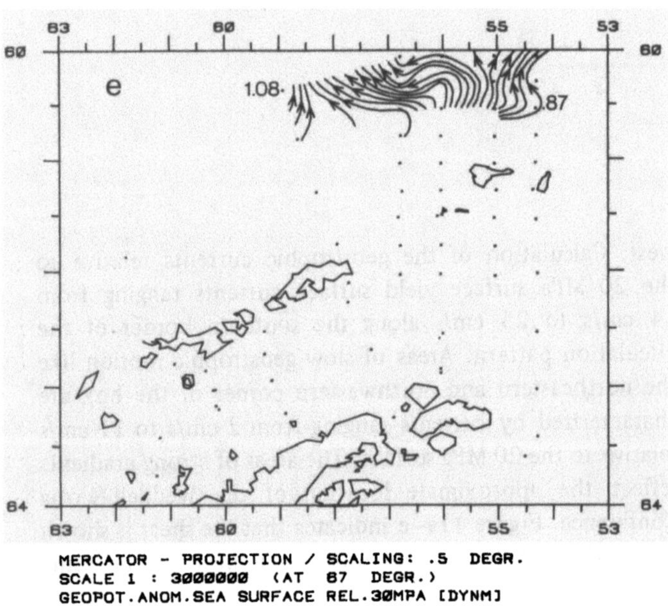

Fig. 11a–e. Geopotential anomaly of the sea surface relative to the 3, 5, 10, 20, and 30 MPa surface; October 22 to November 10, 1983

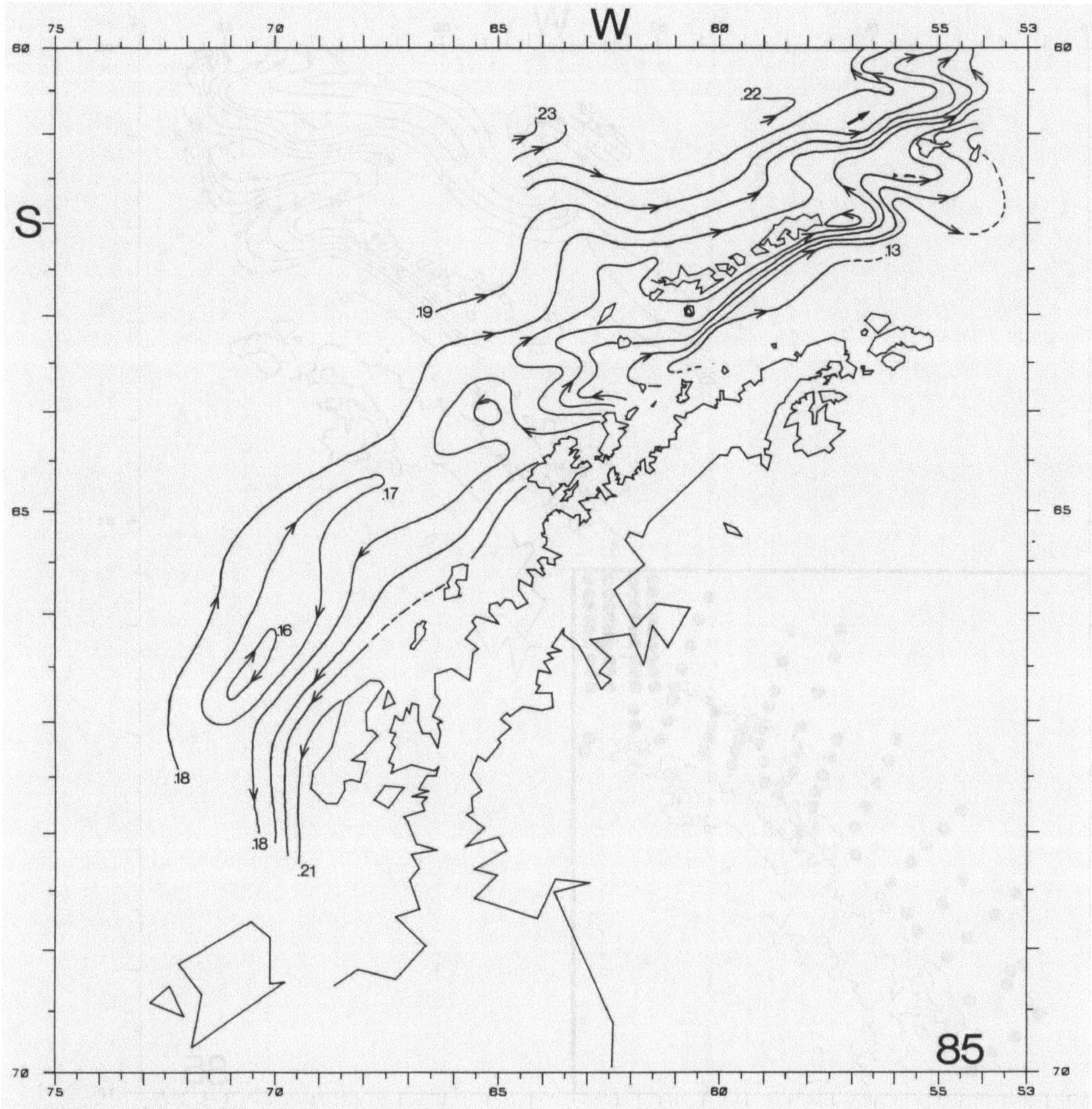

Fig. 12. Geopotential anomaly of the sea surface relative to the 3 MPa surface; March 7 to April 16, 1985. The *arrow* denotes the mean current direction according to Wittstock and Zenk (1983); station grid see *insert* Fig. 13

sense. Throughout all depths, the characteristic meandering of the stream lines north of Clarence Island, 61°15′S, 54°W is visible. As noted already from the 1977/78 and the FIBEX observations, this feature seems to indicate topographic steering on the flow of water masses. Accordingly, the curvature of the stream lines in the west of Elephant Island reflects steering influenced by the Shackleton Fracture Zone (see below).

6.2 The Autumn Situation

From March 19, 1985 to April 8, 1985 FRV *Walther Herwig* worked along a station grid from Elephant Island to Adelaide Island (Figs. 12, 13). The general circulation pattern referenced to the 3 MPa surface (Fig. 12) is consistent with the spring 1983 circulation shown in Fig. 11a. North of the South Shetland Islands the flow pattern indicates a steady current to the northeast. Between King George Island and Elephant Island, the Loper Channel, the contour lines show

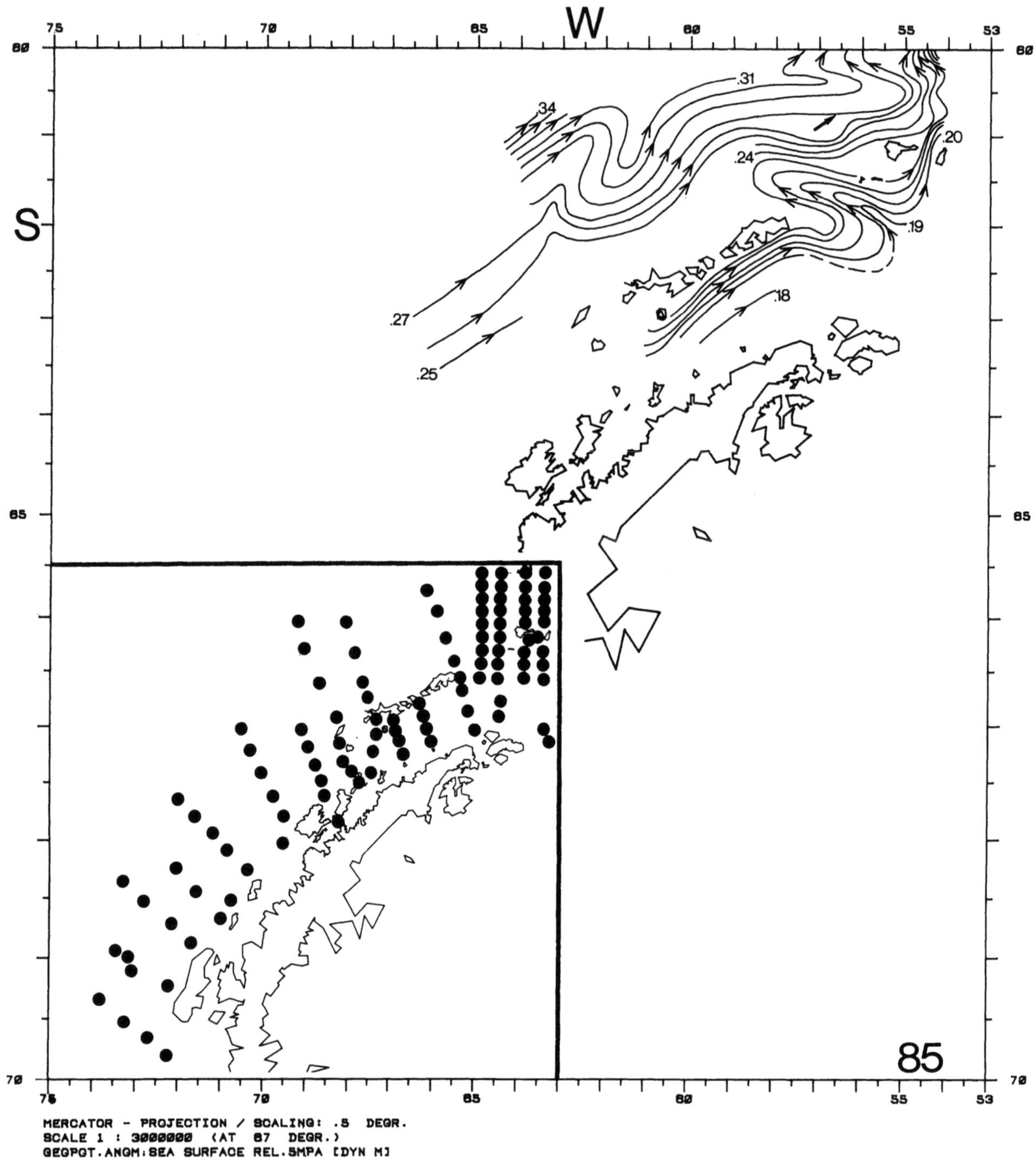

Fig. 13. Geopotential anomaly of the sea surface relative to the 5 MPa surface; March 7 to April 16, 1985. The *arrow* denotes the mean current direction according to Wittstock and Zenk (1983); station grid see *insert*

a bending to the northwest. This is consistent with the findings of the First Post-FIBEX Hydrographic Data Interpretation Workshop (BIOMASS 1982) where a northward movement of water off the north coast of King George Island was reported. Northwest of Elephant Island at 60°54.6′S, 57°06.0′W mooring 270 as published by Wittstock and Zenk (1983) was deployed from December 2, 1980 until January 21, 1981. The mean direction of the current measured at 439 m depth is given in Fig. 12 as arrow. North of this position the Shackleton Fracture Zone (SFZ) is situated, which is a submarine ridge covering about half of the Drake Passage. The SFZ rises from the abyssal plane at about 4000 m depth to depths less than 2000 m. The SFZ influences the flow through Drake Passage by a distinct curvature of the stream-lines to the northwest (Reid and Nowlin 1971; Sievers and Nowlin 1984; Stein 1986), a trend which is also detectable from the 1983 data (Fig. 11) and the 1985 data (Figs. 12, 13).

Off Adelaide Island, at about 66°30′S, 71°W, a cyclonic eddy is seen in the geopotential anomaly field. This eddy is seaward of a southwest-setting flow, which seems to begin in the vicinity of Anvers Island, 65°S, 64°W. Similar results are reported by Witek et al. (1981) and Kock and Stein (1978). This fits in the scheme given by Dietrich et al. (1975). They describe the area west of the Antarctic Peninsula as being under the regime of the southwest-flowing Polar Current or East Wind Drift. The autumn situation referenced to the 500 decibar level (Fig. 13), dynamic height 0/500 db, amplifies the gradients but indicates a similar flow pattern in the Bransfield Strait, southern Drake Passage and around Elephant Island, as outlined in Fig. 12. The flow of water through the Loper Channel is even more expressed than in the 3 MPa referenced geopotential anomaly pattern.

7 Conclusions

The geopotential anomaly of the sea surface (Figs. 2 to 13) referenced to individual pressure levels reflects the spatially averaged near-surface shear of the Antarctic Peninsula and in the southwest Scotia Sea from 1975 to 1985. The consistency of this method with measured mean current vectors is shown for historic data sets originating from the southern Drake Passage.

Variability of the geopotential anomaly field is expressed both on the seasonal as well as on the spatial scale. From the seasonal coverage of the Elephant Island Box during 1977/78, it would appear that divergent current flow and eddying played a major role during the summer and autumn season in the surface layer of the box. In contrast, the spring situation of the surface waters and of the deeper water layers is governed by a steady northeasterly flow.

Current flow features of permanent nature emerge from all geopotential anomaly charts. They reflect the influence of the submarine topography on the upper ocean geostrophic shear pattern. Three of those topographic features, the Shackleton Fracture Zone, the Loper Channel and the gap between the Elephant and Orkney Islands plateaus, mark their influence on the observed patterns.

Largest variation of upper ocean baroclinic shear is observed on the seasonal scale. Due to warming of the upper water layers with increasing solar heat input, the structure of the density field changes. This in turn results in variations of the geostrophic part of the circulation off the Antarctic Peninsula and in the southwest Scotia Sea.

However, there seems to be evidence for large-scale advective processes induced by anomalous wind fields in the area under consideration. Advection of warm or cold water masses on time scales of a week or less may change the density field of the surface waters. These effects might even exceed the seasonal signal of the upper ocean geostrophic shear pattern.

References

Anonymous (1982) Antarktis-Expedition 1981 der Bundesrepublik Deutschland mit FFS *Walther Herwig*. Arch Fischereiwiss 33: 1–176

Anonymous (1986) Antarctic Expeditions of the Federal Republic of Germany with RV *Polarstern* (Legs II/2, 1983 and III/2, 1984) and FRV *Walther Herwig* in 1985. Arch Fischereiwiss 37:1–234

BIOMASS (1982) First Post-FIBEX Hydrographic Data Interpretation Workshop. BIOMASS Rep Ser 30:1–11

BIOMASS (1983) Second Post-FIBEX Hydrographic Data Interpretation Workshop. BIOMASS Rep Ser 31:1–26

Clowes AJ (1934) Hydrology of the Bransfield Strait. Discovery Rep 9:1–64

Dietrich G, Kalle K, Krauss W, Siedler G (1975) Allgemeine Meereskunde, 3rd edn. Bornträger, Berlin Stuttgart

Gordon AL, Nowlin WD Jr (1978) The basin waters of the Bransfield Strait. J Phys Oceanogr 8:258–264

Gordon AL, Molinelli EJ (1982) Southern Ocean Atlas. Columbia Univ Press, New York

Hempel G, Sahrhage D, Schreiber W, Steinberg R (1979) Antarktis-Expedition 1977/78 der Bundesrepublik Deutschland. Arch Fischereiwiss 30:1–119

Kock K-H, Stein M (1978) Krill and hydrographic conditions off the Antarctic Peninsula. Meeresforschung 26:79–95

Nowlin WD Jr, Clifford M (1982) The kinematic and thermohaline zonation of the Antarctic Circumpolar Current at Drake Passage. J Mar Res 40:481–507

Nowlin WD Jr, Whitworth III T, Pillsbury RD (1977) Structure and transport of the Antarctic Circumpolar Current at Drake Passage from short-term measurements. J Phys Oceanogr 7:788–802

Pillsbury RD, Whitworth III T, Nowlin WD Jr, Sciemammano F Jr (1979) Currents and temperatures as observed in Drake Passage during 1975. J Phys Oceanogr 9:469–482

Reid JL, Nowlin WD Jr (1971) Transport of water through the Drake Passage. Deep-Sea Res 18:51–64

Sahrhage D, Schreiber W, Steinberg R, Hempel G (1978) Antarktis-Expedition 1975/76 der Bundesrepublik Deutschland. Arch Fischereiwiss 29:1–96

Schröder M (1983) Beschreibung einer Salzgehaltsfront in der östlichen Bransfield-Straße, dem Ursprungsgebiet der Weddell-Scotia-Konfluenz. Dtsch Hydrogr Z 36:217–235

Sievers HA, Nowlin Jr WD (1984) The stratification and water masses at Drake Passage. J Geophys Res 89:10489–10514

Stein M (1979) Ozeanographische Untersuchungen. In: Antarktis-Expedition 1977/78 der Bundesrepublik Deutschland. Arch Fischereiwiss 40:1–119

Stein M (1981) Thermal structure of the Weddell-Scotia Confluence during February 1981. Meeresforschung 29:47–52

Stein M (1986) Variability of water masses and current off the Antarctic Peninsula during SIBEX. Arch Fischereiwiss 37:25–50

Stein M, Rakusa-Suszczewski S (1983) Geostrophic currents in the South Shetland Islands Area during FIBEX. Proc. BIOMASS Colloquium 1982. Mem Natl Inst Polar Res Spec Issue 27:24–34

Whitworth III T, Nowlin WD Jr, Worley SJ (1982) The net transport of the Antarctic Circumpolar Current through Darke Passage. J Phys Oceanogr 12:960–971

Witek Z, Kalinowski J, Grelowski A, Wonomiejski (1981) Studies of aggregations of krill (*Euphausia superba*). Meeresforschung 28: 228–243

Wittstock R-R, Zenk W (1983) Some current observations and surface T/S distribution from the Scotia Sea and the Bransfield Strait during early austral summer 1980/81. "Meteor" Forschungsergeb Reihe 24:77–86

Zenk W (1981) Detection of overflow events in the Shag Rocks Passage, Scotia Ridge. Science 213:1113–1114

Variation of Positions and Structures of the Oceanic Fronts in the Indian Ocean Sector of the Southern Ocean in the Period from 1965 to 1987

Y. NAGATA[1], Y. MICHIDA[2], and Y. UMIMURA[3]

Summary. Based on oceanographic observations made from Japanese icebreakers on their cruises to and from Syowa Station over the last 22 years, temporal and spatial variations of the oceanic fronts in the Southern Ocean are analyzed. It is concluded that interannual variability is detectable even from observations taken only once a year.

1 Introduction

Systematic oceanographic observations related with the Japanese Antarctic Research Expedition (JARE) were started in December 1965 when the icebreaker *Fuji* (5250 GRT) came into operation. The icebreaker *Fuji* was replaced by *Shirase* (11 000 GRT) in 1983, but the observations have been continued annually along the ship courses from Fremantle, Australia to the Japanese Syowa Station (69.0°S, 39.5°E), and from Syowa Station to Cape Town, South Africa or Port Louis, Mauritius. Serial observations are usually made at several stations during the cruise from Syowa Station, but they are sometimes omitted due to logistic reasons. Here, we analyze temperature data mainly taken by MBT in the earlier years and by XBT in later years, and examine temporal and spatial variability of the positions of the Polar Front, the Sub-Antarctic Front, the Subtropical Front and the other fronts in the Indian Ocean sector of the Southern Ocean. The dates that the icebreaker leaves Fremantle and Syowa Station are a little different year by year; the former ranges between December 3 to 22, and the latter between February 9 to 25. However, for simplicity, the time differences are ignored in this analysis.

2 Interannual Variations

All observation lines in the period from December 1965 to March 1987 are shown in Fig. 1. We draw the temperature cross-section which is projected into longitudinal plane for each observation line for each year. Examples of the temperature cross-section are shown in Fig. 2a and b for the observation lines from Fremantle to Syowa Station in 1966 and 1980. Spatial intervals of the observation points vary greatly from year to year, and positions of the fronts are sometimes difficult to be identified from the positions of high temperature gradient portions in the section in the case of coarse observation intervals. We therefore defined several parameters which represent the positions of the oceanic fronts, by checking the temperature cross-sections in which the observation intervals are relatively small. Thus, for the position of the Polar Front (PF), two parameters, the position of the 3°C isotherm at 100 m depth surface and of the position of the northernmost edge of the 0°C isotherm contour (NE-0), were defined. The position of the 7°C isotherm at 100 m depth surface is selected as a parameter representative for the position of the Sub-Antarctic Front (SAF), and that of the 12°C isotherm at 150 m depth surface for the Subtropical Front (STF). These selected temperature values roughly coincide with those of the surface water inside the corresponding fronts (Lutjeharms and Valentine 1984). Additionally, positions of the front associated with the Leeuwin Current (LF) near Fremantle and that associated with the Agulhas Current (AF) near Cape Town are represented by those of the 17°C isotherm and of the 18°C isotherm at 150 m depth surface, respectively.

Interannual variations of PF and NE-0 are shown in Fig. 3, for the cruises both to Syowa Station, and from Syowa Station to Cape Town and to Port Louis. The edges of the vertical line attached to each data point indicate the positions of nearby observation points, and so the line would indicate a measure of ambiguity in our estimation. Interannual variations of SAF and STF are shown in Fig. 4 in a similar way. Although shorter periods of fluctuations with small amplitudes in these variation curves may be caused by the ambiguity of present estimations, these figures seem to show that some interannual variability can be detected even from data taken only once a year.

The amplitude of the variation for NE-0 is significantly larger than that of PF (Fig. 3). As these parameters may be regarded as measures of the northern extent of the Antarctic Surface Water, this suggests that results would be largely influenced by the manner of selection or definition

1 Geophysical Institute, University of Tokyo, Tokyo, Japan
2 Hydrographic Department, MSA, Tokyo, Japan
3 Tokai University, Shimizu, Japan

Antarctic Ocean and Resources Variability
(ed. by D. Sahrhage)

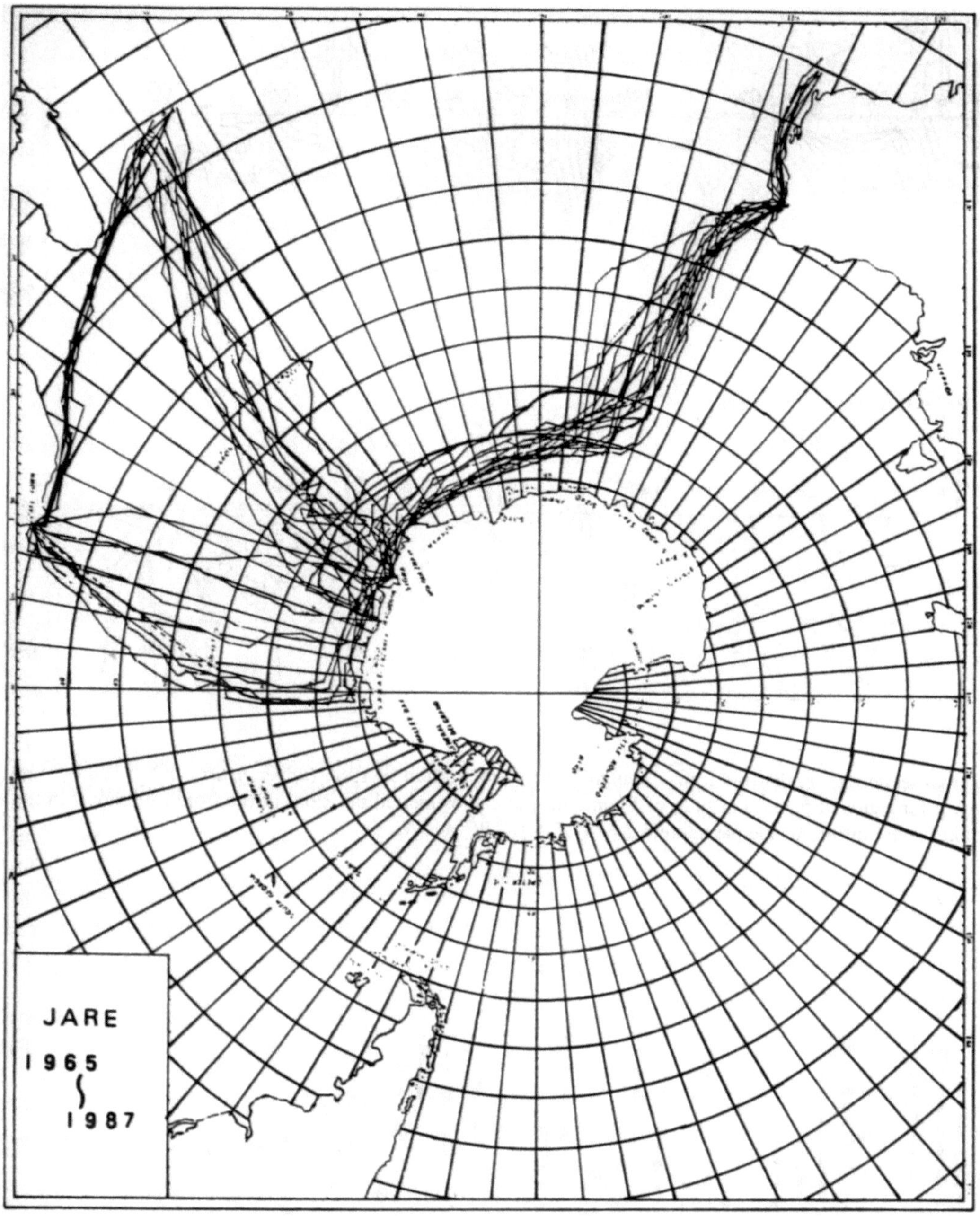

Fig. 1. Oceanographic observation lines occupied in JARE in the period from December 1965 to March 1987

of the parameters. However, long-period fluctuations seen in the four variation curves (at least, three of them) indicate rather similar trends. A northern shift of the front occurs in 1967–1968, 1972–1973, 1979–1981, and 1983–1984, and a southern shift in 1968–1972, 1974–1978, and 1981–1984. On the other hand, no similarity can be recognized for short-period fluctuations. As the longitudes of the courses to and from Syowa Station are quite different from each other, the similarity in these curves suggests that the interannual variation of the position of the Polar Front is independent of longitude, at least within the analyzed sector in comparison with those of the Sub-Antarctic Front and of the Subtropical Front discussed below.

The variation curves of SAF and STF in Fig. 4 exhibit much clearer interannual variations than those of PF and NE-0. However, the southward shift of STF in December 1975 to January 1976 along the course to Syowa Station is doubtful because of the limited observation range in this year and during the following 2 years (the observation line covered only the edge of the Subtropical Front in 1976 and did not cover the Subtropical Front region in 1977 and 1978). Also, a clear tendency that the position of SAF along the course from Syowa Station shifted northward in the period after 1976 would not be considered as an interannual variation, because the port of call was changed from Cape Town to Port Louis after 1976, except in 1980 and in

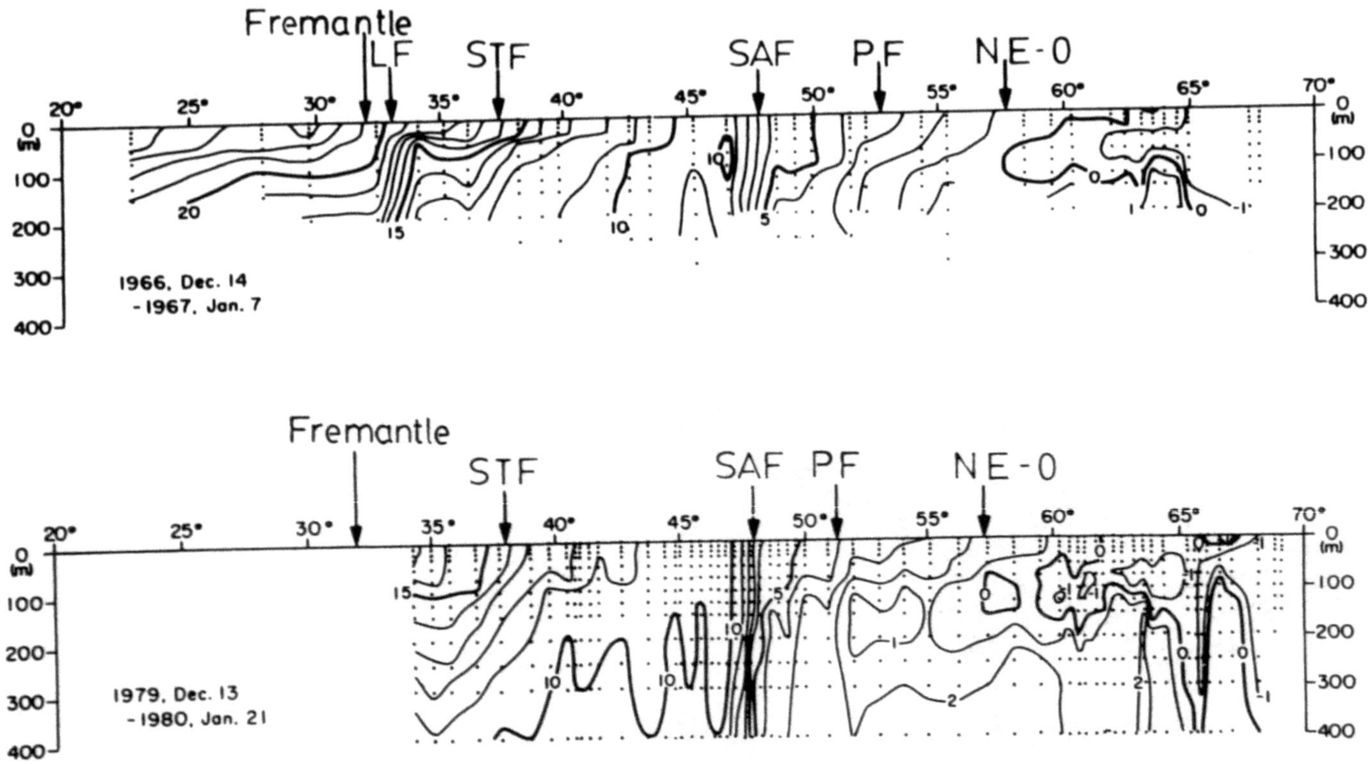

Fig. 2. Examples of temperature cross-section along the line from Fremantle to Syowa Station in 1966 (*upper figure*) and in 1986 (*lower figure*). The sections are projected to longitudinal plane. Latitude of Fremantle and the estimated positions of the fronts are shown in the *upper margin* of each figure. See text for definitions of the parameters; *LF, STF, SAF, PF,* and *NE-0*

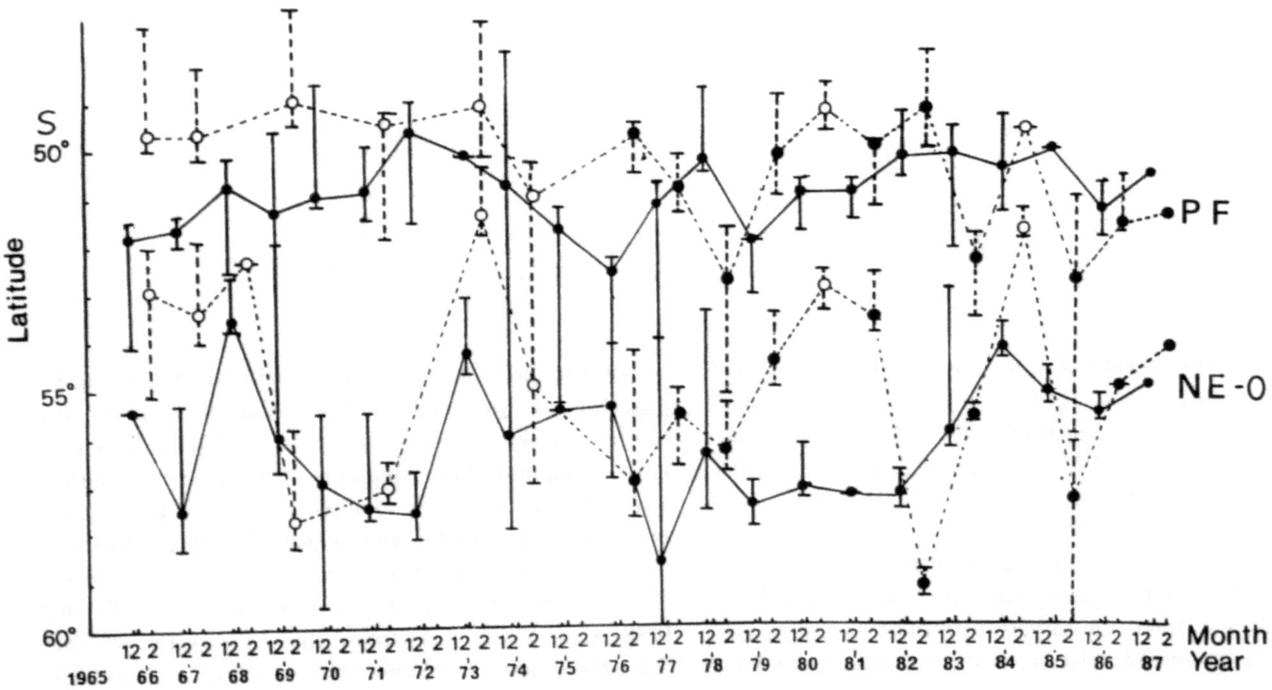

Fig. 3. Interannual variations of PF and NE-0. The data taken in the cruise to Syowa station are shown by *small black circles with full line,* those from Syowa Station to Cape Town by *large white sircles,* and those from Syowa Station to Port Louis by *large black circles with dashed line,* respectively. The edges of the *vertical line* attached to each data point indicate the positions of nearby observation points

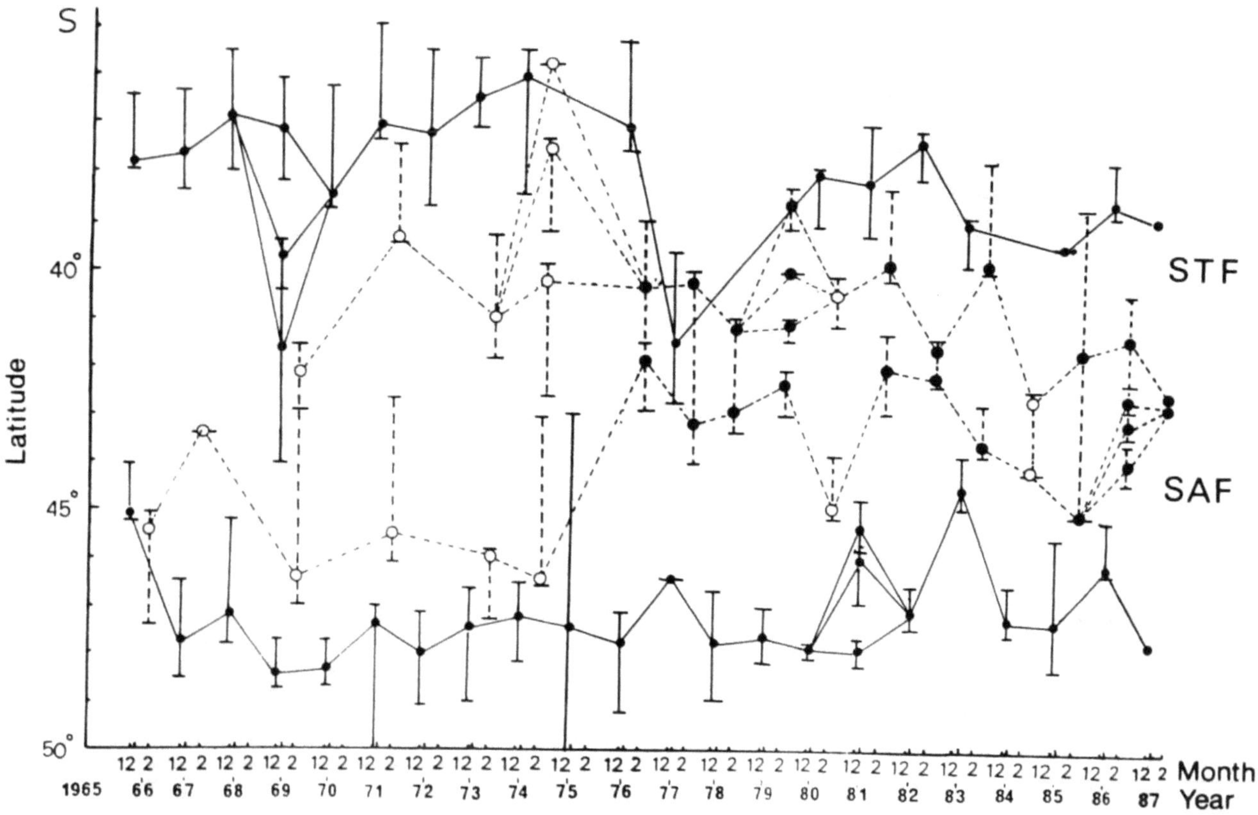

Fig. 4. As in Fig. 3, except for SAF and STF

1984. We will discuss the spatial variations of the front positions in the next section. The multiple positions in Fig. 4 (STF in Dec–Jan 1968 and in Feb–Mar 1978 and SAF in Dec–Jan 1980 and in Feb–Mar 1986) indicate that the front shows a clear double structure and the position of the front defined here becomes multivalued. The double structure may be caused by meandering of the current, but no conclusions can be drawn from our limited data. The details will be discussed in Section 4, but it should be noticed that the variational range of the front position corresponds roughly with the domain where the multivalued positions are found.

3 Spatial Variations

In order to clarify the spatial variation, the positions of all fronts found in the period from 1965 to 1987 are simultaneously plotted in Fig. 5. As seen in this figure, data points may be classified into three groups according to the ship courses, namely, (A) the course from Fremantle to Syowa Station, (B) that from Syowa Station to Port Louis and (C) that from Syowa Station to Cape Town. The position of PF does not show significant differences among the three groups, but that of NE-0 in group C has a slight tendency to be located further north. Conspicuous spatial differences are seen in the positions of SAF and STF: the position of SAF in group B tends to be located northernmost, that in group A is located southernmost and that in group C is located in between. The positions of STF in groups B and C are almost identical, but that in group A is located clearly to the north compared with the other two groups. Consequently, STF is located adjacent to SAF along the observation line from Syowa Station to Port Louis. The identification of these two fronts is very difficult along this line, when the observation interval is very coarse. The mean positions and the ranges of the distribution for each group are shown in Table 1 for STF, SAF, PF, and NE-0, respectively. The result indicates that the interannual variation of SAF seen in Fig. 4 should be attributed to spatial variation of the front position rather than to temporal variation. Therefore, in order to examine the interannual variability of these fronts, the observation line needs to be fixed as far as possible.

4 Apparent Front Migration Due to the Change of Frontal Structure

The temperature cross-sections near the Sub-Antarctic Front taken along the course to Syowa Station are shown in Fig. 6 for 4 years from 1979 to 1982. The cross-section in 1980 shows a clear double structure of the front, and the 7°C isotherm crosses the 100 m depth surface three times. In this case, the exact position of the front is hard to define, and we should rather speak of the Sub-Antarctic

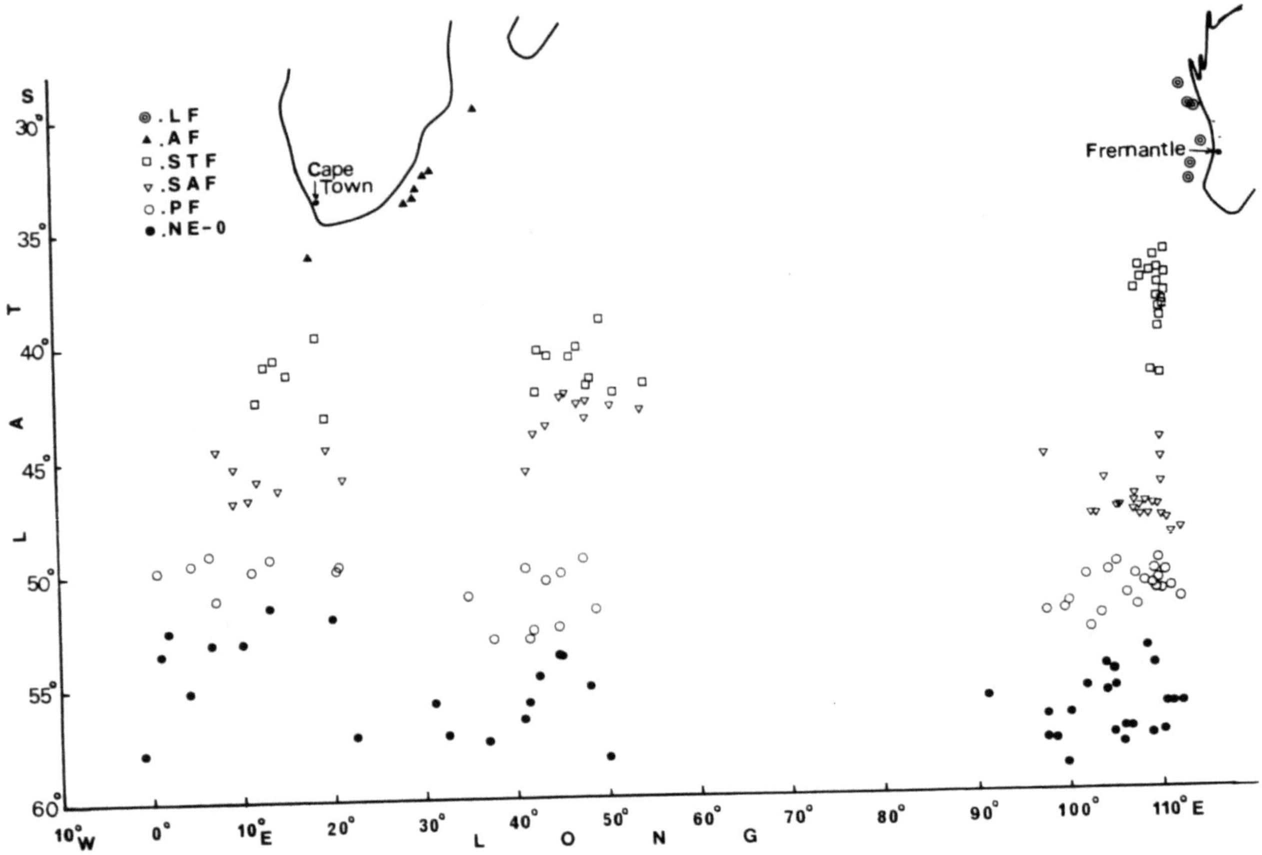

Fig. 5. Spatial distribution of the fronts in the period from 1966 to 1986

Table 1. Mean latitudinal positions and ranges of the distribution of STF, SAF, PF and NE-0 for three groups; A: along the course from Fremantle to Syowa Station, B: from Syowa Station to Port Louis, and C: from Syowa Station to Cape Town

	Subtropical Front (STF)	Sub-Antarctic Front (SAF)	Polar Front (PF)	Polar Front (NE-0)
Fremantle – Syowa Station	38°09′S 36°06′–41°41′S	47°05′S 44°30′–48°20′S	51°04′S 49°50′–52°40′S	56°16′S 53°30′–58°44′S
Syowa Station – Port Louis	40°50′S 38°53′–42°00′S	43°10′S 42°06′–45°25′S	51°09′S 49°22′–52°56′S	56°04′S 54°44′–59°16′S
Syowa Station – Cape Town	41°14′S 39°34′–42°59′S	45°33′S 43°35′–46°44′S	49°45′S 49°06′–51°06′S	53°56′S 51°28′–58°50′S

The upper column indicates the mean position and the lower the range of distribution.

"Frontal Zone" than the "Front". The zone may then be defined by two isotherms which sandwich the double structure of the front, say by the 5°C and 10°C isotherms in the depth range between 100 and 300 m depth.

According to our definition of the parameter SAF (7°C at 100 m), the position of the front migrates north and south in these 4 years. However, the migration appears to be limited within the domain between 44 and 50°S. It is worthwhile to note that this domain corresponds to the "Frontal Zone" of 1980. The observation interval in 1979 was fairly dense, and so the frontal structure would be well resolved. The width of the frontal zone appears very narrow in 1979. However, the complicated configuration of the 10°C isotherm may suggest that a weak double structure exists and that the frontal zone might be traced to about 45°S. Observations in 1981 were rather sparse, but the position of the 7°C isotherm at 100 m depth is well defined. The frontal zone appears very narrow also in this year. However, if we assume that the 5°C and 10°C isotherms in mid-depth indicate the edges of the "frontal zone", the width of this zone is just as broad as in 1980. The structure in 1982 indicates that the sharp temperature contrast exists in a relatively narrow region near 45°S. However, the configurations of the 4°, 5° and 6°C isotherms

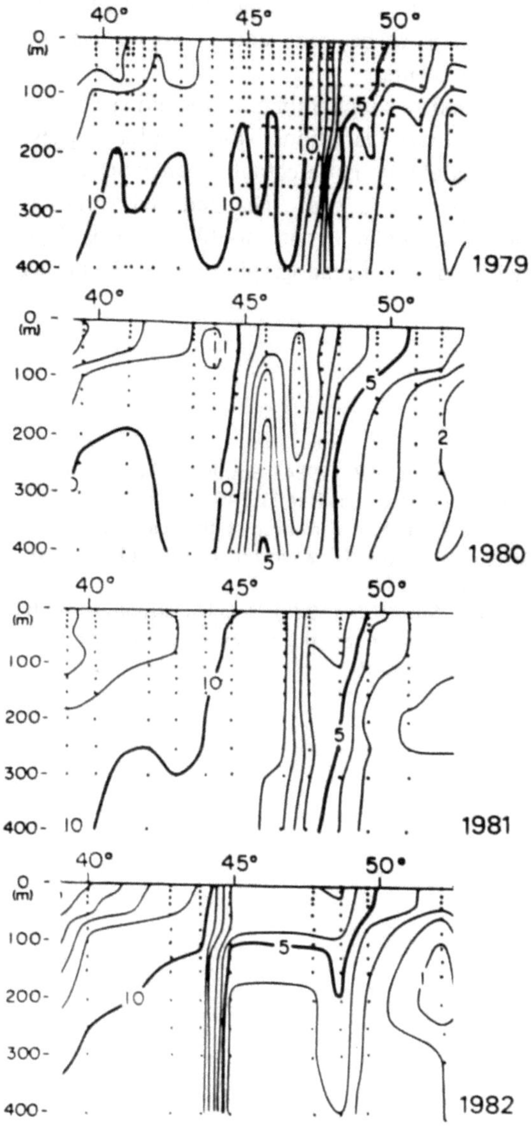

Fig. 6. Temperature cross-sections near the Sub-Antarctic Front along the course to Syowa Station for 4 years from 1979 to 1982

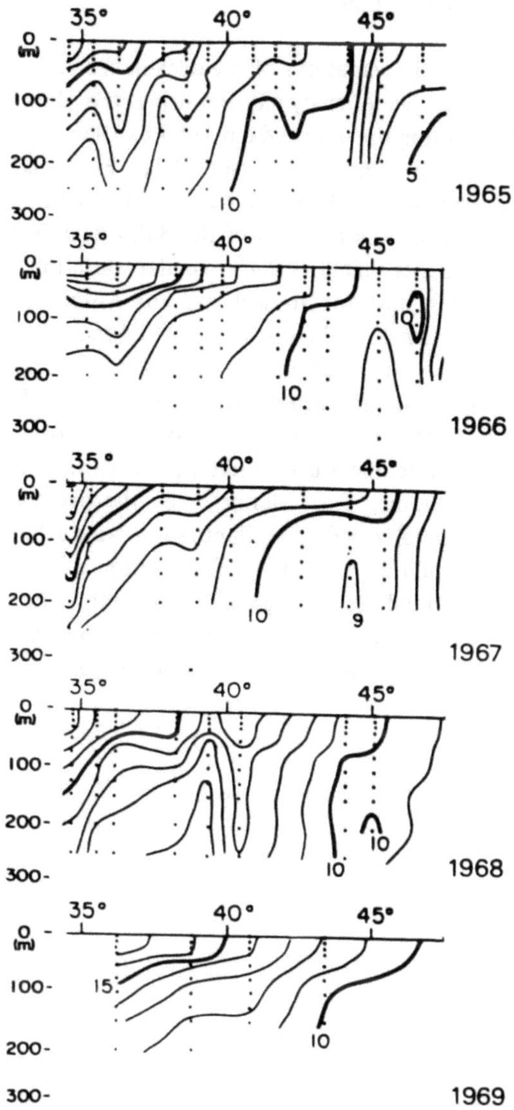

Fig. 7. Temperature cross-sections near the Subtropical Front along the course to Syowa Station for 5 years from 1965 to 1969

suggest that weak double structure of the front exists and that the width and position of the frontal zone are identical with those in 1980.

Another example of the double structure is shown in Fig. 7 (fourth figure from top) for the Subtropical Front along the course to Syowa Station. The interpretation is somewhat difficult due to the relatively coarse observation interval. However, if we consider the "frontal zone" between the 10° and 14°C isotherms, its position and width are almost identical for these sections from 1965 to 1969.

This suggests that, if we define some "frontal zone", the interannual variation of these fronts is much smaller than that expected from the variation of the position of the maximum horizontal temperature gradient. Or, in some cases, the shift of the front defined by the maximum horizontal temperature gradient may be understood as a weakening of one subfront and a strengthening of the other subfront, as seen in the short-period fluctuations of the Kuroshio front in the Tokara Strait (Nagata and Takeshita 1985), and in the sea to the east of Honshu (Nagata et al. 1986).

This may also give a reason why PF exhibits smaller variability than NE-0 in Fig. 3. The definition of PF, the position of the 3°C isotherm at 100 m depth, may give the outer edge of the Polar Front or the measure of the position of overall "frontal zone". Then, it would be much more stable than NE-0 which may give information of the fine structure of the front.

5 Summary and Conclusion

The temporal and spatial variations of the oceanic fronts in the Southern Ocean were analyzed by using the data obtained in JARE during the period from 1965 to 1987.

The variation of the latitude of the Polar Front is comparatively independent of longitude, at least within the analyzed sector, but the position of the Sub-Antarctic and Subtropical Fronts exhibits large spatial variation. The distance between the Subtropical and Sub-Antarctic Fronts is usually very small to the south of Port Louis.

It is suggested that the variational characteristics of the position of the fronts largely depend on the definition of the position itself. If we define the position by local maximum horizontal temperature gradient, the variation would be large. However, if we define it by the overall position of the "frontal zone", the fluctuation would be rather small. Some of the variation of the position of the "front" may be regarded as a change of the fine structure of the "frontal zone". Therefore, in order to detect precise interannual variability, it is recommended that the observation lines are to be fixed as much as possible and that the observation interval is to be small enough to resolve the fine structure of the fronts.

Although there exists some ambiguity including aliasing from unknown shorter variations, the results suggest that interannual variability is detectable even from observations taken only once a year. The deduced variation of the position of the Polar Front should be checked by comparison with other oceanographic data including biological data.

References

Lutjeharms JRE, Valentine HR (1984) Southern Ocean thermal fronts south of Africa. Deep-Sea Res 31:1461–1475

Nagata Y, Takeshita K (1985) Variation of the sea surface temperature distribution across the Kuroshio in the Tokara Strait. J Oceanogr Soc Jpn 41:244–258

Nagata Y, Yoshida J, Shin H-R (1986) Detailed structure of the Kuroshio front and origin of the water in warm-core rings. Deep-Sea Res 33:1509–1526

Part 2
Meso/Large-Scale Variability in the Biota (Related to the Environment)

Seasonal and Interannual Variabilities in Antarctic Phytoplankton with Reference to Krill Distribution

S. Z. EL-SAYED

Summary. The spatial and temporal variabilities in the distribution of Antarctic phytoplankton biomass, based primarily on data collected by the author and other investigators during the past 25 years, are summarized. Mesoscale seasonal and interannual variabilities in phytoplankton distribution and abundance in the Drake Passage/ Scotia Sea, the Bransfield Strait/Elephant Island, and the Ross Sea are examined. The general picture that emerges is one of great variability of the phytoplankton biomass and primary production. This spatial variability (up to two orders of magnitude) tends to overshadow the expression of seasonal differences. The time of peak phytoplankton abundance varies from year to year. Seasonal variability is much more pronounced than any interannual variability.

Of the several factors thought to cause spatial and temporal variability in Antarctic phytoplankton, three factors, namely, nutrient salts, water column stability, and grazing are discussed. The relationship between phytoplankton and krill remains as one of the vexing problems in Antarctic marine ecosystem studies. For better understanding of phytoplankton/krill interrelationships, future studies must resolve finer spatial scales and include time-course measurements to examine underlying mechanisms. Recommendations for future research are discussed.

1 Introduction

The study of Antarctic marine phytoplankton has a long and almost uninterrupted history that dates back nearly a century and a half to the *Erebus* and *Terror* expedition (1839–43). This history continued with the celebrated expeditions of the *Challenger* (1872–76), *Antarctica* (1895), *Belgica* (1897–99), *Valdivia* (1898–99), *Southern Cross* (1899–1900), *Gauss* (1901–03), *Scotia* (1902–04), *Pourquoi Pas?* (1908–10) and the *Discovery* investigations (1925–39) (see El-Sayed 1987 for review). Thanks to the extensive investigations carried out on board the USSR *Ob* and *Vitiaz,* the USNS *Eltanin,* the Argentinian *San Martin* and *Islas Orcadas,* the French *Commandant-Charcot* and *Marion Dufresne,* and the numerous Japanese and other ships throughout the 1960's and 1970's, a large amount of useful data on the geographical and temporal distributions of phytoplankton standing crop and on the magnitude of primary production in the circum-Antarctic waters was obtained (Klyashtorin 1961; Ichimura and Fukushima 1963; Mandelli and Burkholder 1966; El-Sayed 1967, 1968; Fukuchi 1980; Fukuchi and Tamura 1982).

The extensive investigations carried out during FIBEX (First International BIOMASS Experiment 1980–1981) and SIBEX (Second International BIOMASS Experiment 1983–84 and 1984–85) have greatly enriched our data base with regard to the phytoplankton and their primary consumers (zooplankton, especially krill). Although we now have a fair body of knowledge on phytoplankton distribution, abundance, and productivity in the waters surrounding Antarctica, we still have a poor understanding of the factors that regulate phytoplankton production in this area. We know, for instance, very little of the life history, physiology, behavior, and population dynamics of even the most common components of Antarctic phytoplankton, and whether they are well adapted to their environment is still a matter of controversy (Holm-Hansen et al. 1977; El-Sayed and Taguchi 1981). Moreover, the low levels of phytoplankton standing crops and primary production that are characteristic of most of the oceanic regions are puzzling in view of the high inorganic nutrient concentrations reported from these waters.

This chapter reviews the subject of spatial distribution of the phytoplankton biomass and productivity based primarily on data collected at 780 stations by the author and other investigators in the circum-Antarctic waters during the past 25 years (Fig. 1). The mesoscale seasonal and interannual variabilities in phytoplankton distribution and abundance in the three best-studied areas of the Southern Ocean, namely, the Bransfield Strait/Elephant Island, the Drake Passage/Scotia Sea, and the Ross Sea are discussed. The factors thought to govern the spatial and temporal variabilities of Antarctic phytoplankton, in light of data collected during the last decade, are also reviewed. In this review, the pelagic primary producers are featured, leaving the discussion of ice-algae to other contributors to this seminar (see Smith et al. this Vol.).

2 Data Limitations

Despite the large amount of data collected by the author and others since the early 1960's, both the geographical

1 Department of Oceanography, Texas A & M University, College Station, Texas 77843, USA

Antarctic Ocean and Resources Variability
(ed. by D. Sahrhage)

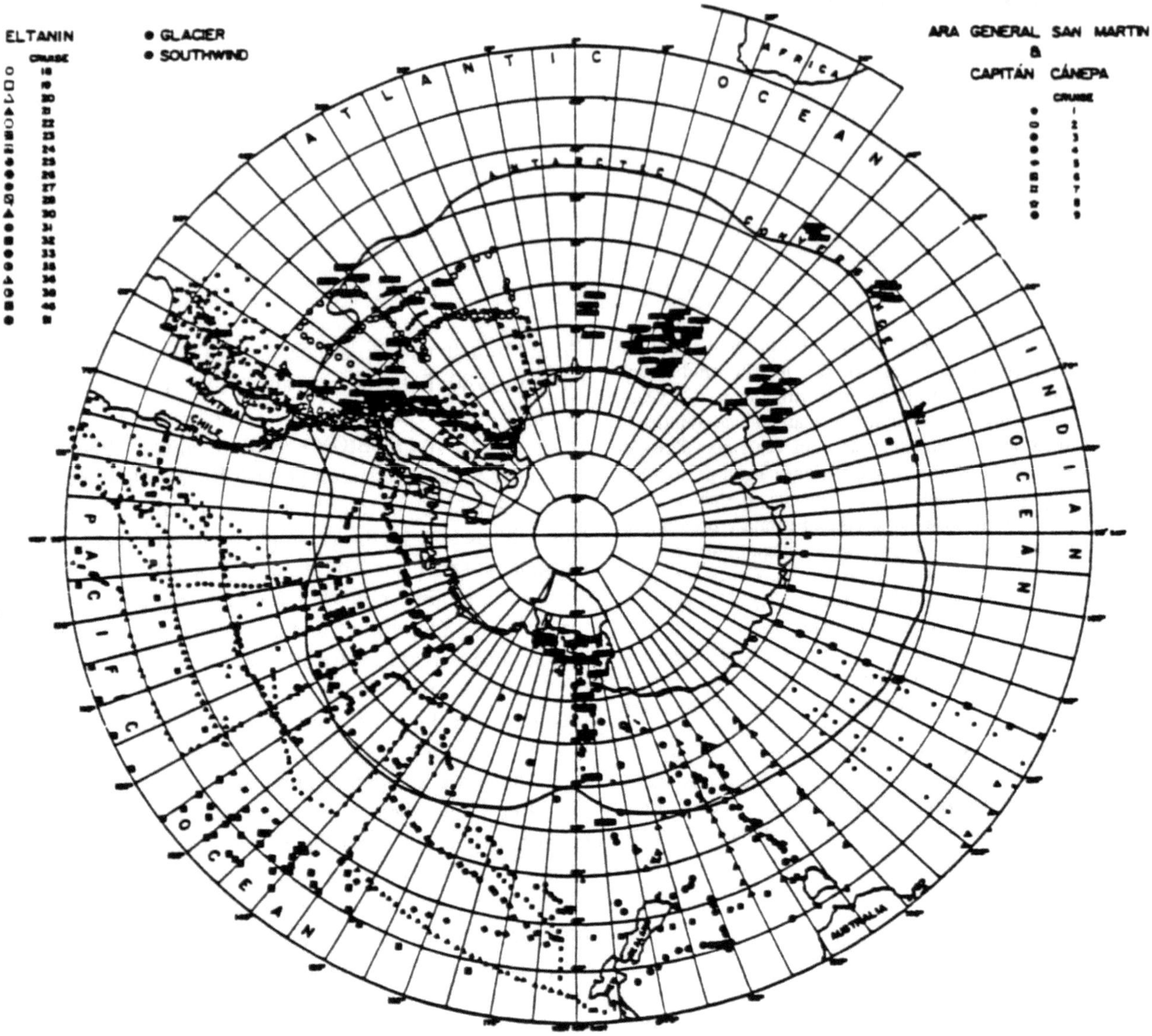

Fig. 1. Distribution of data on phytoplankton biomass and primary productivity discussed in the text. Stations occupied between 1976 and 1987 are indicated with ▬

and temporal coverage of the data is far from complete because of the vastness of the Southern Ocean. Heywood and Whitaker (1984) remind us that observations of standing crop and productivity rates have been made in isolation, both in time and space, as dictated by logistics and the workload of research vessels rather than by the biology of the flora. Most of the expeditions to the Southern Ocean have been limited to the periods of austral spring and summer, thus covering only part of the annual growth cycle, and giving us an incomplete picture of seasonal variability. Moreover, current methods and techniques have many limitations and have changed or been modified during the past two decades, thus making it difficult to compare results. Such limitations should be borne in mind, especially when comparing data obtained from various sources.

3 The Marine Environment

Before details about spatial and temporal variabilities of Antarctic phytoplankton are discussed, it is important to understand the unique physical/chemical settings of the Antarctic seas, since these strongly affect the biology, distribution, and abundance of the planktonic organisms.

The Southern Ocean is an upwelling system whose large-scale physical oceanography is dominated by the Antarctic Circumpolar Current. The unique features of this system are: the presence of pack ice around the continent and the seasonal waxing and waning of the pack-ice zone; the variability of the light regime that alternates between perpetual dankness in winter and continuous daylight during the summer; the extensive cloud cover; the stormy weather; and the Circumpolar Current, which connects

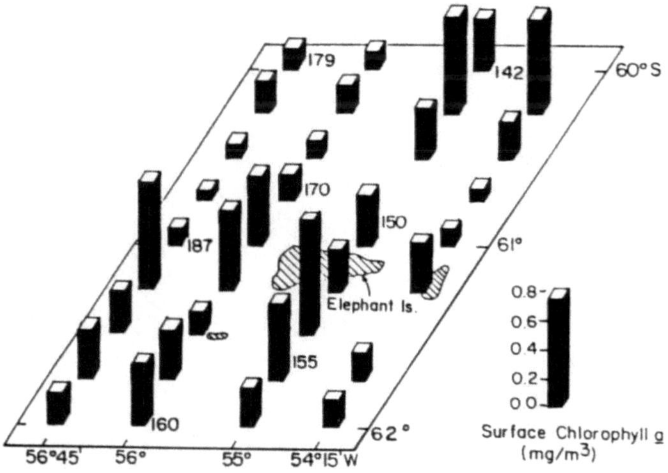

Fig. 2. Distribution and abundance of surface chlorophyll a in waters in the vicinity of Elephant Island during November 1984. (Data from Weber and El-Sayed 1986)

circum-Antarctic waters into a single system, the Southern Ocean, through the transport of nutrients and plankton. The Antarctic organisms have developed their characteristic features in response to the physical/chemical environments, and these are precisely the unique environmental parameters that present a paradox of considerable scientific interest, as is discussed later.

4 Spatial Distribution of Phytoplankton Standing Crop

4.1 Surface Distribution of Chlorophyll a

The bulk of the phytoplankton data shows chlorophyll concentrations in the range of 0.1 to 1.0 mg/m^3, with a mean value of 0.5 mg/m^3 (Saijo and Kawashima 1964; El-Sayed and Mandelli 1965; El-Sayed 1970; Fukuchi 1980). Within the confines of such low values, there is pronounced geographical variability in chlorophyll a concentration, even over relatively small spatial scales. This is clearly shown by the marked spatial heterogeneity of the phytoplankton biomass around Elephant Island (Fig. 2). Similar observations are also reported for around South Georgia (Priddle et al. 1986b), where the highest values of phytoplankton biomass were associated with stations off the northwest of the island and the lowest chlorophyll levels were located to the southeast. Priddle and his coworkers are essentially correct in stating that this patchiness around South Georgia (240 × 180 nm grid) reflects in a microcosm the distribution of phytoplankton throughout the Southern Ocean.

In contrast to the generally low standing crop of phytoplankton in oceanic regions, elevated biomass levels are reported from inshore waters, e.g., west of the Antarctic Peninsula and the Gerlache Strait (El-Sayed 1968), in the southern Ross Sea (El-Sayed et al. 1983), near the Kerguelen and Heard Islands (El-Sayed and Jitts 1973); off the Crozet Archipelago (El-Sayed et al. 1979), and in the inshore waters of Signy Island (Horne et al. 1969). Exceptionally high values (i.e., in excess of 25 mg/m^3) were reported by Mandelli and Burkholder (1966) during a phytoplankton bloom near Deception Island. Elevated values in phytoplankton standing crop and primary production in Antarctic coastal waters and in the vicinity of Antarctic and Sub-Antarctic islands have been attributed to the so-called island-mass effect (El-Sayed et al. 1964; El-Sayed and Jitts 1973; Nast and Gieskes 1986). However, increased concentrations of inorganic salts cannot be invoked here, as is usually the case in tropical or temperate waters, since these nutrients, as discussed later, are in abundant supply.

Phytoplankton blooms also occur near the edge of the seasonally retreating pack-ice (El Sayed and Taguchi 1981; Smith and Nelson 1985; Nelson et al. 1987). The most extensive and richest of such blooms occurred in the southwestern Weddell Sea (El-Sayed 1971a). This bloom, composed entirely of the diatom *Thalassiosira tumida,* covered an area 15 000 km^2, with chlorophyll concentrations of up to 190 mg/m^3.

In addition to inshore waters and the ice edge zone, high chlorophyll concentrations have been frequently associated with frontal structures such as the Polar Front (Allanson et al. 1981; El-Sayed and Weber 1982; Yamaguchi and Shibata 1982; Lutjeharms et al. 1985; Bidigare et al. 1986) and the Weddell/Scotia Confluence (Bidigare et al. 1986; Nelson et al. 1987).

In recent years, the advent of remote sensing has added a new and powerful tool to the study of geographical variability of phytoplankton biomass. Such variability is graphically documented in a satellite (Nimbus 7 Coastal Zone Color Scanner) image from the region of the Polar Front in the Drake Passage (Fig. 3). This image is a striking example of the interaction of mesoscale physics and phytoplankton biomass (El-Sayed and Hofmann 1986). The phytoplankton-poor waters of the Drake Passage and the cold core rings are clearly visible. The image further shows generally low pigment values in the Polar Front, a region of relatively high pigment concentration squeezed between bands of low pigment concentration. Pigment values north of the Polar Front are also generally low, but isolated regions of concentrations in excess of 0.1 mg Chl a/m^3 are observed. The reasons for these enhanced concentrations are unknown.

4.2 Vertical Distribution of Phytoplankton Biomass

The vertical distribution of phytoplankton biomass (in terms of chlorophyll a) generally exhibits maximum values at subsurface (50–70 m) depths. Following these maxima, there is a gradual decrease in the chlorophyll values to a depth of 200 m (Fig. 4), below which chlorophyll con-

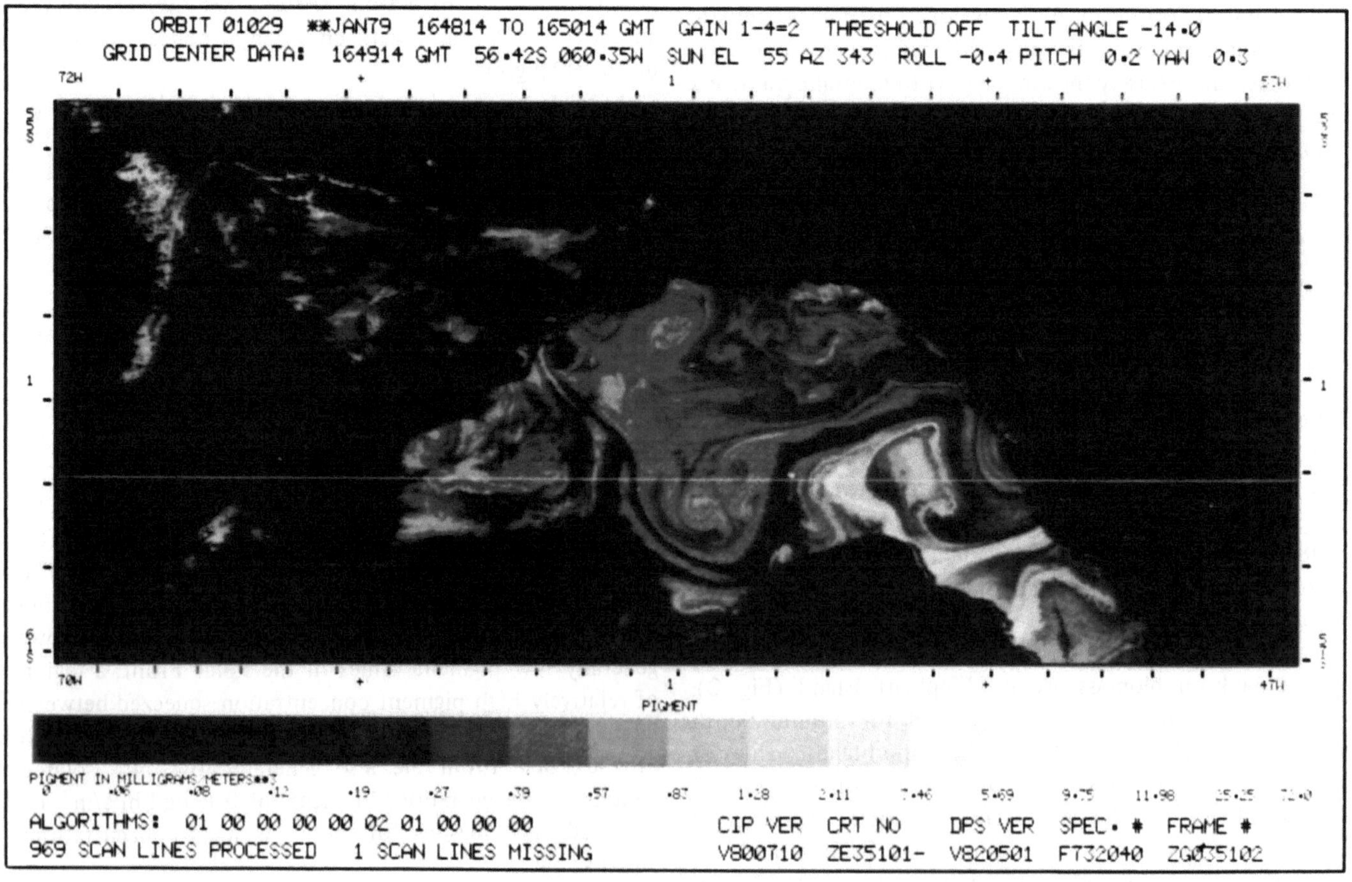

Fig. 3. CZCS pigment image of the Drake Passage (Nimbus-7 Orbit 1029, 6 January 1979). (El-Sayed and Hofmann 1986)

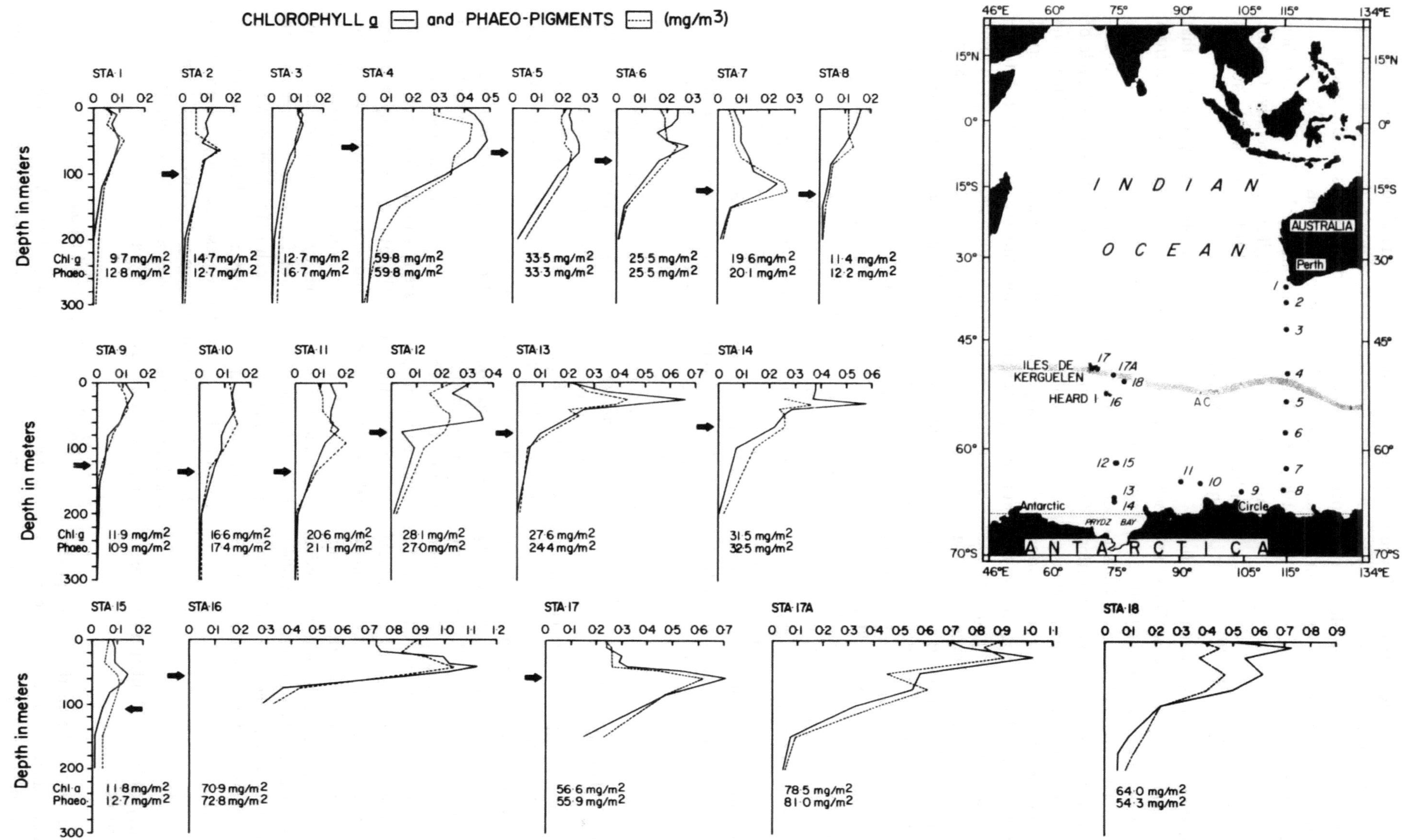

Fig. 4. Vertical distribution of chlorophyll a and phaeopigments (mg/mg^3) at stations occupied during *Eltanin* Cruise 46. Position of arrows indicate depth of euphotic zone. (El-Sayed and Jitts 1973)

centration is usually negligible (El-Sayed 1970; El-Sayed and Turner 1977; El-Sayed and Weber 1982). It is not uncommon for substantial amounts of chlorophyll to be found below the euphotic zone (i.e., below the depth to which 1% of surface light penetrates).

4.3 Spatial Distribution of Primary Production

The magnitude of primary production is a function of the phytoplankton biomass and the growth rates. Therefore, it is not surprising that most primary productivity data (estimated by the ^{14}C uptake method of Steemann Nielsen 1952) from the Southern Ocean show good correlation with the distribution of the phytoplankton standing crop. For instance, low values of primary production are reported for the Drake Passage, the Bellingshausen Sea, and in the oceanic waters in general. The open-ocean system has rates of production typical of oligotrophic regions (< 0.1 g C/m^2/day). On the other hand, high values are often found in coastal regions and in the vicinity of Antarctic and Sub-Antarctic islands. For example, El-Sayed (1968) recorded 3.2 g C/m^2/day in the Gerlache Strait; Mandelli and Burkholder (1966) reported 3.62 g C/m^2/day near Deception Island; and Horne et al. (1969) found a peak productivity of 2.8 g C/m^2/day in the inshore waters of Signy Island in the South Orkney Islands. Such values, which are comparable to those from the upwelling systems off Peru, southeast Arabia, Somalia, and southwest Africa, have no doubt perpetuated the belief in the proverbial "richness" of the Antarctic waters.

4.4 Vertical Distribution of Primary Production

As to the vertical distribution of primary productivity in the circum-Antarctic waters, maximum photosynthetic activity generally occurs at depths corresponding to between 50 and 10% of surface light intensity, corresponding to a photosynthetically available radiation (PAR) between 500 and 100 μE/m^2/s (von Bodungen et al. 1986). Carbon fixation, has been measured in samples from well below the classical euphotic zone. For example, El-Sayed and Taguchi (1981) and El-Sayed et al. (1983) report that primary production below the euphotic zone in the Ross and Weddell Seas was nearly one-fourth of the total production in the water column. However, in other studies, ^{14}C uptake below the euphotic zone did not exceed 5 to 10% of that in the entire water column (El-Sayed and Jitts 1973; El-Sayed and Weber 1982).

5 Seasonal Variability of Phytoplankton Biomass and Primary Production

As to the seasonal variability in the distribution of phytoplankton, unfortunately, there is a relative dearth of long-term phytoplankton data from any of the Antarctic regions. In this respect Antarctic biological oceanographers are not as fortunate as, for example, their British counterparts, who have long-term data records for the North Sea and the English Channel from as early as the 1920's (Cushing 1966; Russell et al. 1971; Southward 1974, 1980), or their American colleagues, whose California Cooperative Oceanic Fisheries Investigation (CALCOFI) spans nearly 40 years (Brinton and Reid 1986).

The problem is further complicated by the fact that seasonal cycles based on data from single stations, or from single depths, may not be representative. For example, deep chlorophyll maxima will be missed entirely if surface samples are used to reconstruct the seasonal cycle of phytoplankton biomass (Harris 1986). Further, large-scale advection may also modify the seasonal cycle as observed at a single point if the seasonal biomass cycle differs between water masses. Such complications should be kept in mind when examining the seasonal variability of phytoplankton biomass in the Bransfield Strait and neighboring waters, which are noted for their complex hydrographic structure (Stein this Vol.; Sievers and Nowlin this Vol.).

Despite the nonquantitative methods used by Hart (1934, 1942), his is still the most complete data set on seasonal variability of Antarctic phytoplankton to date. Hart demonstrated that the onset of the period of maximum production changes from early spring to late summer or early fall with increasing latitude (Fig. 5). He was able to show that the period of maximum production decreased in duration with increasing latitude and, further, that annual differences in the hydrographic conditions could affect the timing and magnitude of these peaks and the species that are active during the period. Hart's observations have been corroborated by the observations of many investigators. Figure 5 illustrates the marked resemblance between El-Sayed's (1970) observations and those Hart made in "the Northern Zone".

In this section we will discuss the seasonal and interannual variabilities of the standing crop of phytoplankton in the Drake Passage/Scotia Sea region, the Bransfield Strait and contiguous waters, and the Ross Sea. We will discuss how these data fit into the generally accepted patterns of geographical and seasonal productivity of the Antarctic waters. Again, it is worth repeating (cf. Introduction) that seasonal biomass cycles cannot be adequately defined by mounting three or four cruises between austral spring and late summer in such large bodies of water as the Scotia Sea or the Bransfield Strait. Further, it is widely recognized that in order to describe the physicochemical fluctuations that drive year-to-year variations in the phytoplankton biomass, it is necessary to have very long data sets, covering at least 20–30 years.

5.1 Drake Passage/Scotia Sea Region

The two cruises of the ARA *Islas Orcadas* (Cruise 17 – 2. September–14. October 1978 and Cruise 19 – 22. Feb-

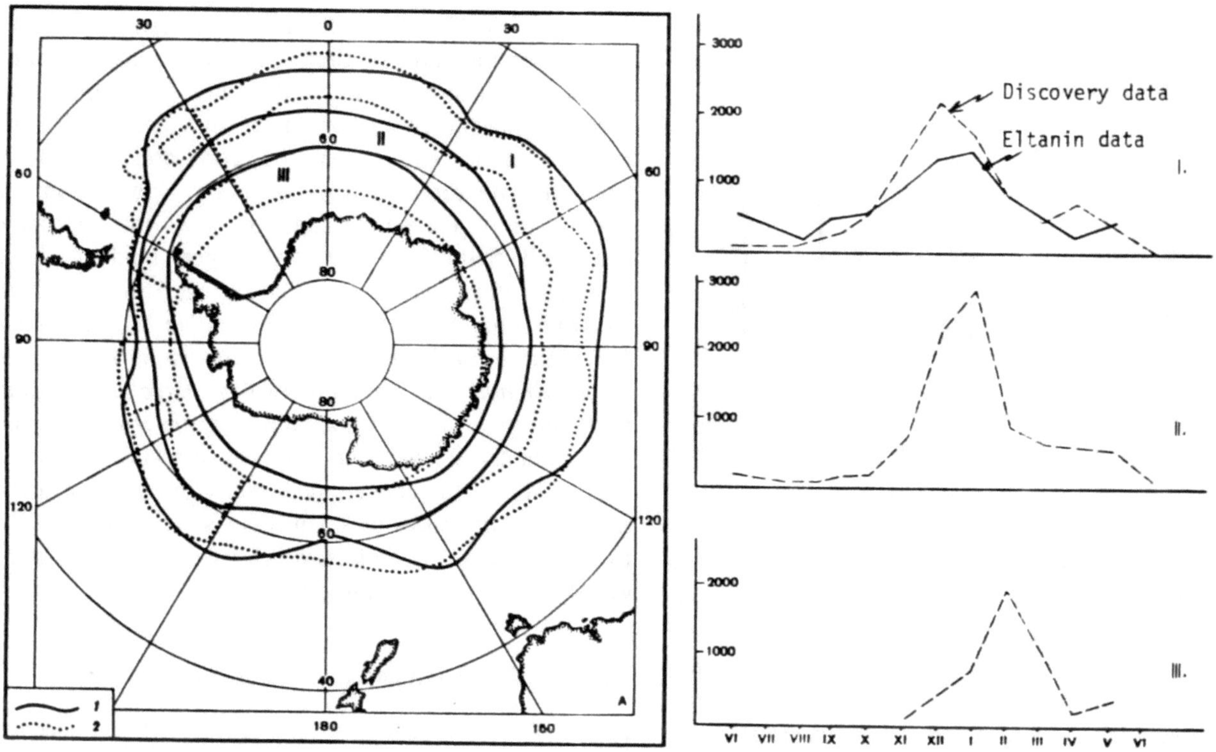

Fig. 5. Seasonal variation in plant pigment concentrations with latitude in the Southern Ocean. (Hart's 1942 *Discovery* data). *Eltanin's* chlorophyll a data are plottet for comparison (El-Sayed 1971a). *I* Northern Zone (ca. 50°–55° 0.50′S); *II* Intermediate Zone (ca. 55°–66° 00′S); *III* Southern Zone (above 66° 00′S). (After Voronina 1984)

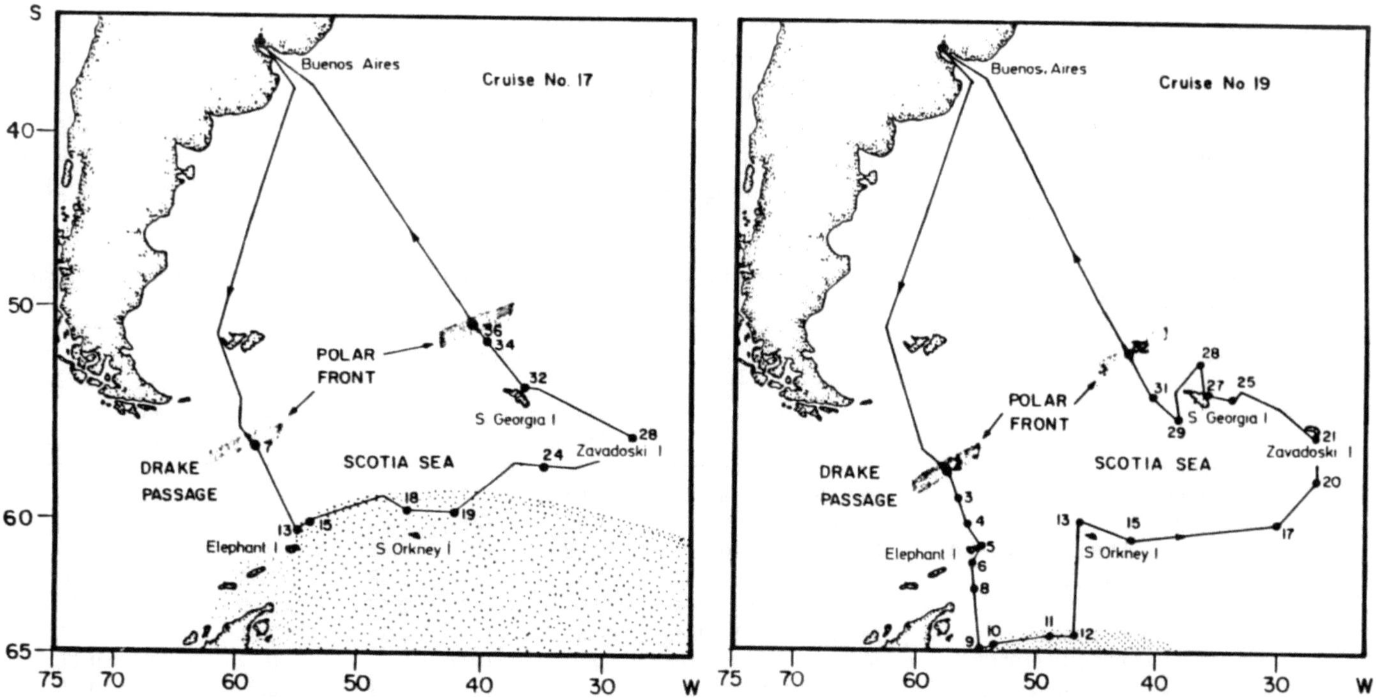

Fig. 6. Cruise tracks of the *Islas Orcadas* Cruise 17 and 19 in the Southwest Atlantic and the Scotia Sea. (El-Sayed and Weber 1982)

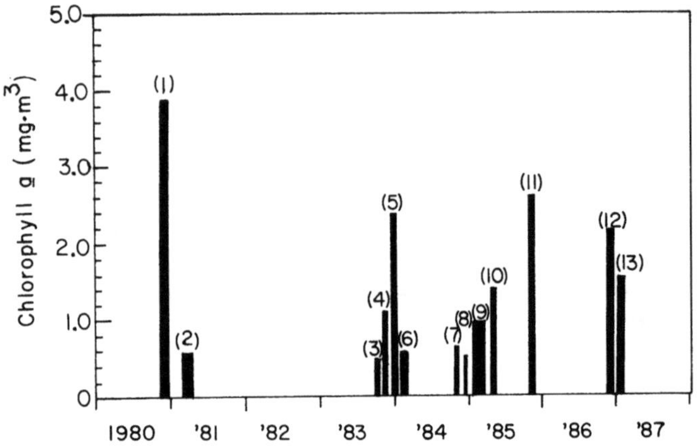

Fig. 7. Seasonal distribution of chlorophyll a in the Bransfield Strait between 1980 and 1987. (For explanation of numbers shown see Table 1)

ruary–9. April 1979) had as their major objectives the studies of the biological productivity and the metabolic activities of selected Antarctic organisms at two different seasons, namely late winter/early spring and late summer/ early fall. Study of the seasonal variability was made possible by occupying stations along nearly identical cruise tracks (Fig. 6).

The general picture that emerges from the investigation carried out during the two cruises of the ARA *Islas Orcadas* (see El-Sayed and Weber 1982 for details) is one of great variability of the phytoplankton biomass and primary productivity between low values, typical of oligotrophic waters, and high values characteristic of eutrophic regions. *This spatial variability (up to two orders of magnitude) tends to overshadow the expression of seasonal differences.*

Surprisingly, when discrete and integrated values for primary productivity, chlorophyll, and phaeopigments, at all stations occupied in or south of the Polar Front during Cruises 17 and 19 are compared, only the phaeopigment values are significantly different. When stations from the two cruises are paired on the basis of geographical location, two stations from the late winter/early spring cruise have order of magnitude higher values of chlorophyll a and primary productivity than the corresponding stations from the late summer/early fall cruise. However, in considering all of the paired stations, values for chlorophyll and productivity from Cruise 17 appear randomly larger or smaller than values measured during Cruise 19. *Thus, temporal variability is clearly discernible at small spatial scales but is easily masked when comparing mean values of chlorophyll and productivity from large areal regions.* It should be noted that maximum phytoplankton biomass and productivity have been observed to occur in December (Hart 1942) in the area studied, indicating that the lack of significant differences between seasonal values may be due to the timing of the cruises, which fall on both sides of the vernal austral peak (see Fig. 5).

5.2 Bransfield Strait and Contiguous Waters

In Fig. 7 we compare surface chlorophyll a concentrations measured between 1980 and 1987 in the Bransfield Strait during the cruises of the West Germany's *Polarstern* and *Walther Herwig,* Chile's *Itsumi,* Poland's *Professor Siedlecki,* and Brazil's *Professor Besnard.* Dates, number of stations occupied during these cruises, and average surface chlorophyll values are given in Table 1. It is noted that chlorophyll a concentrations in the Bransfield Strait are generally high when compared to average oceanic values for the Southern Ocean. It should be realized that data averaged over the periods and geographic ranges of individual cruises will conceal significant small-scale spatial and temporal differences. Also, it is risky to compare

Table 1. Sources of data on average surface chlorphyll a concentrations (mg/m^3) in the Bransfield Strait and Elephant Island region during 1980–1987. (n) = number of stations

Number[a]	Investigator	Dates	Chlorophyll a values					
			Bransfield Strait			Elephant Island		
			Mean	(n)	Median	Mean	(n)	Median
1	von Bodungen (1986)	Nov. 20–Dec. 14, 1980	4.63	(14)	3.83	2.44	(2)	2.44
2	Lipski (1981)	Feb.–Mar. 1981	0.77	(21)	0.53			
3	Gieskes and Elbrächter (1986)	Oct. 24–Nov. 16, 1983	0.28	(17)	0.25	0.29	(33)	0.19
4	von Bodungen (1986)	Nov. 26–Dec. 21, 1983	2.20	(9)	1.18	1.94	(1)	1.94
5	Lipski (1985)	Dec. 1983–Jan. 1984	2.91	(33)	2.37	1.00	(7)	0.35
6	Kutner (pers. commun.)	Jan. 21–Feb. 9, 1984	0.51	(19)	0.37	0.21	(7)	0.15
7	Weber and El-Sayed (1986)	Nov. 21–Dec. 4, 1984	0.93	(7)	0.60	0.32	(39)	0.27
8	von Bodungen (pers. commun.)	Jan. 6–12, 1985	1.52	(3)	0.51	0.95	(1)	0.95
9	Kutner (pers. commun.)	Jan. 27–Mar. 15, 1985	4.64	(20)	0.85	0.29	(10)	0.27
10	Elbrächter (pers. commun.)	Mar. 20–Apr. 5, 1985	3.07	(15)	1.20	1.57	(33)	0.80
11	von Bodungen (pers. commun.)	Nov. 15–28, 1985	2.44	(13)	2.50			
12	Weber and El-Sayed (in prep.)	Jan. 2–19, 1987	2.44	(11)	2.09	1.24	(36)	1.01
13	Ochocki (pers. commun.)	Jan. 26–Feb. 22, 1987	1.46	(27)	1.52	0.91	(12)	0.66

[a] Corresponds to Numbers in Figs. 7 and 8.

average values from different cruises if the proportions of neritic and oceanic stations are different. Recognizing these caveats, we nevertheless suggest that this limited data set shows seasonal variability in chlorophyll a values within individual study years. Within a given year, average chlorophyll values differ by an order of magnitude, depending on the month of sampling. However, the period of peak phytoplankton abundance varies considerably from year to year. For the 5 years of observation, highest chlorophyll values occurred anywhere between November (1980) and March (1985). Considering the large month-to-month differences, interannual variability in the phytoplankton standing stock appears to be small in the Bransfield Strait.

5.3 The Elephant Island "Box"

For the past several years, West German scientists have been studying, on an almost annual basis, an area of 36 000 km^2 around the Elephant Island (referred to as the Elephant Island Box, see Fig. 2). This area, deliniated by 60°–62°S and 54°15′–56°45′W, was covered with a grid of 36 stations along 4 meridional transects, 45 nm and 60 nm apart. The distance between stations on each transect is 15 nm. One objective of this mesoscale investigation was to assess the phytoplankton and krill standing stocks in the study area in an effort to improve our understanding of the macroscale processes.

Average results from the Elephant Island investigations are given in Table 1 and plotted in Fig. 8. Chlorophyll values in this region are, with few exceptions, similar to typical values (<0.5 mg/m^3) reported for oceanic waters, and thus much lower than the Bransfield Strait values. As in the case for the Bransfield Strait, approximate order of magnitude differences in chlorophyll concentrations are noted within individual years, and the period of peak phytoplankton abundance also varies from year to year. Again, seasonal variability is much more pronounced than any interannual variability.

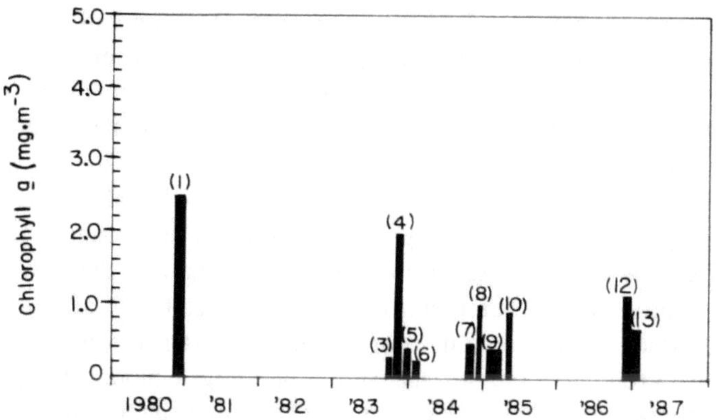

Fig. 8. Seasonal distribution of chlorophyll a in the Elephant Island Box between 1983 and 1987. (For explanation of numbers shown see Table 1)

5.4 The Ross Sea

The average surface chlorophyll values for the USNS *Eltanin* Cruises 27 (January/February 1967) and 51 (January/ February 1972) and USCGC *Glacier* (December 1977/ January 1978) for all the stations occupied south of the Antarctic Circle in the Ross Sea are plotted in Fig. 9. Based on these three data sets (collected at approximately the same time of year), interannual variability of phytoplankton biomass in the Ross Sea is minimal (less than a factor or two). Unfortunately, seasonal data are not available for this region.

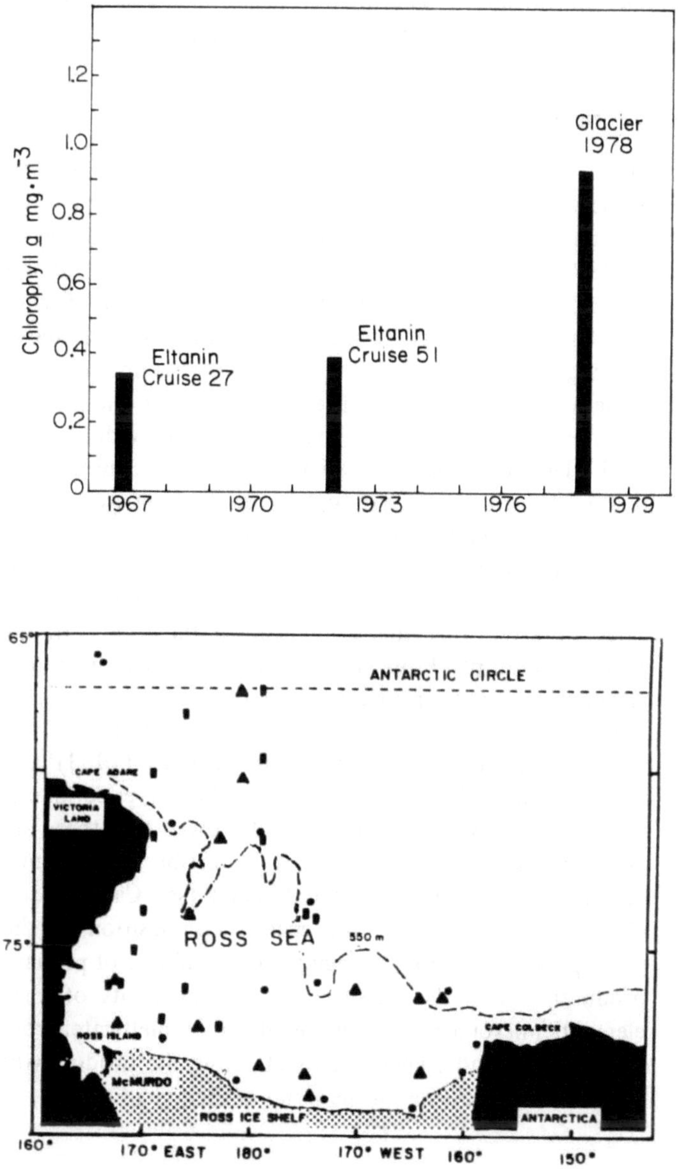

Fig. 9. Seasonal distribution of surface chlorophyll a in the Ross Sea during the austral summers of 1967, 1972 and 1977/78. Position of stations occupied are shown in *inset* below

6 Factors Governing the Distribution and Productivity of Antarctic Phytoplankton

Since the publication of Hart's (1934 and 1942) works, the factors governing the distribution and productivity of Antarctic phytoplankton have occupied the attention of several investigators (see Fogg 1977; Jacques 1983; Heywood and Whitaker 1984; Sakshaug and Holm-Hansen 1984; Priddle et al. 1986a; El-Sayed 1987 for reviews). Of the numerous physical, chemical, and biological factors that might limit the productivity of Antarctic marine phytoplankton, large-scale variation in radiation climate, water column stability, nutrients, and grazing are generally concluded to be the most important. Unfortunately, the evidence provided by investigators is often contradictory and unsubstantiated. In this section we will discuss three of these factors: nutrient salts, water column stability, and grazing, and the possible effects that each has on spatial and temporal distributions of Antarctic phytoplankton.

6.1 Nutrient Salts

The numerous observations of the nutrient salts in Antarctic waters clearly show that these salts generally exceed phytoplankton requirements (Hart 1942; Holm-Hansen et al. 1977; El-Sayed 1984; Hayes et al. 1984; Heywood and Whitaker 1984). Even at the peak of phytoplankton growth, the concentration of nutrient salts remains well above limiting values (El-Sayed 1971a). Thus, it is unlikely that these nutrient salts are sufficiently low at any one time to become limiting factors to phytoplankton growth. Priddle et al. (1986b), using multivariate analysis to examine interrelationships between phytoplankton and environmental variables around South Georgia, found a relatively strong, inverse relationship between phytoplankton biomass and the concentrations of a combination of dissolved nutrients. They observed that the diminution of the total nutrient pool by phytoplankton growth was clearly demonstrated, although no single nutrient could be identified as being important in determining phytoplankton variation.

With regard to trace elements, Jacques (1983) and Hayes et al. (1984) carried out enrichment experiments using Zn, Mo, Co, Mn, and Fe, and showed that these are not limiting factors. However, it is possible that organic factors, e.g., vitamin B_{12} and thiamin (see Carlucci and Cuhel 1977) may alter the species composition of the phytoplankton without changing the overall rate of primary production. It is also possible that the availability of trace elements/micronutrients may be altered by meltwater from the pack ice and icebergs, thus affecting the productivity or species composition of the water in their vicinity.

6.2 Water Column Stability

Braarud and Klem (1931), Gran (1931), and Sverdrup (1953) were among the early investigators to draw attention to the importance of the stability of the water column in controlling primary production and, hence, phytoplankton biomass. The length of time that phytoplankton cells remain within the euphotic zone depends, in part, on the extent of vertical mixing. Sverdrup (1953) showed theoretically that phytoplankton blooms will occur only when the depth of the mixed layer is less than the critical depth. The critical depth in the Southern Ocean is generally between 130–200 m (Jacques and Minas (1981). This is deeper than the usual summer mixed layer (60–95 m, Jacques and Minas 1981; 60–160 m, Priddle et al. 1986b; 12–48 m, Miller and Monteiro this Vol.), which generally corresponds to the depth of the euphotic zone. Such data imply that Sverdrup's critical depth hypothesis cannot explain the low phytoplankton production (Jacques 1983). However, turbulence within the euphotic zone may play a significant role in preventing large increases in phytoplankton biomass. Sakshaug and Holm-Hansen (1984) report that for ten stations in the Scotia Sea where chlorophyll a was >2 mg/m^3, the pycnocline was located at 20–40 m. They speculate that 50 m may be the maximum pycnocline depth for a bloom to develop. Thus, the presence of an homogeneous (i.e., isothermal) water column reaching to depths of 50–100 m during most of the year hinders the development of blooms and contributes to the low primary production of Antarctic waters. In this respect, it is interesting that the *Discovery* investigations found that the depth to which wind action was generally effective in overturning the water column in the open ocean ranged from 60–80 m, especially after storms. Heywood (1984) observed the discontinuity layer at 112 m after a 2-day storm in latitude 56°S (in the sector between 20° and 30°E), and Nast and Gieskes (1986), reported a deep mixed layer (to 200 m) north of Elephant Island as a result of high wind forces.

Bottom topography may, on a regional basis, play an important role in inducing vertical mixing. This was first suggested by Hart (1934) and corroborated by the findings of El-Sayed and Jitts (1973) at Station 12 (61°35′S, 75°12′E) of the *Eltanin* Cruise 46 which lies in proximity to the Banzare Bank. This station was noted for its extremely high standing stocks of phytoplankton and zooplankton. A 1.0-m No. 3 zooplankton net towed vertically from 500 m to the surface yielded 3600 ml of wet volume that was predominantly composed of the diatom species *Thalassiothrix antarctica.*

The vertical stability induced by meltwater has also been suggested as a significant process in initiating and sustaining near-ice blooms. According to this suggestion, the low salinity of meltwater contributes to the stability of the near-ice water column, thus helping to retain the phytoplankton near the surface and promoting a bloom. Corroborative evidence that this mechanism is important in the initiation of ice-edge blooms in the Antarctic is furnished by Jacobs and Amos (1967), El-Sayed (1971a), Smith and Nelson (1985) and Nelson et al. (1987). Thus, the high phytoplankton standing stock (up to 190 mg Chl a/m^3) recorded in the 1968 bloom in the southern Weddell Sea (El-Sayed 1971a) may have been due to a strong pycnocline produced by the melting of ice at the edge of the Filchner Ice Shelf.

Further, low vertical stability of the water column has been implicated by several investigators as a factor contributing to a low standing crop of phytoplankton in the region of the Polar Front (Hart 1942; Hasle 1956; El-Sayed and Mandelli 1965). However, as noted earlier, higher-than-average values of chlorophyll a and primary productivity have also been noted in the vicinity of the Polar Front (Allanson et al. 1981; El-Sayed and Weber 1982; Yamaguchi and Shibata 1982; Lutjeharms et al. 1985; Bidigare et al. 1986). Allanson et al. (1981) postulate that the Polar Front may be characterized, at the sea surface, by a divergence rather than a convergence. Thus, the physical complexity of this frontal system (Amos 1984) must be examined more closely before we can decide whether the variability in phytoplankton standing crop and primary production at the Polar Front stations can be explained in terms of stability or instability of the water column, or if other factors must be invoked.

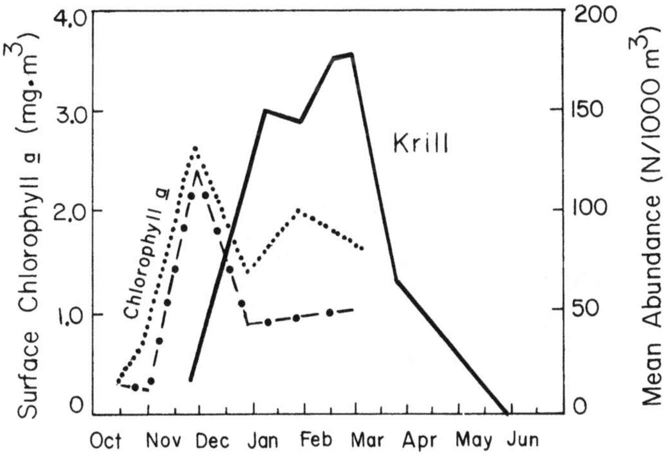

Fig. 10. Mean monthly abundance of krill (N/1000 m^3) for the Bransfield Strait/Elephant Island region (Siegel 1987). Also shown are monthly mean (• • • •) and median (– • –) values of surface chlorophyll a (mg/m^3) for October–March (1980–1987)

6.3 Grazing

To date, much of the importance given to grazing has been by default. That is, the generally low phytoplankton biomass of Antarctic waters cannot be otherwise accounted for, therefore grazing pressure has been assumed to play an important limiting role. In examining the role of grazing in controlling Antarctic phytoplankton biomass and productivity, attention generally focuses on the Southern Ocean krill, *Euphausia superba,* which constitutes a significant (but not necessarily the most abundant) portion of the Antarctic zooplankton biomass (Holdgate 1967; Brinton and Antezana 1984) and serves as a key link between marine primary producers and top predators. Although *E. superba* is regarded by some investigators as the dominant herbivorous species in the Antarctic waters, one should not overlook the significance of two other main components of Antarctic zooplankton (i.e., salps and copepods) as important herbivores affecting phytoplankton density.

The seasonal rise and fall of the krill population appears to lag behind that of the phytoplankton biomass. This is demonstrated in Fig. 10, where mean (and median) concentrations of surface chlorophyll a and mean values of krill abundance for the Bransfield Strait/Elephant Island region are plotted by month. Krill abundance values include data from 1977 to 1985 (Siegel this Vol.), while phytoplankton data were collected between 1980 and 1987.

Early investigators observed that areas of high krill concentration are usually noted for their low standing crop of phytoplankton (Hardy and Gunther 1935; Hart 1942). Polish (Rakusa-Suszczewski 1982) and Chilean (Uribe 1982) investigators found that areas of dense krill concentration in the central parts of the Bransfield Strait exhibited low chlorophyll a values at the surface (< 0.5 mg/m^3) and in the water column (< 50 mg/m^2). According to Uribe (1982), the poverty of the phytoplankton was due not to nutrient limitation but most likely to intensive krill feeding. Figure 11, taken from Nast and Gieskes (1986), demonstrates the inverse relationship between phytoplankton and krill in the waters around Elephant Island during early November 1983. Similarly, an inverse relationship was observed with time when an area north of the Elephant Island was sampled prior to and during the occurrence of a super swarm of krill (Holm-Hansen and Huntley 1984). The phytoplankton biomass was very high in January (633 mg Chl a/m^2 at Station 16), but had been reduced more than tenfold (55 mg Chl a) by March (Fig. 12). On the other hand, the very low zooplankton biomass in January was replaced in March by a dense swarm of furcilia, which extended for 200 km. Paden et al. (1981) reported a chlorophyll a concentration of 3 mg/m^3 upstream of the Elephant Island krill swarm, while downstream, chlorophyll a values were only about 10% of those upstream.

It should be pointed out that coincidence of high phytoplankton standing crop and high krill density has also been reported (Weber and El-Sayed 1985; Mujica and Asencio 1983). Nast and Gieskes (1986) conclude that: "Whereas the negative correlation between chlorophyll a and krill abundance seems to be valid in the mesoscale, a positive correlation between the high krill mass in the Scotia Sea and enhanced phytoplankton food sources may also exist."

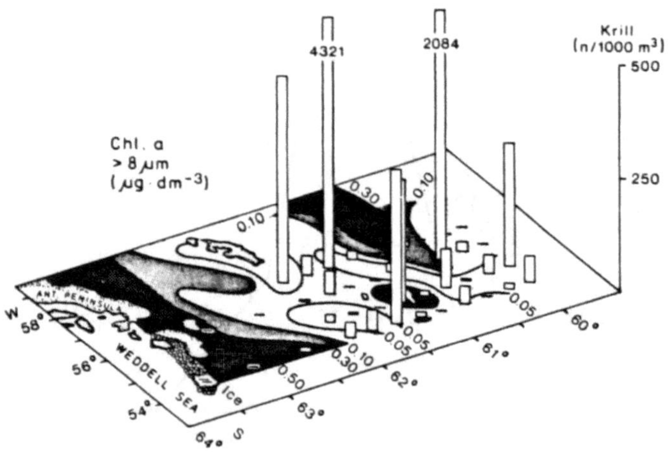

Fig. 11. Surface distribution of chlorophyll a in particles <8 µm (mg/m³) around Elephant Island, November 1983. (Nast and Gieskes 1986)

7 Phytoplankton/Krill Distributional Relationships

Although grazing pressure has been proposed as a factor that controls the abundance of Antarctic phytoplankton, it was only recently (during the 10 February–20 March 1981 cruise of the S.A. *Agulhas*) that the availability of phytoplankton data taken concurrently with acoustic estimates of krill biomass has allowed, perhaps for the first time, a statistical examination of the distributional relationship between these two important components of the ecosystem (Weber and El-Sayed 1985). Because of the significance of the *Agulhas* results to our understanding of the phytoplankton/krill distributional relationship, we will discuss them in some detail.

The waters surveyed during the *Agulhas* cruise were entively within the Antarctic Zone except for the southernmost section near the continent, which lay within the Continental Water Mass (Fig. 13). Thus the discussion relates to an open-ocean region of the Antarctic where frontal mixing and bottom topography play a very limited role in determining phytoplankton and krill distribution patterns. Surface seawater temperature over the entire study area varied only between −2° and +2°C. Temperatures during any single daytime transect were always within ±0.2°C of the transect mean. Extracted chlorophyll a concentrations were low, averaging only 0.12 mg/m³. The detected krill biomass was concentrated in 1034 aggregations, having a mean diameter of only 13 m (Hampton 1985).

Station data were examined statistically using simple correlation, multivariate regression, and canonical correlation analyses. Parameters considered include chlorophyll a, phaeopigments, primary productivity, nutrient concentrations, light, temperature, salinity, sigma-t, water column stability, and krill. Krill parameters were estimated acoustically and included (a) mean krill density (g/m²), (b) number of swarms, (c) mean radius of swarms, and (d) the maximum swarm biomass. Each of these krill parameters was evaluated for the 10 km before and after each station. In the simple correlation matrix (Table 2) chlorophyll a is correlated (positively) more strongly with maximum swarm biomass than with any other variable. In both the multiple regression and canonical correlation analyses, the

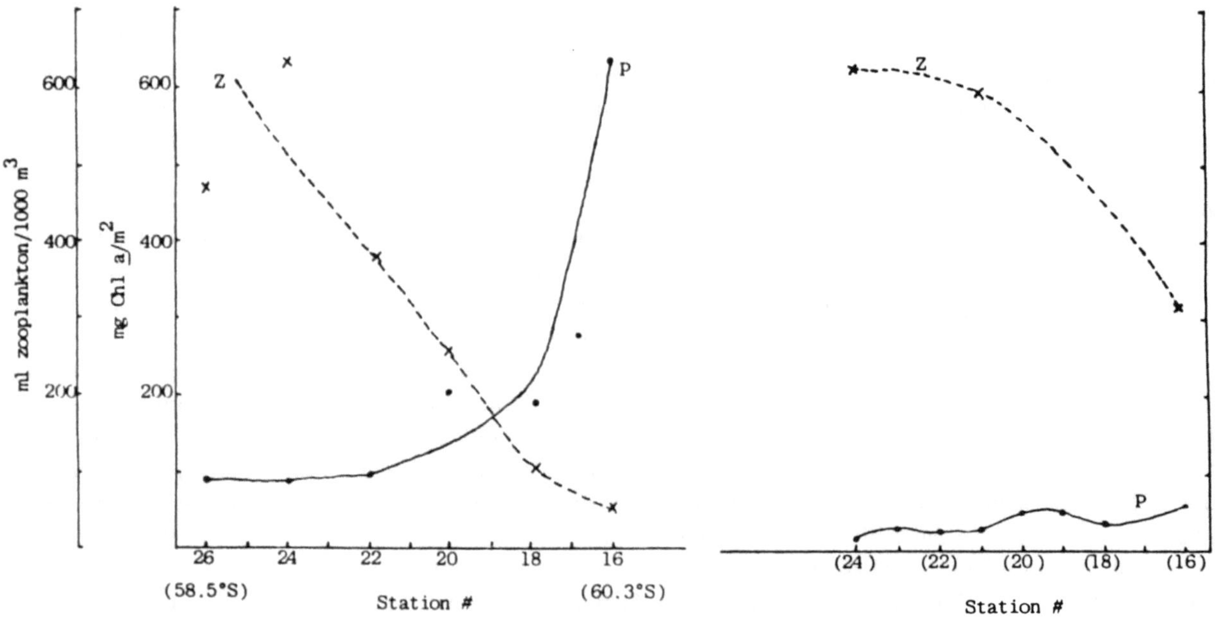

Fig. 12. Total zooplankton biomass and chlorophyll a in the upper 200 m of the water column of a north-south transect to the northwest of the South Orkney Islands. The same stations were sampled during January and again during March 1981. The cruise track lay close to longituide 46° 30′W. (Holm-Hansen and Huntley 1984)

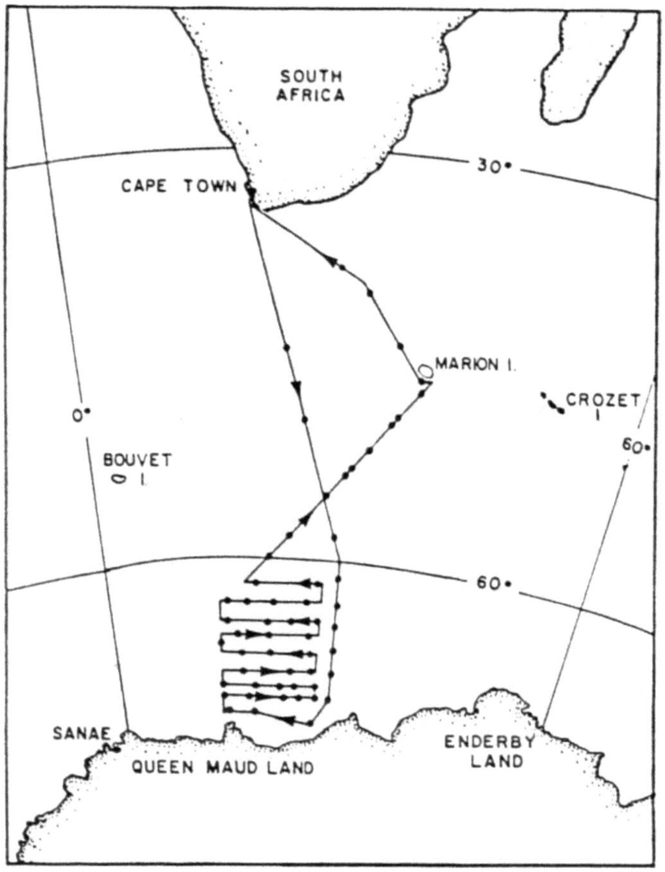

Fig. 13. Cruise track of the M.V. S.A. *Agulhas* during FIBEX, February–March 1981. (Weber et al. 1986)

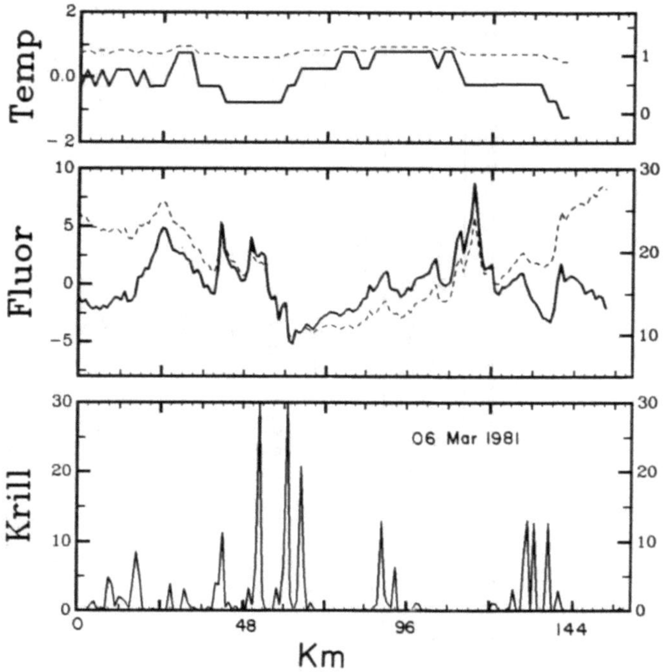

Fig. 14. Horizontal profiles of integrated krill density (g/m²), surface in vivo fluorescence (relative units), and surface seawater temperature 0°C) from the 6 March 1981 transect of the *Agulhas* cruise. Both the original (– – – –; *right-hand scale*) and detrended (———; *left-hand scale*) data are plotted. (Weber et al. 1986)

relation between phytoplankton biomass and maximum swarm biomass remains strongly positive. On the other hand, although the simple correlation between chlorophyll a and mean krill density is positive, an inverse relationship is noted when other parameters are statistically held constant via multivariate regression. Although the absence of time course measurements makes interpretation difficult, it appears that krill are somehow cueing in on areas of local phytoplankton abundance. Hampton (1985) suggests that the larger swarms may have been nonfeeding aggregations that were migrating or foraging. In support of the conclusion of Meyer and El-Sayed (1983) concerning the size-dependent selectivity of krill feeding, the amount of phaeopigments relative to net chlorophyll concentration is correlated positively with each of the four krill parameters in the multiple regression models.

Representative continuous profiles of integrated krill biomass, surface in vivo fluorescence and surface seawater temperature from a single daytime transect are shown in Fig. 14. Three- to fivefold differences in in vivo fluorescence values over the course of the daily transects (79–245 km) suggest strong, though fairly large-scale phytoplankton patchiness. In contrast, krill exhibit intense patchiness over very short scales.

Interrelationships between the krill, phytoplankton, and the physical environment were examined by applying the techniques of power spectral analysis to the suite of continuous daytime profiles of acoustically derived krill biomass, surface in vivo fluorescence and seawater temperature as functions of spatial scale are similar, suggesting that the distributional patterns of phytoplankton biomass are largely determined by physical processes. However, consistent coherence between the fluorescence and krill profiles suggests that predator-prey interactions are also important in determining patterns of phytoplankton distribution.

8 Discussion

Patterns of spatial and temporal variabilities in Southern Ocean phytoplankton distribution, abundance, and productivity are complex. As a whole, there is considerable heterogeneity in the distribution of the phytoplankton density and a great difference between the low values of phytoplankton biomass and primary productivity in oceanic regions and the high values of the inshore waters. This geographical variability tends to overshadow the expression of seasonal differences as the analysis of the data collected during Cruises 17 and 19 of the ARA *Islas Orcadas*

Table 2. Simple correlation coefficients for all pairs of phytoplankton, krill, and physicochemical variables. Only r values ⩾ 0.25 (with P-values ⩽ 0.05) are included. (Weber and El-Sayed 1985)

	Temp.	Salinity	Sigma-t	DO	Silicate	Nitrate	Latitude	Stabil-ity	Max fons	Nano prod	Net prod	Nano AT	Nano Chl	Net Chl	Nano phago	Net phago	Nano AN	Net AN	Net AN	
Light	0.38	−0.40	−0.43							0.30							0.39			
Temp.		−0.62	−0.74		−0.44	−0.31	−0.83	−0.44							−0.36	−0.45			0.36	
Salinity			*0.98*	−0.57	0.53	0.55		−0.38		−0.25					0.58	0.50			−0.45	
Sigma-t				−0.53	0.56	0.54	0.39								0.57	0.53			−0.44	
DO					−0.32	−0.34									−0.58	−0.48			0.38	
Phosphate					0.30	0.29	−0.52	−0.35		−0.26									−0.25	
Silicate						0.60	0.44								0.42	0.36			−0.29	
Latitude								0.39											0.39	
Integ. krill					−0.49				0.66					0.48	0.47					
Mn radius									0.50			0.50								
Nano AT					−0.31			0.46	*0.60*		0.25		049	0.41				0.34	0.41	
Net AT		0.41	0.39	−0.26		0.26			*0.60*			0.32	0.58	0.60	0.70	0.30	−0.37	−0.31	−0.29	0.33
Nano EC																			−0.26	−0.26
Nano Chl	−0.35	0.44	0.52			0.31			*0.73*					0.63	0.40		−0.39	−0.28		0.28
Net Chl		0.26	0.26						*0.80*						0.36		−0.28	−0.33		0.47
Nano Phaeo						0.48			*0.60*							0.67	−0.28	−0.26	*−0.76*	
Net Phaeo						0.32													0.58	*0.39*
Nano AN		−0.36	−0.36							0.69								0.36		
Net AN											0.59									0.35
Nano pig rat						−0.30														0.35

clearly demonstrated. Although the open ocean regions are noted for their impoverished phytoplankton standing crop and low productivity, the fact that they constitute 12 times the combined neritic and upwelling areas explains why they supply the bulk of primary production of the Southern Ocean (Heywood and Whitaker 1984).

The seasonal variability of the phytoplankton biomass is just as complicated as that of the spatial variability. Although it is nearly half a century since Hart's (1942) publication (see Fig. 6) and despite the growing interest in recent years in the study of all aspects of Antarctic phytoplankton, our knowledge of the seasonal variation of Antarctic phytoplankton is still limited. As was pointed out earlier, observations from single stations, or from single depths, may not be representative of a body of water. Further, average seasonal values could be misleading in that they conceal temporal or spatial variability, or both. Further, because of the scarcity of data from open waters during late austral fall and winter, it is difficult to construct a complete year-round picture of the seasonal variability or to describe the physicochemical fluctuations which drive seasonal and year-to-year variations in phytoplankton biomass.

Although much useful information on the physiological ecology of phytoplankton has been gathered during the last two decades, it is humbling to state that an explanation of the factors which govern phytoplankton distribution and production has continued to defy sceintific inquiry. Investigators are confounded by the fact that the ocean south of the Polar Front contains a circumpolar phytoplankton population living in a fairly uniform environment with sufficient light for photosynthesis (at least during the austral spring and summer), that is more or less adapted to cold temperature and has access to abundant nutrient salts, and yet that population is not only patchy, but its density is very low, approaching that of oligotrophic regions. The nutrient levels should be able to support a phytoplankton biomass of at least 25 mg/m^3 chlorophyll a (Holm-Hansen and Huntley 1984). However, Antarctic phytoplankton seldom achieve the high growth rates suggested by the high nutrient levels, except for occasional blooms of phytoplankton such as those reported by El-Sayed (1971a), Mandelli and Burkholder (1966), and others.

In seeking an explanation for the spatial variability in primary production, Fogg (1977) contends that variability in the waters south of the Polar Front cannot be accounted for by differences in incident radiation, water temperature, or concentrations of nitrate, phosphate, and silicate, as each of these parameters has similar values over the entire area in question. After noting that environmental factors act synergistically, Priddle et al. (1986a) conclude that "no one single factor is likely to be identified as ecologically important from a subjective comparison of species and environemental data, even though this method may be effective over large time- and space-scales (e.g., Hart 1942)."

We are still left with the puzzling questions as to why phytoplankton biomass achieves only a fraction of its potential and why such a small part of the Antarctic Ocean is highly productive, while its vast expanses are much less productive. While Sverdrup's (1953) theory of critical depth is essentially sound, it cannot, by itself, provide sufficient explanation of this paradox, as Jacques (1983) has clearly shown. Perhaps an explanation lies in considering light limitation mediated by variation in vertical mixing. To test this hypothesis we need to enlist the cooperation of the physical oceanographers to clarify once and for all the relationship between vertical mixing and primary production. The results of this joint effort could have global implications considering the vast areas in the world ocean where such study is relevant.

Such cooperation may also yield an explanation of the generally enhanced production during most of the production season in the waters off the Antarctic Peninsula and around the Antarctic and Sub-Antarctic islands. This island effect is generally explained in temperate and tropical waters by the input of nutrients from depth as a result of turbulence around the islands (Simpson et al. 1982). In the Antarctic, on the other hand, the nutrients are in abundant supply at the surface, so that turbulence-induced variations in the light regime may provide the answer for the above paradox.

Study of the phytoplankton/krill distributional relationship still remains one of the most vexing problems in marine Antarctic ecosystem studies. Although many hypotheses/explanations have been advanced to elucidate the observed inverse and coincident relationships, for the most part they remain unconvincing or unsubstantiated. Despite its limitations, we feel that the investigation of Weber et al. (1986) of the spatial variability in surface in vivo fluorescence carried out simultaneously with krill assessment by the hydroacoustic method is a step in the right direction.

The positive coherence observed in their study between phytoplankton and krill agrees with the spectrally derived positive correlations between phytoplankton and zooplankton in the Mediterranean Sea and over Georges Bank (Mackas 1977) and in the North Pacific (Star and Mullin 1981), but is in contrast to the inverse correlation found between phytoplankton and zooplankton in the North Sea (Mackas 1977; Mackas and Boyd 1979). Moreover, the positive correlation between in vivo fluorescence and krill biomass observed at scales of 2–20 km is in contrast to the negative correlations usually reported between these two entities on the basis of random, discrete samples separated by much larger distances (Hardy and Gunther 1935; Hart 1942; Witek 1979; Witek et al. 1982; Rakusa-Suszczewski 1982; Uribe 1982; Weber and El-Sayed 1985; Nast and Gieskes 1986).

As a result of their investigation, Weber et al. (1986) identified several problems in the study of phytoplankton/krill relationships that need further elucidation. The most critical need is to resolve finer spatial scales, that is hundreds of meters instead of kilometers. In terms of interpreting results, the most serious problem in their data set stems from the fact that in vivo fluorescence values yield estimates of surface phytoplankton concentrations, while acoustic estimates of krill density give water column values integrated from near-surface to a depth of 100 m. During the day, krill were generally concentrated near the top of the thermocline at a depth of 40–60 m; whereas at night they were found near the surface and could not be detected acoustically (Hampton 1985). Although there is a fair correlation between surface and integrated water column values of chlorophyll, the different sampling strategies for krill and in vivo fluorescence result in an implicit time lag between these two parameters. Moreover, the lag period varies as a function of the time of sampling. Thus, the use of daytime transects (necessary because the downward-looking acoustics could not "see" nighttime surface concentrations of krill) confounds temporal and spatial variability. Side-scan sonar (which was not available on the *Agulhas* cruise) could adequately sample the upper water column and thus provide reliable nighttime acoustic estimates of krill.

These nighttime estimates, together with simulatneous records of surface in vivo fluorescence, could then be analyzed by cross-spectral analysis without having the contend with the problem of implicit temporal variability. Additional advantages of using nighttime transects would be the relatively larger signal-to-noise ratio for the in vivo fluorescence, as well as a decrease in the amount of data manipulation necessary to detrend the profiles.

The application of multivariate regression and power spectral analysis procedures to the suite of data obtained during the *Agulhas* FIBEX cruise (Weber and El-Sayed 1985; Weber et al. 1986) has generated several hypotheses concerning phytoplankton/krill interrelationships. In order to test these hypotheses, future studies must resolve finer spatial scales and include time-course measurements in order to examine the underlying mechanisms.

9 Conclusions and Recommendations

Despite more than 50 years of intensive field work in the Southern Ocean and the expenditure of millions of dollars, we are still unable to answer some of the most fundamental questions in Antarctic marine ecology, for example, why is such a small part of the Antarctic Ocean so much more productive than its vast expanses? However, now researchers more than ever before realize that the distribution and abundance of Antarctic phytoplankton and the rates of primary production are intimately related to the physical oceanographic processes of horizontal circulation and vertical mixing. These relationships operate on time scales ranging from hours to months and over spatial scales of meters to hundreds of kilometers.

Until recently, primary productivity studies have often consisted of piecemeal projects directed at problems of specific interest to individual investigators and limited in scope to localized areas and short time periods. Although data from such projects provide a wealth of information relevant to the localized areas at the time of study (the Elephant Island "Box", for example), we still lack an overall description of the productivity of the Southern Ocean. This seriously limits our ability to assess the significance of the results of many individual research projects and perpetuates the considerable uncertainties regarding estimates of the overall primary production of the Southern Ocean. The unavoidable lesson learned from half a century of Antarctic marine research is that the field season in the Southern Ocean is too short and the logistic support is too expensive for ship-based studies ever to provide a proper perspective of the productivity of the marine ecosystem. Continued application of conventional research techniques and sampling strategies will no longer advance our quest for better understanding of the temporal and spatial variabilities of the biota of the Southern Ocean. Bold, imaginative new approaches are needed in the way we study the Southern Ocean ecosystem. I offer a few suggestions as to the directions our research should take in the next decade.

Research satellites have demonstrated a remarkable ability to measure ocean variables (e.g., sea surface temperature, sea-ice concentrations and ocean color). Measurements made with the Coastal Zone Color Scanner (CZCS) on the Nimbus-7 satellite launched in 1978 have shown that we can now estimate chlorophyll concentrations and thus infer the amount and distribution of the phytoplankton over large areas of the oceans. Now that the CZCS sensor is no longer operational, NASA is investigating the deployment of a satellite-borne Ocean Color Imager (OCI) in 1990. The primary goal of OCI will be to measure visible and near-infrared ocean radiance in order to derive chlorophyll concentrations as an estimation of ocean productivity. The OCI program will be aimed at providing a new "global" perspective of the distribution, abundance, and variance of phytoplankton in the ocean. Such satellite-based data could enhance understanding of phytoplankton dynamics on space and time scales that have so far eluded quantitative investigations, thus providing fundamental new information on Southern Ocean biology, biogeochemical fluxes, and the complex interactions of phytical and biological processes.

A second technological achievement in recent years is the successful use of moored instruments to obtain long-term (months to years), uninterrupted records of currents, temperature, and fluorescence measurements. Here again we should take advantage of this new technology to advance our knowledge of the scales and intensity of both biological and physical processes and to gain an understanding of the causes behind the seasonal and interannual fluctuations of the phytoplankton.

Finally, there is a great need to organize the widely scattered body of knowledge on Antarctic phytoplankton distribution and abundance collected during the past 25 years into an atlas form. Similar efforts have been made by our physical oceanography colleagues (e.g., Wyrtki's 1971 Indian Ocean Atlas, Bainbridge's 1980 Atlantic GEOSECS Atlas, and more recently, Gordon and Molinelli and Baker's 1982 Southern Ocean Atlas). In the latter, the enormous wealth of data on the physics and chemistry of the Southern Ocean is neatly summarized and attractively displayed. Although we cannot aspire to match the quantity of data contained in the physical oceanographic atlases, nonetheless the rapidly growing data set on Antarctic productivity is currently being collated into a Southern Ocean Productivity Atlas. Upon completion, the productivity atlas will be useful in establishing correct overall patterns of Southern Ocean productivity and in providing background information against which spatial and temporal variations in phytoplankton data can be interpreted.

References

Allanson BR, Hart RC, Lutjeharms JRE (1981) Observations on the nutients, chlorophyll and primary production of the Southern Ocean south of Africa. S Afr J Antarct Res 10/11:3–14

Amos AF (1984) Distribution of krill (*Euphausia superba*) and the hydrography of the Southern Ocean: large-scale processes. J Crustacean Biol 4:306–329

Bainbridge AE (1980) Atlantic GEOSECS expedition, vol 2. Sections and profiles. Int Decade of Ocean Explor. Nat Sci Found, Washington DC

Bidigare RR, Frank TJ, Zastrow C, Brooks JM (1986) The distribution of algal chlorophylls and their degradation products in the Southern Ocean. Deep-Sea Res 33:923–937

Braarud T, Klem A (1931) Hydrographical and chemical investigations in the coastal waters off More and in the Romsdalfjord. Hvalradets Skr 1:1–88

Brinton E, Antezana T (1984) Structures of swarming and dispersal popularion of krill (*Euphausia superba*) in Scotia Sea and South Shetland waters during January–March 1981, determined by bongo nets. J Crustacean Biol 4:45–46

Brinton E, Reid JL (1986) On the effects of interannual variations in circulation and temperature upon euphausiids of the California Current. In: Pelagic biogeography, Proceedings of an International Conference, The Netherlands, 29 May–5 June 1985. UNESCO Tech Pap Mar Sci 49:25–34

Carlucci AF, Cuhel RL (1977) Vitamins in the south polar seas: distribution and significance of dissolved and particulate vitamin B_{12}, thiamin, and biotin in the Southern Indian Ocean. In: Llano GA (ed) Adaptations within Antarctic ecosystems (Proceedings of the Third SCAR Symposium on Antarctic Biology). Gulf, Houston, pp 115–128

Cushing DH (1966) Biological and hydrographic changes in British Seas during the last thirty years. Biol Rev 41:221–258

El-Sayed SZ (1967) On the productivity of the southwest Atlantic Ocean and the waters of the Antarctic Peninsula. In: Schmitt W, Llano G (eds) Biology of the Antarctic Seas, vol III. Antarct Res Ser, vol II. Am Geophys Un, Washington DC, pp 15–47

El-Sayed SZ (1968) Primary productivity of the Antarctic and Subantarctic. In: Bushnell V (ed) Primary productivity and benthic marine algae of the Antarctic and Subantarctic. Folio 10. Antact Map Folio Ser. Am Geogr Soc, New York, pp 1–6

El-Sayed SZ (1971a) Observations on phytoplankton bloom in the Weddell Sea. In: Llano GA, Wallen IE (eds) Biology of the Antarctic Seas IV, vol 17. Am Geogr Un, Washington DC, pp 301–312

El-Sayed SZ (1971b) Dynamics of trophic relationships in the Southern Ocean. In: Quan LO (ed) Research in the Antarctic. Am Assoc Adv Sci, Washington DC, pp 73–91

El-Sayed SZ (1984) Productivity of Antarctic waters – a reappraisal. In: Holm-Hansen O, Bolis L, Gilles R (eds) Marine phytoplankton and productivity. Springer, Berlin Heidelberg New York Tokyo, pp 19–34

El-Sayed SZ (1987) Biological productivity of the Antarctic waters: Present paradoxes and emerging paradigms. In: El-Sayed SZ, Tomo A (eds) Regional symposium on recent advances in Antarctic aquatic biology with special reference to the Antarctic Peninsula. San Carlos de Bariloche, Argentina, 6–10 June 1983. BIOMAS Sci Ser 7 SCAR/Scott Polar Res Inst, Cambridge

El-Sayed SZ, Hofmann E (1986) Drake Passage and Western Scotia Sea (Antarctica). In: Hovis WA (ed) Nimbus-7 CZSC coastal zone color scanner imagery for selected coastal regions. NASA, pp 97–99

El-Sayed SZ, Jitts H (1973) Phytoplankton production in the Southwestern Indian Ocean. In: Zeitschel B (ed) Biology of the Indian Ocean. Proc Symp Univ Kiel, FRG, March 31–April 6, 1971. Springer, Berlin Heidelberg New York, pp 131–544

El-Sayed SZ, Mandelli EF (1965) Primary production and standing crop of phytoplankton in the Weddell Sea and Drake Passage. In: Llano GA (ed) Biology of the Antarctic Sea II, vol 5. Antarct Res Ser. Nat Acad Sci, Nat Res Counc Publ No 1297, Washington DC, pp 87–106

El-Sayed SZ, Taguchi S (1981) Primary production and standing crop along the ice-edge in the Weddell Sea. Deep-Sea Res 28: 1017–1032

El-Sayed SZ, Turner JT (1977) Productivity of the Antarctic and tropical subtropical regions: A comparative study. In: Dunbar MJ (ed) Proc SCOR/SCAR Polar Oceans Conf, Montreal, Canada, May 1974. Arct Inst North Am, Calgary, Alberta, Canada, pp 463–504

El-Sayed SZ, Weber LH (1982) Spatial and temporal variations in phytoplankton biomass and primary productivity in the southwest Atlantic and Scotia Sea. Polar Biol 1:83–90

El-Sayed SZ, Mandelli EF, Sugimura Y (1964) Primary organic production in the Drake Passage and Bransfield Strait. In: Lee MO (ed) Biology of the Antarctic Seas I, vol 1. Am Geophys Un, Washington DC, pp 1–110

El-Sayed SZ, Stockwell DA, Reheim HR, Taguchi S, Meyer MA (1979) On productivity of the southwestern Indian Ocean. In: Arnaud PM, Hureau JC (eds) CNFRA, Campagne océanographique MD 08/Benthos aux iles Crozet, Marion et Prince Edward: premiers résultats scientifiques, pp 83–110

El-Sayed SZ, Biggs DC, Holm-Hansen O (1983) Primary productivity, standing crop of phytoplankton and ammonium chemistry of the Ross Sea. Deep-Sea Res 30:871–886

Fogg GE (1977) Aquatic primary production in the Antarctic. Philos Trans R Soc London Ser B 279:27–38

Fukuchi M (1980) Phytoplankton chlorophyll stocks in the Antarctic Ocean. J Oceanog Soc Jpn 36:73–84

Fukuchi M, Tamura S (1982) Chlorophyll a distribution in the Indian sector of the Antarctic Ocean in 1975–1979. Antarct Rec 14:143–162

Gieskes WW, Elbrächter M (1986) Abundance of nanoplankton-size chlorophyll-containing particles caused by diatom disruption in surface waters of the Southern Ocean (Antarctic Peninsula Region). Neth J Sea Res 20:291–303

Gordon AL, Molinelli EJ, Baker TN (1982) Southern Ocean Atlas. Columbia Univ Press, New York

Gran NH (1931) On the conditions for the production of plankton in the sea. J Cons Cons Perma Int Explor Mer 75:37–46
Hampton I (1985) Abundance, distribution and behavior of *Euphausia superba* in the Southern Ocean between 15°E and 30°E during FIBEX. In: Siegfried WK, Condy PR, Laws RM (eds) Antarctic nutrient cycles and food webs. Springer, Berlin Heidelberg New York Tokyo, pp 294–303
Hardy AC, Gunther ER (1935) The plankton of the South Georgia whaling grounds and adjacent waters, 1926–1927. Discovery Rep 11:1–456
Harris GP (1986) Phytoplankton ecology: Structure, function and fluctuation. Chapman and Hall, New York, 384 pp
Hart TJ (1934) On the phytoplankton of the southwest Atlantic and the Bellingshausen Sea, 1929–1931. Discovery Rep 8:1–268
Hart TJ (1942) Phytoplankton periodicity in Antarctic surface waters. Discovery Rep 21:261–356
Hasle GR (1956) Phytoplankton and hydrography of the Pacific part of the Antarctic Ocean. Nature (London) 177:616–617
Hayes PK, Whitaker TM, Fogg GE (1984) The distribution and nutrient status of phytoplankton in the Southern Ocean between 20° and 70°W. Polar Biol 3:153–165
Heywood RB (1984) British Antarctic Survey offshore biological programme: Cruise 4 (27 July–29 September 1983) Rep
Heywood RB, Whitaker TM (1984) The Antarctic marine flora. In: Laws RM (ed) Antarctic ecology, vol II. Academic Press, London New York, pp 373–419
Holdgate MW (1967) The Antarctic ecosystem. Philos Trans R Soc London Ser B 252:363–383
Holm-Hansen O, Huntley M (1984) Feeding requirements of krill in relation to food sources. J Crustacean Biol 4:156–173
Holm-Hansen O, El-Sayed SZ, Franceschini GA, Cuhel R (1977) Primary production and the factors controlling phytoplankton growth in the Southern Ocean. In: Llano GA (ed) Adaptations within Antarctic ecosystems. Proc 3rd SCAR Symp Antarct Biol. Gulf, Houston, pp 11–50
Horne AJ, Fogg GE, Eagle DJ (1969) Studies in situ of primary production of an area of inshore Antarctic sea. J Mar Biol Assoc UK 49:393–405
Ichimura S, Fukushima H (1963) On the chlorophyll content in the surface water of the Indian and the Antarctic Oceans. Bot Mag 76:395–399
Jacobs SS, Amos A (1967) Physical and chemical oceanographic observations in the Southern Ocean. Tech Rep 1-Cu-1-67. Lamont-Doherty Geol Observ, 287 pp
Jacques G (1983) Some eco-physiological aspects of the Antarctic phytoplankton. Polar Biol 2:27–34
Jacques G, Minas M (1981) Production primaire dans le secteur indien de l'Océan Antarctique en fin d'été. Oceanol Acta 4:33–41
Klyashtorin LB (1961) Pervichnaya produktsiya v Alanticheskon i Yuzhom okranakh po dannym pyatogo Antarkticheskogo reisa dizel-elektrokhoda. Ob Dokl Akad Nauk USSR 141:1204–1207
Lipski M (1981) The distribution of chlorophyll a in relation to the water masses in the Southern Drake Passage and the Bransfield Strait (BIOMASS FIBEX, February–March 1981). Pol Polar Res 3(3/4):143–152
Lipski M (1985) Chlorophyll a in the Bransfield Strait and the southern part of Drake Passage during BIOMASS FIBEX (December 1983–January 1984). Pol Polar Res 6(1/2):21–30
Lutjeharms JRE, Walters NM, Allanson BR (1985) Oceanic frontal systems and biological enhancement. In: Siegfried WR, Condy PR, Laws RM (eds) Antarctic nutrient cycles and food webs. Springer, Berlin Heidelberg New York Tokyo, pp 11–21
Mackas DL (1977) Horizontal spatial variability and covariability of marine phytoplankton and zooplankton. PhD Diss, Dalhousie Univ
Mackas DL, Boyd CM (1979) Spectral analysis of zooplankton spatial heterogeneity. Science 204:62–64
Mandelli EF, Burkholder PR (1966) Primary productivity in the Gerlache and Bransfield Straits of Antarctica. J Mar Res 24: 15–27
Meyer MA, El-Sayed SZ (1983) Grazing of *Euphausia superba* on natural phytoplankton populations. Polar Biol 1:193–197
Mujica A, Asencio V (1983) Distribution and abundance of krill larvae (*Euphausia superba* Dana). Ber Polarforsch 4:21–29
Nast F, Gieskes W (1986) Phytoplankton observations relative to krill abundance around Elephant Island in November 1983. Arch Fischereiwiss 37:73–94
Nelson DM, Smith WO Jr, Gordon LI, Huber B (1987) Early spring distributions of density, nutrients and phytoplankton biomass in the ice-edge zone of the Weddell/Scotia Sea. J Geophys Res 92:7181–7190
Paden CA, Hewes CD, Neori A, Holm-Hansen O, Weaver E, Kiefer DA, Sakshaug E (1981) Phytoplankton studies in the Scotia Sea. Antarct J US 16:163–164
Priddle J, Hawes I, Ellis-Evans JC, Smith TJ (1986a) Antarctic aquatic ecosystems as habitats for phytoplankton. Biol Rev 61: 199–238
Priddle J, Heywood RB, Theriot E (1986b) Some environmental factors influencing phytoplankton in the Southern Ocean around South Georgia. Polar Biol 5:65–79
Rakusa-Suszczewski S (1982) Feeding of *Euphausia superba* Dana under natural conditions. Pol Polar Res 3:289–297
Russell FS, Southward AJ, Boalch GT, Butler EI (1971) Changes in biological conditions in the English Channel off Plymouth during the last half century. Nature (London) 234:468–470
Saijo Y, Kawashima T (1964) Primary production in the Antarctic Ocean. J Oceanogr Soc Jpn 19:190–196
Sakshaug E, Holm-Hansen O (1984) Factors governing pelagic production in polar oceans. In: Holm-Hansen O, Bolis L, Gilles R (eds) Marine phytoplankton and productivity. Springer, Berlin Heidelberg New York Tokyo, pp 1–18
Simpson JH, Terr PB, Argote-Espinoza ML, Edwards A, Jones HJ, Savidge G (1982) Mixing and phytoplankton growth around an island in a stratified sea. Cont Shelf Res 1:15–31
Smith WO, Nelson DM (1985) Phytoplankton bloom produced by a receding ice edge in the Ross Sea: spatial coherence with the density field. Science 227:163–166
Southward AJ (1974) Changes in the plankton community of the Western English Channel. Nature (London) 249:180–181
Southward AJ (1980) The Western English Channel – an inconstant system? Nature (London) 285:361–366
Star JL, Mullin MM (1981) Zooplanktonic assemblages in three areas of the North Pacific as revealed by continuous horizontal transects. Deep-Sea Res 28:1303–1322
Steemann Nielsen E (1952) The use of radioactive carbon (^{14}C) for measuring organic production in the sea. J Cons 18:117–140
Sverdrup HU (1953) On conditions for the vernal blooming of phytoplankton. J Cons 18:287–295
Uribe F (1982) Influence of the phytoplankton and primary production of the Antarctic waters in relationship with the distribution and behavior of krill. Inst Antart Chileno Sci Ser 28:147–163
von Bodungen B (1986) Phytoplankton growth and krill grazing during spring in the Bransfield Strait, Antarctica – implications from sediment trap collections. Polar Biol 6:153–160
von Bodungen B, Smetacek VS, Tilzer MM, Zeitzschel B (1986) Primary production and sedimentation during spring in the Antarctic peninsula region. Deep-Sea Res 33(2):177–194
Voronina NM (1984) Pelagic Ecosystems of the Southern Ocean. Acad Sci USSR. Nauka, Moscow, 206 pp

Weber LH, El-Sayed SZ (1985) Spatial variability of phytoplankton and distribution and abundance of krill in the Indian sector of the Southern Ocean. In: Siegfried WR, Condy PR, Laws RM (eds) Antarctic nutrient cycles and food webs. Springer, Berlin Heidelberg New York Tokyo, pp 284–293

Weber LH, El-Sayed SZ (1986) Phytoplankton data from the November–December cruise of the FS *Polarstern* to the Bransfield Strait/Elephant Island region of the Southern Ocean. Dep Oceanogr, Texas A & M Univ, 25 pp

Weber LH, El-Sayed SZ, Hampton I (1986) The variance spectra of phytoplankton, krill and water temperature in the Antarctic Ocean south of Africa. Deep-Sea Res 33:1327–1343

Witek Z (1979) Phytoplankton distribution and some aspects of the biology of Antarctic krill *Euphausia superba* Dana. ICES CM 1979/L:14

Witek Z, Grelowski A, Kalinowski J (1982) Formation of Antarctic krill concentrations in relation to hydrodynamic processes and social behavior. ICES CM 1982/L:59

Wyrtki K (1971) Oceanographic Atlas of the International Indian Ocean Expedition. Nat Sci Found, Washington DC

Yamaguchi Y, Shibata Y (1982) Standing stock and distribution of phytoplankton chlorophyll in the Southern Ocean south of Australia. Trans Tokyo Univ Fish 5:111–128

Scales of Interaction Between Antarctic Krill and the Environment

E. J. MURPHY, D.J. MORRIS, J. L. WATKINS, and J. PRIDDLE[1]

Summary. A conceptual framework to allow a constructive approach to the analysis of the Antarctic marine ecosystem taking into account the variability of the system was developed. The temporal and spatial scale relationships of biotic and abiotic processes were found to be important in the investigation of ecosystem structure. These processes were regarded as forming hierarchies over temporal and spatial scales. The hierarchies define which processes interact over common scales and can also be used to consider which interactions can occur. Different levels of krill aggregation change the possible interactions in the system, and the scales of operation of these aggregations have important consequences for the investigation of ecosystem structure. The framework allows specific interactions, such as grazing and predation, as well as more general aspects such as the formation of sea ice to be considered and allows an overview of ecosystem structure and function.

1 Introduction

Investigations of the biology of Antarctic krill (*Euphausia superba* Dana) have been extensive, especially over the last decade. These studies have produced much detailed information on biological processes ranging from physiological to interannual variability and attempts to synthesize this information have resulted in a series of comprehensive reviews (Everson 1977; Lubimova et al. 1980; Bengtson 1985; Miller and Hampton ms). The quantity and range of the information generated on krill biology in particular, and the Antarctic marine ecosystem in general, makes it very difficult to consider the structure and functioning of the system in the absence of some unifying approach. We have therefore produced a framework to allow this large body of information to be integrated so that a more constructive and cohesive approach can be developed. This should lead to a greater understanding of krill dynamics and its role in the Southern Ocean marine ecosystem.

The complexity of the Southern Ocean food web is now being increasingly recognized and ideas of the simple food chain of "diatoms-krill-consumers" (see, for example, Tranter 1982 and Fig. 1) have been reassessed (Clarke 1985). Despite this, krill are clearly a major link in the transfer of energy from primary producers to larger organisms in the system. They are the characteristic zooplankton of the Southern Ocean and one of the key features of their biology is an extremely heterogeneous distribution. They can occur at very high densities in localized areas, for example "superswarms" (Macaulay et al. 1984), but they can also be almost completely absent; the dominant macrozooplankton herbivores are then copepods or salps (Clarke 1985).

As variability in time and space is a key feature of the Antarctic marine ecosystem, we consider in this Chapter the implications of that variability. We have expanded the ideas expressed by Southwood (1977) on the variability of ecosystem structure and by Haury et al. (1978) on the importance of time and space scales in the distribution of plankton. In addition, we draw on work on the dynamics of planktonic systems (Legendre and Demers 1984; Harris

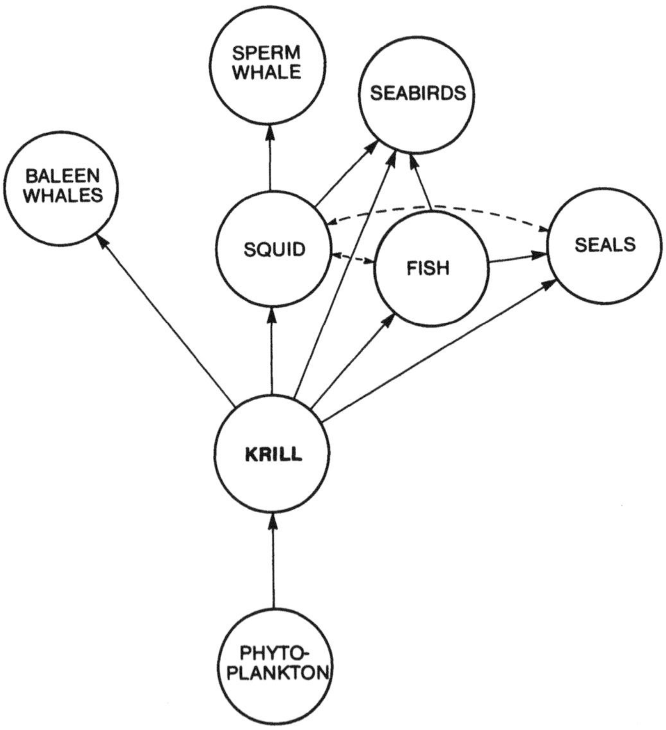

Fig. 1. Simplified representation of part of the Southern Ocean food web involving krill

1 British Antarctic Survey, Natural Environment Research Council, High Cross, Madingley Road, Cambridge CB3 OET, UK

Antarctic Ocean and Resources Variability
(ed. by D. Sahrhage)

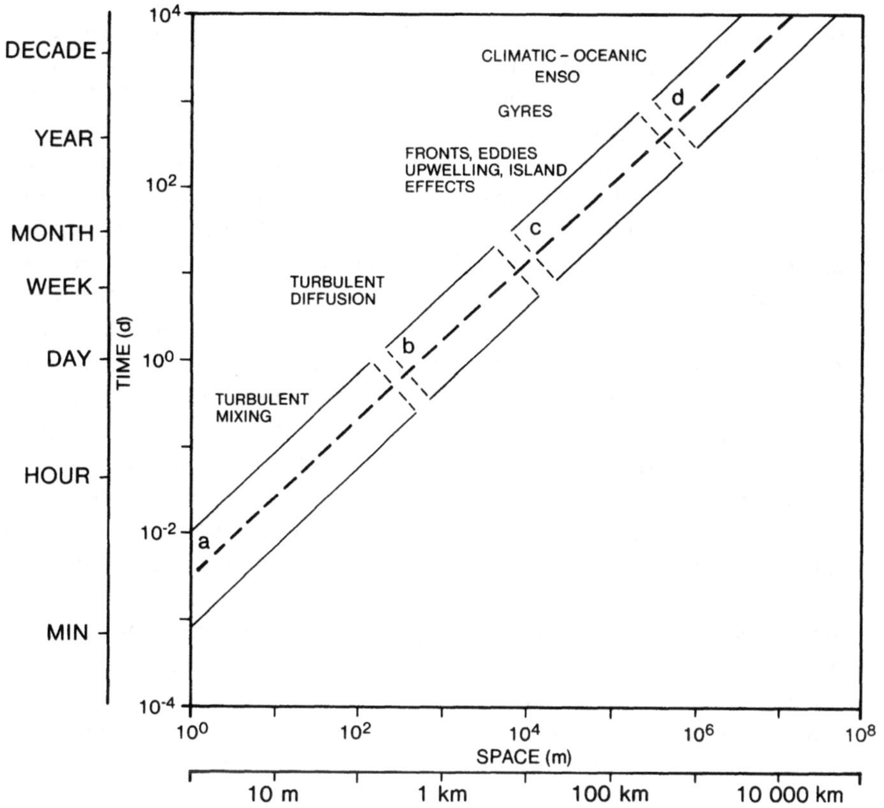

Fig. 2. The relationship between time and space scales for mixing processes. The *broken line* represents the general relationship established by Bowden (1970). The four blocks (*a–d*) represent groups of hydrodynamic processes operating at similar scales in the ocean

1986) and hierarchical ecosystem theory (Allen and Starr 1982) to allow the analysis of not just the planktonic part but the whole of the Southern Ocean ecosystem. To do this we utilize the concept that the highly variable hydrodynamics of the Southern Ocean comprise a set of interacting processes operating at different time and space scales. We then consider the ability of an organism to respond to variation in the environment and apply these ideas to the Southern Ocean food web. The importance of the role of krill in the ecosystem is investigated from this viewpoint of time-space interaction. This procedure is regarded as a preliminary development of a conceptual framework for the study of the Southern Ocean ecosystem.

2 Time and Space Scales in the Southern Ocean Marine Ecosystem

A major point emphasized in discussions of the hydrodynamic environment of lakes and oceans is the distinct relationship between temporal and spatial scales of mixing (Bowden 1970; Harris 1980, 1986; Legendre and Demers 1984). Thus, for particular hydrodynamic processes there are corresponding temporal and spatial scales (Fig. 2).

The hydrodynamic environment of surface waters, at scales of hours and less in time and shorter than a kilometre in space, is dominated by vertical diffusion processes (e.g. Langmuir vortices) resulting in turbulent mixing (Thorpe 1985) (Fig. 2a). At the scale of days and kilometers, however, horizontal turbulent diffusion dominates (Legendre and Demers 1984; Thorpe 1985; Harris 1986; Priddle et al. 1986) (Fig. 2b). An important additional feature of the Southern Ocean is the presence and formation of sea ice, which modifies the hydrodynamic environment over these scales by removing wind-induced mixing effects, resulting in reduced turbulent mixing and diffusion.

Events with temporal scales of several months to a year have corresponding spatial scales of tens to hundreds of kilometres. In the Southern Ocean these probably include such features as seasonal thermal stratification on continental shelf regions (e.g. Bransfield Strait – Heywood and Priddle 1987), frontal systems (e.g. the Ross Sea shelf-break front – Amos 1984; see also Makarov 1972 and Lutjeharms et al. 1985), areas of upwelling (as observed in the Scotia Sea region – Mackintosh 1972), eddies (Bogdanov et al. 1969; Bryden 1983) and the seasonal fluctuations in ice cover (Tranter 1982) (Fig. 2c).

In the longer term and on a larger spatial scale, changes in ocean circulation patterns are influenced by climatic variation (Amos 1984). Such variations are probably at or above the scale of variations in the Antarctic Circumpolar Current (Legeckis 1977; Colton and Chase 1983). Thus, the effects of El Niño Southern Oscillation (ENSO) events are characteristic of this scale in the Pacific Ocean and may be important in the Antarctic region (Carleton 1988). The extent of the ice cover over the Southern Ocean is also likely to be affected by processes operating over these large scales (Fig. 2d).

Together, these hydrodynamic processes can be regarded as forming a hierarchy over temporal and spatial scales

(Harris 1980, 1986; Legendre and Demers 1984), producing variation from large to small scales.

3 Responses of an Organism to Variability of the Environment in Time and Space

Biological structures can also be considered as elements in a hierarchy (organelles, cells, tissues, organs, individuals, populations, communities), as can biological processes (biochemistry, physiology, population growth, changes in species composition) (Webster 1979; Allen and Starr 1982; Harris 1986). Particular biological processes have characteristic time and space scales. An organism's response to environmental variation at particular scales will be through biological processes operating at corresponding scales. The range of these responses in time and space thus allows an organism to exploit particular parts of the ecosystem and it is therefore adapted to exploit variation over particular scales. Southwood (1977) and Harris (1986) thus emphasized that it is the way in which organisms perceive the environment that is important (see also Caddy 1986). A complete hierarchy, describing possible biological responses of an organism, would also define its niche.

Where these ideas have been previously applied to phytoplankton ecology, a close coupling between the hydrodynamic processes and the biological responses has been shown (Legendre and Demers 1984; Harris 1986). It is therefore possible to describe the relationships between the scales of biological processes and those of environmental variability. These relationships will show a similar relationship in time and space to that shown in Fig. 2 for the hydrodynamic processes (see also Priddle et al. 1986, Fig. 10). Before considering in more detail the biological responses of phytoplankton to variation of the environment in time and space in the Southern Ocean, we need to define what possible biological responses can occur and how these are positioned in time and space; that is, we need to define the niche. This will then allow us to expand these ideas to other elements of the ecosystem.

4 The Scales of the Population Growth Process as the Fundamental Units of Species

We need to identify a fundamental time unit by which we can compare the responses of a wide range of organisms to environmental variation in time and space. Generation time is a key element in determining the ability of an organism to respond to environmental variation and is closely related to the intrinsic rate of population increase (Tranter 1976). Phytoplankton growth rate is commonly expressed in terms of doubling time and is of the order of 1 to 10 days (Sheldon et al. 1972; Harris 1986). For sexually reproducing animals, however, the definition of doubling time as expressed for phytoplankton is inapplicable. We have therefore taken the mean time between reproductive events over the whole of the life-span of an individual as the population growth period (this can be considered as a standardized doubling time). Thus, for a large baleen whale with a life span of 30 years, reproducing every 2 years after a pre-maturity period of 10 years, the population growth period is 3 years (10 events in 30 years). This concept can also be applied to krill, which may breed a few times in a restricted season each year. Assuming krill breed twice each year, with a prematurity phase of 2 years and an average lifespan of 4 years, then the population growth period is 1 year. With a lifespan of 6–8 years, the population growth period is reduced to 9 or 8 months. Environmental variation on the time scale of the population growth period therefore elicits a response by affecting the rate of change of the population in numbers. Environmental changes at larger or smaller scales necessitate other biological responses. Population growth period can therefore be regarded as defining a fundamental time unit for a species (Harris 1986).

Having established a time scale for the population growth process, we need a corresponding spatial unit in order to be able to compare the responses of organisms to environmental variation. A possible approach to this is the relationship between the population growth period and organism size, which is commonly used in a variety of forms to describe ecosystems (Bonner 1965; Sheldon et al. 1972; Calder 1985; Platt 1985) following the size spectrum approach of Elton (1927). Organism size has other profound implications for biological response and ecological interactions (see Peters 1983), such as metabolic rate (Platt 1985), dispersal ability (Harris 1986) or trophic level (Steele 1978). Although such a relationship (Fig. 3a) provides a useful description of ecosystem structure, it contains no direct information on the spatial scale of any biological process. We can, however, produce a relationship between population growth period and its associated spatial scale. At one extreme, population growth period for phytoplankton has a spatial scale of approximately 1 km, defined by hydrographic processes. Whales, however, are more or less independent of such influences and disperse actively over distances of 10^3 km. Other organisms occupy intermediate positions on such a curve (Fig. 3b), which is similar to the relationship between time and space scales for hydrodynamic processes (Fig. 2). These spatial scales are, therefore, likely to reflect the level of independence of an organism from water movement and thus the amount of "new water" occupied.

We have now defined fundamental time and space units for organisms in the Southern Ocean food web by identifying the temporal and spatial scales of the population growth process (Fig. 3b). It is also possible to describe the scales of other processes which define an organism's niche and thus the responses that can occur to environmental variation of those scales.

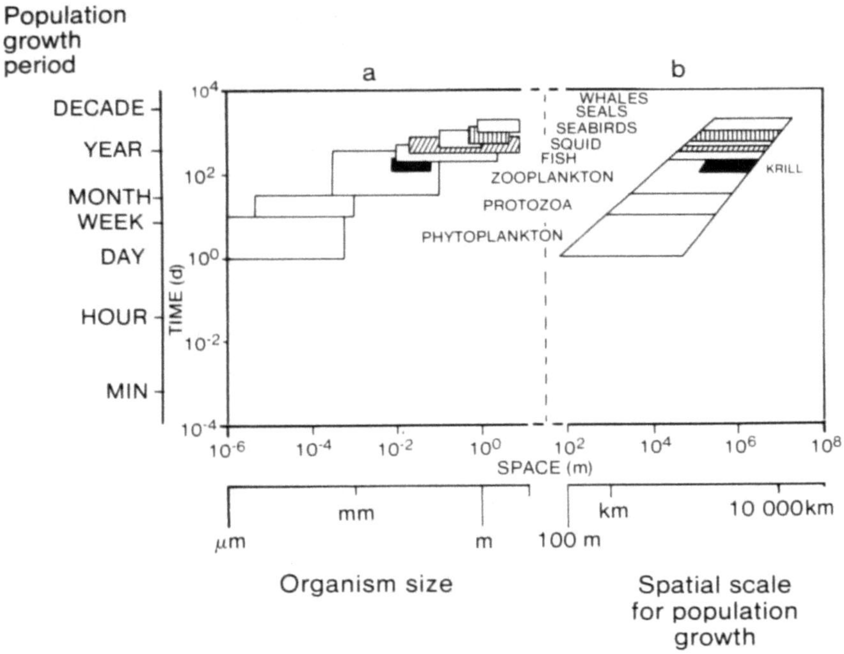

Fig. 3a,b. The relationship of the population growth period to a organism size and **b** the spatial scale of population growth of organisms in the Southern Ocean

5 Scales of Biological Processes

In a simplified view of the Southern Ocean ecosystem, krill interact with the hydrodynamic environment (described in Sect. 2, Fig. 2), the food supply (represented here by phytoplankton) and predators (a grouping of the organisms that eat krill). Utilizing the position of the population growth process in the time and space scale framework (Fig. 3b), other processes through which organisms respond to environmental fluctuations can be placed on the time-space relationship for each of these general trophic levels.

The relationship for phytoplankton (Fig. 4) is based on the definitions of Harris (1980, 1986) and Legendre and Demers (1984). Thus, physiological responses occur when environmental fluctuations are at scales less than that of the process of population growth. Changes in biomass result from variation in population numbers, so they occur at the scale of the population growth process. Changes in distribution and abundance thus occur over scales of weeks to months and tens of kilometres, while the overall community structure changes over months (on a seasonal basis) and hundreds to thousands of kilometres. At scales longer and larger than this, changes are seen in the biogeography and eventually the evolutionary history (Haury et al. 1978).

For the predators (Fig. 5) the position of the biological responses can only be defined approximately because of the range of organisms covered (Fig. 3b). Thus, the process of population growth, which is the fundamental temporal and spatial unit, is at a longer and on a larger scale for the predators than for phytoplankton, and this results in corresponding shifts in the scales of other processes. Behavioral changes are also included here because the predators can respond in this manner to environmental variation over certain scales. In contrast to phytoplankton, changes in predator biomass can be observed as a result of individual growth so these can also occur below the scales of the population growth process. The position of these processes in the time-space framework will depend not only on the species of predator but also on the age (and thus size) of the organism.

A similar procedure was carried out for krill (Fig. 6), where physiological and behavioral responses occur below scales of days and kilometres. Changes in biomass occur over scales of days and kilometres and for the zooplankton community structure changes will be mainly seasonal to interannual and over hundreds to thousands of kilometres. However, before considering some of the possible interactions in this trophic-environmental, time-space framework, there is a further aspect of krill biology which should be taken into account. The highly heterogeneous distribution of krill, particularly the formation of aggregations, is an important feature of krill biology and has been a major focus of studies (Mackintosh 1967; Everson 1977; Mauchline 1980; Watkins 1986; Watkins et al. 1986; see also Miller and Hampton ms, for a review). Attempts to classify these aggregations and identify specific types (Mauchline 1980; Kalinowski and Witek 1985) have been mainly in terms of size, krill density or persistence in time and are, therefore, attempts to define particular time and space scales over which peaks occur in the variability of the krill distribution.

To aid the investigation of the time-space interactions of krill with the environment we identify five levels of organisation in krill populations (Fig. 7). We do not attempt to completely formalize the definition of these aggregations, but identify overall types of aggregation based on a scale perspective. We thus suggest a hierarchical organiza-

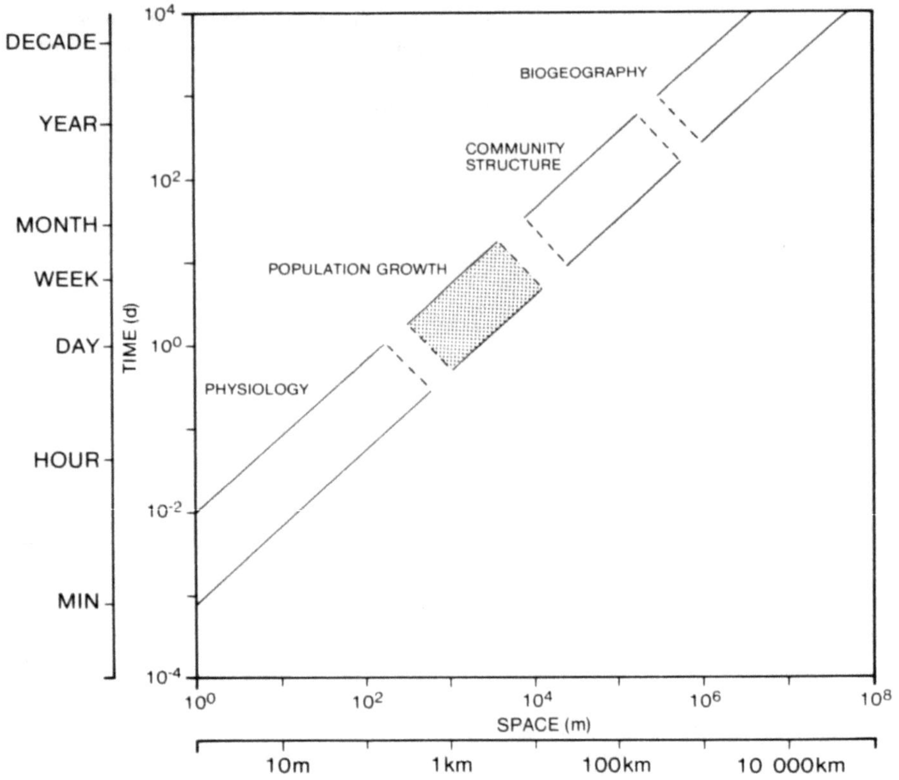

Fig. 4. Schematic representation of the relationship between temporal and spatial scales of phytoplankton responses to environmental variation. Here and in Figs. 5 and 6 the process of population growth is indicated by *shading*

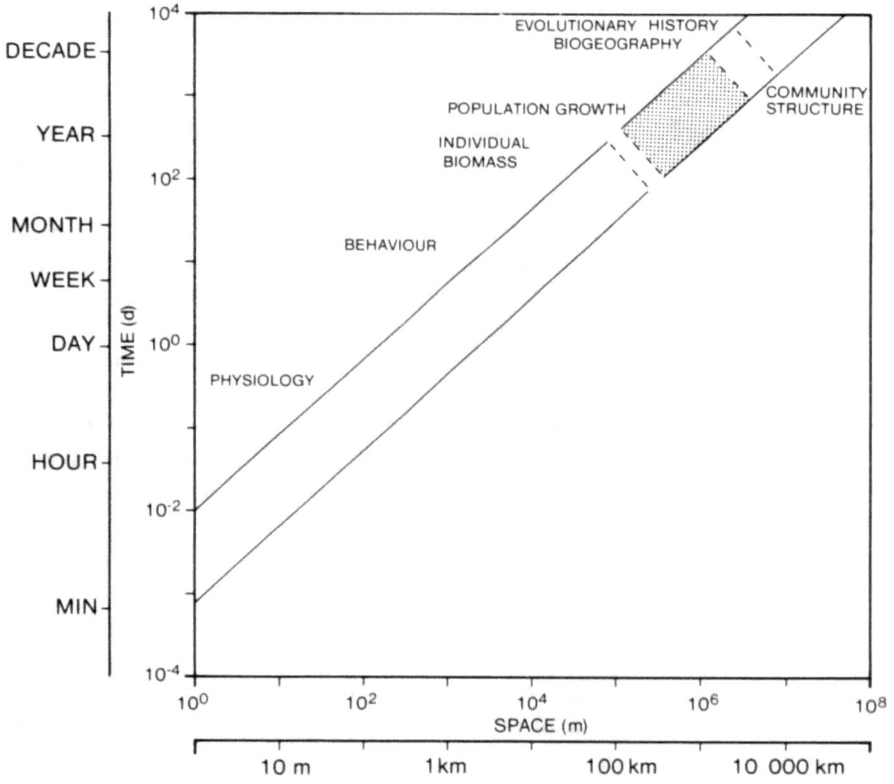

Fig. 5. Schematic representation of the relationship between temporal and spatial scales of predators' responses to environmental variation. *Shading* see Fig. 4

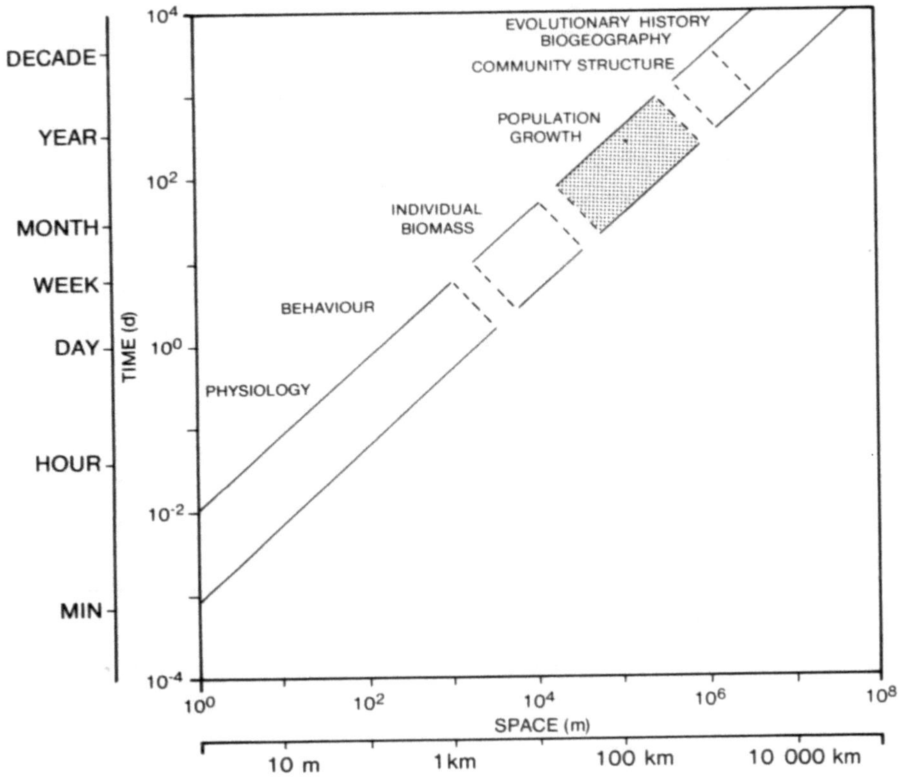

Fig. 6. Schematic representation of the relationship between temporal and spatial scales of krill responses to environmental variation. *Shading* see Fig. 4

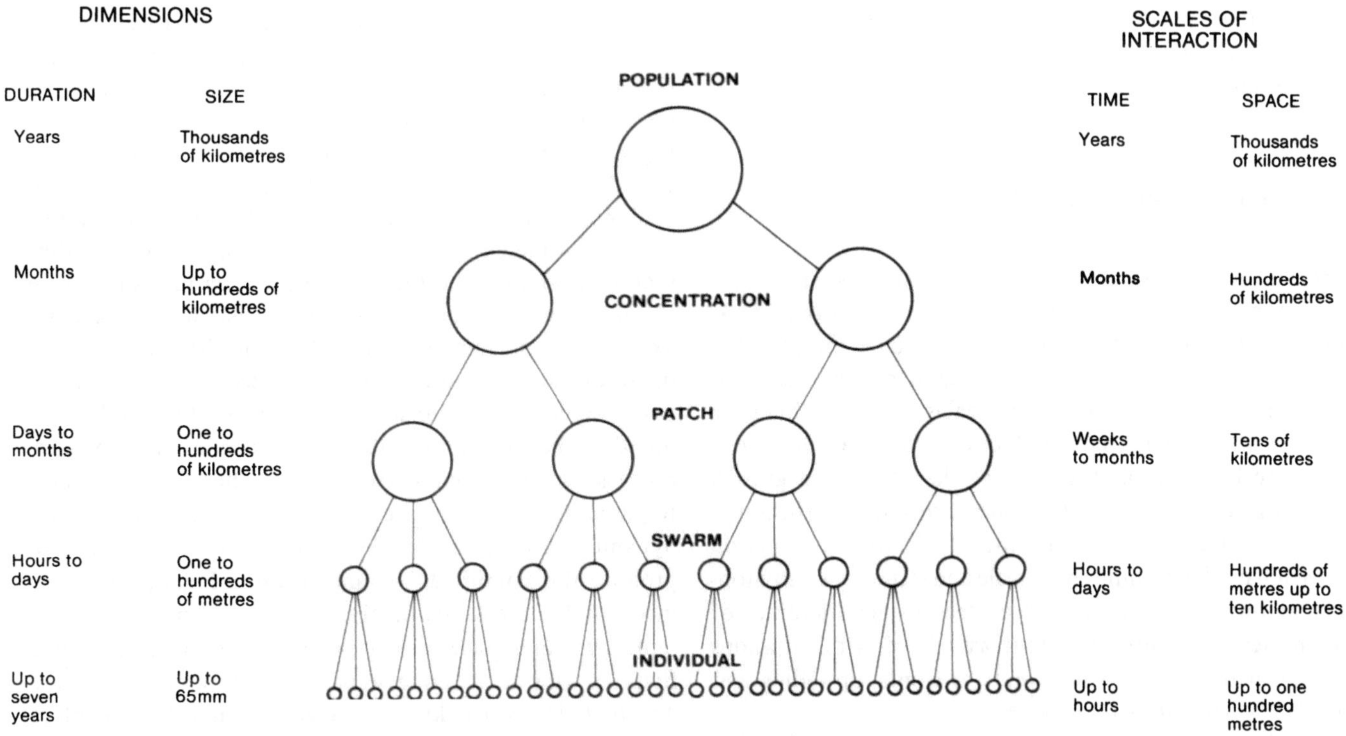

Fig. 7. Diagram showing a hierarchy of krill organisation based on successive levels of aggregation. The dimensions of the units and the scales over which they interact with the environment are shown

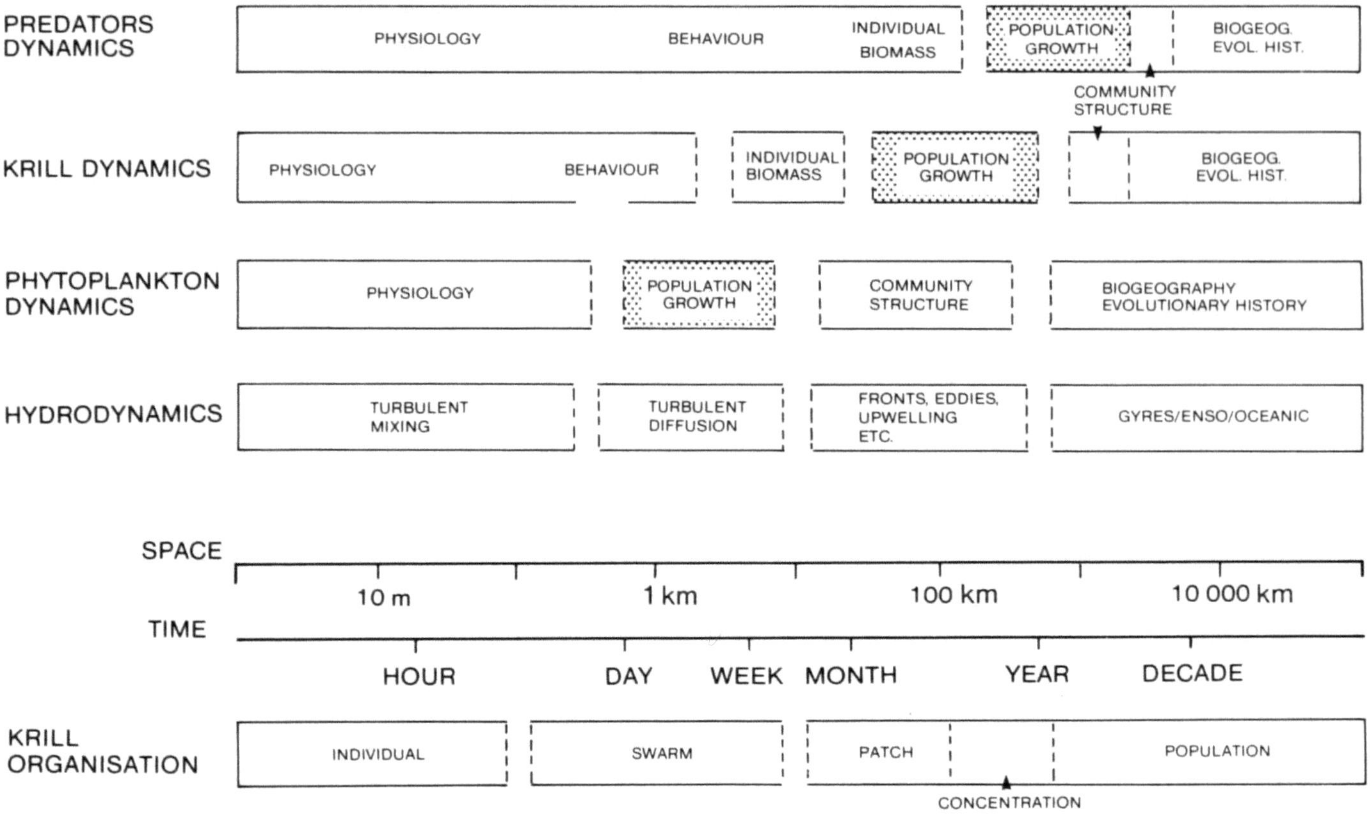

Fig. 8. Comparison of the scales of operation of processes in the hydrodynamic environment and three groupings of the biological components (cf. Figs. 2 and 4–6). The corresponding scales for five levels of krill organization (Fig. 7) are also shown

tion: the individuals forming swarms, which then occur in patches, these may then become concentrations in particular geographical regions; the population is composed of all these levels.

6 Interactions in Time and Space

Interactions can occur between processes operating over similar scales and the major point to note from our breakdown of the ecosystem (Figs. 2 and 4–6) is that, for different organisms, a particular biological process has characteristic temporal and spatial scales and can therefore interact with environmental fluctuations occurring at different scales. For phytoplankton, the population growth process is associated with the scale of turbulent diffusion in the hydrodynamic environment. For krill, population growth has a larger time-space scale and responds to environmental fluctuations at a scale of seasons and hundreds of kilometres. This process is thus associated with variations due to such hydrodynamic events as fronts, upwelling and eddies occurring over these scales.

Furthermore, for a particular scale, organisms of different size will interact through biological processes appropriate to that scale. For example, the process of krill population growth corresponds to the scale of changes in the phytoplankton community structure. Bearing in mind the importance of aggregations in the biology of krill, we need to incorporate our hierarchy of krill population organization (Fig. 7) in the time-space framework. We have already assigned time and space scales to the levels of krill population organization, each of which can interact with environmental fluctuations occurring at corresponding scales. If we map the time and space-scale relationship for the different components of the ecosystem on to a single axis, we can examine the interactions of different processes operating at the same scale. This axis is defined by the curve describing the relationship between time and space, noted as being common to hydrodynamic and biological processes (Figs. 2 and 4–6). We have, therefore, been able to combine the time-space scale relationships for hydrodynamic, phytoplankton, krill and predator processes to produce the simplified diagram shown in Fig. 8, which also incorporates the levels of organization of krill. This represents our time-space scale view of the ecosystem. We will first use this to describe some of the interactions of the different levels of krill organization and then explore specific processes such as grazing, predation and the modifying effects of sea ice.

6.1 Different Levels of Krill Organization

Individual krill can probably respond to environmental variations occurring over minutes to an hour and up to tens of metres in space, so turbulent mixing processes are likely to dominate their hydrodynamic environment at this scale with the responses of krill being physiological or behavioral. At the swarm level, the major hydrodynamic processes are turbulent diffusion and the biological responses of krill over this scale are mainly behavioral. Thus, the formation and persistence of krill swarms probably result from a combination of hydrodynamics and krill behavior.

At the levels of patches and concentrations, interactions occur at or just above the scale of the krill population growth process, and are associated with the scale of fronts, upwelling and eddies in the hydrodynamic environment. This suggests that such physical features may be particularly important in the formation and maintenance of patches and concentrations of krill. This is consistent with the attempts to classify krill aggregations (Mauchline 1980; Kalinowski and Witek 1985), and the different forms of aggregation observed may be due to the particular features they are associated with. So patches associated with eddies in the Scotia Sea may have a very different form to those associated with upwelling events or frontal systems.

The krill population level in the hierarchy of organization is associated with interannual and longer-term events and is thus likely to respond to the hydrodynamic processes varying at these scales, such as climatically driven systems. Thus, changes at this level may be reflected in changes in zooplankton community structure and longer-term processes. So although the population growth process is affected at the level of patch and concentration, changes in the overall population abundance occur above this scale. Below the scale of the population growth process, behavioral responses occur with environmental fluctuations; above it, changes occur in the biomass and zooplankton community structure. Therefore, hydrodynamic processes are likely to be the major determinant of krill biology above this scale (seasonal to annual and hundreds to thousands of kilometres). This scale and above will, therefore, be of critical importance in monitoring changes in the biomass of krill and in the rates of change of the population.

6.2 Grazing Interactions

To exploit variation in the food supply, krill must be able to detect the pattern of variation. There is a critical scale for phytoplankton patch formation where net growth overcomes the dispersive action of horizontal turbulent diffusion (Wroblewski and O'Brien 1976; Legendre and Demers 1984). This scale greatly exceeds that for the responses of individual krill, but is associated with the scale of krill swarm formation. This is therefore the scale at which major interaction between krill and its food supply occurs. Thus, at the swarm level, integration of the responses of the individual allow the unit (the swarm) to detect fluctuations in the biomass of the food supply. A behavioral response is possible at the swarm level to allow searching for food in a patchy environment. This is consistent with the view of Miller and Hampton (ms), who suggested that swarming was a group-feeding strategy (see also Hamner et al. 1983; Weber and El-Sayed 1985; Weber et al. 1986). The behavioral responses elicited might include vertical and horizontal migrations or the breakdown of swarm structure. Below the swarm scale krill as individuals will also encounter food supply fluctuations; however, hydrodynamic factors and the associated scale of the phytoplankton responses will be perceived as noise rather than as an interpretable signal.

6.3 Predator-Prey Interactions

Krill is a significant prey item for a wide variety of predators in the Southern Ocean. A major point from our time-space framework is that interactions between krill and its different groups of predators will occur at different scales. However, for predators of the same size, the scales of interaction will be similar. Thus, we would expect similar-sized predators to exploit prey over scales of the same order of magnitude. However, sperm and baleen whales are a similar size, yet they appear to exploit animals with very different fundamental time units, squid and krill respectively. This is resolved in our treatment of ecosystem structure where they are feeding over the same scales. It is krill swarms rather than individuals which are exploited by baleen whales and swarms occur over similar scales to squid and are therefore available within the feeding scales of whales.

In the early days of ideas of simple food chains, the true role of krill and the structure and complexity of the system was not recognized. However, krill are a central element of the Southern Ocean food web and the above consideration of predator-prey interactions in time and space explains why. It is not just that krill occur in large numbers as individuals, but also that it is available as a prey item over a wide range of scales because of the formation of aggregations (see also Brodie et al. 1978). This has important implications for the possible interactions within the ecosystem. For example, the release of predator pressure at one level may not make that resource fully available to predators operating at another scale. The ability to exploit the resource at the level released will depend on the degree of niche overlap between predators. So the decrease in large baleen whale numbers (Gambell 1985) may have allowed those predators with the greatest niche overlap, exploiting krill at a similar scale, to exploit the resource most successfully (cf. Payne 1977; Laws 1984, 1985; Bengtson 1985; Croxall et al. this Vol.). Our scale view predicts that such predators would be larger animals such as minke whales and seals rather than smaller animals such as fish and squid.

In the case of investigations of seabird biology, the importance of scale has already been recognized in analyses of animal distribution and correlations with variation in the environment (Schneider and Piatt 1986; Hunt and Schneider 1987). Schneider and Piatt (1986) found significant scale-dependent correlations between seabirds and schooling fish and a similar approach in the Southern Ocean would be of value in analyses of predator-krill interactions.

6.4 Effects of Sea Ice on Structure and Functioning of the Ecosystem

Sea ice is a distinctive feature of the polar oceans and it has a profound influence on the hydrography of the Southern Ocean. The presence of sea ice modifies the pelagic environment in two fundamental ways: as a surface layer it isolates the hydrodynamic environment under it, while at the ice edge there is a transition zone with unique hydrodynamic and biological characteristics.

Under the ice, wind-forced turbulence is damped, isolating the upper water column from the atmosphere and changing the scale of turbulent motion in the surface water. In contrast to the open ocean, where individual krill are unable to detect gradients in the food supply, the under-ice habitat has very different scale characteristics. Here, individual krill may exploit patches of microalgae on the underside of the ice for long enough to detect real differences in food availability. Furthermore, the physical processes implicated in the formation of patches and concentrations of krill may be absent. If this is so, then we would expect to find different mechanisms for the formation of aggregations, if indeed they occur at all. The possible absence of the large-scale hydrodynamic influences on the formation of patches or concentrations of krill may result in the absence of krill swarms beneath the ice.

The ice-edge zone is a region of great variability at the northerly margin of the sea ice cover, up to 200 km in width (Maykut 1985), which is often associated with a frontal system (see Niebauer and Alexander 1985). Biological activity in this zone is enhanced and phytoplankton can attain a high biomass in the water newly exposed by the retreating ice edge (El-Sayed and Taguchi 1981; Nelson et al. 1985; Wilson et al. 1986). The hydrodynamic environment associated with the ice-edge zone will have scale characteristics which differ from those of both the under-ice and ice-free systems. As when under the ice, individual krill may be able to detect the steeper environmental gradients which are also characteristic of the ice-edge zone.

The ice-edge zone is a highly dynamic area and can be envisaged as a frontal system which tracks between high and low latitudes of the Southern Ocean on a seasonal cycle. This means that approximately 50% of the area of the Southern Ocean is affected by the transit of the ice-edge zone each year. This has important implications for interactions within the ecosystem which occur at this scale. The seasonal catch data for pelagic whaling (Mizroch et al. 1985) indicated that areas of high baleen whale abundance followed the ice-edge southwards during the summer recession of the ice cover. On a larger scale there is interannual variation in the extent of the ice cover (NOCD 1985) which will affect the higher levels of organization such as population levels, community structure and biogeography.

7 Implications of the Scale Approach

In this chapter we have produced a conceptual framework for the analysis of the Southern Ocean ecosystem in terms of time and space. An important point is that it is possible to consider individual species and interactions within such a framework (cf. Southwood 1977).

It is also possible to consider what processes are associated at particular scales and can thus interact, and consequently what processes can be investigated by observation at any particular scale. Once a system of interest has been defined there are constraints on survey design (Haury et al. 1978; Steele 1978; Platt et al. 1981; Legendre and Demers 1984; Field et al. 1985; Harris 1986). The maintenance of a scale view of the ecosystem in any survey design is extremely important. Furthermore, experimental attempts to gain an understanding of krill population dynamics should be aimed at scales above the fundamental time unit for krill (population growth period), i.e. at the patch level of organization and above where hydrodynamic processes act as major determinants of krill distribution and abundance. This is consistent with the phenomenon discussed by Priddle et al. (this Vol.), where hydrodynamic features of seasonal to annual scale affect overall distribution and abundance of krill. An important implication of the scale concept is that localized predator population processes will not provide an overall index of krill abundance unless the predators forage over sufficiently large scales. This scale view of the ecosystem has also, therefore, important consequences for resource management with its requirements to monitor changes within the ecosystem. Harvesting particular species, such as whales, which exploit prey items over particular scales, may have resulted in changes in ecosystem structure and functioning which would not be predicted by simple ideas of the food web (May et al. 1979; Beddington and May 1980, 1981). Development of models which take the scale of interactions into account will provide a greater understanding of ecosystem structure and function and allow prediction of the effects of disturbance of the system. To do this, further detail is required on the scales of variation in the Antarctic marine ecosystem. Such data can be provided by spectral analysis of transect data over small scales (Weber et al. 1986), while larger scale synoptic data are more easily derived from remote sensing and long series of in situ measurements.

8 Conclusions

1. The relationships between temporal and spatial scales are relevant to the investigation of ecosystem structure.
2. For a process at a particular scale, interactions with the environment or another process occur at similar scales.
3. For different-sized organisms a particular biological process occurs at different scales.
4. An environmental gradient can only be detected by organisms, or levels of organization of those organisms, which operate at or above that scale. For instance, krill swarms can detect gradients in the food supply that individuals cannot.
5. The presence of sea-ice changes the characteristics of the environment over the complete range of scales.
6. Krill occupy a central role in the Southern Ocean food web, not only because of their large numbers but also because of their availability to predators over a wide range of scales. This is due to their occurrence in aggregations which can be regarded as levels of organization different from that of the individual.

References

Allen TFH, Starr TB (1982) Hierarchy: perspectives for ecological complexity. Univ Chicago Press, Chicago, 310 pp

Amos AF (1984) Distribution of krill (*Euphausia superba*) and the hydrography of the Southern Ocean: large-scale processes. J Crustacean Biol 4 (Spec No 1):306–329

Beddington JR, May RM (1980) Maximum sustainable yields in systems subject to harvesting at more than one trophic level. Math Biosci 51:261–281

Beddington JR, May RM (1982) The harvesting of interacting species in a natural ecosystem. Sci Am 247:42–49

Bengtson JL (1985) Review of Antarctic marine fauna. Rep SC-CAMLR-III/BG/13 to CCAMLR Sci Comm, 226 pp

Bogdanov MA, Oradvoskiy SG, Solyankin V, Khvatskiy NV (1969) On the frontal zone in the Scotia Sea. Oceanology 9:777–783

Bonner JT (1965) Size and cycle, an essay on the structure of biology. Princeton Univ Press, Princeton, 219 pp

Bowden KF (1970) Turbulence II. Oceanogr Mar Biol Annu Rev 8:11–32

Brodie PF, Sameoto DD, Sheldon RW (1978) Population densities of euphausiids off Nova Scotia as indicated by net samples, whale stomach contents, and sonar. Limnol Oceanogr 23:1264–1267

Bryden HL (1983) The Southern Ocean. In: Robinson AR (ed) Eddies in marine science. Springer, Berlin Heidelberg New York, pp 265–288

Caddy JF (1986) Modelling stock-recruitment processes in Crustacea: some practical and theoretical perspectives. Can J Fish Aquat Sci 43:2330–2344

Calder WA (1985) Size and metabolism in natural systems. In: Ulanowicz RE, Platt T (eds) Ecosystem theory for biological oceanography. Can Bull Fish Aquat Sci 213:65–75

Carleton AM (1988) Antarctic sea ice–atmosphere signal of the Southern Osciallation in the Weddell Sea, Antarctica. J Climat 1:379–388

Clarke A (1985) Energy flow in the Southern Ocean food web. In: Siegfried WR, Condy PR, Laws RM (eds) Antarctic nutrient cycles and food webs. Springer, Berlin Heidelberg New York Tokyo, pp 573–580

Colton MT, Chase RRP (1983) Interaction of the Antarctic Circumpolar Current with bottom topography: an investigation using satellite altimetry. J Geophys Res 88:1825–1843

El-Sayed SZ, Taguchi S (1981) Primary production and standing crop of phytoplankton along the ice-edge in the Weddell Sea. Deep-Sea Res 28:1017–1032

Elton C (1927) Animal ecology. Macmillan, New York, 207 pp

Everson I (1977) The living resources of the Southern Ocean. South Ocean Fish Serv Programme. Food and Agriculture Organisation of the United Nations, Rome. FAO GLO/SO/77/1, 156 pp

Field JG, Wolff FV, Allen PM, Fasham MJR, Flos J, Frontier S, Kay JJ, Silvert W, Trainor L (1985) Ecosystem theory in relation to unexploited marine ecosystems. (Working group report). In: Ulanowicz RE, Platt T (eds) Ecosystem theory for biological oceanography. Can Bull Fish Aquat Sci 213:241–247

Gambell R (1985) Birds and mammals – Antarctic whales. In: Bonner WN, Walton DWH (eds) Key environments – Antarctica. Pergamon, Oxford, pp 223–241

Hamner WM, Hamner PP, Strand SW, Gilmer RW (1983) Behavior of Antarctic krill, *Euphausia superba:* chemoreception, feeding, schooling and molting. Science 220:433–435

Harris GP (1980) Temporal and spatial scales in phytoplankton ecology. Mechanisms, methods, models, and management. Can J Fish Aquat Sci 37:877–900

Harris GP (1986) Phytoplankton ecology: structure, function and fluctuation. Chapman and Hall, London, 384 pp

Haury LR, McGowan JA, Wiebe PH (1978) Patterns and processes in the time-space scales of plankton distributions. In: Steele JH (ed) Spatial pattern in plankton communities. Plenum, New York, pp 277–328

Heywood RB, Priddle J (1987) Retention of phytoplankton by an eddy. Cont Shelf Res 7:937–955

Hunt GL, Schneider DC (1987) Scale dependent processes in the physical and biological environment of marine birds. In: Croxall JP (ed) Seabirds: feeding ecology and role in ecosystem. Cambridge Univ Press, Cambridge, pp 7–41

Kalinowski J, Witek Z (1985) Scheme for classifying aggregations of Antarctic krill. BIOMASS Handb 27:9 pp

Laws RM (1984) Seals. In: Laws RM (ed) Antarctic ecology. Academic Press, London New York, pp 617–711

Laws RM (1985) The ecology of the Southern Ocean. Am Sci 73: 26–40

Legeckis R (1977) Ocean polar front in the Drake Passage – satellite observations during 1976. Deep-Sea Res 24:701–704

Legendre L, Demers S (1984) Towards dynamic biological oceanography and limnology. Can J Fish Aquat Sci 41:2–19

Lubimova TG, Makarrov RR, Maslennikov VV, Shevtsov VV, Shust KV (1980) Results of Soviet research into assessment of stocks, ecology and the role of krill in the Antarctic ecosystem. Trudy VNIRO, Moscow, 87 pp (in Russian)

Lutjeharms JRE, Walters NM, Allanson BR (1985) Oceanic frontal systems and biological enhancement. In: Siegfried WR, Condy PR, Laws RM (eds) Antarctic nutrient cycles and food webs. Springer, Berlin Heidelberg New York Tokyo, pp 11–21

Macaulay MC, English TS, Mathisen OA (1984) Acoustic characterisation of swarms of Antarctic krill (*Euphausia superba*) from Elephant Island and Bransfield Strait. J Crustacean Biol 4 (Spec No 1):16–44

Mackintosh NA (1967) Maintenance of living *Euphausia superba* and frequency of moults. Nor Hvalfangst Tid 5:97–102

Mackintosh NA (1972) Life cycle of Antarctic krill in relation to ice and water conditions. Discovery Rep 36:1–94

Makarov RR (1972) Life cycle and distribution pattern of *Euphausia superba* Dana. Trudy VNIRO 77:85–92 (in Russian)

Mauchline J (1980) Studies on patches of krill, *Euphausia superba* Dana. BIOMASS Handb 16:36 pp

May RM, Beddington T, Clark CW, Holt SJ, Laws RM (1979) Management of multispecies fisheries. Science 205:267–277

Maykut GA (1985) The ice environment. In: Horner RA (ed) Sea ice biota. CRC Press, Boca Raton, pp 21–82

Miller DGM, Hampton I (ms) A review of the biology and status of the Antarctic krill (*Euphausia superba* Dana). Sea Fish Res Inst, Roggebaai, S Afr, 261 pp

Mizroch SA, Rice DW, Bengtson JL, Larson SW (1985) Preliminary atlas of balaenopterid whale distribution in the Southern Ocean based on pelagic catch data. Rep SC-CAMLR-IV/BG/21 to CCAMLR Sci Comm, 9 pp

Nelson DM, Gordon LI, Smith WO (1985) Phytoplankton dynamics of the marginal ice zone of the Weddell Sea, November and December 1983. Antarct J US 19(5):105–107

Niebauer HJ, Alexander V (1985) Oceanographic frontal structure and biological production at an ice edge. Cont Shelf Res 4: 367–388

NOCD (1985) Sea ice climate atlas, vol 1. Antarctic. Naval Oceanography Command Detechment, Asheville, 131 pp

Payne MR (1977) Growth of a fur seal population. Philos Trans R Soc London Ser B 279:67–79

Peters RH (1983) The ecological implications of body size. Cambridge Univ Press, Cambridge, 329 pp

Platt T (1985) Structure of the marine ecosystem: its allometric basis. In: Ulanowicz RE, Platt T (eds) Ecosystem theory for biological oceanography. Can Bull Fish Aquat Sci 213:55–64

Platt T, Mann KH, Ulanowicz RE (1981) Mathematical models in biological oceanography. UNESCO, Paris, 157 pp

Priddle J, Hawes I, Ellis-Evans JC, Smith TJ (1986) Antarctic ecosystems as habitats for phytoplankton. Biol Rev 61:199–238

Schneider DC, Piatt JF (1986) Scale-dependent correlation of seabirds with schooling fish in a coastal ecosystem. Mar Ecol Prog Ser 32:237–246

Sheldon RW, Prakash A, Sutcliffe WH (1972) The size distribution of particles in the ocean. Limnol Oceanogr 17:327–340

Southwood TRE (1977) Habitat, the templet for ecological strategies. J Anim Ecol 46:337–365

Steele JH (1978) Some comments on plankton patches. In: Steele JH (ed) Spatial pattern in plankton communities. Plenum, New York, pp 1–20

Thorpe SA (1985) Small-scale processes in the upper ocean boundary layer. Nature (London) 318:519–522

Tranter DJ (1976) Herbivore production. In: Cushing DJ, Walsh JJ (eds) The ecology of the seas. Blackwell, London, pp 186–224

Tranter DJ (1982) Interlinking of biological processes in the Antarctic Ocean. Oceanogr Mar Biol Annu Rev 20:11–35

Watkins JL (1986) Variations in the size of Antarctic krill, *Euphausia superba* Dana, in small swarms. Mar Ecol Prog Ser 31:67–73

Watkins JL, Morris DJ, Ricketts C, Priddle J (1986) Differences between swarms of Antarctic krill and some implications for sampling populations. Mar Biol 93:137–146

Weber LH, El-Sayed SZ (1985) Spatial variability of phytoplankton and the distribution and abundance of krill in the Indian sector of the Southern Ocean. In: Siegfried WR, Condy PR, Laws RM (eds) Antarctic nutrient cycles and food webs. Springer, Berlin Heidelberg New York Tokyo, pp 284–293

Weber LH, El-Sayed SZ, Hampton I (1986) The variance spectra of phytoplankton, krill and water temperature in the Antarctic Ocean south of Africa. Deep-Sea Res 33:1327–1343

Webster JR (1979) Hierarchical organisation of ecosystems. In: Halfon E (ed) Theoretical systems ecology. Academic Press, London New York, pp 119–129

Wilson DL, Smith WO, Nelson DM (1986) Phytoplankton bloom dynamics of the western Ross Sea ice edge. 1, Primary productivity and species-specific production. Deep-Sea Res 33: 1375–1387

Wroblewski JS, O'Brien JJ (1976) A spatial model of phytoplankton. Mar Biol 35:161–175

Interannual Variability in Estimated Primary Productivity of the Antarctic Marginal Ice Zone

W. O. SMITH Jr.[1], N. K. KEENE[2], and J. C. COMISO[3]

Summary. The significance of marginal ice zones as loci for primary productivity has been firmly established in recent years, yet little data exist on the variations within these areas over long time periods. Interannual variations in primary productivity of the marginal ice zone were calculated by combining actual field data on primary productivity, remote sensing data on ice position and concentrations, and a conceptual model of the dynamics of ice-edge phytoplankton blooms. The Ross Sea, the Weddell Sea, and the entire Southern Ocean were analyzed for the years from 1973–1986. Interannual variations in primary productivity within the marginal ice zone are shown to be substantial, with the maximum productivity being 50% greater than the minimum. In view of the importance of the marginal ice zone as a site of biological production and its temporal variations, an understanding of the causes of interannual variability of the ice-edge location and of biological processes within marginal ice zones is needed to assess the impact of these time/space variations on the food webs of the Southern Ocean.

1 Introduction

The marginal ice zone (the region in which a transition from completely ice-covered waters to waters uninfluenced by effects of pack ice) is now widely recognized as a major site of biological activity. Large accumulations of phytoplankton (Alexander and Niebauer 1981; Rey and Loeng 1985; S.L. Smith et al. 1985; W.O. Smith and Nelson 1985), zooplankton (Marr 1962), seabirds (Ainley and Jacobs 1981; Fraser and Ainley 1986) and marine mammals (Laws 1985) are associated with the marginal ice zone. Accumulations of higher trophic levels (e.g. birds, marine mammals) most likely are attributable to increased food resources and a reduction in predation provided by the physical presence of ice (Ainley and Jacobs 1981). Increased phytoplankton biomass at the ice edge is regulated by environmental factors, which are a function of the specific geographic location. However, in the Southern Ocean the dominant feature controlling phytoplankton growth and accumulation is water column stability (W.O. Smith and Nelson 1985).

The blooms of phytoplankton at ice edges in the Antarctic are in large part created by the vertical stability induced by the seasonal melting and retreat of the pack ice (W.O. Smith and Nelson 1985). Given that pack ice in the Southern Ocean is ca. 1 m thick, complete melting of the ice at the surface would reduce the surface salinity from 35 per mil to 33.62 per mil (assuming that the pack ice salinity is 6 per mil and the meltwater is distributed throughout the upper 20 m; Ackley et al. 1979). At −1°C the decrease in sigma-t of 1.12 units would create a highly stable surface layer which would be greatly resistant to vertical mixing, thereby providing a near optimal environment for phytoplankton growth. This is especially true since nutrients never appear to be limiting to growth in the Southern Ocean (Nelson and W.O. Smith 1986) and the daily integrated radiant flux during the Antarctic summer is greater than that experienced at the equator (Holm-Hansen et al. 1977). Therefore, during the austral summer, phytoplankton growth in stabilized regimes can be expected to be near the maximum predicted on the basis of temperature alone (Eppley 1972).

Because of the latitudinal range of the area covered and uncovered by the seasonal ice retreat and advance in the Southern Ocean, it is impractical to design observational studies to quantify the annual cycle of biogenic matter production in the marginal ice zones. This is especially true in view of the fact that marginal ice zones are mesoscale features which often are not well represented in large-scale oceanographic investigations. In one study conducted in the austral summer, the average productivity of a Ross Sea ice-edge bloom was established to be 962 mg $C/m^2/d$ (Wilson et al. 1986). The enhanced carbon assimilation was confined to the region of reduced density created by the retreating ice, which extended ca. 250 km from the ice edge. Nelson et al. (1987) and W.O. Smith and Nelson (unpublished) investigated the growth and productivity within the marginal ice zone in the Weddell Sea during austral spring, and found that primary productivity averaged 572 mg $C/m^2/d$. The enhanced biomass again extended ca. 250 km from the ice edge. Other data have been collected within the marginal ice zone, but no other studies have systematically described the spatial variations which occur within the region.

1 Botany Department and Graduate Program in Ecology, University of Tennessee, Knoxville, TN 37996, USA

2 Graduate Program in Ecology, University of Tennessee, Knoxville, TN 37996, USA

3 Laboratory for Oceans, NASA Goddard Space Flight Center, Greenbelt, MD 20771, USA

Antarctic Ocean and Resources Variability
(ed. by D. Sahrhage)

In order to predict the magnitude of the contribution of ice-edge phytoplankton blooms to the annual productivity cycle of the Southern Ocean, W.O. Smith and Nelson (1986) proposed a simple model to estimate the impact of blooms in the marginal ice zone both in one localized area and for the entire Southern Ocean. The model used the measured productivity values for the Ross and Weddell Seas, and calculated overall productivity as a function of ice retreat. Their results suggested that estimates of primary productivity for the entire Southern Ocean should be increased by 60% to include ice-edge blooms. Furthermore, they found that ice-edge-related productivity in the Ross Sea accounted for approximately 74% of the annual productivity at one location. In the Weddell Sea a similar calculation showed that 67% of the yearly productivity was associated with the ice edge. These figures agree with the nutrient depletion model described by Jennings et al. (1984) using data from the Weddell Sea. Despite the lack of quantitative descriptions of ice-edge-related processes, it is clear that marginal ice zones are very important features in the annual cycle of biogenic matter production over the ice-covered waters of the Southern Ocean.

Significant interannual variability occurs in all ecosystems which are coupled tightly to local physical processes. Examples from marine systems include upwelling centers (Barber and R.L. Smith 1981), intertidal communities (Dungan et al. 1982), the Bering Sea (Niebauer 1980), the Fram Strait (W.O. Smith et al. 1987) and coral reefs (Woodley et al. 1981). However, little information exists on the year-to-year variations in primary productivity within the Southern Ocean. More specifically, nothing is known concerning the yearly variations of phytoplankton growth within marginal ice zones. Because significant interannual variations occur in the ice cover and concentration in all sectors of the Southern Ocean (Zwally et al. 1983a,b), it can be expected that similar variations would occur in biogenic matter production in marginal ice zones. In this paper we estimate the extent of interannual variations of primary productivity within the Antarctic marginal ice zone using a variation of the method of W.O. Smith and Nelson (1986) in conjunction with observed interannual variations of the Antarctic ice cover. Predicted variations in the productivity of the entire Southern Ocean as well as the Ross and Weddell Seas are computed to estimate the impact of interannual variability within these regions.

2 Methods and Data

2.1 Assumptions Within the Model

To predict the magnitude of interannual variations in primary productivity within the marginal ice zone of the Southern Ocean, we calculated the ice-edge-related productivity using the method of W.O. Smith and Nelson (1986). This procedure combines knowledge of temporal variations in ice concentration and distribution with estimates of areal productivity to calculate the productivity in large regions of the Antarctic. A number of assumptions are inherent in this model. The first is that *productivity within Southern Ocean marginal ice zones is controlled by the density regime, which in turn is regulated by the input of low density meltwater from the receding ice* (W.O. Smith and Nelson 1985, 1986). Since the Southern Ocean is never nutrient-limited (Sakshaug and Holm-Hansen 1984; Nelson and W.O. Smith 1986), light is of over-riding importance in determining the rate of phytoplankton growth; therefore, at the ice edge growth and accumulation can occur at maximal rates, since vertical mixing has been largely reduced by the low density meltwater input. The horizontal extent of the stabilized region is likely to be a function of the rate of meltwater input (which is creating stratification) and local wind conditions, horizontal flow, etc. (which dissipate the stratification). Two mesoscale studies in the Southern Ocean have determined that the stabilized region extends ca. 250 km from the solid pack ice (W.O. Smith and Nelson 1985; Nelson et al. 1987). However, in other marginal ice zones the spatial extent can be very different (e.g. Bering Sea; Alexander and Niebauer 1981; Fram Strait: S.L. Smith et al. 1985). In addition, spatial variations in phytoplankton biomass along the ice edge have been noted (Nelson et al. 1987), indicating that the effects of ice melt can be altered by other hydrographic conditions. In this study the horizontal extent of ice-edge blooms is assumed to be a constant 250 km; however, we recognize that variations undoubtedly occur and that they will impact the calculated productivity. Further observational work is needed to determine the spatial extents of ice-edge blooms in the Antarctic and their seasonal and geographic variations in a manner (and scale) similar to the variations of ice cover and concentration observed by Comiso and Zwally (1984).

A second assumption in this method is that *ice retreat occurs at a uniform rate throughout the austral spring and summer.* Clearly this assumption is not totally justified, in that ice retreat is minimal at the beginning and end of the austral summer (Zwally et al. 1983b; Comiso and Zwally 1984) and exhibits a large degree of local variation (Comiso and Sullivan 1986). But to a large degree the large-scale rates of retreat for the various sectors and the Southern Ocean as a whole are constant (and maximal) for most of the melt-back period (Zwally et al. 1983b; Comiso and Zwally 1984), and the influence of the early and late periods (when the rates of retreat are decreased relative to the entire season) is minimal.

2.2 Ice Concentration Data

The ice concentration data used in the model calculations were determined from US Navy-NOAA ice extent maps which have been produced on a biweekly basis since 1973. We used digitized values of areal ice extent from 1973

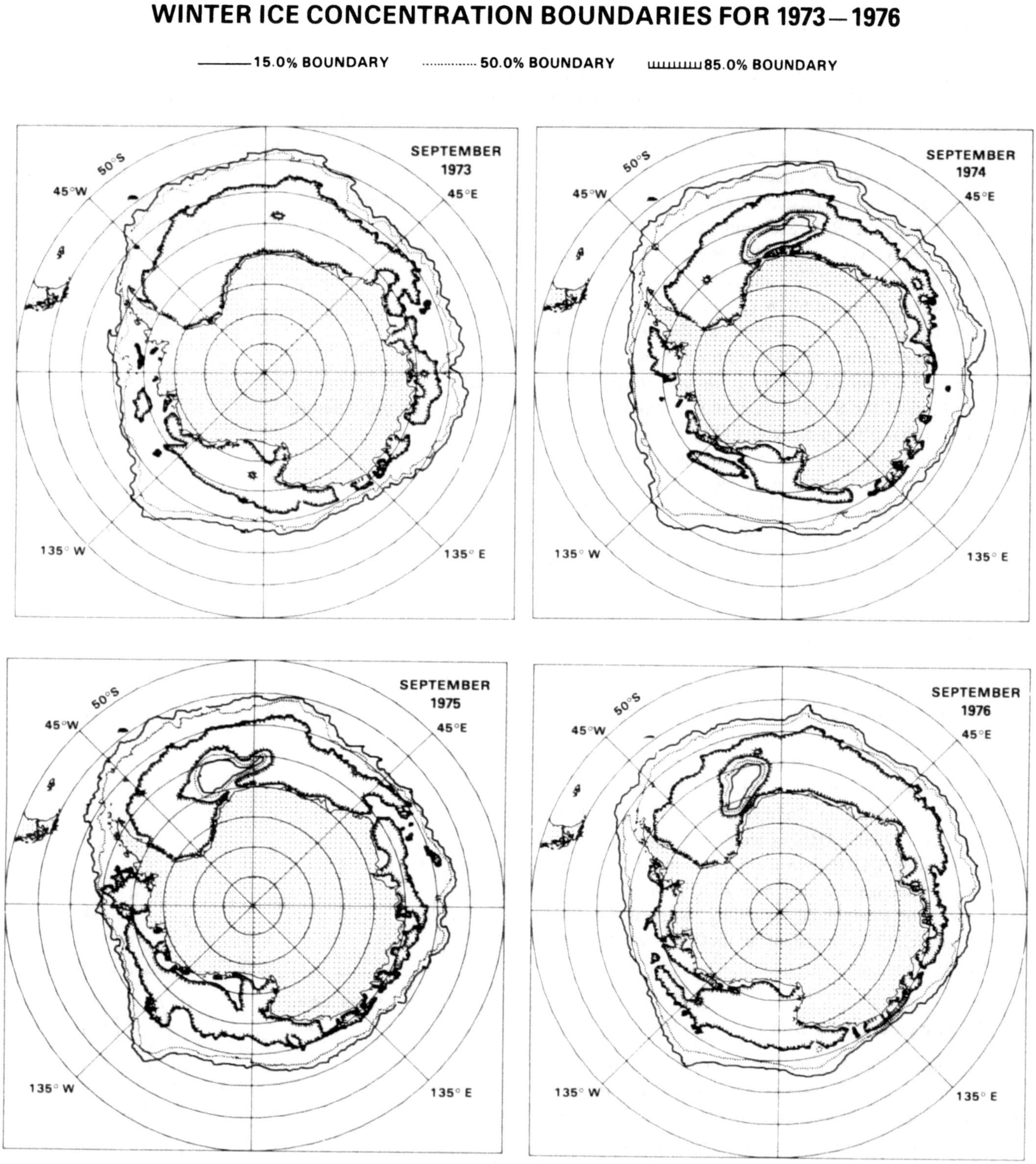

Fig. 1. Variations in ice cover from 1973–1976. Note the formation of a large polynya in the Weddell Sea area which dominated the interannual variations in ice cover. Polynyas of this magnitude have been absent since 1976

through 1981. These data have been described by Zwally et al. (1983b). The data were generated from various sources, including the NOAA Advanced Very High Resolution Radiometer (AVHRR), the Electronically Scanning Microwave Radiometer (ESMR), the Scanning Multifrequency Microwave Radiometer (SMMR), and the visible wavelength data from Landsat satellites and ship reports. Such a combined data set has the advantage of providing a longer temporal sequence than any one data source alone. A comparison of ESMR data with the Navy-NOAA maps for the periods when both were available showed little systematic difference between the two (Zwally et al. 1983a). In addition, data from SMMR images for the years 1978 through 1986 were used to analyze the more recent interannual ice extent variations. The latter data set is a significant improvement of ESMR because the variations due to physical temperature and emissivity are better taken into account. Also, it is derived more consistently through time than the Navy-NOAA images.

In this report three regions were analyzed: the entire Southern Ocean, the Ross Sea, and the Weddell Sea. The Ross Sea sector was arbitrarily defined as the region between 160°E and 130°W, and the Weddell Sea as the region between 20°E and 60°W (Zwally et al. 1983a). The Southern Ocean area was defined as the entire oceanic region surrounding the Antarctic continent which is covered by ice at its maximum extent. It should be noted that this definition of the Southern Ocean encompasses a smaller area than that often used (e.g. El-Sayed 1978). The area considered included only that encompassed by the retreat and advance of the seasonal pack ice, and did not include the effects of polynyas except those associated with the large Weddell Sea polynyas which occurred in the 1970's (see Fig. 1). In general, the total area which was influenced by polynyas was less than 10% of the total area covered by ice, except when both the polynyas and ice edge areas were increasing due to seasonal ice melt (usually only in December and January). The polynya-associated ice edge was not included in this analysis, so that computations of ice-edge-related productivity are conservative estimates of the total marginal ice zone biogenic production. A complete description of the seasonal and interannual variability of the Southern Ocean and its five sectors for the years 1975–1976 is given in Comiso and Zwally (1984) and Zwally et al. (1983b). Data of ice extent from 1978–1986 processed from SMMR data were also used (Fig. 2). Such data show conclusively that the 1970's were unusual in the amount of interannual variability, whereas the 1980's have been relatively constant in both ice concentration and extent.

2.3 Primary Productivity Data

Few mesoscale studies have been completed in the Southern Ocean to directly assess the productivity associated with the ice-edge phytoplankton bloom. Wilson et al. (1986) determined the mean productivity within the stabilized region of an ice edge in the Ross Sea to be 962 mg C/m^2/d, with a range from 31 to 1750 mg C/m^2/d. The median productivity for this study was 906 mg C/m^2/d. This indicates the magnitude of variations inherent within the ice-edge bloom. However, the greatest productivity and accumulations of phytoplankton biomass occurred in regions with the strongest vertical stability. W.O. Smith and Nelson (in preparation) studied the primary productivity associated with a retreating ice edge in the Weddell Sea and found a mean carbon assimilation rate of 571 mg C/m^2/d; again, the productivity within the bloom varied by two orders of magnitude (from 17 to 2089 mg C/m^2/d, with the median value being 447 mg C/m^2/d). In this analysis the two means are used to calculate sector-wide productivity, but it should be emphasized that given the wide range of productiyity within a bloom, a degree of uncertainty is introduced by using the mean values. An average between the Ross and Weddell means was used to calculate Southern Ocean ice-edge productivity; although we realize this is only a crude approximation, no data exist for other Antarctic marginal ice zones.

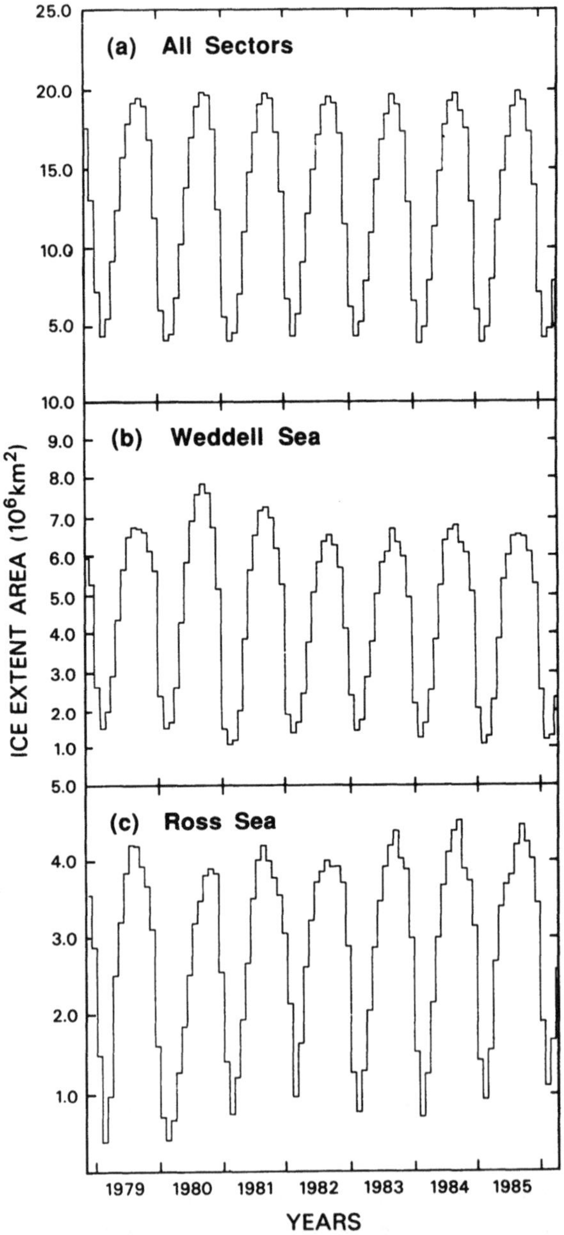

Fig. 2. Ice extent in the Southern Ocean, Ross Sea, and the Weddell Sea from 1979–1985 as determined from SMMR imagery. These years were characterized by interannual variations which were much smaller than those of the previous decade. Ice extent defined by the region of ⩾ 15% ice cover

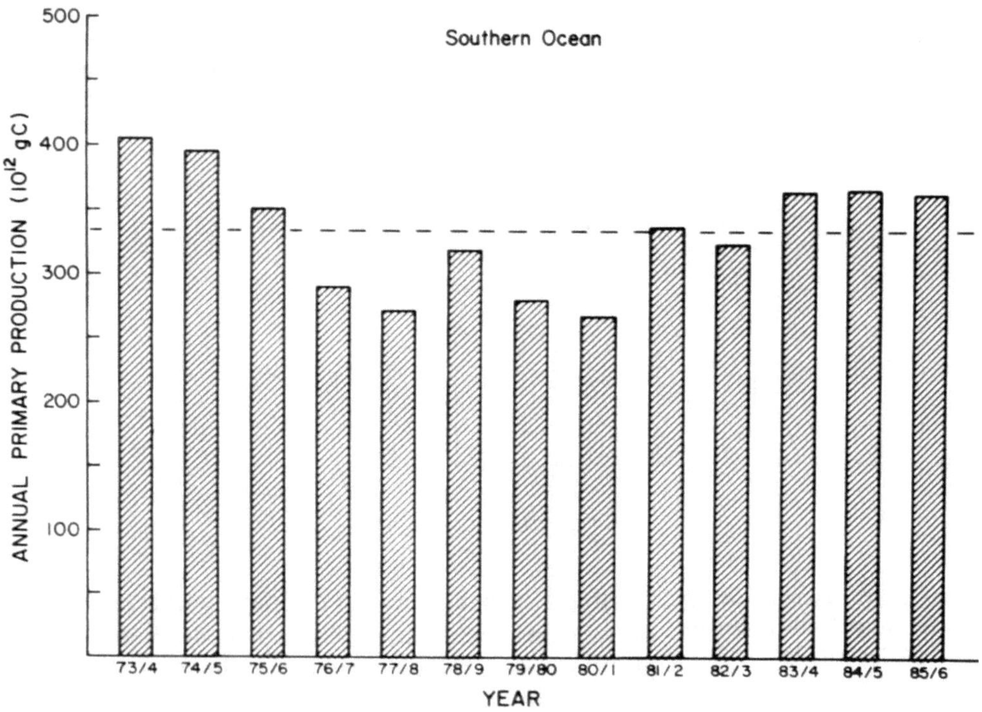

Fig. 3. The calculated yearly primary productivity within the marginal ice zone for the Southern Ocean. The *dashed line* represents the mean for the entire period

3 Results

A large degree of interannual variation in ice-edge primary productivity was observed from 1973 to 1986. The Southern Ocean as a whole had a calculated ice-edge productivity of 333.2 x 10^{12} g C, which is comparable to the 383.1 x 10^{12} g C estimated from the original model of W.O. Smith and Nelson (1986). What is noteworthy is the extent of interannual variability in the primary productivity within the marginal ice zone (Fig. 3). The maximum productivity occurred in the 1973/74 austral growing season (405.2 x 10^{12} g C), whereas a minimum was noted in the 1980/81 season (268.4 x 10^{12} g C). The coefficient of variation for the mean annual production was 13.7%. Given that the maximum productivity was 50% greater than the minimum, it is clear that such large variations in biogenic matter production could have a significant impact on the transfer of energy within the localized food web.

Variations within the Ross and Weddell sectors were also predicted. The maximum productivity in the Ross Sea occurred in the 1974/75 growing season (ice-edge related primary production equaled 125.3 x 10^{12} g C), whereas the minimum occurred in 1980/81 (71.0 x 10^{12} g C; Fig. 4). Mean annual productivity from 1973 to 1982 in the marginal ice zone was 99.7 x 10^{12} g C, with a coefficient of variation of 15.0%, and the maximum productivity was 76% greater than the minimum. For the Weddell Sea, maximum and minimum values occurred in the 1973/74

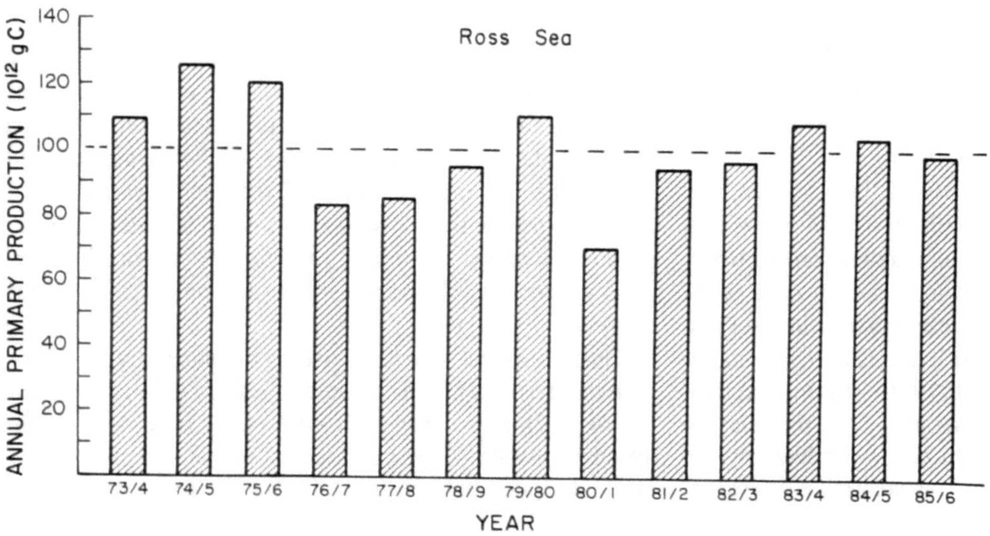

Fig. 4. The calculated yearly primary productivity within the marginal ice zone for the Ross Sea. The *dashed line* represents the mean for the entire period

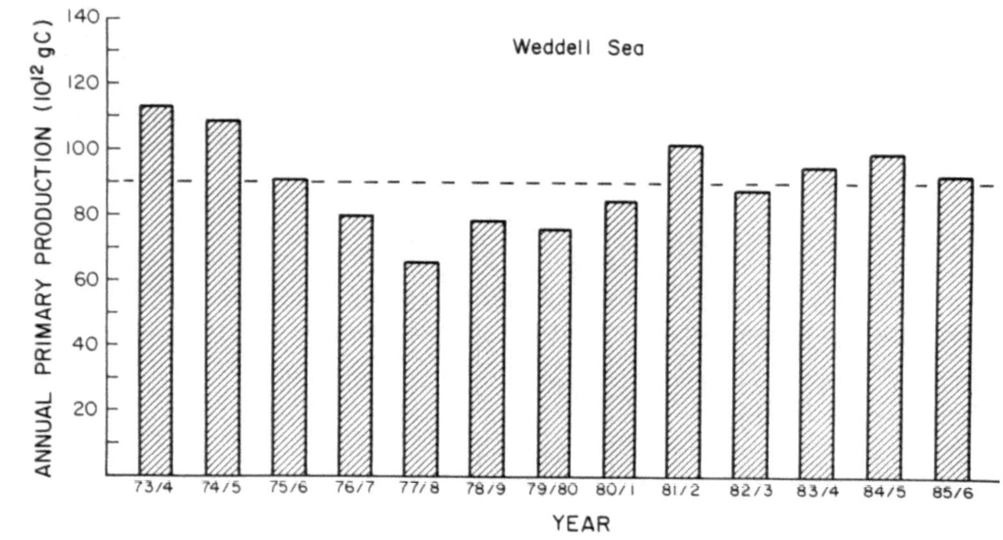

Fig. 5. The calculated yearly primary productivity within the marginal ice zone for the Weddell Sea. The *dashed line* represents the mean for the entire period

Table 1. Calculated interannual variations in primary productivity associated with the marginal ice zone for the entire Southern Ocean, and the relative contribution to the total productivity by the Ross Sea and Weddell Sea sectors

Year	Southern Ocean primary production (10^{12} g C)	Ross Sea contribution (%)	Weddell Sea contribution (%)
1973/74	405.2	26.9	27.8
1974/75	395.4	31.7	25.0
1975/76	349.6	34.6	25.9
1976/77	289.7	28.5	27.7
1977/78	271.7	30.9	24.4
1978/79	319.5	29.8	24.7
1979/80	280.8	39.1	27.1
1980/81	268.4	26.5	31.7
1981/82	338.0	28.2	30.2
1982/83	323.9	29.6	27.3
1983/84	363.6	30.1	26.1
1984/85	364.2	29.0	27.1
1985/86	361.9	27.5	25.5
Mean	333.3	30.1	26.9
SD	45.6		

and 1977/78 growing seasons and had productivities of 112.6 and 66.4 g C, respectively (Fig. 5). Therefore, the maximum was 70% greater than the minimum. The mean ice-edge productivity for the 8-year period was 86.1 × 10^{12} g C (coefficient of variation 16.8%).

The calculated yearly productivity of the Ross Sea, the Weddell Sea, and the entire Southern Ocean were not necessarily correlated. For example, maximal productivity was observed in the 1973/74 growing season for the Weddell Sea and the total Southern Ocean, whereas in the Ross Sea the maximum occurred 1 year later. The Ross Sea also showed a large increase in the 1979/80 season, but the Weddell Sea did not. This increase, as well as the relative contribution of each sector to the entire productivity of the Southern Ocean, was due not to a greater ice retreat in the Ross Sea, but to the fact that our available information on ice-edge productivity indicates that the Ross Sea marginal ice zone is more productive per unit area than that of the Weddell Sea. The percentage that each sector contributes to total productivity also varies interannually (Table 1); the mean contributions of the Ross Sea and Weddell Sea were 30.9 and 26.7%, respectively.

A strict error analysis, such as that described by Zwally et al. (1983a), cannot be constructed for these results, since the calculations are not the result of a mathematical description of a physical process. However, the uncertainty of our results can be described by using the variations observed within ice-edge blooms in the Weddell and Ross Seas and computing the expected annual productivity of the marginal ice zone. This was done for the year of maximum productivity of the Southern Ocean (1974). When one standard deviation above and below the mean (±47%) was used for the Ross Sea, the productivity of the ice edge ranged from 55.0 to 163.4 × 10^{12} g C/yr (the mean ice-edge contribution was 109.2 × 10^{12} g C/yr). Similarly, in the Weddell Sea the average ice-edge contribution was 112.6 × 10^{12} g C/yr, whereas when one standard deviation above and below the mean was used, the productivity ranged from 60.0 to 165.1 × 10^{12} g C/yr. Because productivity was greatest in 1974, these error calculations represent the largest potential error introduced by using mean productivity values. However, further work designed to increase our understanding of the spatial variations of primary productivity within ice-edge blooms and to produce a model to predict productivity within the marginal ice zone would greatly reduce the uncertainty in this analysis.

One additional source of error is the means by which ice melts and breaks up in the austral spring. Often ice retreat is not a simple, linear melt back (i.e. from a northern ice-edge towards the continent), but is punctuated by internal breaks and melting episodes of varying dimensions (see, for example, seasonal ice melt of Ross Sea; Zwally et al. 1983a). Such variations are not included in this analysis and result in underestimates of primary productivity by this method.

4 Discussion

Our results indicate that there exists a significant potential for interannual variability of primary production within the Southern Ocean's marginal ice zone. The variability is directly related to the variations observed in the extent and retreat of the seasonal pack ice (Zwally et al. 1983a,b). The maximum primary productivity associated with the marginal ice zone occurred in the 1973/74 growing season and was more than 50% larger than the minimum in ice-edge productivity (1980/81). Such a large variation in biogenic production undoubtedly plays an important role in regulating the flux and transformation of biogenic material. The exact role of this interannual variability within food web dynamics of the Southern Ocean, however, remains to be determined.

The interannual variability we estimated for the Southern Ocean (and for the Ross and Weddell Seas) does not appear to be as great as that observed in some pelagic systems. For example, Barber and R.L. Smith (1981) and Chavez and Barber (1987) compiled primary productivity data from the Peruvian coastal upwelling system for 10 separate years. We calculate a coefficient of variation for areal productivity of 37.1% from these data, much greater than that calculated for the marginal ice zone. The productivity off the Peruvian coast is regulated directly by the persistence of a basin-wide slope of the nitracline, the depth of which is in turn modulated by global air-ocean interactions. These interannual variations are dramatic in that large failures of commercially important species occur with direct economic effects (Barber and Chavez 1983). These anomalies, which result in decreases in productivity, are called ENSO (El Niño-Southern Oscillation) events; they greatly reduce the area of high productivity, although they do not necessarily reduce the maximum areal production which occurs (Chavez and Barber 1987). ENSO events are the physical forcing mechanism whereby interannual variability in biological processes can be analyzed. These periods also are strongly correlated with the ice cover in the Bering Sea (albeit with a substantial time lag; Niebauer 1984) and undoubtedly create interannual variations in the productivity of the Bering Sea. The teleconnection of the tropical ocean with the Bering Sea occurs via a movement of the Aleutian low pressure system which deepens and moves southward during the northern hemisphere winter in response to an ENSO event. Unfortunately, it is unknown what the nature (if any) and strength of a teleconnection between the tropic region's coupled atmosphere-ocean system and that of the Antarctic might be. If such a connection occurs, then it should have a direct impact on ice extent and areal coverage, which ultimately should influence primary productivity and the flux of material within the Southern Ocean.

A major difference between the marginal ice zone during its period of active retreat and the pelagic zone of the Southern Ocean is the degree of new production (Eppley and Peterson 1979) which occurs. Although few studies have been conducted, the ice edge appears to be an active site of new production during periods of stability (Nelson and W.O. Smith 1986; W.O. Smith and Nelson in preparation) with "f-ratios" in excess of 0.5. Such high ratios indicate that a large fraction of the primary productivity is nitrate-based and thus available for export from the euphotic zone in the form of higher trophic level biomass or particles which sink to the benthos. In contrast, open-water areas have "f-ratios" of approximately 0.1, indicating that very little of the material produced can be removed. Therefore, not only are marginal ice zones of the Southern Ocean a quantitatively significant source of biogenic production, they function so as to channel much of the productivity into an exported form. Thus not only are ice-edge systems important quantitatively, they are also extremely important qualitatively to the trophic dynamics of the Southern Ocean.

It remains to be determined what effect, if any, the interannual variations in marginal ice zone primary productivity have on higher trophic levels, particularly krill stocks and reproduction. Cushing (1981) commented that for upwelling systems "differences in production of the higher parts of the ecosystem may well be related to differences in primary production from year to year". This implies that the life strategies of herbivores in upwelling regions are structured to utilize the primary productivity pulses or events associated with the physical circulation (Cushing 1976). Data from a number of upwelling regions seem to support such a hypothesis (e.g. Parrish et al. 1981; Sherman et al. 1984). However, ice-edge phytoplankton blooms and the associated ice-edge secondary producers may be significantly different from those of upwelling systems. For example, *Euphausia superba,* which is generally considered to be the major herbivore in the Southern Ocean, is planktonic and less able to "follow" the ice edge as it (and the associated bloom) undergoes its seasonal retreat than are the major herbivores in upwelling areas, which are phytophagous fishes which can navigate against the local currents and maintain themselves at specific locations. The ice-edge bloom apparently lasts 2 months or less at one location, depending on the rate of ice retreat (W.O. Smith and Nelson 1986), whereas the upwelling "season" in tropical coastal upwelling systems lasts 6 months or more. Given that the increased phytoplankton biomass in the marginal ice zone is restricted both in space and time, it is possible that the biogenic production is not utilized in situ, but remains intact and sinks from the euphotic zone. The material then would be consumed and remineralized at depth, either within the water column or at the water/benthos boundary. There is evidence that this occurs in at least two marginal ice zones. The production of biogenic silica (diatomaceous material) within the Ross Sea ice-edge bloom is nearly quantitatively delivered to the sediments (W.O. Smith 1987), and substantial amounts of organic production also appear to be exported from the euphotic zone via passive sinking. In the Bering Sea, the ice retreat occurs so rapidly that the duration of any phytoplankton bloom is limited to a period of a few weeks (W.O. Smith 1987). The material is not utilized by the dominant copepod herbivores but accumulates in the sediments on the shallow (ca. 30–80 m) continental shelf and supports an active benthic food web.

Further evidence for the lack of utilization of production within a bloom comes from studies of temperate blooms. Malone et al. (1983) found that approximately 90% of the material produced during the spring diatom bloom (February through April) in the mid-Atlantic Bight was in large part not directly consumed by zooplankton, but transported off the continental shelf to the continental slope. Walsh (1981) has argued that in a number of continental shelf regions (mid-Atlantic Bight, Bering Sea, and Oregon coast) up to 50% of the annual primary production is exported as phytodetritus (noningested phytoplankton) to the upper slope. It is certainly possible that the biogenic production in marginal ice zones of the Southern Ocean is also exported to deeper waters rather than utilized directly within the upper ocean. If true, the effects of interannual variations of ice-edge production on the food web would be smoothed; however, such large-scale temporal variations would have a marked impact on benthic processes as well as on biogeochemical cycles. In order to accurately assess the impact of such variations, the temporal and spatial patterns of distribution of euphausids within marginal ice zones must be clearly established. Furthermore, the flux of biogenic material from the surface layers should be measured concurrently in order to determine the coupling among production, consumption and export. Those data will then allow a quantitative analysis of the role of the interannual variability of production processes of Southern Ocean marginal ice zones on living resources in the Antarctic.

Acknowledgments. This research was supported by the National Science Foundation (DPP-8420213) and the Oceanic Processes Program of the National Aeroneutic Space Agency (NASA). We thank Dr. Richard Barber for his helpful comments.

References

Ackley SF, Buck KA, Taguchi S (1979) Standing crop of algae in the sea ice of Weddell Sea region. Deep-Sea Res 26A:269–281

Ainley DG, Jacobs SS (1981) Seabird affinities for ocean and ice boundaries in the Antarctic. Deep-Sea Res 28A:1173–1185

Alexander V, Niebauer HJ (1981) Oceanography of the eastern Bering Sea ice-edge zone in spring. Limnol Oceanogr 26:1111–1125

Barber RT, Chavez FP (1983) Biological consequences of El Niño. Science 222:1203–1210

Barber RT, Smith RL (1981) Coastal upwelling ecosystems. In: Longhurst AR (ed) Analysis of marine ecosystems. Academic Press, London New York, pp 31–68

Chavez FP, Barber RT (1987) An estimate of new production in the equatorial Pacific. Deep-Sea Res 34:1229–1243

Comiso JC, Sullivan CW (1986) Satellite microwave and in situ observations of the Weddell Sea ice cover and its marginal ice zone. J Geophys Res 91:9663–9681

Comiso JC, Zwally HJ (1984) Concentration gradients and growth/decay characteristics of the seasonal ice cover. J Geophys Res 89:8081–8103

Cushing DH (1976) Marine ecology and fisheries. Cambridge Univ Press, Cambridge, 278 pp

Cushing DH (1981) Temporal variability in production systems. In: Longhurst AR (ed) Analysis of marine ecosystems. Academic Press, London New York, pp 443–472

Dungan ML, Miller TE, Thomson DA (1982) Catastrophic decline of a top carnivore in the Gulf of California rocky intertidal zone. Science 216:989–991

El-Sayed SZ (1978) Primary productivity and estimates of potential yields of the Southern Ocean. In: McWhinnie MA (ed) Polar research: to the present, and the future. AAAS Select Symp 7. Westview Press, New York, pp 141–160

Eppley RW (1972) Temperature and phytoplankton growth in the sea. Fish Bull 70:1063–1085

Eppley RW, Peterson B (1979) Particulate organic matter flux and planktonic new production in the deep ocean. Nature (London) 282:677–680

Fraser WR, Ainley DG (1986) Ice edges and seabird occurrence in Antarctica. BioScience 36:258–263

Holm-Hansen O, El-Sayed SZ, Franceschini GA, Cuhel RL (1977) Primary production and the factors controlling phytoplankton growth in the Southern Ocean. In: Llano GA (ed) Adaptations within Antarctic ecosystems. Proc 3rd SCAR Symp Antarct Biol. Smithsonian Inst, Washington DC, pp 11–50

Jennings J, Gordon L, Nelson D (1984) Nutrient depletion indicates high primary productivity in the Weddell Sea. Nature (London) 309:51–54

Laws RM (1985) The ecology of the Southern Ocean. Am Sci 73: 26–40

Malone TC, Hopkins TS, Falkowski PG, Whitledge TE (1983) Production and transport of phytoplankton biomass over the continental shelf of the New York Bight. Cont Shelf Res 1: 305–337

Marr JWS (1962) The natural history and geography of the Antarctic krill (*Euphausia superba* Dana). Discovery Rep 32:33–464

Nelson DM, Smith WO Jr (1986) Phytoplankton bloom dynamics of the western Ross Sea ice edge. II. Mesoscale cycling of nitrogen and silicon. Deep-Sea Res 33:1389–1412

Nelson DM, Smith WO Jr, Gordon LI, Huber B (1987) Early spring distributions of density, nutrients and phytoplankton biomass in the ice-edge zone of the Weddell/Scotia Sea. J Geophys Res 92:7181–7190

Niebauer HJ (1980) Sea ice and temperature variability in the eastern Bering Sea and the relation to atmospheric fluctuations. J Geophys Res 85:7507–7515

Niebauer HJ (1984) On the effects of El Niño events in Alaskan waters. Bull Am Meteorol Soc 65:472–473

Parrish RH, Nelson CS, Bakun A (1981) Transport mechanisms and reproductive success of fishes in the California Current. Biol Oceanogr 1:175–203

Rey F, Loeng H (1985) The influence of ice and hydrographic conditions on the development of phytoplankton in the Barents Sea. In: Gray JS, Christiansen ME (eds) Marine biology of polar regions and effects of stress on organisms. Wiley, Chichester, pp 49–63

Sakshaug E, Holm-Hansen O (1984) Factors governing pelagic production in polar oceans. In: Holm-Hansen O (ed) Lecture notes on coastal and estuarine studies. Springer, Berlin Heidelberg New York Tokyo, pp 1–18

Sherman K, Smith W, Morse W, Berman M, Green J, Ejsymont L (1984) Spawning strategies of fishes in relation to circulation, phytoplankton production, and pulses in zooplankton off the northeastern United States. Mar Ecol Prog Ser 18:1–19

Smith SL, Smith WO Jr, Codispoti LA, Wilson DL (1985) Biological observations in the marginal ice zone of the East Greenland Sea. J Mar Res 43:693–717

Smith WO Jr (1987) Phytoplankton dynamics in marginal ice zones. Oceanogr Mar Biol Annu Rev 25:11–38

Smith WO Jr, Nelson DM (1985) Phytoplankton bloom produced by a receding ice-edge in the Ross Sea: spatial coherence with the density field. Science 227:163–167

Smith WO Jr, Nelson DM (1986) The importance of ice-edge phytoplankton blooms in the Southern Ocean. BioScience 36:251–257

Smith WO Jr, Baumann ME, Wilson DL, Aletsee L (1987) Phytoplankton biomass and productivity in the marginal ice zone of the Fram Strait during summer, 1984. J Geophys Res 92: 6777–6786

Walsh JJ (1981) Shelf-sea ecosystems. In: Longhurst AR (ed) Analysis of marine ecosystems. Academic Press, London New York, pp 159–196

Wilson DL, Smith WO Jr, Nelson DM (1986) Phytoplankton bloom dynamics of the western Ross Sea ice edge. I. Primary productivity and species-specific production. Deep-Sea Res 33:1375–1387

Woodley, JD, Chornesky EA, Clifford PA, Jackson JB, Kaufman LS, Knowlton N, Lang J, Pearson M, Porter J, Rooney M, Rylaarsdam K, Tunnicliffe V, Wahle C, Wulff J, Curtis A, Dallmeyer M, Jupp B, Koehl M, Neigel J, Sides E (1981) Hurricane Allen's impact on Jamaican coral reefs. Science 214:749–755

Zwally HJ, Comiso JC, Parkinson CL, Campbell WJ, Carsey FD, Gloersen P (1983a) Antarctic sea ice, 1973–1976: Satellite passive-microwave observations. NASA Spec Publ 459, Washington DC, 206 pp

Zwally HJ, Parkinson CL, Comiso JC (1983b) Variability of Antarctic sea ice and changes in carbon dioxide. Science 220:1005–1012

Effects of Pack Ice on the Composition of Micronektonic Communities in the Weddell Sea

D.G. AINLEY[1], W.R. FRASER[1], and K.L. DALY[2]

Summary. Using seabirds as sampling devices on two cruises in the northwestern Weddell Sea, we investigated the species and length-frequency distribution of micronekton in surface waters. Our micronekton samples are among the first ever collected simultaneously in both ice-covered and adjacent ice-free waters in the Antarctic using the same technique. Changes in the abundance or presence of certain crustacean, squid, and fish species were related to the presence of ice, and to distance from the ice edge. Length-frequency distributions representing age- or sex-classes of fish and krill were also affected by proximity to the ice edge. Our data indicate that the proximity of sampling to pack ice may affect our perception of the composition of micronektonic communities, and that the marginal ice zone is an important area for juvenile age-classes of pelagic fish and krill.

1 Introduction

An understanding of trophodynamics at levels of complexity greater than the organismic level requires knowledge of the spatial and temporal variation in the abundance of organisms, and the causes for that variation in the system being studied. Perceptions of variability are dependent on real patterns in the systems and the technologies and sampling regimes employed. In the case of one organism, the Antarctic krill *Euphausia superba,* whose role in the trophodynamics of the Southern Ocean has received a great deal of attention, studies such as those by Brinton and Antezana (1984) and Siegel (1986) have indicated an extremely complex pattern in distribution, abundance, and age structure relative to the meso- and macroscales of oceanic features. The former study emphasized use of the same sampling technique in different seasons and areas; the latter study discussed some of the difficulties in relating samples collected by different techniques.

Despite many formal and informal attempts at international coordination of research, the distribution and abundance of krill and other micronekton continues to be studied with a wide array of sampling techniques. For example, in two recent symposia volumes, Crustacean Biology (Special No. 1, 1984) and Siegfried et al. (1985), among eight studies published on krill distribution six different sampling techniques were employed. Furthermore, studies of the diets of organisms higher in the food web have indicated the presence of certain krill length-frequency classes or other faunal components where more typical sampling techniques (i.e., nets) failed to detect them (Croxall et al. 1985; Ainley et al. 1986).

If we are to understand the temporal and spatial variability in the distributional and abundance patterns of organisms in the Southern Ocean, some standardization of the sampling methodologies will be necessary. We will also need more information on what the different techniques actually sample (e.g., Everson and Bone 1986). In the case of some important organisms in marine food webs, for example cephalopods, typical methodologies of marine biological research do not suffice, and alternate ones must be used or found (e.g., Clarke 1980; Clarke and Prince 1981).

One aspect of sampling which has been standardized among almost all studies to date on the distribution and abundance of Antarctic micronekton or macrozooplankton is the choice of study areas away from pack ice. Such a sampling regime has resulted from problems with towing nets or acoustic devices through ice-covered seas. The degree to which this aspect of Antarctic research has biased our perceptions of micronektonic and macrozooplanktonic communities remains to be determined. We present data here to indicate that understanding the spatial and, perhaps, temporal variability in the abundance of these organisms will require knowledge of the recent pack ice conditions in study areas. In the results presented here, unlike most other studies on the distribution of planktonic/nektonic organisms in the Antarctic near pack ice, we used the same sampling technique in both open and ice-covered waters: analysis of prey eaten by seabirds.

On AMERIEZ (Antarctic Marine Ecosystem Research in the Ice Edge Zone) cruises, micronekton were investigated by towed acoustic array, Tucker trawl (RMT) and bongo net in open waters, and (from a stationary ship) by a vertically deployed acoustic array, plummet net, and SCUBA-assisted observation in ice-covered water, as well as by seabird stomach contents. The latter technique was the only one that was relatively consistent in both habitats. If

1 Point Reyes Bird Observatory, 4900 Shoreline Highway, Stinson Beach, CA 94970, USA
2 School of Oceanography, University of Washington, Seattle, WA 98195, USA

Antarctic Ocean and Resources Variability
(ed. by D. Sahrhage)

inconsistencies exist in the seabird-derived sampling, between habitats, they should largely be a function of two factors.

First, the seabird species differed to some degree between deep pack, pack ice edge and open water habitats, as well as between open waters in the Weddell Sea versus those of the Scotia-Weddell Confluence (Fraser and Ainley 1986). Except for the relatively small number of penguins which were sampled only in the ice, however, all other birds were essentially surface feeders with similar representation of predator sizes in open-water versus ice-covered habitats. Thus, controlling for penguin samples, which we did, equilibrated sampling with regard to predator size and feeding depth.

Second, the success of our micronekton sampling technique is predicated on the assumption that seabirds are opportunistic in their feeding, and thus take prey that are most abundant and most easily available to them. The only major study of diet overlap (prey species and size) among seabirds at sea in pelagic waters (i.e., the situation in the present study) was conducted in the Ross Sea by Ainley et al. (1984). One of the major findings of the latter study (and a trend in the present study, based on preliminary inspection of the data) was that in the same habitat, different seabird species take the same prey, and that when between-species differences in diet occur they are a function of the species' tendencies to exploit different habitats. This particular pattern, observed in the Ross Sea study, is the basis for using seabirds to sample nekton in the present study. Land-based studies of seabird diet (i.e., analysis of prey returned to chicks at the nest) often indicate little overlap in diet between certain species (i.e., Croxall and Prince 1980; Prince 1980), which would argue against the validity of studies such as the present one. Land-based studies, however, almost never know in what habitats the birds sampled had been feeding. On the other hand, consideration of energetics dictates that, given a choice, seabirds should eat the prey which provide the most energy per unit effort of catch, e.g., one 5-cm-long fish instead of five 5-cm euphausiids, or one 5-cm euphausiid instead of 15 1-cm amphipods. Whether or not seabirds make such a selection is largely a subject of conjecture. As with other micronekton sampling techniques, especially before more information is available on seabird feeding selectivity at sea, abundance or size-class comparisons within a group of organisms are most likely more valid than between groups. Recognizing that a potential exists for some biases to occur in the seabird samples, nevertheless, it is possible that seabirds offer one of the better means to understand occurrence patterns of certain organisms at the sea surface because disturbance by the ship is not a problem, as it is with any type of towed device. Future analysis of all the AMERIEZ data, i.e., those derived from birds, nets, and acoustics, should shed additional light on the sampling selectivity of seabirds.

2 Methods

Our studies were conducted on AMERIEZ cruises in the northwestern Weddell Sea, between 57 and 63°S, 36 and 51°W, during November–December 1983 and March 1986. During the spring period, the ice pack had just begun to retreat rapidly from its maximal winter extent, and the orientation of the ice edge was east-west. During the late autumn period, the pack was at its minimal extent, had just begun to show signs of refreezing, but was not yet advancing; orientation of the ice edge then was north-south. No sampling was conducted over the continental shelf. Other information on the general conditions, cruise tracks, and station plans are contained in Ainley and Sullivan (1984) and Sullivan and Ainley (1988).

We collected or pumped the stomachs of 327 seabirds in 1983 and 246 in 1986. Species sampled included Emperor Penguin *Aptenodytes forsteri* (n = 18, during 1986 only), Adélie Penguin *Pygoscelis adeliae* (11, during 1986 only), Antarctic Petrel *Thalassoica antarctica* (71), Cape Petrel *Daption capense* (64), Snow Petrel *Pagodroma nivea* (111), Kerguelen Petrel *Pterodroma brevirostris* (23), Blue Petrel *Halobaena caerulea* (46), Antarctic Prion *Pachyptila vittata* (66), White-Chinned Petrel *Procellaria aequinoctialis* (11), Wilson's Storm Petrel *Oceanites oceanicus* (61), Antarctic Tern *Sterna vittata* (7), and Arctic Tern *S. paradisaea* (22). These species take prey mainly during the night or darkest period of the day, and within 10 m of the surface (Ainley et al. 1984), except for the Adélie, which probably feeds to about 70 m (assuming similarity with its congener, *P. antarctica,* Lishman and Croxall 1983) and the Emperor which feeds to more than 250 m (Kooyman et al. 1971). Samples were taken just after dawn, and thus stomach contents were fresh; items found in the gizzard are not included in this analysis unless they corresponded to items in the stomach, e.g., otoliths in gizzard and a partly digested fish in the stomach.

Prey sizes were measured directly, using total length for crustaceans and standard length for fish, or were calculated from mathematical relationships between length (L) and certain body parts. In the case of krill, we used a formula based on the diameter of the eye (D), where $\ln D = 0.738 \ln L + 3.262$ (n = 163, r = 0.945); and in the case of the myctophid fish *Electrona antarctica,* we used a formula based on the largest diameter of an otolith (D), where $L = 3.974 D - 1.047$ (n = 101, r = 0.980). For cephalopods, we measured the lower rostral length (LRL; see Clarke 1986) only of intact (including membranous edges), hence relatively fresh beaks. Few data are available to relate LRL to mantle length for the species we encountered, and thus only LRL's are presented. Sexual maturity stages of krill were determined according to criteria in Makarov and Denys (1981).

3 Results

3.1 General Patterns

As a result of the 1983 cruise, Ainley et al. (1986) observed a strong effect of the pack ice on the presence/absence of certain micronektonic organisms in surface waters of the study area. A similar phenomenon was observed in 1986 (Fig. 1), with some differences in the species involved. During both seasons, krill were most prevalent in seabird diets in open water immediately adjacent to the pack ice edge. During spring, krill were less prevalent in seabirds sampled in the pack ice and least prevalent in those sampled in open water 200 km north of the ice edge, just south of the Scotia-Weddell Confluence (at about 58°S). A few individuals of *Thysanoessa macrura* were eaten by birds in the pack ice, as well as in waters north of the Confluence. *E. crystallorophias* was encountered, but only rarely and only in the pack ice. During autumn, krill also were less predominant in seabird diets sampled in the pack ice and least prevalent in the diets of seabirds of open waters 150 km from the ide edge.

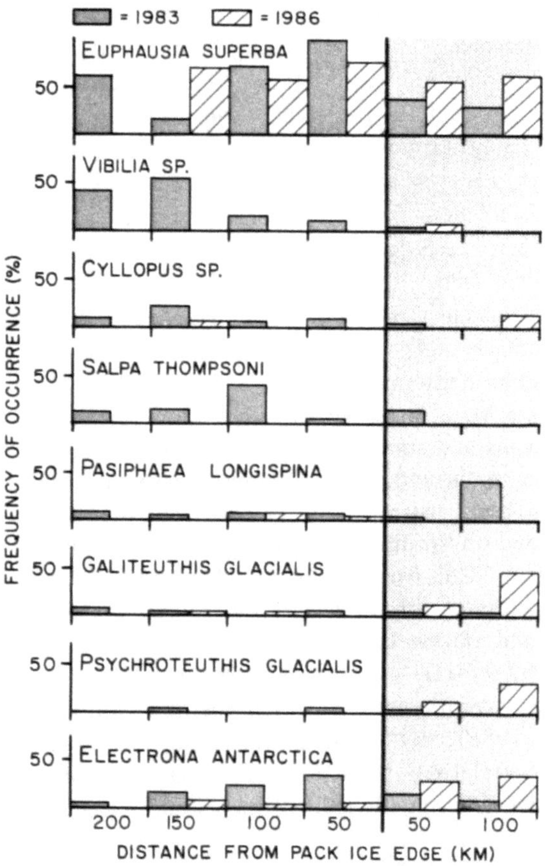

Fig. 1. Frequency of occurrence of prey species in the stomach contents of seabirds as a function of distance from the pack ice edge, with pack ice indicated by *shading*, spring 1983 and fall 1986

Two amphipods, *Vibilia* sp. and *Cyllopus* sp., were especially abundant in the diets of seabirds in open water during spring 1983, particularly in the vicinity of the Scotia-Weddell Confluence, but were virtually absent in the diets during autumn 1986. In neither year were these prey important within the pack ice. These crustaceans are obligate symbionts of salps (e.g., Madin and Harrison 1977). Thus, it is not surprising that their prevalence closely paralleled that of *Salpa thompsoni.* Another crustacean, *Pasiphaea longispina,* was important in the diets of pack ice seabirds in 1983 (Ainley et al. 1986), but was encountered much less frequently in 1986.

Squid were rarely eaten by the seabirds sampled in 1983, and the same would be true for 1986, except that in the latter but not the former year we sampled Emperor Penguins and only within the pack ice. As a result, *Galiteuthis glacialis* and *Psychroteuthis glacialis,* which dominated the Emperor's diet, appear to be important components in the diets of pack ice birds. Two other squid, *Kondakovia longimana* and *Gonatus antarcticus,* showed a similar pattern. Thus, the apparent distributions of squid were an artifact of the seabird species we sampled, far more so than any other micronektonic organisms discussed here (see Introduction). Squid did, however, occur in the diet of pack ice petrels, and only in the pack did we see schools of squid at the surface.

During both 1983 and 1986, myctophid fishes were prevalent in seabird diets. *Electrona* dominated, but *Gymnoscopelus braueri, G. opistomelas* and *Protomyctophum bolini* also occurred. *P. bolini* was eaten mainly in the vicinity of the Scotia-Weddell Confluence. Another mesopelagic fish, *Notolepis coatsi,* occurred in the diets, especially of birds sampled in the pack ice. In 1983, *Electrona* reached a peak in prevalence immediately adjacent to the pack ice edge; in 1986, peak prevalence occurred within pack ice-covered waters.

3.2 Length-Frequency Patterns

Three size-classes of *E. superba* were encountered, with modes at 20, 35 and 50–55 mm in spring 1983, and with modes at 25, 45 and 55–60 mm in autumn 1986 (Figs. 2, 3). It is possible that these represent three year-classes of krill (Brinton 1984), but recent work has suggested that assignment of krill to ages, based on length frequency, is not straightforward (Rosenberg et al. 1986; Siegel 1986). A few larval krill were also eaten, especially by Antarctic Prions north of the Confluence, but these were too few to discuss further. Year I krill predominated in the vicinity of the Confluence (< 59°S), as well as in the pack ice near its edge (61°S). This year-class was largely absent from the diets of seabirds frequenting open waters adjacent to the ice edge, as well as in the pack ice interior. In the latter two habitats, year III or older krill predominated almost exclusively. In 1986, year III (or older) individuals were eaten largely in open waters adjacent to the ice edge (45°–47°W)

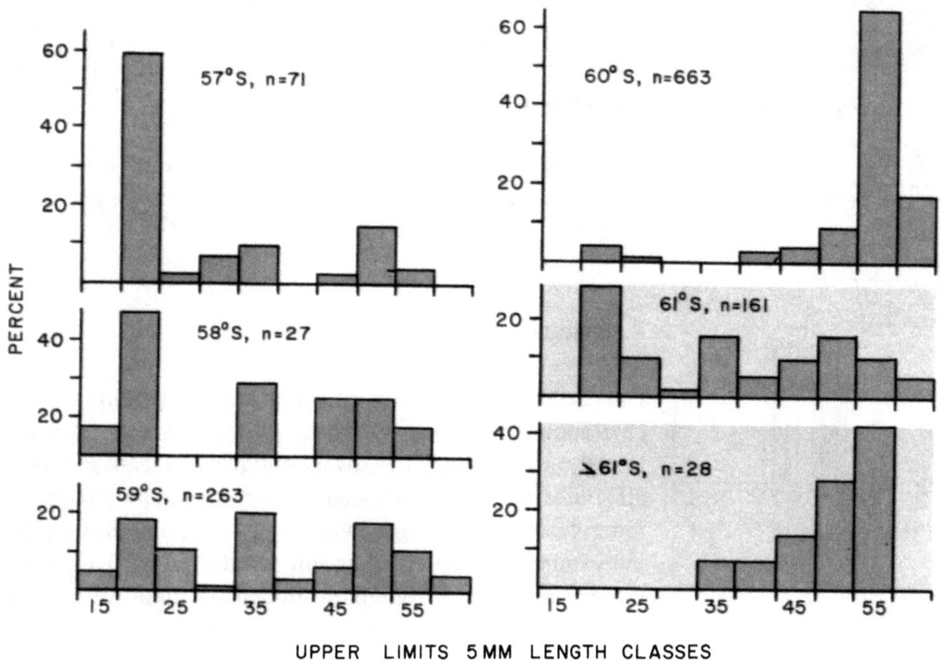

Fig. 2. Length frequencies of krill in seabird diets as a function of latitude, and, thus, distance from the pack ice edge at about 61°S, in 1983; pack ice indicated by *shading*

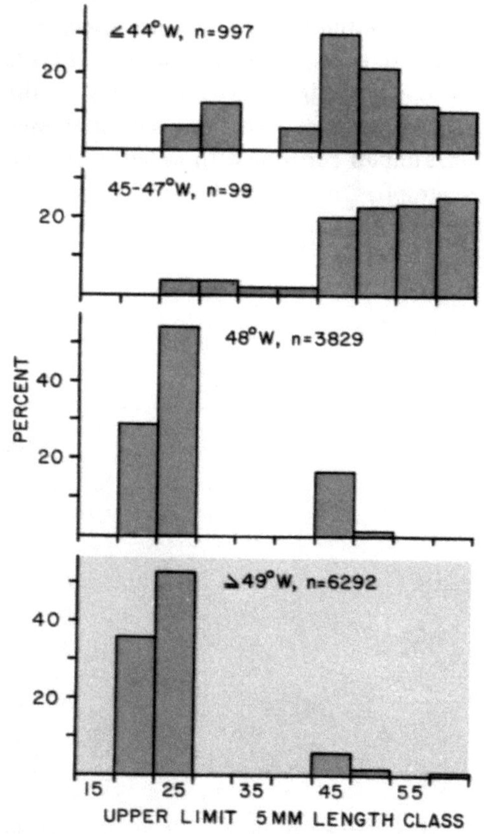

Fig. 3. Length frequencies of krill in seabird diets as a function of longitude, and, thus, distance from the pack ice edge at about 48° W, in 1986; pack ice indicated by *shading*

and year I individuals were eaten largely in ice-covered waters. Year II krill were taken by birds in open as well as ice-covered waters but reached greatest frequency at distance from the ice.

For seven stations where seabirds were sampled and krill were eaten in 1983, male *E. superba* dominated the diets, ranging between 70 and 100% of the adults eaten. A similar pattern was apparent in 1986, where at seven of eight stations, male krill comprised 60 to 100% of the adults eaten, while at the eighth females predominated (55%).

Individuals of *Pasiphaea longispina* eaten by seabirds ranged between 50 and 120 mm in length (Fig. 4). A bimodal frequency distribution was evident, but whether or not it is biologically meaningful requires more information to become available on this little-known species. There appears to be a relationship, however, between size and habitat. Combining both years, individuals eaten in open water ranged between 50 and 95 mm ($\overline{x}$ = 74, s = 16, n = 11), but 64% were less than 80 mm long; in contrast, individuals eaten in ice-covered water ranged between 50 and 120 mm ($\overline{x}$ = 88, s = 19, n = 20), but 75% were longer than 80 mm (difference between means significant, t = 2.12, $p < 0.05$).

The squid eaten by seabirds in 1983, although fewer in number, were consistently smaller in size than those eaten in 1986 (Fig. 4). Squid are known to have extremely fast growth rates, and thus this pattern could be explained in terms of when collections were made. The small squid in 1983 were collected during the spring, and the larger squid in 1986 were collected during the autumn, 4 months later. The change in size between 1983 and 1986 could also be an artifact of sampling: most of the squid in 1986 were taken from Emperor Penguins (very large predators), which were not sampled in 1983.

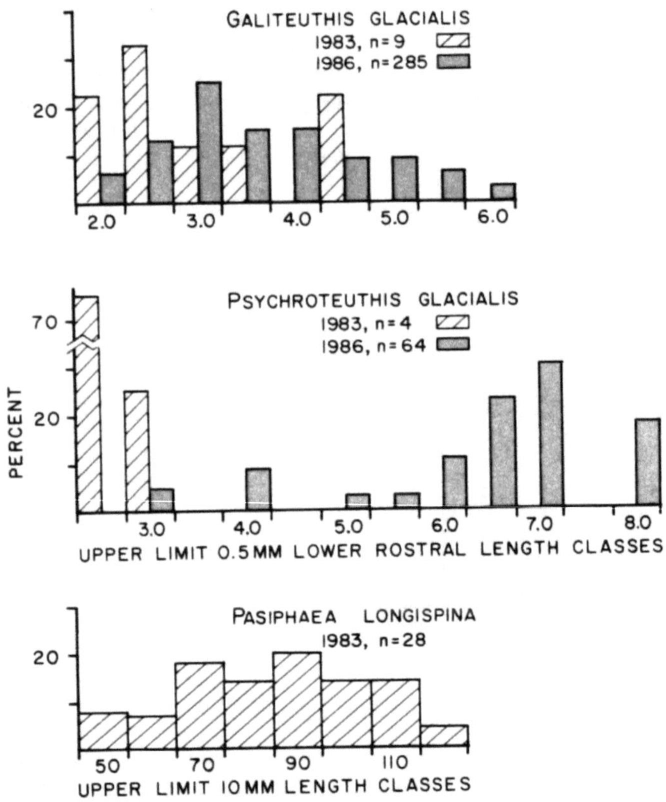

Fig. 4. Frequency patterns of total length in *Pasiphaea* and LRL in two squid species; prey eaten by seabirds in 1983 and 1986

A bimodal frequency distribution was clearly evident in the size of *Electrona antarctica* eaten by seabirds (Fig. 5). One group of fish centered around 61–70 mm and the other around 81–90 mm, with the former likely being year IV and the latter year V individuals (Robison and Hopkins 1983, personal communication). Interestingly, during both years fish of the 90 mm size class (as well as ones even larger than 81–90 mm) outnumbered fish of the smaller size-class by as much as 7:1 in the diets of seabirds frequenting waters away from the ice. In ice-covered waters of the marginal ice zone the smaller size-class (as well as still smaller individuals) became more prevalent, with the ratio at 2:1 or 1:1.

4 Discussion

Our data on the distribution of *Salpa thompsoni* in 1983 correspond well with Brinton's (1984), which were collected using a bongo net towed obliquely in ice-free waters. He observed highest concentrations in waters 50 km or more from the ice edge, with a slight decline in densities in waters immediately adjacent to the edge. This was a pattern apparent in our data, but in our data we can also see that the high densities of salps did not reach very far into the pack, an observation supported by the plummet net samples (Daly and Macaulay unpublished data).

Similarly, some patterns exhibited by *Euphausia superba* in the seabird data also coincide with those in Brinton's data (Table 1). The bird data, however, extend those patterns into ice-covered waters. Brinton observed adult krill (>35 mm) mainly between 60°S and the ice edge (ca. 61°S), as we did, but our data indicated that this higher prevalence of adults continued into the pack ice. Both data sets also detected a preponderance of juvenile krill in the vicinity of the ice edge, but ours indicated that juveniles were concentrated far into the pack ice as well. The plummet net samples confirmed the high abundance

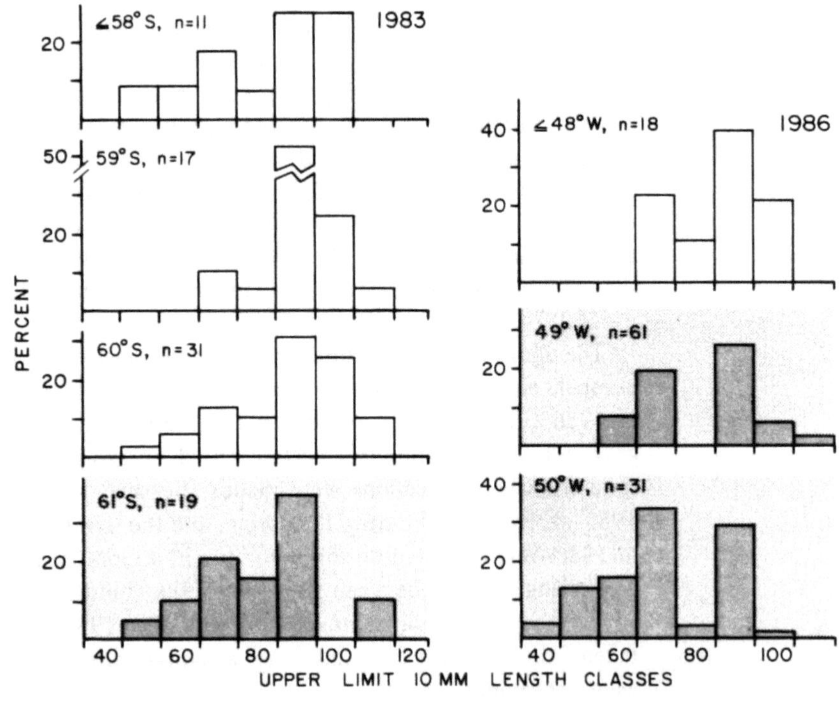

Fig. 5. Length frequencies of *Electrona* in seabird diets as a function of latitude in 1983 and longitude in 1986, and thus, as a function of distance from the pack ice (see legends Figs. 2, 3)

Table 1. A qualitative comparison of where three different sampling techniques found adult (>45 mm) and juvenile (<30 mm) krill in abundance, by ice-related habitat south of the Scotia-Weddell Confluence in 1983

	Pack ice	Pack edge	Open water
Adult krill			
Bongo net	0	*	*
Plummet net			0
Birds	*	*	*
Juvenile krill			
Bongo net	0	*	
Plummet net	*	*	0
Birds	*	*	

0 = not sampled; * = krill abundant.

of juveniles in the marginal ice habitat, but the samples contained few adults (Daly and Macaulay, 1988). On the other hand, the bird sampling revealed the presence of juveniles in the vicinity of the Scotia-Weddell Confluence, whereas Brinton's bongo samples did not. Both the bird and bongo data sets detected the same three year-classes of krill, with the year I and year III (and older) individuals predominating overall.

There are few data on the depth distribution of krill by sex, but the predominance of males in the seabird diet indicate a sexual difference in behavior making males more vulnerable to predation, or a greater preponderance of males near the sea surface. A similar prevalence of male krill has been observed in the diet of penguins at King George Island (W. Trivelpiece personal communication).

Currently, only preliminary comparisons are possible between our 1986 data and those of other researchers. The lower prevalence of salps in 1986 compared to 1983 was detected by all AMERIEZ researchers sampling micronekton. The bird data, however, seem to indicate higher densities of krill in 1986 than detected by other investigators who were most likely sampling lower in the water column. A strong association of juvenile krill with the pack ice was again apparent in 1986.

Given the emphasis on the trophodynamics of krill in Antarctic marine food web research (e.g., Siegfried et al. 1985, and references therein), we were surprised that myctophid fishes occurred so frequently in the diets of seabirds during both 1983 and 1986. It appears that in pelagic waters, these fishes are as important to predators as is *Pleuragramma antarcticum* in neritic waters (cf. Ainley et al. 1984; Plötz 1986). Although *Pleuragramma* is considered an important component in various treatments of Antarctic faunas (e.g., Everson 1984; Hempel 1985), the importance of myctophids, and especially *Electrona*, has apparently been overlooked.

While squid are recognized as important components of marine food webs in lower latitudes of the Antarctic (Clarke 1980; Lubimova 1985; Okutani and Clarke 1985), they, too, have been overlooked at higher latitudes. The data presented here, and in Ainley et al. (1984) and Offredo et al. (1985), extend the recognized ranges of several squid species appreciable distances to the south, and, moreover, indicate them to be exceedingly abundant and important components of the high latitude pelagic fauna. Also, whereas Lubimova (1985) considered the species we observed in seabird diets to reach "lower epipelagic depths", occurrence in the diets of surface-feeding predators and other observations of squid schools indicate that these species also reach the sea surface, at least in the pack ice.

Why smaller fish became more prevalent in diets near to the ice is another pattern where further information will be necessary before an explanation may be possible. In this regard, smaller fish also seemed more prevalent in another "frontal area", the Scotia-Weddell Confluence. Thus, the energetics of growth and regions of higher production may be involved; younger individuals may find better feeding conditions in these regions.

In many of the patterns we have reported, as well as in those reported by Ainley al. (1986), strong correspondence to the presence/absence of pack ice has been obvious Because of this, we suggest that for future biological oceanographic studies, the recent cryographic history and characteristics of a region should be considered when interpreting results. Otherwise, spatial and temporal variation in the observed abundance and distributional patterns of micronekton may be incompletely explained. In addition, by continuing to concentrate studies in ice-free waters, the importance of several micronektonic organisms may continue to be underestimated.

Acknowledgments. We wish to thank the US Antarctic Research Program, and the officers and crew of R/V *Melville*, USCGC *Westwind*, and USCGC *Glacier* for logistical support. L. Quetin provided krill for measurements, and T. Lancraft and B. Robison provided measurements of myctophids. Assistance in the field was given by R. Ferris, E. O'Connor, P. Pietz, C. Strong and G. Wallace. Our work benefited greatly from discussions about micronekton patterns with E. Brinton, T. Hopkins, T. Lancraft, M. Macaulay, and J. Torres. E. Brinton and P. Slattery helped with the identification of crustaceans, as did T. Lancraft with fish. Helpful comments on the ms were provided by J.P. Croxall, T. Hopkins, and B. Robison. The research was funded by the National Science Foundation. This is contribution no. 360 of the Point Reyes Bird Observatory.

References

Ainley DG, Sullivan CW (1984) AMERIEZ 1983: a summary of activities on board the R/V *Melville* and USCGC *Westwind*. Antarct J US 18(5):100–103

Ainley DG, O'Connor EF, Boekelheide RL (1984) The marine ecology of birds in the Ross Sea, Antarctica. Am Ornithol Un Ornithol Mongr No 32:97p

Ainley DG, Fraser WR, Sullivan CW, Torres JJ, Hopkins TL, Smith WO (1986) Antarctic mesopelagic micronekton: evidence from seabirds that pack ice affects community structure. Science 232: 847–849

Brinton E (1984) Observations of plankton organisms obtained by bongo nets during the November–December 1983 ice-edge investigations. Antarct J US 19(5):113–115

Brinton E, Antezana T (1984) Structures of swarming and dispersed populations of krill (*Euphausia superba*) in Scotia Sea and South Shetland waters during January–March 1981, determined by bongo nets. Crustacean Biol 4 (Spec No 1):45–66

Clarke MR (1980) Cephalopoda in the diet of sperm whales of the Southern Hemisphere and their bearing on sperm whale biology. Discovery Rep 37:1–324

Clarke MR (1986) A handbook for the identification of cephalopod beaks. Clarendon Press, Oxford, 273 pp

Clarke MR, Prince PA (1981) Cephalopod remains in regurgitations of Black-browed and Grey-headed Albatrosses at South Georgia. Br Antarct Surv Bull 54:1–7

Croxall JP, Prince PA (1980) Food, feeding ecology and ecological segregation of seabirds at South Georgia. Biol J Linn Soc 14: 103–131

Croxall JP, Prince PA, Ricketts C (1985) Relationships between prey life-cycles and the extent, nature and timing of seal and seabird predation in the Scotia Sea. In: Siegfried WR, Condy PR, Laws RM (eds) Antarctic nutrient cycles and food webs. Springer, Berlin Heidelberg New York Toyko, pp 516–533

Daly KL, Macaulay MC (1988) Abundance and distribution of krill in the ice edge zone of the Weddell Sea, austral spring 1983. Deep-Sea Res 35:21–41

Everson I (1984) Fish biology. In: Laws RM (ed) Antarctic ecology, vol 2. Academic Press, London New York, pp 491–532

Everson I, Bone DG (1986) Effectiveness of the RMT8 system for sampling krill (*Euphausia superba*) swarms. Polar Biol 6:83–90

Fraser WR, Ainley DG (1986) Ice edges and seabird occurrence in Antarctica. BioScience 36:258–263

Hempel G (1985) Antarctic marine food webs. In: Siegfried WR, Condy PR, Laws RM (eds) Antarctic nutrient cycles and food webs. Springer, Berlin Heidelberg New York Tokyo, pp 266–270

Kooyman GL, Drabek CM, Elsner R, Campbell WR (1971) Diving behavior of the Emperor Penguin, *Aptenodytes forsteri*. Auk 88: 775–795

Lishman GS, Croxall JP (1983) Diving depths of the Chinstrap Penguin *Pygoscelis antarctica*. Br Antarct Surv Bull 61:21–25

Lubimova TG (1985) Results of Soviet investigations of the distribution and ecology of pelagic squids (Oegopsida) in the Southern Ocean. CCAMLR Select Pap Sci Comm 1985:79–112

Madin LP, Harrison GR (1977) The association of Amphipoda Hyperiidea with gelatinous zooplankton – I. Association with Salpidae. Deep-Sea Res 24:449–463

Makarov RR, Denys CJ (1981) Stages of sexual maturity of *Euphausia superba* Dana. BIOMASS Handb No 11:11 pp

Offredo C, Ridoux V, Clarke MR (1985) Cephalopods in the diets of Emperor and Adélie Penguins in Adélie Land, Antarctica. Mar Biol 86:199–202

Okutani T, Clarke MC (1985) Identification key and species description for antarctic squids. BIOMASS Handb No 21:57 pp

Plötz J (1986) Summer diet of Weddell seals (*Leptonychotes weddelli*) in the eastern and southern Weddell Sea, Antarctica. Polar Biol 6:97–102

Prince PA (1980) The food and feeding ecology of the blue petrel *Halobaena caerulea* and dove prion *Pachyptila desolata*. J Zool 190:59–76

Robison BH, Hopkins TL (1983) Nutrient and energetic dynamics of the southern ocean myctophid, *Electrona antarctica*. EOS (Abstract) 64(52):1099

Rosenberg AA, Beddington JR, Basson M (19860 Growth and longevity of krill during the first decade of pelagic whaling. Nature (London) 324(6093):152–153

Siegel V (1986) Structure and composition of the Antarctic krill stock in the Bransfield Strait (Antarctic Peninsula) during the Second International BIOMASS Experiment (SIBEX). Arch Fischereiwiss 37:51–72

Siegfried WR, Condy PR, Laws RM (eds) (1985) Antarctic nutrient cycles and food webs. Springer, Berlin Heidelberg New York Tokyo, 700 pp

Sullivan CW, Ainley DG (1988) AMERIEZ 1986: summary of activities on board R/V *Melville* and USCGC *Glacier*. Antarct J US 22:(in press)

Patterns of Spatial and Temporal Distribution and Their Variation in Early Life Stages of Antarctic Fish in the Antarctic Peninsula Region

A. KELLERMANN[1] and K.-H. KOCK[2]

Summary. Ichthyoplankton collected during six expeditions to the Antarctic Peninsula area between 1975/1976 and 1984/85 was analyzed for patterns of spatial and temporal distribution and interannual variability of abundance. In some species, e.g., *Nototheniops larseni,* median numbers per 10^3 m^3 varied mostly within one order of magnitude, whereas in others, e.g., *Notothenia gibberifrons,* considerable interannual fluctuation of postlarval abundance was observed. Unusually low abundance of *N. larseni* was recorded during the 1984/85 season and in summer 1982. Larvae and postlarvae of the myctophid *Electrona antarctica* showed an oceanic type of distribution, whereas larvae and postlarvae of *Notothenia kempi* hatched in summer were chiefly confined to slope waters. A large cyclonic gyre north of Adelaide Island was identified as a possible retention area. Early stages of some species, particularly icefish and other commercially harvested fish, could not be recorded in sufficient quantities to assess annual population size. For these, specific survey grids and the combined use of different nets are required.

1 Introduction

Recruitment variability in commercially exploited fish stocks is still poorly understood, although studies on recruitment processes have been a major issue in fishery science for several decades. It is, however, generally accepted that ichthyoplankton and young fish surveys which are carried out over successive years together with measurements of abiotic and biotic environmental factors may provide valuable information toward a better understanding of mechanisms governing year-class strength.

Despite their ecological and economical importance and the steadily increasing research effort in the last decade (e.g., BIOMASS Program), knowledge of the early life history of Antarctic fish and in particular of the seasonal and year-to-year variability in distribution and abundance is still rather fragmentary and confined to a few investigations in the Weddell Sea (Keller 1983; Hubold 1985) and the Antarctic Peninsula region (Kellermann 1986a,b; Slosarczyk 1986). Results of these studies, however, are not readily comparable due to the different nets used. Since the Convention for the Conservation of Antarctic Marine Living Resources (CCAMLR) came into force in 1982, its Scientific Committee stressed the need for recruitment surveys as methods independent of the fishery that could be used to estimate and monitor year-class strength in Southern Ocean's exploited fish stocks, particularly in those which have been heavily depleted in recent years and are now under protection measures (e.g., *Notothenia rossii, Champsocephalus gunnari*) (CCAMLR 1986).

The effect of such surveys, however, will largely depend on the detailed knowledge of the patterns of spatial and temporal distribution and abundance of the target species. Therefore, this paper is an attempt to synthesize the results of ichthyoplankton collections carried out by the Federal Republic of Germany in the Antarctic Peninsula region from 1976 to 1985, and to identify possible sources for the observed variability and to discuss these in the light of future ichthyoplankton surveys planned for the area by CCAMLR.

2 Material and Methods

Ichthyoplankton was collected as by-catch in krill directed investigations in Bransfield Strait and adjacent waters during the course of the Antarctic Expeditions 1975/76, 1977/78, 1980/81, 1983/84 and 1984/85 by the Federal Republic of Germany and the Joint Anglo-German Antarctic Expedition in 1981/82. The net used was the RMT 1+8 (Baker et al. 1973) and its multiple version RMT 1+8 M (Roe and Shale 1979) with mesh sizes of 0.35 mm in the RMT 1 and 4.5 mm in the RMT 8. Due to the coarse mesh, the catches of the RMT 8 underestimated the abundance of early stages of many species of Antarctic fish up to a standard length of ca. 30 mm (Kellermann 1986a). Consequently in Notoheniids and Myctophids only data from the RMT 1 were used to calculate abundance as number of specimens per 10^3 m^3. Analysis of the net selectivity for Channichthyids demonstrated that estimates of abundance per unit volume was similar for the RMT 1 and RMT 8, and so data from both nets could be used. For March 1985 only RMT 8 catches were available.

1 Alfred-Wegener-Institut für Polar- und Meeresforschung, Columbusstraße, 2850 Bremerhaven, FRG
2 Institut für Seefischerei, Bundesforschungsanstalt für Fischerei, Palmaille 9, 2000 Hamburg 50, FRG

Antarctic Ocean and Resources Variability
(ed. by D. Sahrhage)

Table 1. Summary of the time periods of investigations, the numbers of standard hauls, their depth ranges fished and the references for details of cruise tracks and hauling and sorting procedures

Time of season		Period of investigation	Vessel	Number of standard hauls	Depth range of standard hauls	Cruise track, hauling and sorting procedures
Spring	1977	November 15 – December 8	*Walther Herwig*	49	0 – 140	Hempel et al. (1979)
	1983	October 25 – November 16	*Polarstern*	53	200 – 0	Sahrhage (1986), Fütterer (1984)
	1984	November 11 – December 4	*Polarstern*	45	200 – 0	Sahrhage (1986), Hempel (1985)
Summer	1976	January 27 – February 6	*Walther Herwig*	21	0 – 200 – 0	Sahrhage et al. (1978); James and Wörner (1978)
	1978	January 6 – February 13	*Walther Herwig*	44	0 – 140	Hempel et al. (1979)
	1982	February 2–27	*John Biscoe*	36	200 – 0	Hempel and Heywood (1982)
Autumn	1978	March 3–12	*Walther Herwig*	63	0 – 140	Hempel et al. (1979)
	1981	March 10–19	*Walther Herwig*	28	140 – 0	Hempel et al. (1982), Piatkowski and Klages (1983)
	1985	March 10 – April 8	*Walther Herwig*	85	200 – 0	Sahrhage (1986), Hempel (1985)

The details of the standard hauls used are summarized in Table 1. The terminology followed that recommended by BIOMASS (Hureau 1982). Length measurements are standard lengths to the nearest mm below. All types of hauls, including also those fished at other than the standard depth ranges, were taken into account to describe the patterns of spatial and temporal distribution. Statistical processing of abundance estimates for comparison of data, however, was based only on standard hauls (oblique hauls down to 140 m or 200 m).

Both the median ($\tilde{X}$) and the arithmetic mean ($\bar{X}$) were calculated to estimate abundance. Preference was given to the median as in most cases data were not even approximately nomally distributed ($0.9 < \frac{\tilde{X}}{\bar{X}} < 1.1$). Hence, nonparametric statistics were applied for comparison of data between seasons or years. In the case of equal patterns of distribution of data the Mann-Whitney (Wilcoxon) U-test was used, and in all other cases the Friedman's median test was employed.

3 Results

Early life stages (larvae and postlarvae) of 21 species of Notothenioidei and 1 species of Myctophidae were identified (Table 2). Abundance data for the predominant and most frequently occurring species are summarized in Tables 3–5 for spring, summer, and autumn. Some of these were analyzed for pattern of spatial and temporal distribution and variability of abundance (Table 2).

Table 2. Species so far identified from ichthyoplankton collections in the Peninsula region during 1976–1985. The most abundant species are **in bold type**

Notoheniidae

Nototheniops larseni (Lönnberg 1905)
Nototheniops nudifrons (Lönnberg 1905)
Notothenia neglecta (Nybelin 1951)
Notothenia kempi (Norman 1937)
Notothenia gibberifrons (Lönnberg 1905)
Trematomus newnesi (Boulenger 1902)
Trematomus scotti (Boulenger 1907)
Trematomus eulepidotus (Regan 1914)
Pleuragramma antarcticum (Boulenger 1902)

Harpagiferidae

Harpagifer sp.

Artedidraconidae

Artedidraco spp.

Channichthyidae

Chionodraco rastrospinosus (De Witt and Hureau 1979)
Cryodraco antarcticus (Dollo 1900)
Chaenodraco wilsoni (Regan 1914)
Chaenocephalus aceratus (Lönnberg 1906)
Champsocephalus gunnari (Lönnberg 1905)
Neopagetopsis ionah (Nybelin 1947)
Pagetopsis macropterus (Boulenger 1907)
Pagetopsis maculatus (Barsukov and Permitin 1958)

Bathydraconidae

Parachaenichthys charcoti (Vaillant 1906)
Prionodraco evansii (Regan 1914)

Myctophidae

Electrona antarctica (Gunther 1878)

Table 3. Spring 1977, 1983, 1984. Total number of specimens caught in the RMT 1+8 (n), percentage of the total number of specimens (%), median abundance per 10^3 m^3 and its 95% confidence interval, ratio of median vs. arithmetic mean, maximum observed abundance and frequency of occurrence as absolute number of standard stations (FO) for the most abundant species. Data from standard stations only

	n RMT 1+8	n (%)	Median abund.	95% conf. interval	Median/ mean	Max. obs. abund.	FO
	November 15 – December 12, 1977						
Nototheniops larseni	510	72.7	6.6	3.7–11.2	0.34	153.7	30
Trematomus newnesi	39	5.6	1.4	0.1–14.6	0.26	14.6	7
Chionodraco rastrospinosus	46	6.6	0.2	0.1– 0.6	0.29	4.0	19
Electrona antarctica	55	7.8	5.8	2.5–10.2	0.75	28.6	13
Remaining species	52	7.4	–	–	–	–	23
Total	702	100					
	October 25 – November 16, 1983						
Nototheniops larseni	80	58.8	0.6	0.1–12.1	0.27	12.1	7
Trematomus newnesi	12	8.8	0.5	0.5– 1.7	0.63	1.7	5
Chionodraco rastrospinosus	14	10.3	0.1	0.1– 0.3	0.50	0.3	8
Electrona antarctica	2	1.5	–	–	–	0.7	2
Remaining species	28	20.6	–	–	–	–	14
Total	136	100					
	November 21 – December 4, 1984						
Nototheniops larseni	56	42.4	1.5	0.7– 4.0	0.65	7.8	16
Trematomus newnesi	32	24.2	2.0	0.1– 7.6	0.71	7.6	5
Chionodraco rastrospinosus	12	9.1	0.1	0.1– 0.6	0.50	0.6	7
Electrona antarctica	11	8.3	0.4	0.1– 0.8	1.00	0.8	10
Remaining species	21	15.9	–	–	–	–	15
Total	132	100					

Table 4. Summer 1976, 1978, 1982. Total number of specimens caught in the RMT 1+8 (n), percentage of the total number of specimens (%), median abundance per 10^3 m^3 and its 95% confidence interval, ratio of median vs. arithmetic mean, maximum observed abundance and frequency of occurrence as absolute number of standard stations (FO) for the most abundant species. Data from standard stations only

	n RMT 1+8	n (%)	Median abund.	95% conf. interval	Median/ mean	Max. obs. abund.	FO
	January 27 – February 6, 1976						
Nototheniops larseni	59	14.4	2.0	1.1– 7.5	0.59	9.1	13
Notothenia kempi	20	4.9	2.2	0.6–12.0	0.49	12.0	5
Notothenia gibberifrons	38	9.3	1.9	1.1– 2.5	0.61	15.0	9
Pleuragramma antarcticum	252	61.6	8.0	6.5–47.6	0.51	47.6	5
Chionodraco rastrospinosus	10	2.4	0.1	–	0.50	0.5	5
Electrona antarctica	3	0.7	–	–	–	1.0	1
Remaining species	27	6.6	–	–	–	–	6
Total	409	100					
	January 6 – February 13, 1978						
Nototheniops larseni	184	64.3	3.7	0.9– 5.8	0.42	98.4	17
Notothenia kempi	27	9.4	6.1	2.9– 9.9	1.03	9.9	6
Notothenia gibberifrons	–	–	–	–	–	–	–
Pleuragramma antarcticum	–	–	–	–	–	–	–
Chionodraco rastrospinosus	12	4.2	0.2	–	1.00	0.3	5
Electrona antarctica	50	17.5	6.7	2.3–21.1	0.57	56.8	13
Remaining species	13	4.6	–	–	–	–	8
Total	286	100					

Table 4 (continued)

	n RMT 1+8	n (%)	Median abund.	95% conf. interval	Median/ mean	Max. obs. abund.	FO
	February 2 – 28, 1982						
Nothothenips larseni	142	14.8	1.0	0.5– 1.5	0.67	8.1	18
Notothenia kempi	227	23.6	0.9	0.4–10.7	0.19	28.7	11
Notothenia gibberifrons	–	–	–	–	–	–	–
Pleuragramma antarcticum	362	37.6	1.1	0.5– 8.8	0.22	32.6	14
Chionodraco rastrospinosus	12	1.3	0.1	–	1.00	0.1	8
Electrona antarctica	112	11.6	1.5	0.5– 3.4	0.71	8.3	14
Remaining species	107[a]	11.1	–	–	–	–	24
Total	962	100					

[a] Mostly *Notolepis* sp. (n = 63).

Table 5. Autumn 1978, 1981, 1985. Total number of specimens caught in the RMT 1+8 (n) (RMT 8 only in 1985), percentage of the total number of specimens (%), median abundance per 10^3 m^3 and its 95% confidence interval, ratio of median vs. arithmetric mean, maximum observed abundance and frequency of occurrence as absolute number of standard stations (FO) for the most abundant species. Data from standard stations only

	n RMT 1+8	n (%)	Median abund.	95% conf. interval	Median/ mean	Max. obs. abund.	FO
	March 3 – 12, 1978						
Nototheniops larseni	22	13.4	2.1	1.2– 3.6	0.95	4.5	12
Notothenia kempi	15	9.2	1.7	0.8– 2.1	0.77	9.0	10
Pleuragramma antarcticum	2	1.2	–	–	–	2.2	2
Chionodraco rastrospinosus	1	0.6	–	–	–	–	1
Electrona antarctica	118	72.0	5.7	2.4–11.0	0.58	38.7	27
Remaining species	6	3.7	–	–	–	–	6
Total	164	100					
	March 10 – 19, 1981						
Nototheniops larseni	191	33.0	3.4	0.1–23.2	0.25	102.2	11
Notothenia kempi	116	20.0	15.4	0.1–48.1	0.59	65.8	9
Pleuragramma antarcticum	215	37.1	2.7	1.8–90.0	0.13	90.0	7
Chionodraco rastrospinosus	6	1.0	–	–	–	–	2
Electrona antarctica	6	1.0	–	–	–	0.1	4
Remaining species	45	7.8	–	–	–	–	14
Total	579	100					
	March 10 – April 8, 1985						
Nototheniops larseni	5	2.0	–	–	–	0.4	3
Notothenia kempi	43	17.1	0.2	0.1– 0.3	0.67	0.8	17
Pleuragramma antarcticum	162	64.3	0.3	0.1– 1.5	0.27	7.7	15
Chionodraco rastrospinosus	5	2.0	–	–	–	–	5
Electrona antarctica	4	1.6	–	–	–	0.1	4
Remaining species	33	13.1	–	–	–	–	18
Total	252	100					

3.1 Nototheniops larseni

3.1.1 Spatial and Temporal Occurrence

Larvae and postlarvae occurred in the whole area investigated (Figs. 1–3). Distribution in spring in the eastern Bransfield Strait and adjacent waters to the east was similar in 1977 and 1984 (the western Strait was not sampled in 1984), but was confined chiefly to the shelves north of D'Urville, Joinville, and Elephant Islands in 1983 (Fig. 1). Length compositions were unimodal in 1977, 1983, and 1984 with means at 10.4 mm (95% confidence interval 10.2–10.6 mm), 10.9 mm (95% confidence interval 10.2–11.6 mm), and 10.3 mm (10.0–10.6 mm), respectively. The portion of yolk sac stages was 69% in 1977, 45% in 1983, and 38% in 1984.

Distribution in summer 1978 had changed little when compared with spring. Postlarvae were also frequently observed over the shelf of Joinville Island where the pack ice had retreated (Fig. 2b). Distribution in 1976 and 1982 (Fig. 2a,c) was comparable to that during 1978.

In autumn (March) 1978 the center of postlarval distribution was off D'Urville and Joinville Islands. Early stages of *N. larseni* were scarce in Bransfield Strait and postlarvae were completely absent around Elephant Island (Fig. 3a). In contrast, during 1981 postlarvae were abundant near Elephant Island, while occurring rarely in the Strait (Fig. 3b). In 1985 only single specimens were obtained from near Adelaide Island to the western Bransfield Strait, but these levels of abundance may be underestimates due to the selectivity of the RMT 8.

3.1.2 Abundance

Abundance data of *N. larseni* from Tables 3–5 were plotted versus time of the season to demonstrate year to year variability (Fig. 4). Median numbers per 10^3 m^3 at a given season of different years varied mostly within one order of magnitude; however, statistical analysis revealed significant differences in spring and summer. Abundance (median) in spring 1977 was significantly higher than in 1983 and 1984 (Mann-Whitney U-test, $p < 0.05$). It was about four times higher than during the same time interval in 1984 ($p < 0.001$), and statistical significance is maintained even if the different extent of station grids in the two years is taken into account and the comparison restricted to the area east of 58° W ($p < 0.01$, cf. Fig. 2a,c). Maximum abundance during 1977 exceeded that in 1984 by two orders of magnitude (Fig. 4). Median numbers per 10^3 m^3 during 1983 and 1984 were not statistically different and also maximum values were about at the same level.

In summer there was no difference between (median) abundance levels in 1978 and 1976, statistically. However, there were significantly fewer postlarvae of *N. larseni* in 1982 than in both earlier years (Fig. 4; Mann-Whitney U-test, $p < 0.05$). In autumn 1978 abundances were approximately normally distributed and numbers per 10^3 m^3 evenly low (Table 5). Comparison of data from 1978 with 1981 did not reveal a statistical difference in median levels, but maximum abundance recorded in 1981 (off Elephant Island) was much higher (Fig. 4).

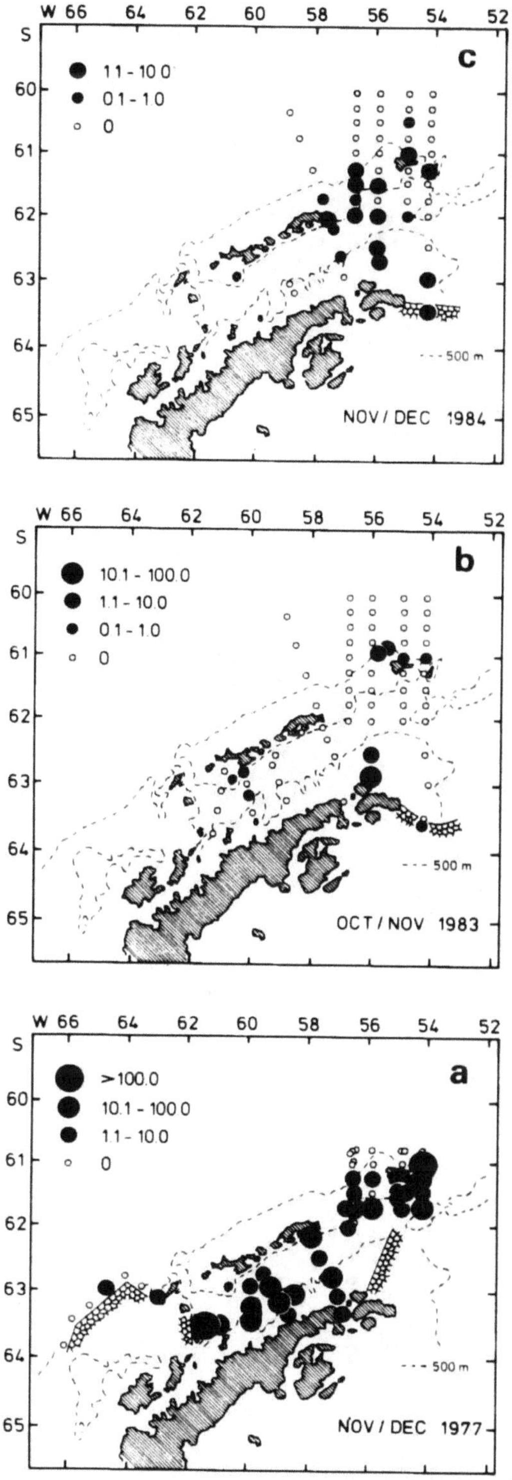

Fig. 1a–c. Quantitative distribution (number per 10^3 m^3) of larval and postlarval *Nototheniops larseni* in spring 1977 (a), 1983 (b) and 1984 (c)

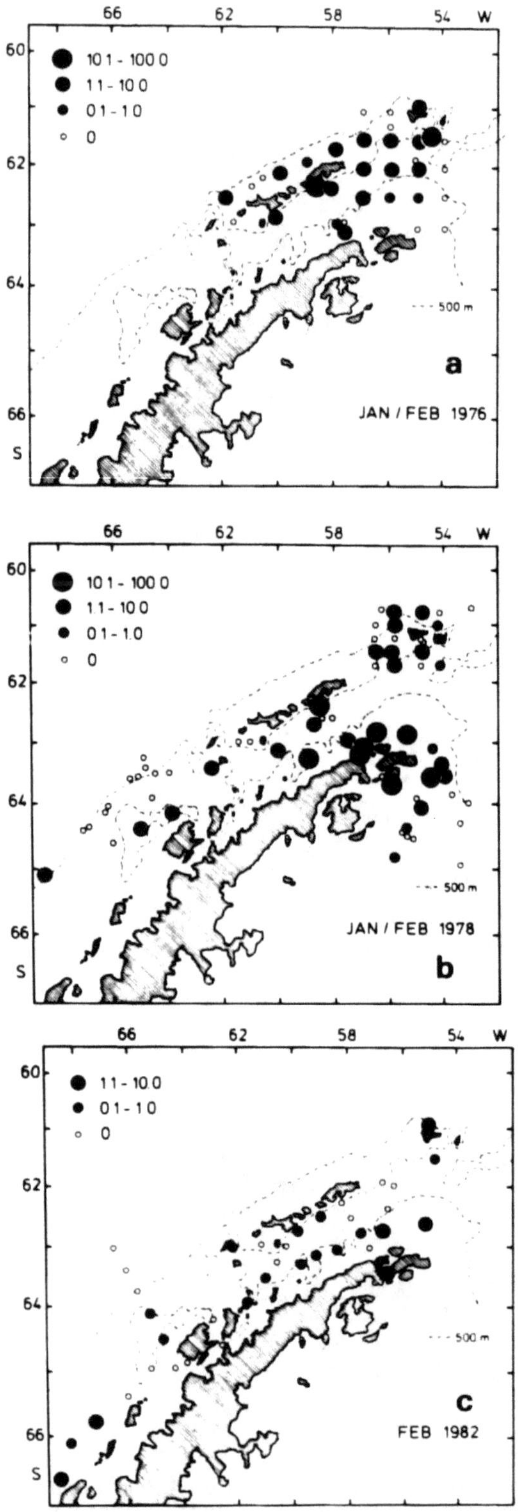

Fig. 2a–c. Quantitative distribution (number per 10^3 m^3) of post-larval *Nototheniops larseni* in summer 1976 (a), 1978 (b) and 1982 (c)

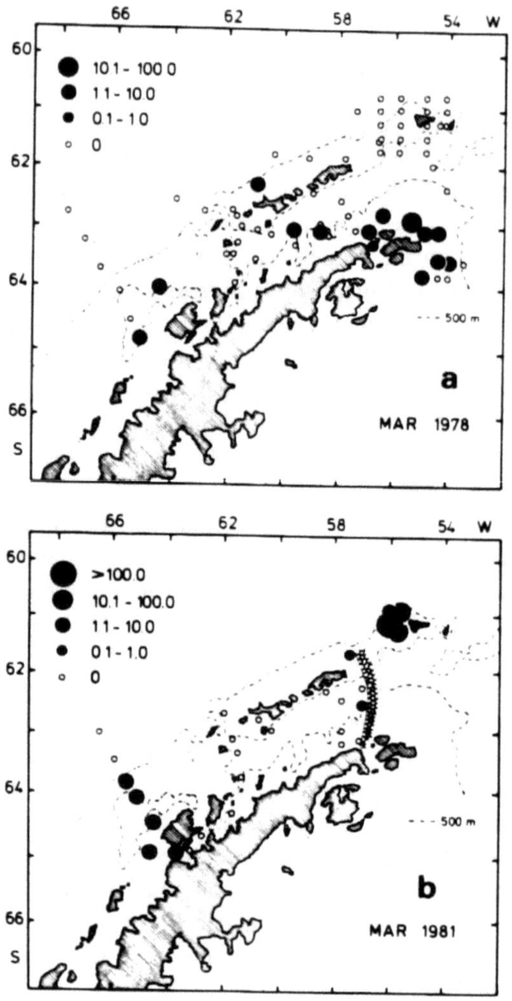

Fig. 3a,b. Quantitative distribution (number per 10^3 m^3) of post-larval *Nototheniops larseni* in autumn 1978 (a) and 1981 (b)

3.1.3 Seasonal Differences in Abundance

The logarithmic decrease of median numbers per 10^3 m^3 (Fig. 4) and a tendency toward a less skewed positive distribution of data (cf. Tables 3–5) suggest an overall decline in abundance of *N. larseni* from November 1977 (spring) to March 1978 (autumn). The seasonal decrease of values was significant only between spring and summer (Friedman's median test, $p < 0.05$), whereas no statistical significance was revealed for the apparent decline between summer and autumn.

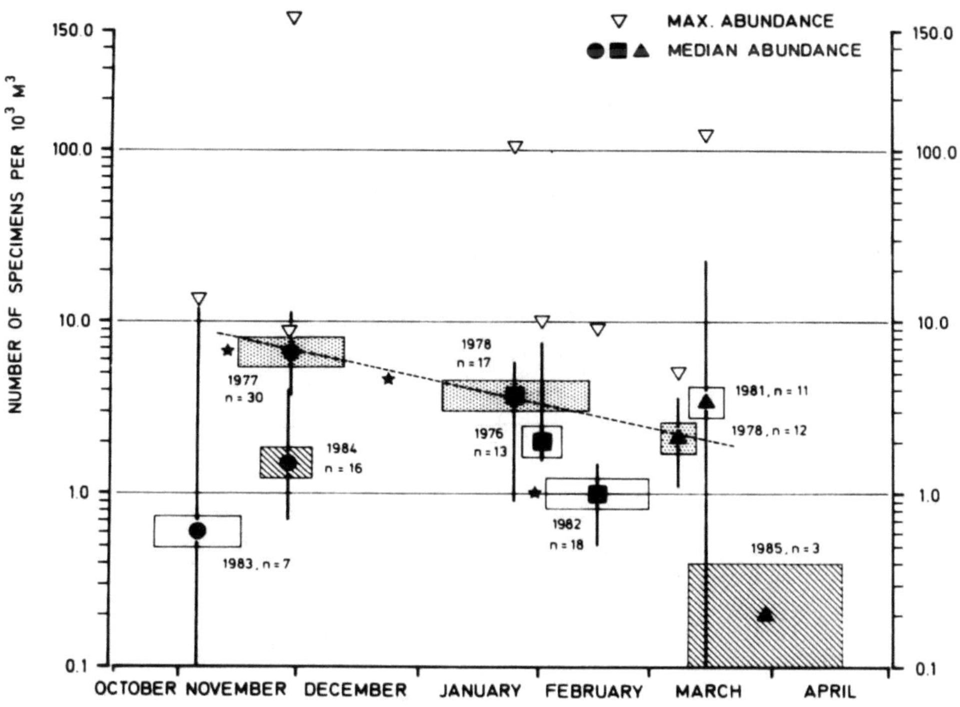

Fig. 4. *Nototheniops larseni.* Median and maximum abundances per 10^3 m^3 of seawater recorded during spring (*dots*), summer (*squares*) and autumn (*filled triangles*) of years investigated. *Rectangles* represent time periods covered during individual cruise legs. The area was sampled repeatedly during the 1977/78 and 1984/85 seasons. The *hatched line* was fitted by eye to the decline in abundance during 1977/78. *Vertical bars* are 95% confidence intervals and *asterisks* denote statistical significant high or low abundances. *n* is the number of observations, i.e., the FO in Tables 3–5

3.2 Nothothenia kempi

3.2.1 Spatial and Temporal Occurrence

In summer 1976 catches were confined to the area east of King George Island and around Elephant Island (Fig. 5a). In summer 1978 postlarval *N. kempi* were exclusively caught close to the 500-m depth contour (Fig. 5b). Most of these records were made in January. Distribution in 1982 was similar, but the species was also observed over the shelf south of the Palmer Archipelago (Fig. 5c). Postlarvae caught over the shelf break were significantly smaller than those from coastal waters (Kellermann 1986a). In January 1978 50% of larvae had yolk remains, whereas in February 1976 and 1982 only postlarvae were encountered.

Occurrence in autumn closely resembled that in summer: postlarval *N. kempi* were chiefly present close to the shelf break and on the shelf south of the Palmer Archipelago. Maximum abundance was observed off Anvers Island both in 1978 and 1981. In 1985, postlarvae occurred only rarely around Elephant Island (Fig. 6a–c).

3.2.2 Abundance

Despite the higher (median) abundance of *N. kempi* in summer 1978 (Table 4), statistical analysis of data did not reveal significant differences between years (Friedman's median test, $p < 0.05$). Similarly, maximum abundance was about at the same level in summer of each year. In autumn (median) abundance in 1981 was greater than that in 1978 by one order of magnitude (Table 5), but no statistical significance was indicated by the Mann-Whitney U-test ($p < 0.05$). Low abundanoe observed in 1985 was most probably a result of the selectivity of the RMT 8.

3.2.3 Seasonal Difference in Abundance

Confidence intervals for estimates of (median) abundance during summer and autumn 1978 do not overlap (Tables 4, 5), providing evidence for a statistically significant decrease in numbers per 10^3 m^3. Higher abundance levels in January/February and the skewed negative distribution in comparison to March support the assumption that larvae of *N. kempi* hatch during summer.

3.3 Pleuragramma antarcticum

3.3.1 Spatial and Temporal Occurrence

Larvae, which should have hatched by mid-November (observations from the southern Weddell Sea, Hubold personal communication) were not present in Bransfield Strait in spring. Postlarvae were not observed until late summer (end of January 1976). Distribution in late summer and autumn (February 1982, March 1981, March/April 1985, Fig. 7a–c) was similar: *P. antarcticum* occurred almost exclusively over the shelf of the Peninsula from Joinville down to Adelaide Island. Records were rare in the central and northern Bransfield Strait and off Elephant Island.

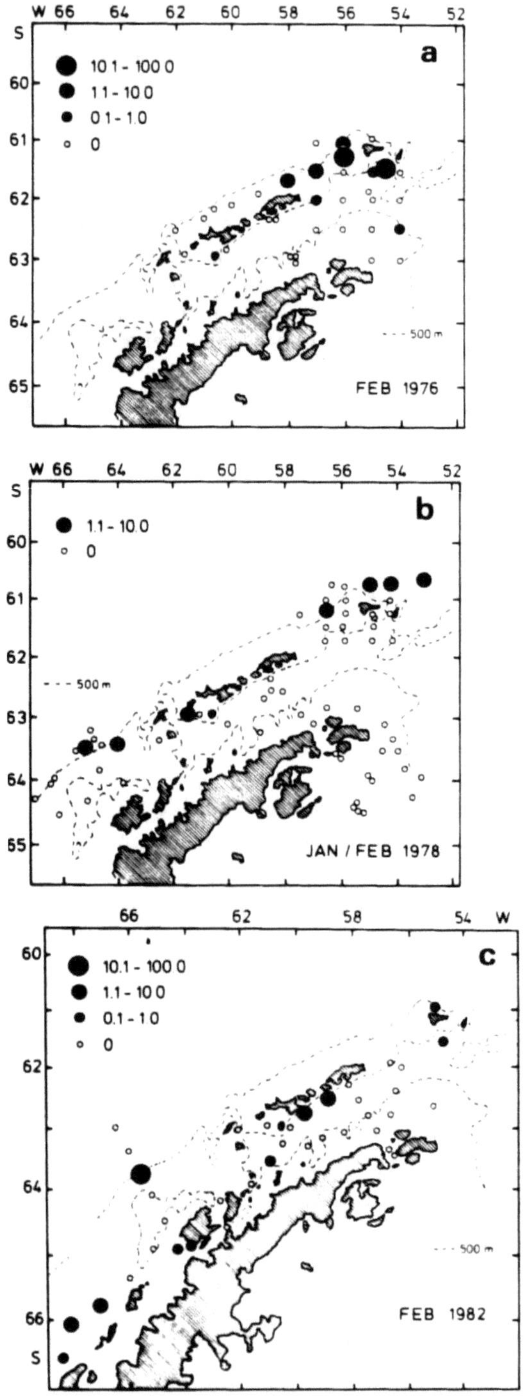

Fig. 5a–c. Quantitative distribution (number per 10^3 m^3) of early stage *Notothenia kempi* in summer 1976 (a), 1978 (b) and 1982 (c)

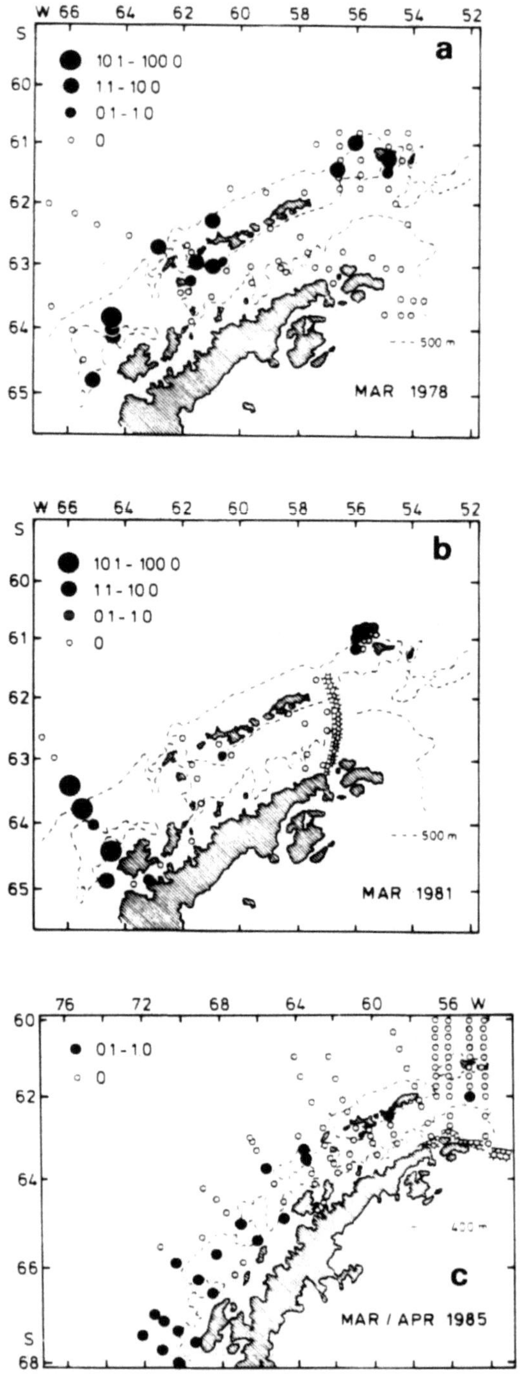

Fig. 6a–c. Quantitative distribution (number per 10^3 m^3) of post-larval *Notothenia kempi* in autumn 1978 (a), 1981 (b) and 1985 (c RMT 8 catches only)

3.3.2 Abundance

The almost complete lack of postlarvae in summer 1978 was striking (see Kellermann 1986b). Abundance (median) was significantly higher in summer 1976 (Friedman's median test, $p < 0.05$) than in 1982 (Table 4) whereas maximum values were similar. However, the southwestern shelf of the Peninsula was not sampled in 1976, and highest numbers per 10^3 m^3 were observed off Joinville Island in both years.

In autumn abundance (median and maximum) was much higher in 1981 than in 1985 (Table 5). This difference, however, was statistically not significant (Friedman's median test, $p < 0.05$), even if analysis was restricted to standard stations east of 66°W (cf. Fig. 7b,c). Hence, allowing for underestimation of abundance in 1985 (RMT 8 catches only), abundance levels may well have been the same in both years.

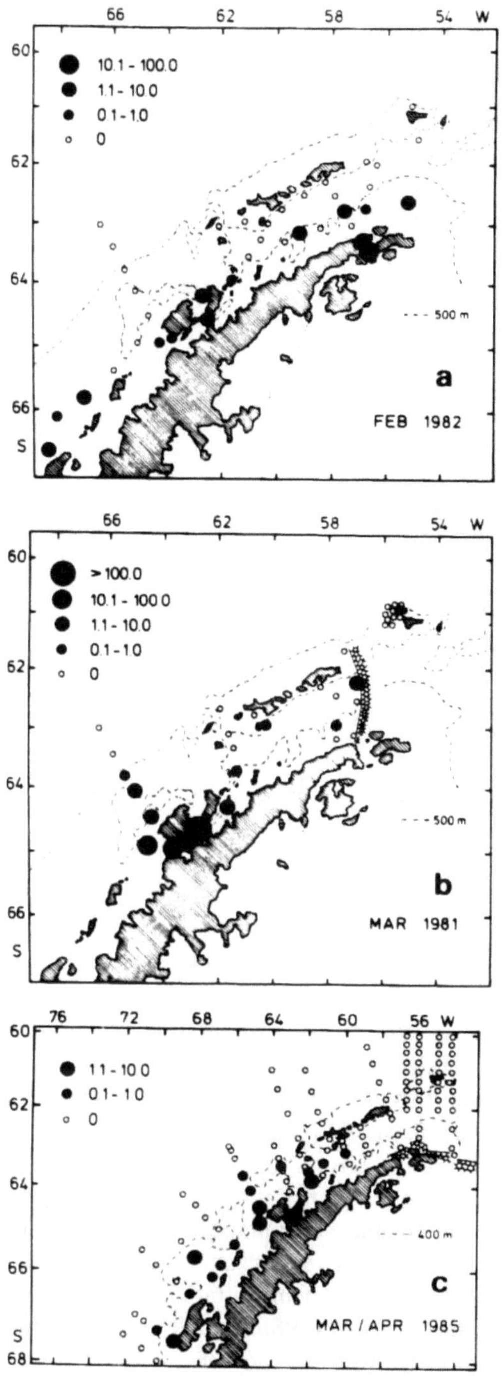

Fig. 7a–c. Quantitative distribution (number per 10^3 m^3) of postlarval *Pleuragramma antarcticum* in summer 1982 (a), and in autumn 1981 (b) and 1985 (c, RMT 8 catches only)

3.4 Electrona antarctica

3.4.1 Spatial and Temporal Occurrence

In spring distribution was confined mostly to waters beyond the shelfbreak west of the South Shetland Islands and north of Elephant Island (Fig. 8a,b). In 1977 early stages were also found in the north-eastern approaches of Bransfield Strait at the 500 m depth contour. In 1983 they were only observed on two stations in the north-east of the Elephant Island box.

In summer (January/February 1978) the overall pattern had changed little compared with that earlier in the season (Fig. 8c). Occurrence in autumn was very variable; in March 1978 *E. antarctica* was spread over most of the shelf around Elephant Island and was also fairly common in Bransfield Strait and its western entrance.

In other years (1976, 1981) only a few specimens were obtained in the whole area, except for February 1982. Here, occurrence was similar to summer 1978, but early stages were also encountered over the shelf south of Anvers Island. They were most frequently observed in the 135–200 m depth layer and only scarcely in the top 70 m of the water column (Kellermann unpubl. data).

3.4.2 Abundance

Only in spring 1977 and 1984 and in summer 1978 and 1982 was the number of records sufficiently high to allow for statistical comparison. Abundance of *E. antarctica* in 1977 was significantly higher (Mann-Whitney U-test, $p < 0.05$) by one order of magnitude and maximum abundance exceeded that in 1984 by two orders of magnitude (Table 3). Median number per 10^3 m^3 in summer 1978 was significantly higher than in summer 1982 (Mann-Whitney U-test, $p < 0.025$).

3.4.3 Seasonal Difference in Abundance

Median abundance estimates and corresponding confidence intervals were almost similar throughout the whole season 1977/78 (Tables 3–5) and there were no statistical differences between median levels (Mann-Whitney U-test, $p < 0.05$). Slightly increased numbers per 10^3 m^3 (median and maximum) in January/February were not statistically significant.

3.5 Other Abundant Species

Early stages of *Notothenia gibberifrons* were found only in January/February 1976 and in March 1981. In summer 1976 it was among the most abundant species (Table 4), and postlarvae occurred mostly south of Elephant Island and in the northeastern part of Bransfield Strait, whereas in autumn 1981 only single specimens were obtained from the whole area (see Kellermann 1986a).

Another abundant Nototheniid in spring of each year was *Trematomus newnesi* (Table 3). Larval and postlarval stages in spring were confined to shelf waters where they were observed regularly in the southern Bransfield Strait, but less consistently east of the Strait and around Elephant Island. Hatching occurs most likely around October in shallow waters (Radtke et al. 1988). Consequently, allowing for a time lag between hatching and dispersal into open

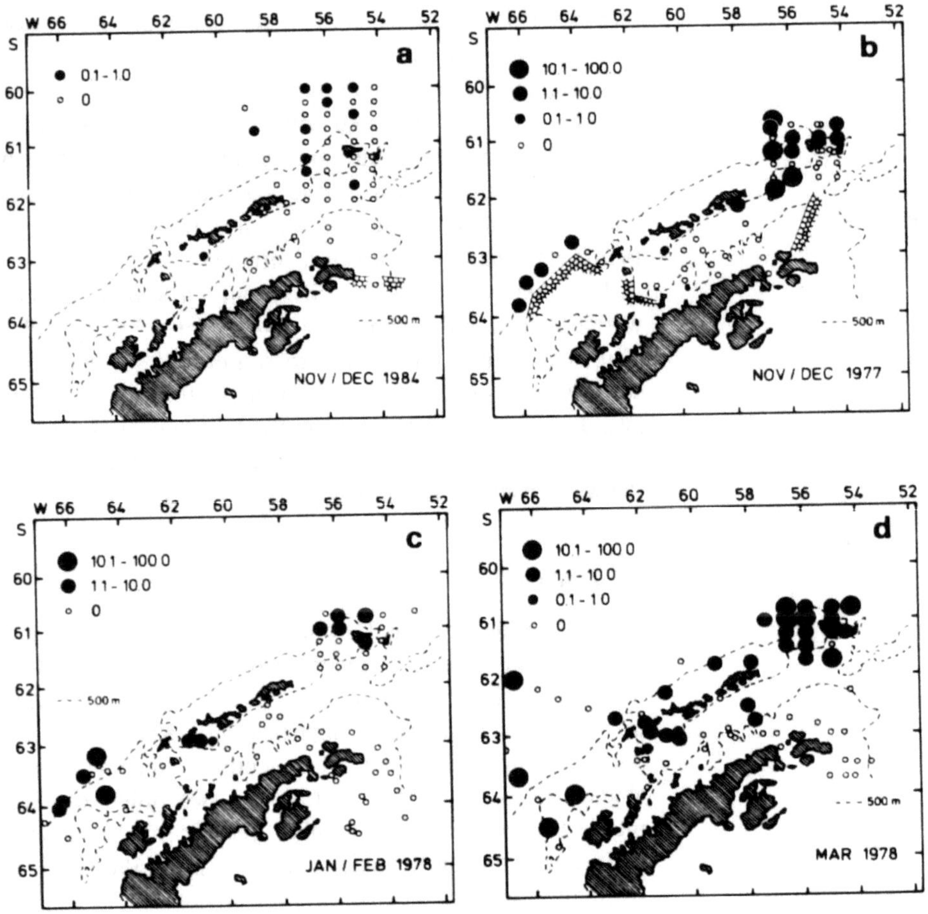

Fig. 8a–d. Quantitative distribution (number of 10^3 m^3) of early stage *Electrona antarctica* in spring 1977 (a) and 1984 (b), and in summer (c) and fall (d) 1978

waters, lowest abundance was recorded in 1983 when the area was fished earliest in the season (Table 3).

The most abundant species of Channichthyids was *Chionodraco rastrospinosus.* Early stages were present in the whole area investigated in spring of each year. They were common in Bransfield Strait and its southeastern approaches in all years with a pattern of distribution which closely resembled that of *N. larseni.* Co-occurrence of both species, however, varied from 83% (1977) to 25% (1983). Abundance levels in spring were mostly below one specimen per 10^3 m^3 (Table 3), and there were no statistical differences between years.

4 Discussion

Survey results are in general subject to two types of error: systematic errors primarily affecting the variance between surveys and random errors. Due to the sampling scheme being designed to investigate krill distribution and ice conditions changing from year to year, systematic errors may have obscured patterns of spatial and temporal distribution to a larger extent than surveys orientated towards ichthyoplankton would have done. Among these systematic errors, the uneven extent to which the area was covered from year to year, and the variation in maximum fishing depth are probably the most significant. Moreover, shallow and inshore waters where early stages of some Notothenioid species (e.g, *Trematomus newnesi*) are likely to occur were under-represented by the station grids. The rigorous standardization and control of survey methods since the introduction of the SIBEX transects has probably reduced systematic errors, but this does not imply that these are the optimum approach for ichthyoplankton surveys.

Another systematic error, though constant between most surveys, is the efficiency of the net to sample a multispecies ichthyoplankton community adequately. The insufficiency of the RMT 8 for quantitative estimates (mesh escapement) has been demonstrated for Nototheniid postlarvae by Kellermann (1986a), and he also found evidence that already postlarval icefish at about 50 mm length may be able to avoid the RMT 8. Therefore, in the present study abundance estimates are based on RMT 1 catches except for 1985, and only the early stages of icefishes of up to 45 mm in length were analyzed.

Besides systematic errors, the observed variability is assumed to reflect responses of early life stages to seasonal and annual variability in abiotic and biotic conditions. Among these conditions, transport by water masses and currents, as well as food availability and predation, are likely to be the most important factors affecting distribution and abundance. Both sources of variance and their relevance to particular species will be discussed in conjunction with some conclusions on the reproduction of the most abundant species.

Early life stages of most of the more than 40 coastal and mesopelagic species known from the Peninsula region (Permitin 1977; Kock 1982; Nast et al. this Vol.) have been identified from ichthyoplankton collections in the area (Kellermann and Slosarczyk 1984; Slosarczyk 1986). Most striking is, however, that only a few occurred frequently (e.g., *Nototheniops larseni, Notothenia kempi*) whereas others, though abundant as adults, were scarcely encountered in our catches. This applies especially to the commercially exploited *Champsocephalus gunnari, Chaenocephalus aceratus* and to some extent also to *N. gibberifrons*. Spawning of *C. gunnari* and *Ch. aceratus* in inshore waters and fjords as around South Georgia and the fact that larvae and postlarvae occur predominantly in coastal waters and/or near-bottom layers might explain the paucity of both species in our catches. However, off South Georgia early stage *C. gunnari* were frequently caught over the shelf in the upper part of the water column (Kellermann in preparation). This might indicate that the low numbers observed are not merely a sampling artifact. Postlarvae of *N. gibberifrons* were caught in larger numbers only in 1976. They were found occasionally in 1981 (Kellermann and Kock 1984), and in summer of the 1984/85 season (Sinque et al. 1986). It seems, however, doubtful that variability in spawning success alone should account for the substantial difference between 1976 and other years.

Nototheniops larseni was the most frequently observed Notothenioid in our catches. Juveniles and adults of this species are common down to 400–500 m depth in the whole area investigated. Spawning probably takes place in July/August. Spawning grounds have not yet been located due to the inaccessability of the area in winter, but the wide distribution of early life stages indicates that *N. larseni* spawn over most of their geographical range.

The unusually high abundance of early stage *N. larseni* in spring 1977 appears to indicate a strong year-class. However, postlarval concentrations subsequently recorded during the 1977/78 season were comparable to those observed in other years. Moreover, the standard haul depth of 140 m used in 1977/78 may overestimate numbers per 10^3 m^3 by about 15% when compared with a haul depth of 200 m (Kellermann 1986a). The abundance level in spring 1984 was not different from that recorded earlier in the season in 1983. Hence, the assumption of relatively low abundance in spring 1984 seems more likely. Further, extremely low postlarval numbers were observed in autumn of the 1984/85 season, even if net selectivity by the RMT 8 is taken into account. The lowest proportion of yolk sac stages in spring 1984 among years may point to a less favorable nutritional environment in that year.

Possible sources for a seasonal decline of postlarval abundance of *N. larseni* as observed in 1977/78 may be predation and dispersal. Starvation seems to be less likely, since high feeding incidences were found during that time (Kellermann 1986a). Predation by larval and postlarval *Chionodraco rastrospinosus* is probably one of the major causes of mortality in the early life of *N. larseni*, since these form major portions of the diets of the early stage icefishes until the summer months (Kellermann 1986a). Dispersal by the easterly currents in Bransfield Strait in spring (Clowes 1934; Stein 1986) was apparent, but there was also evidence for transport and accumulation of postlarval *N. larseni* with Weddell Sea Water. In February 1982, surface water from the Weddell sea appeared to be more widespread in the Strait than in summer 1981 (Heywood 1985); however, overall abundance during 1982 was nevertheless low. Most striking was the complete lack of postlarvae off Elephant Island in March 1978. This coincided with the reinforced advection of oceanic (Bellingshausen Sea/Drake Passage) water to the area (Kock et al. 1980), that apparently had either swept away the postlarval population or substantially affected the nutritional environment in that area.

Notothenia kempi, which is the only Antarctic Nototheniid lacking antifreeze glycoproteins (DeVries and Lin 1977), has a similar geographical range, but concentrations of adults have been found only in shelf and upper slope waters below 300 m depth (Sosinski and Skora 1979; Nast et al. this Vol.), which are under the influence of Warm Deep Water. Spawning takes place in spring and occurrence of early life stages was primarily over the slope in January/February, indicating that spawning is confined to the Warm Deep Water layer. Only traces of this water mass are found in Bransfield Strait (Clowes 1934), which could explain the scarcity of postlarval records there.

West of the Peninsula, larvae and postlarvae hatched over the slope may meet either with the Antarctic Circumpolar Current and are drifted to the northeast, or with the southwesterly flowing coastal Polar current (East Wind Drift) and the cyclonic eddy north off Adelaide Island (Stein this Vol.). This eddy is likely to transport early stages further on to the shelf and retain them in the area. In March 1981 postlarvae occurred on the transect toward Anvers Island in the northeasterly current over the slope and in the coastal current over the shelf area (Stein 1982). The larger mean size of postlarvae on the shelf in 1982 may indicate that postlarvae are in fact older there or that growth conditions are more favorable over the shelf.

Records of adult *Pleuragramma antarcticum* are scattered over the whole area (Kock unpublished). Concentrations, however, have been encountered only off the Biscoe Islands (Sosinski and Skora 1979). The lack of early stages in Bransfield Strait in November/December suggests that this species does not spawn in the Strait. Concentrations of postlarvae off the northern tip of the Peninsula and in southern Bransfield Strait in February most probably originate from the northeastern Weddell Sea (Kellermann 1986b). Weddell Sea Water flowing along the northern shelf of the Peninsula to the southwest could be traced as far south as the northern Gerlache Strait in some years and might have accounted for a transport of postlarvae from the Weddell Sea to west of the Peninsula. However, the velocity of this current is low and is probably in the order of 5–10 cm/s (Stein 1979; Wittstock and Zenk 1983). Further, mixing of this water mass occurs in a northeasterly flowing current in the meander between Deception Island and Tower Island (Anonymous 1982, 1983). The more or less simultaneous occurrence of larger numbers of postlarvae west of the Peninsula (Gerlache Strait, off Biscoe and Adelaide Is.) argues for a separate origin of early life stages in both areas. Variability in abundance in Bransfield Strait and its eastern approaches may thus partly reflect seasonal and annual variation in the transport of Weddell Sea Water to the area.

Chionodraco rastrospinosus is the most abundant icefish in the survey area (Gubsch 1982; Kock unpublished) and was even subject of a large-scale fishery in the seasons 1978/79 and 1979/80 (Kock et al. 1985). Abundance of adults was clearly reflected by the predominance among Channichthyids in the ichthyoplankton. Spawning to the north of the South Shetland Islands is most likely (Gubsch 1982). Early life stages, however, have been primarily encountered in Bransfield Strait and waters to the east (see also Kellermann and Kock 1984; Slosarczyk and Rembiszewski 1982; Slosarczyk 1986).

Early stages of the Myctophid *Electrona antarctica* 6–12 mm in length were present in the ichthyoplankton from November to March. This gave further evidence that the species is a batch spawner and spawning is extended over a long period of time (Lisovenko 1980; Lubimova et al. 1982). Larvae and postlarvae were predominantly caught in oceanic and slope waters beyond the 500-m contour (Efremenko 1972) and rarely observed over the shelf. The pattern of distribution was thus clearly different from the shelf-bound pattern of most Notothenioid species. The increase in occurrence toward autumn 1978 around Elephant Island coincided with the reinforced advection of oceanic water to the area (Kock et al. 1980).

It is obvious from our results that knowledge of the spatial and temporal distribution of early life stages of fish in the Peninsula region, particularly of those of the commercially exploited species, is still not satisfactory enough for the successful implementation of a regular ichthyoplankton survey under the auspices of CCAMLR. Further efforts will be necessary to reduce variance associated with spatial variability, which may be done by refinement of stratification and station allocation on the basis of the existing SIBEX grid system to allow for compatibility with older data. However, for some species (e.g., *Notothenia kempi*) the specific pattern of spatial distribution has to be considered, whereas for others (e.g., Channichthyid species) the efficiency of sampling has to be improved by directing surveys to particular life stages such as postlarvae and juveniles, and by the combined use of different gears, as was recommended by Kellermann and Slosarczyk (1984).

Acknowledgments. We thank crew and colleagues who helped during the sampling and sorting procedures on board of RV *Polarstern* and FRV *Walter Herwig*. This publication has the contribution number 49 of the Alfred-Wegener-Institut für Polar- und Meeresforschung, Bremerhaven.

References

Anonymous (1982) 1st Post-FIBEX Hydrographic Data Interpretation Workshop, Hamburg FRG, Sept 20–26, 1982. BIOMASS Rep Ser 30:12 pp

Anonymous (1983) 2nd Post-FIBEX Hydrographic Data Interpretation Workshop, Hamburg FRG, May 16–20, 1983. BIOMASS Rep Ser 31:27 pp

Baker A de, Clarke MR, Harris MJ (1973) The N.I.O. Combination Net RMT (1 + 8) and further developments of Rectangular Midwater Trawls. J Mar Biol Assoc UK 53:167–184

CCAMLR (1986) Rep 5th Meet Sci Comm, Hobart, Australia

Clowes AJ (1934) Hydrology of the Bransfield Strait. Discovery Rep 9:1–64

DeVries AL, Lin Y (1977) The role of glycoprotein antifreezes in the survival of Antarctic fishes. In: Llano GA (ed) Adaptations within Antarctic ecosystems. Gulf, Houston, Texas, pp 439–458

Efremenko FN (1972) Morphological peculiarities of *Electrona antarctica* Günther and *Gymnoscopelus braueri* Lönnberg and their distribution in the Scotia Sea. Trudy VNIRO 90:120–130

Fütterer D (1984) Die Expedition ANTARKTIS II mit FS *Polarstern* 1983/84. Ber Polarforsch 18:92 pp

Gubsch G (1982) Zur Verbreitung und zur Biologie der Eisfische (Chaenichthyidae) im atlantischen Sektor der Antarktis. Fischerei-Forschung 20:39–47

Hempel G (1985) Die Expedition ANTARKTIS III mit FS *Polarstern* 1984/85. Ber Polarforsch 25:209 pp

Hempel G, Heywood RB (1982) Joint Biological Expedition on RRS *John Biscoe*, February 1982. Ber Polarforsch 5:39 pp

Hempel G, Sahrhage D, Schreiber W, Steinberg R (1979) Antarktis-Expedition 1977/78 der Bundesrepublik Deutschland. Arch Fischereiwiss 30:119 pp

Hempel G, Kock KH, Stein M (1982) Fahrtverlauf. Antarktis-Expedition 1981 der Bundesrepublik Deutschland mit FFS *Walther Herwig*. Arch Fischereiwiss 33:17–26

Heywood RB (1985) Environmental conditions in the Antarctic Peninsula Area of the Southern Ocean during the Anglo-German Joint Biological Expedition, February 1982. Meeresforschung 30:220–239

Hubold G (1984) Spatial distribution of *Pleuragramma antarcticum* Pisces: Nototheniidae) near the Filchner- and Larsen Ice Shelves (Weddell Sea, Antarctica). Polar Biol 3:231–236

Hubold G (1985) The early life history of the High-Antarctic silverfish *Pleuragramma antarcticum*. In: Siegfried WR, Condy PR, Lews RM (eds) Antarctic nutrient cycles and food webs. Springer, Berlin Heidelberg New York Tokyo, pp 445–451

Hureau JC (1982) Methods for studying early life history stages of Antarctic fishes. Cybium 6:2–11

James R, Wörner FG (1978) Results of the sorting of the micronekton and zooplankton material sampled by the German Antarctic Expedition 1975/76. Ber Inst Meereskd Kiel 59: 53 pp

Keller R (1983) Contributions to the early life history of *Pleuragramma antarcticum* Boulenger 1902 (Pisces, Nototheniidae) in the Weddell Sea. Meeresforschung 30:10–24

Kellermann A (1986a) On the biology of early life stages of Notothenioid fishes off the Antarctic Peninsula. Ber Polarforsch 31: 149 pp

Kellermann A (1986b) Geographical distribution and abundance of postlarval and juvenile *Pleuragramma antarcticum* (Pisces, Notothenioidei) off the Antarctic Peninsula. Polar Biol 6:111–119

Kellermann A, Kock KH (1984) Postlarval and juvenile notothenioids (Pisces, Perciformes) in the southern Scotia Sea and northern Weddell Sea during FIBEX 1981. Meeresforschung 30: 82–93

Kellermann A, Slosarczyk W (1984) Distribution of postlarval and juvenile Notothenioidei in the Atlantic sector of the Southern Ocean during FIBEX 1981. BIOMASS Rep Ser 36:27 pp

Kock KH (1982) Fische aus RMT 8- und Krillschwimmschlepp-netzfängen während FIBEX 1981. Arch Fischereiwiss 33:97–112

Kock KH, Nast F, Stein M (1980) Some interactions between hydrography, krill and fish near Elephant Island in 1977/78. ICES C.M. 1980/L:3. Biol Oceanogr Comm, 12 pp

Kock KH, Duhamel G, Hureau JC (1985) Biology and status of exploited Antarctic fish stocks. BIOMASS Sci Ser 6:143 pp

Lisovenko LA (1980) Some data on the reproduction of *Electrona antarctica* from the Scotia Sea. ONTI VNIRO 1980:33–38 (in Russian)

Lubimova TG, Shust KV, Troganovskij FM, Semenov AB (1982) The ecology of mass species of myctophids from the Atlantic Antarctic. Antarktika 22:99–106 (in Russian)

Permitin YU YE (1977) Species composition and zoogeographical analysis of the bottom fish fauna of the Scotia Sea. J Ichthyol 17:710–726

Piatkowski U, Klages N (1983) German Antarctic Expedition 1980/81 with FRV *Walther Herwig* and RV *Meteor*. First International BIOMASS Experiment (FIBEX). Data of micronekton and zooplankton hauls. Ber Polarforsch 15:59 pp

Radtke RL, Targett TE, Kellermann A, Bell J, Hill K (1988) Antarctic fish growth: profile of *Trematomus newnesi*. Mar Ecol Progr Ser (in press)

Roe HSJ, Shale DH (1979) A new multiple rectangular midwater trawl (RMT 1 + 8) and some modifications to the Institute of Oceanographic Science RMT 1 + 8. Mar Biol 50:283–288

Sahrhage D (1986) Participation of the Federal Republic of Germany in SIBEX and CCAMLR activities on Antarctic marine living resources. Arch Fischereiwiss 37:3–24

Sahrhage D, Schreiber W, Steinberg R, Hempel G (1978) Antarktis-Expedition 1975/76 der Bundesrepublik Deutschland. Arch Fischereiwiss 29:96 pp

Sinque C, Koblitz S, Costa LM (1986) Distribution of larval and postlarval Antarctic fishes around Elephant Island and Bransfield Strait – Antarctica. Neritica (Pontal do Sul) 1(3):103–110

Sosinski J, Skora K (1979) New species of fishes of potential industrial value in the Antarctic region. Biul Morsk Inst Ryb 4/54:12–15 (in Polish)

Slosarczyk W (1986) Attempts at a quantitative estimate by trawl sampling of distribution of postlarval and juvenile notothenioids (Pisces, Perciformes) in relation to environmental conditions in the Antarctic Peninsula region during SIBEX 1983-84. Mem Natl Inst Polar Res Spec Issue 40:299–315

Slosarczyk W, Rembiszewski JM (1982) The occurrence of juvenile Notothenioidei (Pisces) within krill concentrations in the region of the Bransfield Strait and the southern Drake Passage. Pol Polar Res 3:299–312

Stein M (1979) Stratification and currents off Elephant Island in the early February 1976. Meeresforschung 27:75–87

Stein M (1982) Fischereiozeanographische Untersuchungen während FIBEX 1981. Arch Fischereiwiss 33:35–51

Stein M (1986) Variability of water masses and currents off the Antarctic Peninsula during SIBEX. Arch Fischereiwiss 37:25–50

Wittstock RR, Zenk W (1983) Some current observations and surface T/S distribution from the Scotia Sea and the Bransfield Strait during early austral summer 1980/81. "Meteor" Forschungsergeb Reihe A/B 24:77–86

The Pelagic Ecosystem of the Bransfield Strait, Antarctica: An Analysis of Microbiological, Planktological and Chemical Characteristics by Multivariate Analyses

M. BÖLTER[1], B. VON BODUNGEN[2], G. LIEBEZEIT[3], and M. MEYER[1]

Summary. A data set composed of biological, chemical, and physical descriptors was treated with a number of statistical techniques, including correlation and cluster analyses. We aimed at establishing reasons for the observed variability – both horizontal and vertical – of planktonic and microbial biomass and activity. Despite an observed zonation in phytoplankton populations, cluster analysis on the complete data set indicates overlap between these zones. The results indicate that for an understanding of system dynamics it is necessary to include not only biological parameters. Thus, descriptors of the organic fraction, both absolute and ratios, provide additional information for elucidation of water mass history, for example. In this context derived parameters appear to be more important than actually obtained data for a comprehensive interpretation of the original data base. Water masses distinguished by physical parameters cannot necessarily be regarded as being similarly distinct in a biological sense. This stems, e.g., from vertical transport of euphotic zone-produced material into deeper water layers by krill fecal pellets or other particulate organic material.

1 Introduction

The Bransfield Strait and its adjacent areas are characterized by a complex hydrographic structure and a high potential for primary production during austral spring and summer. During November/December 1980 a comprehensive study was carried out in the pelagic environment of this area which was designed to follow the development of early stages of phytoplankton blooms in relation to the hydrographic features (cf. Zeitzschel and Zenk 1981).

Hence, parameters of primary production, phytoplankton standing stock and its sedimentation, inorganic nutrients, dissolved and particulate organic material, as well as microbial standing stock and activity parameters, were monitored. Various results of this cruise have been published in previous reports (Bölter and Dawson 1982; Liebezeit 1984; Bodungen et al. 1986; Liebezeit und Bölter 1986; Bölter 1987).

These reports, however, deal with the specific approaches of the different disciplines and the data sets are mainly discussed by using established methods. This chapter comprises a full data set of 29 stations and is analyzed by non-parametric statistics and multivariate methods at different levels of the lower organism's food web in order to establish biologically determined zonations in this environment. It is assumed that different water masses exhibit biological and chemical signals as multivariate properties which can be used to describe the biological character as well as past history or actual processes at the autotrophic/heterotrophic level.

2 Material and Methods

The field study was carried out during November/December 1980 on board the German research vessel *Meteor,* cruise 56/2. Water samples were taken in sterilized all-glass containers for microbiological and organic chemical parameters or by multisamplers (General Oceanics Co.) for phytoplankton data and inorganic nutrients. All methods are described in detail in previous reports (microbiology and organic chemistry: Bölter and Dawson 1982; Liebezeit 1984; Liebezeit and Bölter 1986; primary production, inorganic nutrients, particulate organic carbon and nitrogen: Bodungen et al. 1986); all parameters used in this study are listed in Table 1. More detailed information about the individual parameters and their statistical properties is available upon request from the authors. Figure 1 depicts a map of the stations mentioned in this study.

Statistical analysis follows the program scheme of Bölter and Meyer (1980) with further information given by Bölter and Meyer (1986). The total data set cromprises 67 parameters measured in 64 water samples at 29 stations/depth profiles, as presented in Fig. 2.

This data matrix is analyzed in two ways, namely (a) comparing and clustering parameters, and (b) comparing and clustering stations.

The procedures of the cluster analysis are based on the correlation matrices of the parameters and stations, respectively. For step (a) the cluster algorithms single linkage, average linkage, and complete linkage were adopted; for

1 Institut für Polarökologie der Universität Kiel, Olshausenstraße 40/60, 2300 Kiel 1, FRG
2 Institut für Meereskunde an der Universität Kiel, Düsternbrookerweg 20, 2300 Kiel 1, FRG
3 Geolog.-Paläontolog. Institut der Universität Hamburg, Bundesstr. 55, 2000 Hamburg 13, FRG

Antarctic Ocean and Resources Variability
(ed. by D. Sahrhage)

Table 1. List of the parameters analyzed

No.	Specification	No.	Specification
1	Total bacterial number	35	Isoleucine
2	Bacterial biomass	36	Phenylalanine
3	Mean cell volume of bacteria	37	Serine
4	Turnover time of glucose	38	Tyrosine
5	Actual uptake of glucose	39	Valine
6	Bacterial C-Production based on glucose	40	Asparagine
7	Remineralization rate of glucose	41	Glutamine
8	Respiration of glucose (%)	42	Arginine
9	Turnover time of amino acids	43	Histidine
10	Actual uptake of amino acids	44	Lysine
11	Bacterial C-Production based on amino acids	45	Ornithine
12	Remineralization rate of amino acids	46	Sum of amino acid-C
13	Respiration of amino acids	47	Amino acid-C of POC (%)
14	Ammonia	48	Sum of amino acid-N
15	Nitrite	49	Amino acid-N of PON (%)
16	Nitrate	50	Amino acid-N/Chl a
17	Silicate	51	Amino acid-C/amino acid-N
18	Phosphate	52	Asparagine/Glycine
19	Amino acids (dissolved)	53	Glucose
20	Carbohydrates (dissolved)	54	Galactose
21	C/N	55	Fructose
22	Chlorophyll a	56	Mannose
23	Particulate organic N	57	Sum of hexoses
24	Particulate organic C	58	Arabinose
25	Dry weight	59	Xylose
26	Oxygen	60	Ribose
27	Total particulate P	61	Sum of pentoses
28	Carotenoids	62	Fucose
29	Carotenoids/chlorophyll a	63	2-d-Ribose
30	POC/chlorophyll a	64	Other sugars
31	Primary production	65	Sum of deoxy-sugars
32	Alanine	66	Sum of carbohydrates
33	Glycine	67	Carbohydrates of POC (%)
34	Isoleucine		

The parameters 1–13, 19, 20 are regarded as descriptors for microbiological properties.
The parameters 14–18, 21–31 are regarded as descriptors for phytoplankton properties.
The parameters 32–67 are regarded as descriptors of the dissolved (19, 20) and particulate organic matter.

step (b) we used only the complete linkage procedure in order to obtain strong separations of the created groups. Further, a graph theoretical approach is used to establish overall groups of stations and parameters. This method and all other statistical procedures are described in detail by Bölter and Meyer (1986). All computational procedures were run on the PDP-10 (DEC-System) at the computer center of Kiel University.

3 Results

3.1 Comparison of Parameters

Most of the parameters analyzed show high variation coefficients ($V_k \geqslant 33$) (Sachs 1984), indicating large variability of the individual parameters. This is mainly true for those parameters which describe biological activities, whereas concentrations of individual amino acids, silicate, phosphate, oxygen, and some parameter ratios appear to be less variable.

The cluster analysis of the correlation matrix of the parameters reveals nine groups at the 5% significance level (average linkage method). Table 2 shows the result of this approach. Although not all of these connections between parameters can be interpreted satisfactorily at this stage, it can be stated that phytoplankton parameters, microbial parameters, and those of organic nutrients are generally separated. These clusters remain consistent even when other clustering algorithms are applied. Only minor changes take place, especially when applying the complete linkage method, which tends to split up some of the larger clusters obtained from the average linkage method.

3.2 Comparison of Stations

This analysis is done in two ways, by using the total data set as the basic information and by extracting parameters

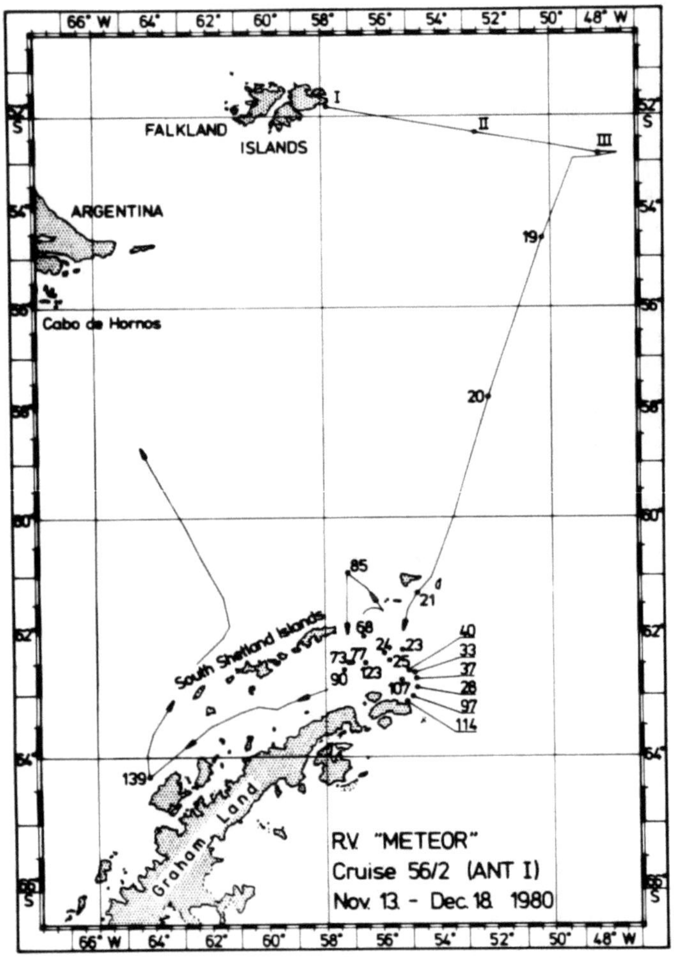

Fig. 1. Map of the stations samples during cruise 56/2 of R/V *Meteor* during November/December 1980

Table 2. Groups of parameters of the cluster analysis of the complete data set. Average linkage method, threshold of acceptance of groups is the 5%-significance level. The numbers refer to Table 1. Parameters not mentioned remain unclustered by this approach

Group 1:	18, 21, 22, 23, 24, 25, 26, 27, 28, 31, 46, 48, 62, 63.
Group 2:	4, 5, 6, 7, 8, 28.
Group 3:	16, 32, 33, 40, 52, 53.
Group 4:	1, 2, 3, 15, 34, 39, 41.
Group 5:	54, 57, 58, 59, 61, 64, 65.
Group 6:	47, 49, 51, 67.
Group 7:	9, 10, 11, 12, 13, 14.
Group 8:	19, 35, 36, 50.
Group 9:	30, 42, 56, 66.

concerning plankton, microbiology, organic chemistry, and those which formed cluster 1 during the analysis of parameters (cf. Table 2).

3.2.1 The Total Data Set

The transposed matrix of the original data set is analyzed by cluster analysis and the graph theoretical approach. The cluster analysis separates the surface samples (Fig. 2) into four groups, which can be distinguished significantly by their absolute correlation coefficients and the sign of the coefficient. Figure 3 shows the result of this procedure, which can be summarized as follows:

A. The region of the Drake Passage, represented by Stations 21, 85, and 139. Further, this cluster includes stations of the South Atlantic (I, II) and some stations of the inner part of the Bransfield Strait (124, 127, 131).

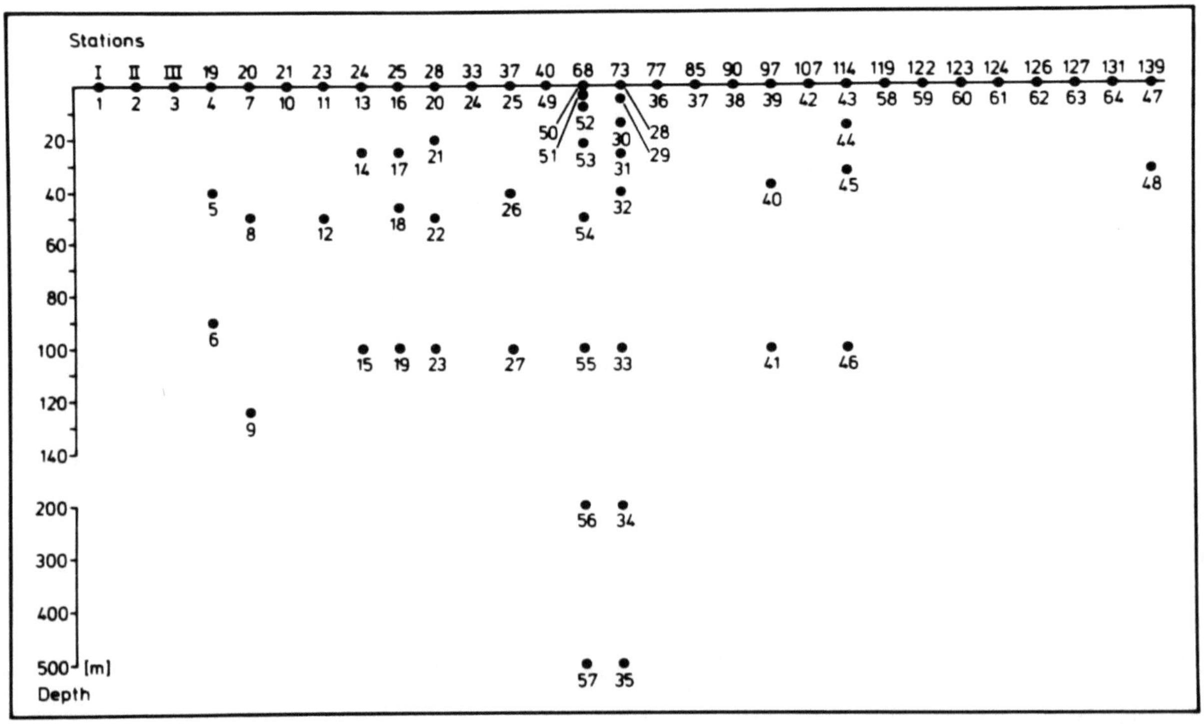

Fig. 2. Samples/depth profiles of Stations 1–139. The samples are marked with regard to their sampling depth by a *dot* and the numbers *1–64*

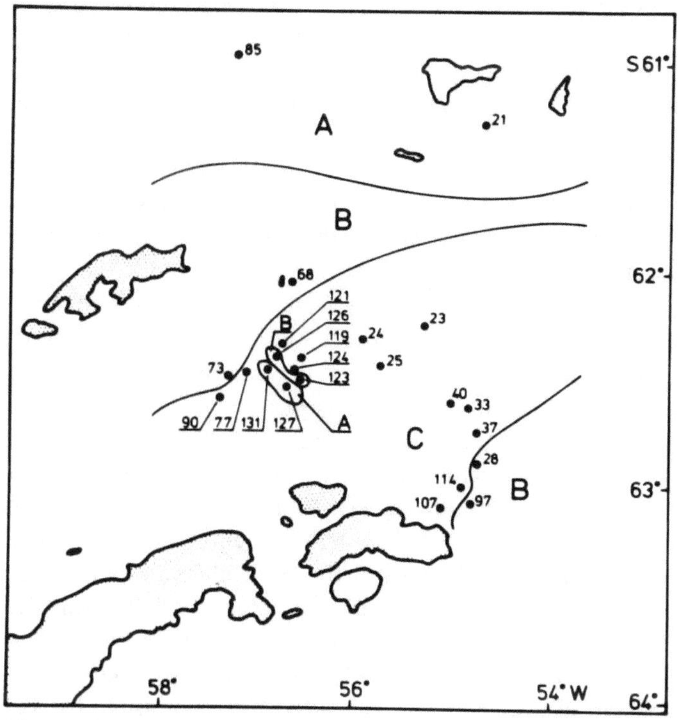

Fig. 3. Zonation of the stations samples in the outer Bransfield Strait. For locations of Stations I–III, 19 and 139 refer to Fig. 1

B. Stations of the outer Bransfield Strait (28, 68, 97) and some of the central Bransfield Strait (73, 123, 126).
C. Most of the stations of the Bransfield Strait.
D. Stations III and 19, located in the northern Drake Passage. For the locations of Stations I–III, 19, 21, and 139 cf. Fig. 1.

The results of the cluster analysis show that the depth profiles of the stations have individual structures which reflect different patterns of the biological and/or chemical situations. Station 25, for example, can be subdivided into three different layers: (1) 0 m, 25 m; (2) 45 m; (3) 100 m, whereas other stations (e.g., Stn. 28) can be regarded as rather homogeneous. The connections between surface samples on one hand and deeper layer samples of other stations on the other hand (e.g., Stns. 37, 73, 114) illustrate the complex structure of the ecosystem which will be discussed below.

3.2.2 Extracted Parameters

More detailed information about these separations of samples by different sets of parameters is available when the original data set is split into parts. For this approach we used the parameters describing the plankton/inorganic nutrients, microbiology, and organic chemistry. The result of a cluster analysis of parameters was used as a further aggregation of parameters. Table 3 lists the members of the different clusters.

Table 3. List of samples which are summarized by the cluster analyses. Threshold for the acceptance of clusters is the 1%-significance level. Surface samples are mentioned in *italics*. Numbers in parenthesis indicate samples of different depth profiles, cf. Fig. 2

A. Extracted parameters: Microbiology

1) *7*, (*11*, 12) 15, (*16*, 17), *20*, *24*, 27, (32, 35), 41, 42, (*43*, 44, 46), *49*, *58*, *59*.
2) (*13*, 14), 18, 22, (*28*, 29, 30, 31, 34), *37*, *39*.
3) *1*, *2*, (21, 23), *25*, *36*, *38*, (51, 53).
4) (*4*, 5, 6), 9, *10*, 26, 33, 40, 45, 48, *60*, *61*, *62*, *63*.
5) *3*, 8, 19, (52, 54, 55, 56, 57), *64*.
6) *47*, *50*.

B. Extracted parameters: Planktology

1) *13*, (*20*, 21, 22, 23), *24*, (*28*, 29), 30, *36*, *39*, *42*, (*43*, 44, 45, 46), (*50*, 51, 52, 53, 54), *60*, *62*.
2) (*4*, 5, 6), (*7*, 8, 9), 15, (*16*, 17, 18), 33, (*47*, 48), 57.
3) 12, 31, 34, 35, 40, 41, *49*.
4) *11*, 14, 19, (*25*, 26, 27), 32, (55, 56).

C. Extracted parameters: Organic chemistry

1) (*20*, 21, 22), (*28*, 29, 30), (*39*, 40, 41), (*43*, 44), (*50*, 51, 52).
2) (*7*, 8, 9), (*11*, 12), 23, 35, 53.
3) (*4*, 5, 6), (*13*, 14, 15), (*16*, 17, 18, 19), (26, 27), (31, 32, 33, 34).

D. Extracted parameters: "Cluster 1" (cf. Table 2)

1) (*4*, 5, 6), (*11*, 12), 15, *16*, 27, (31, 33, 35), (40, 41), (*47*, 48), 49.
2) 14, (17, 19), (*20*, 21), *24*. (*25*, 26), (*28*, 29, 30, 32, 34), *36*, *39*.
3) (22, 23), *42*, (*43*, 44, 45, 46), (*50*, 51, 52, 53, 54), *60*, *62*.

The results can be described as follows:

a) When using the parameters describing microbiological properties, the complete linkage clustering reveals six groups of samples (cf. Fig. 2). Although the picture of the microbial descriptors shows some resemblance to that given in Fig. 3, it should be regarded as an independent system of interactions at this level.

b) The results of the parameters describing phytoplankton properties indicate a higher degree of homogeneity with regard to the depth profiles of some stations; e.g., 20–23, 28–30, 42–46, 50–54, 4–9, 16–18, 40–41, 25–27 (cf. Fig. 2). The first cluster concentrates on most of the surface samples. It shows no intrusions of other clusters in depth profiles. Furthermore, the samples of this cluster do not lie underneath samples from other clusters. The second group comprises stations outside the Bransfield Strait (Stns. 19, 20, 139), although Station 25 is located in its central part and also samples 24 and 33 are from this region. Samples of the third cluster are from deep layers of Stations 23, 73, 97 and only sample 49 is derived from a surface sample (Stn. 40). The last cluster combines mainly samples from Station 37 (# 25–27) and the deep layers of Station 68 (# 55, 56) and Station 73 (# 32).

c) The cluster analysis of samples described by parameters of organic chemistry, especially particulate amino acids and carbohydrates, shows only three groups. These clusters are the most comprehensive: Cluster (1) combines surface samples of Stations 28, 68, 73, 97, and 114. Cluster (2) concentrates Stations 20 and 23 and on two deep layer samples of Stations 28 and 73. Cluster (3) combines samples of Stations 24, 25, and 37 as well as deep layer samples of Station 73.

d) When clustering the stations based on the parameters obtained from the first cluster analysis of parameters (step A, Sect. 2), the result is rather similar to that found during the step (B), i.e., the phytoplankton properties. However, it is more homogeneous with regard to the members of the clusters. Three groups of samples could be established. The first cluster combines Stations 19, 20, 21, 23, and 113 completely and another patch samples from the deep layers of Stations 37, 68, 73, and 97. The second cluster represents mixed samples from surface and deep layers of Stations 24, 25, 28, 33, 37, 73, 77, and 97. The last cluster combines all samples taken at Stations 107, 114, 119, 122, 123, 126 and the surface sample from Station 68 as well as two deep layer samples from Station 28.

A graph theoretical approach (not represented here) resulted in a similar structure of the data. General aspects will be considered below in the discussion.

4 Discussion

Bodungen et al. (1986) described three major zones with regard to phytoplankton data and sedimentation properties. To some extent, they are comparable to the zones established here by analyzing the extended data set (cf. Fig. 3). Liebezeit and Bölter (1986) showed significant differences in the distribution of particulate amino acids between the euphotic zone and the aphotic zone. Bölter (1987) refers to small-scale variations in phytoplankton data and microbiological properties in relation to oceanic variability monitored at transects in the central Bransfield Strait (cf. Wittstock and Zenk 1983). These results point to variable structures of the ecosystem described by the parameters analyzed, although some similar overall aspects can be evaluated from all these studies. The reasons for this situation of diverging or converging sets of ecological properties will be discussed below. Unfortunately, we have no complete data set about temperature and salinity concomitant to the data set analyzed here. Hence, the discussions must be restricted to data from literature and to some obvious figures which are generally relevant in this area.

The comparison and grouping of the parameters analyzed – the first aspect of this study – points to several levels of the biological/chemical systems. A view on these results shows no direct links (significant correlations) between the major compartments, i.e., phytoplankton, microbes, organic material. For instance, Liebezeit (1984), showed a drastic decrease with depth in particulate carbohydrates, especially glucose, which were reduced by 70% from the surface to 100 m. The respective microbial activity was not found. Locally, vertical mixing of different water masses may be responsible for the distribution of particulate carbohydrates rather than microbes and their activities.

Two reasons can be responsible for the lack of interrelationships: (a) the complex hydrography of this region, and (b) the different scales of the activity of the organisms as well as the biochemical processes.

The hydrography pf the Bransfield Strait and adjacent areas is rather complex and influenced by mixing processes of water masses flowing east from Bellingshausen Sea and those coming with northwesterly currents from the Weddell Sea. This particular area of the Weddell-Scotia Confluence is well known for its complex structure, both horizontally and vertically (cf. Patterson and Sievers 1980; Stein 1981, 1986; Wittstock and Zenk 1983). Hence, such hydrographical processes will be partly also responsible for the biological pattern received by the cluster analysis.

The mismatch of the zonations which resulted from cluster analyses of stations by the different sets of parameters, those of the graph theoretical approach, and those of other papers (e.g., Bodungen et al. 1986) leads to the problem of the different biological and/or chemical scales of activity. It is difficult to establish individual parameters which are able to characterize groups of samples and stations. Indeed, Bodungen et al. (1986) gave averaged properties of the euphotic zone in order to evaluate three major zones, described by temperature, primary production, POC, chlorophyll a, C/N, and C/P. However, the thresholds set for separations by these authors are not sufficient to explain to what degree differences within and between these zonations were shaped by differences in the interactions at the lower levels of trophic stages.

An analysis of water samples mirrors both the actual processes which occur at different temporal and spatial scales and biochemical parameters which reflect also the past history of the environment. The nature of the interacting parts of a system can be regarded as fractals (Mandelbrot 1982). The example given in Fig. 4 can be extended to other parts of the food web or individual biochemical processes. For instance, dependent on the physical and biological background, processes regulating the concentration of particulate matter, polymeric, oligomeric, and monomeric compounds may each be linked to quite different parts of the system. Although these processes may occur concomitantly, they are not necessarily geared to each other.

Direct interrelationships between the subsystems of the phytoplankton, microbes, and organic material can only be assumed in synchronized systems. Such systems may develop following the onset of the phytoplankton growth season and are most pronounced in regenerating summer systems (Smetacek et al. 1987). Episodal events, such as physical mixing or grazing, may act as separators of these linked processes. We assume that the congruence of clusters of samples gained from the analysis of different data sets, i.e., between microbiological, phytoplankton, and biochemical data, refers to a high degree of synchronization between these compartments.

For the stations analyzed in the Bransfield Strait, such a high degree of synchronization was found for only very few stations. It appears that in the Bransfield Strait area the pelagic system is not necessarily in balance, as for most of the stations the subsystems act independently at different time and space scales where the individual processes are decoupled.

The degree of synchronization of a system, i.e., the degree to which autotrophic and heterotrophic processes and the resulting biochemical compounds are geared, is dependent on the frequencies of episodical disturbances in the pelagic systems. In the Bransfield Strait intermittent vertical and horizontal mixing, as well as locally intense krill grazing, may represent such disturbance.

The frequently increased input of external energy not only leads to downward mixing of substances produced in surface layers, it may also alter the composition of particulate matter without respective biological processes. For instance, Gieskes and Elbrächter (1986) have shown that after heavy turbulence the abundance of free floating chloroplasts increased without a change in chlorophyll a levels. Concomitant measurements of microbial activity were also much lower than expected from the chlorophyll a levels, i.e., thereof estimated amount of organic matter (Bölter unpublished results).

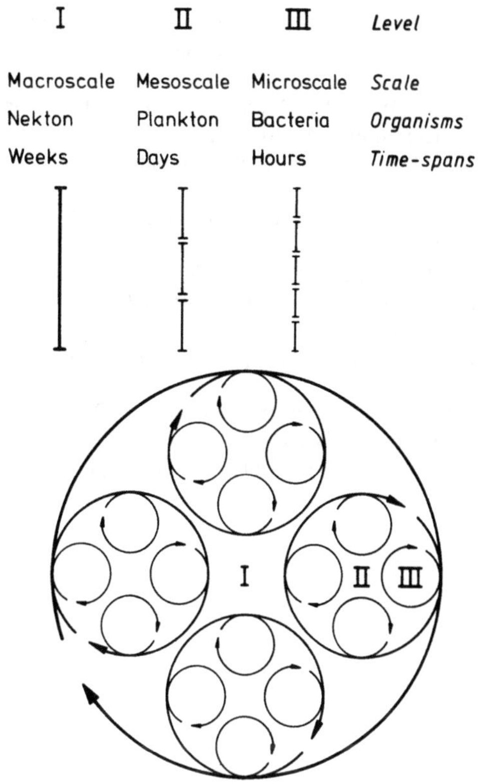

Fig. 4. Presentation of different levels of an ecosystem (levels I–III) by examples of scales, organisms, and time spans of these levels as fractals

5 Conclusions

Analyzing water masses around South Georgia, Priddle et al. (1986) recommended that multidimensional descriptors are needed to describe the heterogeneity of the phytoplankton assemblages and inorganic nutrients, based on ordination techniques. Multivariate techniques, such as cluster analysis and graph theoretical approaches, are useful in establishing groups of similarly behaving parameters or stations (samples). It should be kept in mind that those groups are based on the correlation matrices of the parameters, including to a certain extent the statistical probability. They are based on their internal interrelationships, and, hence, on the ratios of the original characteristics. Although the individual measurements are static, their groupings can be used to describe dynamic processes.

The description and comparison of pelagic systems by a few standing stock parameters, e.g., chlorophyll a levels, POC, nutrients, may be sufficient when comparing seasonal or large-scale geographical differences. For higher resolution and comparison of trophodynamics, standing stocks and rate processes of the different subsystems must be considered. Since it cannot be predicted at which levels the interactions of the subsystems are responsible in shaping the pelagial, multivariate methods are helpful in establishing these levels for further studies.

Acknowledgment. The authors would like to thank Mrs. M. Mempel and Mr. W. Dzomla for skillful technical assistance.

References

Bodungen B v, Smetacek V, Tilzer MM, Zeitzschel B (1986) Primary production and sedimentation during spring in the Antarctic Peninsula region. Deep-Sea Res 33:177–194

Bölter M (1987) Microbiological, planktological and chemical characteristics and their validity for the separation of water masses in the Bransfield Strait, Antarctica. S Afr J Mar Sci 5: 133-143

Bölter M, Dawson R (1982) Heterotrophic utilisation of biochemical compounds in Antarctic waters. Neth J Sea Res 16:315–332

Bölter M, Meyer M (1980) A statistical scheme for structural analysis in marine ecosystems. Ecol Modell 9:143–151

Bölter M, Meyer M (1986) Structuring of ecological data sets by methods of correlation and cluster analysis. Ecol Modell 32: 1–13

Gieskes W, Elbrächter M (1986) Abundance of nanoplankton-size chlorophyll-containing particles caused by diatom disruption in surface waters of the Southern Ocean (Antarctic Peninsula region). Neth J Sea Res 20:291–303

Liebezeit G (1984) Particulate carbohydrates in relation to phytoplankton in the euphotic zone of the Bransfield Strait. Polar Biol 2:225–228

Liebezeit G, Bölter M (1986) Distribution of particulate amino acids in the Bransfield Strait. Polar Biol 5:199–206

Mandelbrot B (1982) The fractal geometry of nature. Freeman, San Francisco, pp 1–468

Patterson SL, Sievers HA (1980) The Weddell-Scotia confluence. J Phys Oceanogr 10:1584–1610

Priddle J, Heywood RB, Theriot E (1986) Some environmental factors influencing phytoplankton in the Southern Ocean around South Georgia. Polar Biol 5:65–79

Sachs L (1984) Angewandte Statistik. Springer, Berlin Heidelberg New York Tokyo, pp 1–552

Smetacek V, Bodungen B v, Bölter M, Bröckel K v, Dawson R, Knoppers B, Liebezeit G, Martens P, Peinert R, Pollehne F, Stegmann P, Wolter K, Zeitzschel B (1987) The pelagic system. In: Rumohr J, Walger E, Zeitzschel B (eds) Seawater-sediment interactions in coastal waters. Lecture notes on coastal and estuarine studies. Springer, Berlin Heidelberg New York Tokyo, pp 32–68

Stein M (1981) Thermal structure of the Weddell-Scotia confluence during February 1981. Meeresforschung 29:47–52

Stein M (1986) Variability of water masses and currents off the Antarctic Peninsula during SIBEX. Arch Fischereiwiss Beih 1 37:25–50

Wittstock R-R, Zenk W (1983) Some current observations and T/S distribution from the Scotia Sea and the Bransfield Strait during early austral summer 1980/81. *Meteor* Forschungsergeb Reihe A/B 24:77–86

Zeitzschel B, Zenk W (1981) Beobachtungen und erste Ergebnisse der *Meteor*-Reise 56 aus der Scotia-See und der Bransfield-Straße im November/Dezember 1980 (ANT I): ein nautischer und wissenschaftlicher Bericht. Ber Inst Meereskd Kiel, 1–73

Part 3
Krill Variability in Relation to the Environment

Large-Scale Fluctuations in Distribution and Abundance of Krill – A Discussion of Possible Causes[1]

J. Priddle[2], J. P. Croxall[2], I. Everson[2], R. B. Heywood[2], E. J. Murphy[2], P. A. Prince[2], and C. B. Sear[3]

Summary. Unusually low abundance of krill may last for periods of several months in the Scotia Sea near South Georgia and in Bransfield Strait. Two longer data sets on krill predators suggest that such events may occur two or three times in a decade, and that the situation normally returns to normal in the following season. It seems most unlikely that these events can be ascribed to features of krill biology. Simple models of recruitment failure or mortality cannot explain the observed changes, and alteration in small-scale distribution is not indicated by the available data. More probable mechanisms must involve large-scale changes in distribution of krill brought about by ocean-atmosphere processes. Whilst natural variation in mesoscale features has an appropriate spatial scale, the likely duration is too short. However, a breakdown of hydrographic structure in the surface water over a large area would drastically decrease the residence time of krill and it would take a longer time to reestablish high krill biomass. A prolonged period of southwards airflow over the Scotia Sea is identified as the likely driving force in this model. Such an airflow has been identified from atmospheric pressure distribution in the winters of 1983 and 1986, and was associated with southwards displacement of both warm surface water and of pack ice in the northern Weddell Sea.

1 Introduction – the Circumpolar Distribution of Krill and Possible Mesoscale Variability

The abundance and distribution of a planktonic animal represents a complex interaction between its biology and both abiotic and biotic factors in the environment. Antarctic krill (*Euphausia superba* Dana) inhabits an environment which exhibits particularly marked variability on some scales. Located at high latitudes, the Southern Ocean is a profoundly seasonal ecosystem which responds to cyclic, within-year variation in incident radiation. This impinges on the physical environment in a number of ways, of which the seasonal change in the areal extent of sea-ice is the most spectacular (see Zwally et al. 1983). Production by photoautotrophic microplankton also follows a large-amplitude seasonal cycle (Priddle et al. 1986a; El-Sayed this Vol.) and this largely determines the pelagic food supply available to krill. However, this paper is concerned with variation at scales larger than these comparatively well-documented seasonal changes. It addresses variation in abundance of krill which has occurred between years, but which has a shorter time-scale than long-term changes (decades) in the Southern Ocean ecosystem described by Laws (1984).

Before describing the character of these interannual changes and exploring possible causes for them, it is necessary to describe the mean spatial pattern of krill distribution. Changes of the time-scale covered here are expected to act over large distance scales (Longhurst 1984; Murphy et al. this Vol.), so that we will be concerned with the gross features of the distribution of krill over the whole of its range, even though the phenomena discussed here were only investigated in a restricted area. The distribution of Antarctic krill is apparently very patchy when examined at this scale. Data for this pattern are derived from a variety of sources. Amos (1984) considered maps of Southern Ocean krill distribution, and its interpretation, by Marr (1962), Jarogov (1969), Mackintosh (1972) and Maslennikov (1980). Few of the more recent data are of additional value in mapping the circumcontinental distribution of krill, as both scientific studies and commercial fisheries have understandably concentrated on areas already thought to offer consistently high krill biomass. The only other relevant supporting information which can usefully be added to that considered by Amos (1984) is the important but circumstantial evidence provided by the distribution of baleen whales between the 1920's and 1950's, as indicated by catches of these whales. Those species migrating to the Southern Ocean in the austral summer to feed on krill can be used as indicators of the distribution of krill (Beklemishev 1960; Chernyy 1966; Mackintosh 1972; Mizroch et al. 1985). Given the limitation of pelagic whaling fleets to ice-free waters, the maps of Blue and Fin Whale catches (Mizroch et al. 1985) show a strong similarity to the suggested distributions of krill (Amos 1984).

These maps inevitably reflect particular theories of circumpolar krill distribution and this must influence representation of range and abundace, especially when interpolating between samples. In spite of this, there is a significant degree of correspondence between them. A synthesis

1 This paper is dedicated to Dr. Richard Laws CBE FRS on the occasion of his retirement from the directorship of the British Antarctic Survey.

2 British Antarctic Survey, Natural Environment Research Council, High Cross, Madingley Road, Cambridge CB3 OET, U.K.

3 Present address: Science Applications International Corporation, Suite 16, Millers Yard, Mill Lane, Cambridge CB2 IRQ, U.K.

Antarctic Ocean and Resources Variability
(ed. by D. Sahrhage)

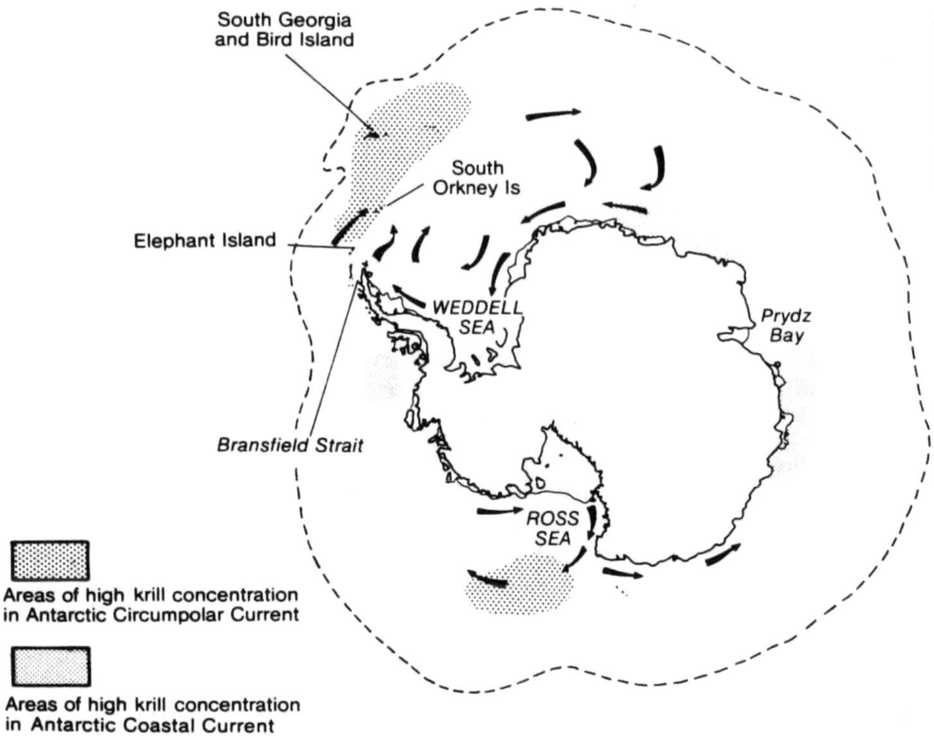

Fig. 1. The circumpolar distribution of Antarctic krill, *Euphausia superba*, showing the presence of areas of high density within the overall northern limits of its range, denoted by the position of the Sub-Antarctic Front. (Data from Marr 1962; Jarogov 1969; Mackintosh 1972; Maslennikov 1980)

of these maps is shown in Fig. 1. Whilst precise delineation of the areas of high krill biomass is not possible, the "hot-spots" can tentatively be divided into two groups on the basis of large-scale hydrography. The first group is found in the Antarctic Coastal Current (East Wind Drift), close to the Antarctic Continent (e.g. Marr 1962, pp. 435 and 444; Everson 1976, 1981; Amos 1984). The pattern of heterogeneity in this zone is unclear, but some correlation with bottom topography can be seen (Everson 1976). The second group of "hot-spots" occurs where the Antarctic Coastal Current enters large gyre systems and interacts with the Antarctic Circumpolar Current (Marr 1962, pp. 435 and 443–444). This is particularly conspicuous in the Scotia Sea region where the Antarctic Circumpolar Current meets the Weddell Sea Gyre.

The presence of "hot-spots" is suggestive of regional "stocks" or populations of krill. However, this is not consistent with the picture of circumpolar distribution of krill being linked intimately with ocean circulation, with its implication of some exchange between areas. A limited amount of work has been carried out on the genetic variability of krill as an indicator of geographic differences. None has had a circumcontinental coverage, most studies being carried out in the Prydz Bay area or in Bransfield Strait and Weddell Sea (Grant 1983; MacDonald and Schneppenheim 1983; Schneppenheim and MacDonald 1983, 1984; Fevolden 1984; MacDonald et al. 1986). It is difficult to estimate the minimum degree of intermixing of originally genetically distinct populations which would lead to blurring of regional stocks (Grant 1983; MacDonald et al. 1986) but there is no evidence so far for discrimination between separate stocks of krill. The concept of a circumcontinental population of *Euphausia superba* is therefore not contradicted by these data.

Whereas this circumpolar distribution pattern appears to be more or less consistent from year to year, interannual mesoscale variation may occur occasionally. Such variation may be extreme, leading to an apparent absence of krill from a region for a period of several months. This paper is concerned with these changes and examines data both on krill abundance and on predators on krill, themselves potential sensitive indicators of local krill biomass. This information is then assessed in the light of present knowledge of krill biology, and in the context of possible meso- and macro-scale interactions with the physical environment.

2 The Data on Long-Term Fluctuations in the Mesoscale Abundance of Krill

Krill are oceanic crustaceans which form an important part of the Southern Ocean food chain. They account for much of the diet of many predators, some of which may rely almost exclusively on krill during their breeding period. Recent studies have provided direct evidence of significant interannual changes in the abundance of krill in certain parts of the Southern Ocean, particularly in the Southwest Atlantic Sector. Parallel data on breeding success of land-based krill predators (breeding seabirds and seals) have provided additional evidence over a longer time scale. Year-to-year changes in local krill abundance, and in the performance of predators reliant on this resource, lead us to suggest that there may be significant variation in the distribution of krill over parts of the Southern Ocean. This is evaluated below, and the appropriate time- and space-scales for mesoscale variation in krill abundance suggested. Three modern data sets have been considered in Sections 2.1–2.3.

2.1 Decrease in the Observed Abundance of Krill in the Scotia Sea and Bransfield Strait, 1983–84

The British Antarctic Survey found that the stock of krill around South Georgia in August 1983 was very much less than expected from comparison with that found when working the same stations during December, 1981 (Heywood et al. 1985 – for South Georgia Zone station grid see Fig. 1 in Priddle et al.1986b). Total biomass can be assumed to be more or less constant for an animal which has a life-span of several years such as krill, but the winter biomass was estimated to be at only 3% of the summer level (based on net haul data, Heywood et al. 1985 – acoustic estimates of biomass give a value of 12%, see Sect. 3.2). In contrast, the winter stocks of other zooplankton approximated to anticipated levels after allowing for seasonal changes (Foxton 1956; Heywood et al. 1985). Krill biomass in September 1983 was also very much lower than in February 1982 along three transects across the Scotia Sea. This reconnaissance included a visit to the area close to Elephant Island which is usually thought to be a region of high krill density (Fig. 1, see also Macaulay et al. 1984). The change in biomass could not be related to any observed physical environmental factor. Microbial biomass, estimated from chlorophyll measurements, was higher than expected for winter samples. This might be explained in part by the reduced grazing pressure resulting from the very much smaller krill population. Gut content analysis of krill caught during the winter cruise showed that they were capable of feeding on phytoplankton at the densities indicated by ambient chlorophyll concentrations (Morris and Priddle 1984).

The SIBEX cruises in the following summer (1983–84) in the Bransfield Strait area also showed very greatly reduced krill biomass (Mujica and Asencio 1985; Witek et al. 1985). Both reports also noted the predominance of salps in these SIBEX samples, possibly indicative of the influx of a different zooplankton community (see also Wormuth this Vol.).

By contrast, biomass of krill in the southern Scotia Sea–northern Weddell Sea appears to have been normal during the 1983–84 summer. Analysis of catch-per-unit effort data from fishery returns (Everson this Vol.) suggests that krill abundance near the South Orkney Islands was as high or even slightly higher in that summer than in adjacent years.

Data for a second SIBEX area for the same season were noted by Heywood et al. (1985) as possibly indicating a similar decrease in krill abundance. The Prydz Bay region also had lower than expected biomass of krill (Miller 1985, 1986; Terazaki and Wada 1986) but the paucity of earlier work in this area makes it more difficult to obtain a "baseline" for these data. Miller (1985) apparently disregards this as a real change, suggesting that the presence of high krill biomass in Prydz Bay may be ephemeral.

2.2 Breeding Success of Avian and Mammalian Predators on Krill – Recent Data

Aspects of the reproductive performance of albatrosses (three species), penguins (two species) and Antarctic Fur Seals (*Arctocephalus gazella* Peters) have been monitored for 10 years at Bird Island, off South Georgia (Croxall et al. this Vol.). Of these species, the Wandering Albatross (*Diomedea exulans* L.) feeds exclusively on carrion, fish and squid while rearing its chicks in winter and has shown a remarkably consistent pattern of reproductive success, with a variation of less than 10% from the mean over a decade of sampling. The remaining bird species and the Antarctic Fur Seal breed in the summer and most rely almost entirely on krill to feed their offspring. Their reproductive success has shown substantial variation over the last decade (described by Croxall et al. this Vol.), although the general pattern has differed between species (Croxall and Prince 1987). Black-Browed Albatrosses (*Diomedea melanophrys* Temminck), Gentoo and Macaroni Penguins (*Pygoscelis papua* Forster and *Eudyptes chrysolophus* Brandt), which are essentially dependent on krill during the breeding season, all had very poor success in 1977–78 and 1983–84. The latter season is notable in that fur seals also showed substantially reduced reproductive performance. Furthermore, in all four species per capita breeding success was greatly improved ("normal") in the following year (i.e. 1978–79 and 1984–85), although the sizes of the breeding populations were significantly reduced in all cases.

The poor breeding success of these birds and seals is attributable to a paucity of krill in the vicinity of Bird Island (Croxall et al. this Vol.). Duration of the breeding season provides a likely minimum time-scale for this phenomenon. Critical dependence on krill probably lasts for at least 2 months for Macaroni Penguins, 3 months for Gentoo Penguins and 4 months for Black-Browed Albatrosses and Antarctic Fur Seals. Krill requirements of these predators are highest during January to March. The need to return to offspring at intervals, varying from 1 day (for penguins) to 4 days (fur seals), largely restricts food capture to the sea overlying the continental shelf, probably within a radius of 100 km of the island. Different species sample different depths. The albatrosses are restricted to the top 1 m of the water column. whilst seals and penguins may dive to 100 m, although they catch most krill at depths of 30–50 m (Croxall et al. 1985; Croxall and Lishman 1987; Croxall et al. 1988).

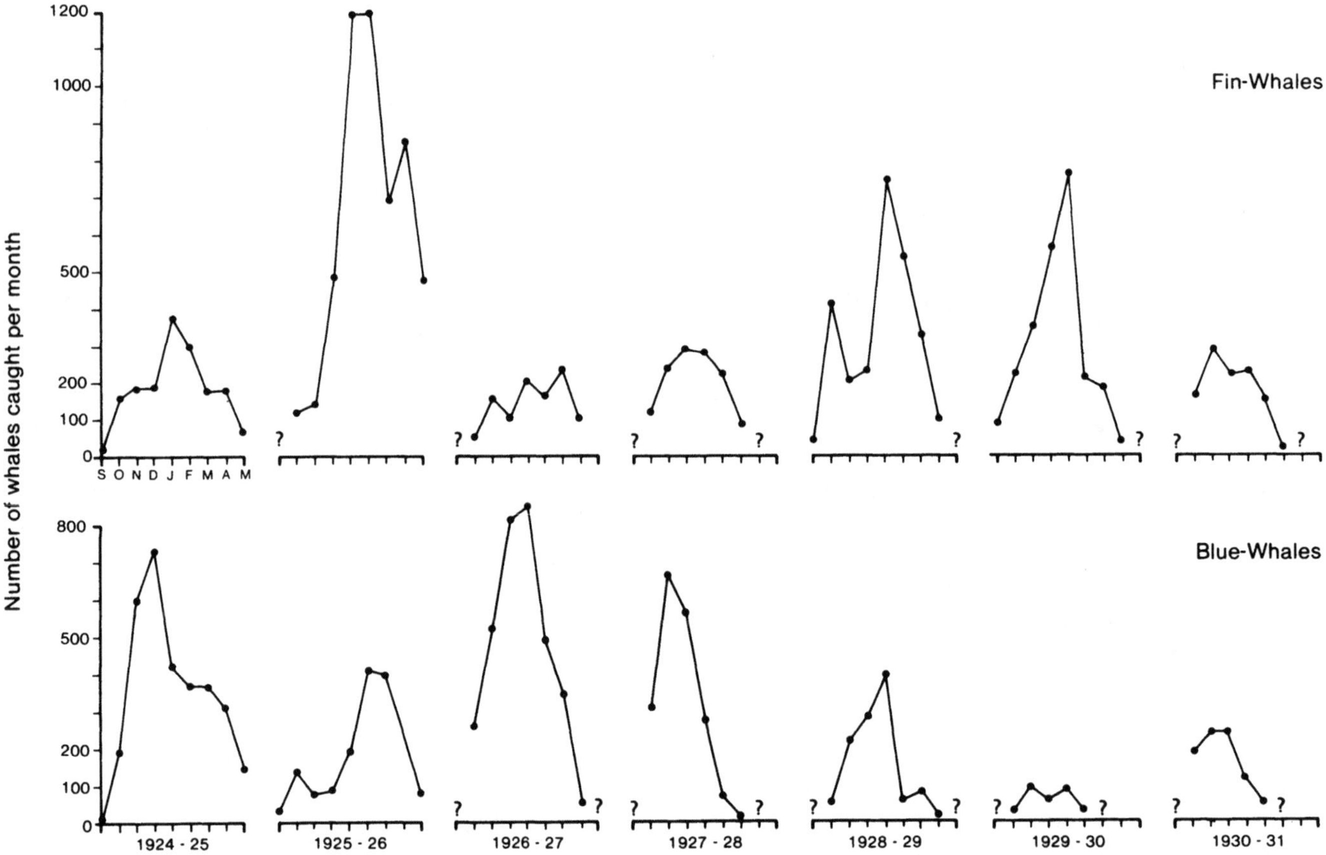

Fig. 2. Variation in the character of catches by the land-based whaling industry at South Georgia during the period 1924–31. (Data from Kemp and Bennett 1932)

2.3 Abundance of Whales near South Georgia, 1924–31

Records of whale catches from the shore-based industry at South Georgia in the 1920's provide an index of the abundance of whales in a defined area. This in turn might be linked to the local abundance of krill. Only a comparatively short data set, that of Kemp and Bennett (1932) for 1924–25 to 1930–31, has been analyzed here. Earlier, the Antarctic whaling industry was developing and interannual comparisons of species are complicated by differing catching effort. Later, of course, the adoption of pelagic whaling meant that whalers could follow their prey and were therefore no longer affected by local variation in whale abundance. The industry finally began also to affect whale stocks, and species composition of catches was therefore then distorted (Beddington and May 1982; Gambell 1985).

Kemp and Bennett (1932) compiled whale catch data for shore stations on South Georgia. These indicate interannual variation between "Blue-Whale years", when this, the preferred species of whale, was abundant within the zone fished by the catchers, and "Fin-Whale years" when Blue Whales (*Balaenoptera musculus* L.) were scarce and the industry switched to taking the less profitable Fin Whales (*B. physalus* L.) (Fig. 2). Harmer (1931) examined the occurrence of Fin-Whale years. He divided the catches into three sets; grouped by the proportion of Blue Whales, by correlations between South Georgia and other whaling grounds, and by the timing of whale population maxima at South Georgia (Harmer 1931, Table 17). He suggested that there were links between these catch classes and both food availability and oceanographic conditions. Harmer (1931, p. 119) observed that Fin Whales taken during the Fin-Whale year of 1925–26 were "exceptionally thin" and that this was typical of such seasons. Thus the pattern of whale catch composition at South Georgia can be taken to be an indicator of total whale abundance in the area and this in turn can be linked to the availability of food. Fin-Whale years can tentatively be identified as seasons with low krill availability in this area. Some support for this suggestion is given by zooplankton abundance data for the corresponding period (Hardy and Gunther 1935, pp. 208–215 and 273–300).

Such changes in the character of whale catches indicated comparatively local changes, whereas later data relate mainly to pelagic whaling and thus are not tied to a small geographical area. The compilation of Antarctic catches of Blue and Fin Whales presented by Mizroch (1984) shows that the proportion of Fin Whales caught over the entire Southern Ocean does not show interannual variation. Thus the variation in South Georgia catch data discussed here can be seen as being local rather than representing larger-scale (e.g circumcontinental) changes.

2.4 Summary – the Character of Long-Term Fluctuations in the Mesoscale Abundance of Antarctic Krill

These three data sets indicate a common phenomenon. Taken together, they point to time- and space-scales which characterize a feature of interannual mesoscale variation in krill abundance. The direct observations in the Scotia Sea and Bransfield Strait during the 1983–84 season suggest that there was a consistently low krill biomass for a period of approximately 9 months. Documentation for Bransfield Strait at this time was especially thorough as part of the BIOMASS SIBEX survey, and provides a useful comparison with a "normal" year in the following phase of SIBEX in 1984–85.

Apparent decrease in the biomass of krill in the Prydz Bay area was cited for the same summer season by Heywood et al. (1985) as possible evidence for a circumpolar decrease in krill abundance in the Southern Ocean. In the absence of background data sufficient to assess the biomass expected in this region, this possibility remains untested.

Data on krill predators provide information on the frequency of such events in the northern Scotia Sea. Records of breeding success of birds and seals at Bird Island indicate unusually low krill availability in 2 out of 10 consecutive years, whilst three "Fin-Whale years" recorded within seven seasons suggest similar paucity. Thus abnormally low krill biomass in the South Georgia area of the Scotia Sea may be expected to occur two or three times in a decade.

Krill biomass in the Bransfield Strait returned to expected levels in the second year (1985–86) of the BIOMASS SIBEX survey after low biomass in the 1984–85 season. Bird and seal breeding data indicate return to normal levels in the years following the two poor breeding years. Two consecutive Fin-Whale years occurred in 1928–29 and 1929–30, but whether this reflects two consecutive years of low krill abundance cannot be tested.

3 Intrinsic Aspects of Krill Biology Which Might Affect Mesoscale Abundance

Given the evidence for interannual variation in the mesoscale abundance of krill, at least in the Scotia Sea and Bransfield Strait area, the first features which should be considered are intrinsic aspects of the biology of krill which might affect its abundance or distribution. These divide into three categories – demographic, small-scale distribution changes, and responses to environmental change.

3.1 Possible Interannual Demographic Changes

Fluctuations in the abundance of krill in a region could be due to variation in mortality rate, recruitment strength or a combination of the two. However, a number of problems are encountered in attempting to assess how these processes operate in krill population dynamics. Here, we attempt to utilize current estimates of population parameters to investigate how these dynamics operate.

Miller and Hampton (ms) have reviewed the studies of krill life-span and they concluded that a range of 4–7 years seemed acceptable. Using length-frequency distributions obtained during the *Discovery* expeditions, Rosenberg et al. (1986) estimated a 6–7-year life-span. This range appears to be a useful starting point for the investigation of the system. Estimates of the instantaneous rate of natural mortality (M) for krill are of the order M = 0.6 to 1.17 for animals older than 2 years (corresponding to annual mortality rates of 45 to 69% respectively) (Miller and Hampton ms). Siegel (1986) obtained a mean value of M = 1.0 for krill (excluding first-year animals). Values of the growth constant, K, of the von Bertalanffy equation are related closely to the rate of natural mortality (Gulland 1983). Siegel (1986) obtained an estimate of K = 0.40, and Rosenberg et al. (1986) a seasonally adjusted value of K = 0.45. Miller and Hampton (ms) suggested that for krill, whose demography probably closely resembles that of clupeoid fish, M would be one to two times greater than K. The values indicate a range of M = 0.8 to 1.35 and are not inconsistent with the more direct estimates. Thus, a value of M = 1.0 does not seem unreasonable for an initial consideration of population dynamics. However, it should be noted that the extreme values of M in the range considered here would produce significant differences in population composition, and that interpretation of population dynamics is dependent on this (see for instance Astheimer 1986). The following examples are offered as preliminary illustrations rather than exact descriptions of krill populations.

Assuming that the mortality rate is constant for each age class in the population we can investigate the effect of recruitment failure on krill abundance. We also assume here that recruitment to the sampled population occurs at 2 years of age. Table 1 shows the stable age structure under conditions of constant recruitment, a constant natural mortality rate (M = 1.0) and potential life span of approximately 7 years, and follows the effect of a failure of recruitment in 1 year on the age structure of the population in that and successive years. The major point to note from this analysis is that the population would be reduced to 36.6% of the stable level with a complete failure in one season.

It is unlikely, given the errors attached to the current estimates of krill abundance (Everson this Vol.; Miller and Hampton ms), that such a decrease would be detected with the present sampling systems. Consistently poor recruitment over a number of seasons could produce a large decrease in abundance and relatively rapid recovery is possible. However, this would be unlikely to appear as a sudden decrease in one season and recovery the next.

Other patterns of recruitment failure can be modelled, based on the same parameters and assumptions. Two successive years of recruitment failure would reduce this theoretical population to 13% of the normal level. Recovery

Table 1. The effect of total failure of recruitment during a single year on age structure and total abundance of a theoretical population of krill with constant mortality rate, M of 1.0 and normal annual recruitment to Age Group 2 of 10 000 individuals

Year	Age group (year)						Total abundance	Total as % of normal
	2	3	4	5	6	7		
Normal age structure	10 000	3679	1353	498	183	67	15 780	100.0
Zero recruitment = "y_{zero}"	0	3679	1353	498	183	67	6 780	36.6
y_{zero} + 1	10 000	0	1353	498	183	67	12 101	76.7
y_{zero} + 2	10 000	3679	0	498	183	67	14 427	91.4
y_{zero} + 3	10 000	3679	1353	0	183	67	15 282	96.8
y_{zero} + 4	10 000	3679	1353	498	0	67	15 597	98.8

over three successive years would result in krill numbers 68%, 88% and 96% of the normal level respectively. The situation where such failure occurs as cyclic behaviour can be envisaged. A 1-year cycle equilibrates to an oscillation between 73% and 27% of the normal population level (which would appear as a decrease to 37% of "normal" in alternate years, based on the revised population level). Similar calculations for a 3-year cycle scarcely differ from the behaviour of a single year's failure – minimum level is 33% (36% of the "new" maximum) and new maximum 91% of normal.

To investigate the effect of such recruitment failure in a single year on the length-frequency distribution of a population, an estimate of the variation in size within a year-class is required. Length-frequency distributions for krill with a single peak and approximately normal distribution, with a mean between 25–35 mm and a standard deviation of 3–6 mm, are often encountered (Marr 1962; Clarke and Morris 1983). From the growth curve of Rosenberg et al. (1986), these samples probably represent a single age group (year 2). Assuming a uniform standard deviation of 5 mm for each age group, and taking the mean length for each age-class from the von Bertalanffy equation of Rosenberg et al. (1986), we can demonstrate the effect of a recruitment failure based on the age structures of Table 1. The length-frequency distribution (Fig. 3) recovers quickly following the recruitment failure, approaching the stable distribution. Mean size of the krill in samples would show an increase in the year of recruitment failure even though abundance would be reduced (Fig. 3), but no significant change in population mean length would be seen 1 year after the failure of recruitment, when Age Group 2 animals recruited to the population would return the mean length to within a millimetre of normal.

Higher mortality rates would lead to a length-frequency distribution dominated by smaller, younger, krill and the overall abundance would be more sensitive to recruitment fluctuations. A similar effect would be observed if krill life-span is significantly shorter than 7 years. Wide fluctuations in recruitment strength are a common feature of shellfish populations (Hancock 1973) and may occur in krill. However, it appears unlikely that this feature alone would explain the observed fluctuations. Mass mortality of the whole population in one season would, of course, result in a large fluctuation in abundance. Complete recovery in the following season would require a mass influx to the area. If that did not occur, replenishment through normal recruitment would result in dominance of the length-frequency curve by small (Age Group 2) animals. Available data do not suggest that the population was dominated by this year-class in summer 1984–85, the season following a year of krill paucity.

3.2 Possible Small-Scale Distribution Changes

The apparent abundance of krill is determined in part by aspects of their local distribution on spatial scales ranging from metres to hundreds of metres. Changes in the behaviour of krill may result in distribution changes which alter their susceptibility to predation or to sampling by nets or detection by hydroacoustics. Two factors likely to affect abundance estimates are changes in vertical distribution and reduction in swarming behaviour.

Significant change in the depth stratum of the water column occupied by krill would have particular impact on their availability to predators feeding in the surface waters. However, the vertical range of hydroacoustic detection employed in krill research has covered the likely full depth range of krill except the top 10 m of the water column (Everson and Bone 1986), and no changes of depth distribution have been observed during a survey in a period of krill paucity (Heywood et al. 1985). Surface net hauls during this survey sampled the water above the upper level of hydroacoustic sensing, and these catches confirmed that krill were not present in concentrations which could account for the paucity in krill biomass estimates based on hydroacoustics and midwater trawls. In any case, occupation of surface layers would not have rendered krill unavailable to predators.

An exaggerated tendency for krill to disperse uniformly in the upper water column, rather than aggregating in swarms, would also affect dramatically their availability

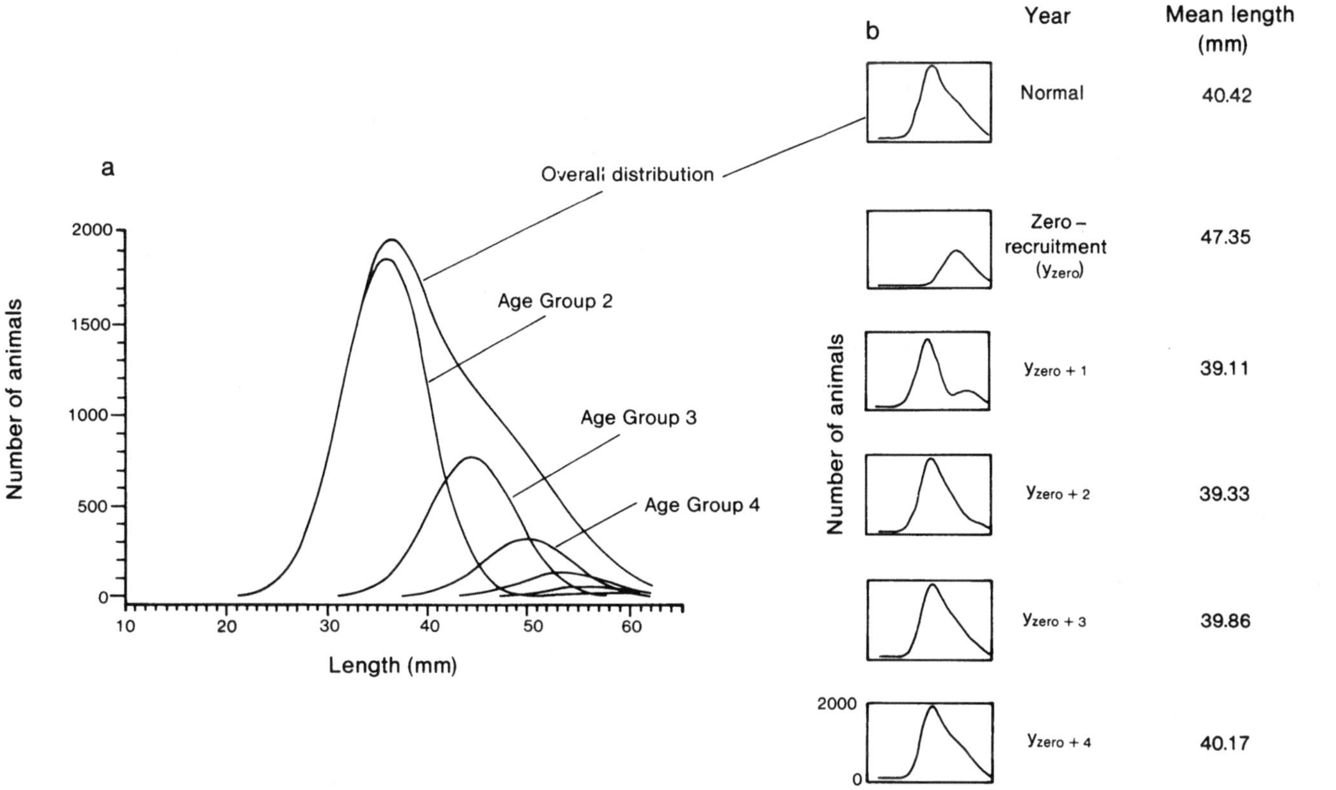

Fig. 3a,b. Theoretical length-frequency distributions for Antarctic krill. **a** Cumulative length-frequency distribution, based on the model presented by Rosenberg et al. (1986) and assuming a standard deviation of 5 mm for each age group. **b** Postulated change in overall length-frequency distribution following total recruitment failure, modelled using the numbers-per-age-class data in Table 1

Table 2. Comparison of the abundance of krill and its variability in a survey area centred on South Georgia, sampled in a "normal" season and during a period of krill paucity, based on RMT hauls

	December 1981	August 1983
Total number of net hauls	71	81
Number containing krill	64 (90%)	46 (57%)
Mean density from net data (number D/m³)	534	3.69
Coefficient of variation	9.5%	143.0%

to large predators (cf. Brodie et al. 1978) and could result in densities undetectable to hydroacoustic sensing. However, the use of large midwater research trawls such as the 8 m^2 Rectangular Midwater Trawl (RMT) should not be so dependent on krill aggregation because the net haul averages krill distribution over kilometre scales. Comparison has been made between two biological surveys in the Scotia Sea on a station grid centred on South Georgia, using both hydroacoustics and RMT hauls. The acoustic survey in December 1981 (taken as a "normal" value) provided an estimate of 224 000 tonnes of krill in the survey area, whilst a similar survey in August 1983 indicated that approximately 12% of this quantity was present in the same area. Comparison data from net hauls from the two surveys confirms this difference (Table 2). If the differences in hydroacoustic estimates of krill abundance for the two surveys had been due to local dispersion in the second season, not only would the net haul data have resembled each other more closely but the proportion of net hauls catching krill would have been expected to increase, whilst the coefficient of variation of density estimates would have decreased. However, the proportion of nets containing krill was lower in the season with krill paucity, and the coefficient of variation was very much higher (Table 2). Thus, the few krill encountered in August 1983 did not appear to exhibit different swarming behaviour from those in December 1981, and we conclude that the low abundance estimated acoustically was a genuine effect.

3.3 Possible Responses to Environmental Changes

Krill are active animals and may respond to various environmental factors which affect their metabolism. It therefore seems possible that disadvantageous environmental conditions might cause krill to seek other, more congenial areas. Indeed, the absence of krill from Bransfield Strait in the 1983–84 summer coincided with higher than usual water temperatures (Rojas 1985), and this was implicated by some investigators as a causal relationship. However, two features of the likely krill responses should be emphasized which suggest that avoidance behaviour by krill is unimportant over large distances. First, consideration of the scales of interaction of krill with their environment (treated in more detail by Murphy et al. this Vol.)

suggests that, whilst krill can perceive features of their local environment, it is unlikely that they can seek actively more advantageous areas over distances suggested by the changes detailed earlier. Secondly, season-to-season changes, for instance in food supply (see El-Sayed this Vol.), may be of comparable importance to interannual variation in environmental quality.

4 Possible Physical Oceanographic Features Influencing the Large-Scale Distribution of Krill

The evidence presented so far indicates that the reduction in the abundance of krill within the Scotia Sea region in some years is a real feature. It cannot be attributed to mass mortality (Sect. 3.1) and must therefore be caused by a redistribution of animals. This does not involve change in vertical range or breakdown of swarming behaviour (Sect. 3.2). Therefore the phenomenon must indicate that krill are redistributed horizontally, outside their usual location. The large scale of the redistribution suggests that active migration, perhaps in response to inclement conditions, cannot be implicated (Sect. 3.3) – krill must move as passive particles in mesoscale circulation even though they can respond actively to their environment over much smaller scales.

4.1 Factors Likely to Influence Mesoscale Distribution of Krill in the Scotia Sea Region

Presence of krill in areas of the open ocean, such as the seas around South Georgia, is dependent largely upon transport of animals to the area in near-surface currents. Thus variation in mesoscale features in hydrography of the Scotia Sea region might be expected to impinge upon krill distribution (see also Maslennikov and Solyankin this Vol.). Circulation in the region is dominated by the interaction of two current systems. The Antarctic Circumpolar Current flows eastwards through Drake Passage and is then deflected northwards on entering the Scotia Sea by the submarine ridge and associated islands of the Scotia Arc, before continuing eastwards beyond South Georgia. Waters of the Weddell Sea Gyre, lying to the south of the Scotia Arc, are similarly deflected (Carmack and Foster 1975). Meridional transport is therefore slowed down within this region of the Southern Ocean, regardless of current speed. Research in Drake Passage, at the western edge of the Scotia Sea, has shown that the Antarctic Circumpolar Current is a multicored stream with bands of geostrophic flow at speeds of 0.025–0.030 m/s, separated by regions of much lower flow (Baker et al. 1977; Georgi 1978; Gordon this Vol.). The relative vorticity between these bands gives rise to numerous eddies (Bryden 1983; Foster and Middleton 1984; see also the review by Nowlin and Klinck 1986). These reduce further the rate of meridional transport because of the circular movement of water within them, conditions which may retain or concentrate planktonic organisms (Angel and Fasham 1983). Thus there is a dual mechanism serving to increase the residence time of krill in the Scotia Sea region – topographically determined deflection of frontal zones combined with eddy genesis.

Influx of krill to the region is a second factor of major importance in determining distribution and abundance of krill. Animals are brought into the Scotia Sea from two sources. The species is found primarily in the Antarctic Coastal Current (East Wind Drift) (Marr 1962; Everson 1976; Makarov et al. this Vol.) and animals enter the study area in water from the Weddell Sea Gyre. These waters also contribute to the population of krill in the Bransfield Strait, as flow from the Weddell Sea enters this area through Antarctic Sound and around Joinville Island. A second input of krill to the Scotia Sea region is derived from the Southeast Pacific Basin, carried through Drake Passage in the Antarctic Circumpolar Current.

The simplest process which could redistribute krill in the region, and give rise to low abundance near South Georgia, is a southward shift of the Polar Frontal Zone under the influence of wind-forcing produced by atmospheric depressions which track across the area. Such large-scale movements of the frontal structure in the Scotia Sea have been observed (for instance Legeckis 1977; Georgi 1978; Heywood unpublished data), with fronts moving up to 100 km in 10 days. Synoptic charts of sea-surface temperature, derived from satellite remote-sensing data (NOAA), show that the Polar Frontal Zone (and hence the Antarctic Circumpolar Current) is a highly mobile feature in the Scotia Sea. The tracks of mesoscale eddies, which may give rise to areas of high krill density, must be similarly influenced. The distance moved by fronts in the Scotia Sea is limited ultimately by bottom topography (Gordon and Goldberg 1970). However, southward displacement could be sufficient to remove krill beyond the foraging ranges of seals and seabirds breeding at Bird Island and away from the BAS survey area around South Georgia. Furthermore, corresponding deflection of the Weddell-Scotia Confluence in the southern Scotia Sea may decrease the input of krill from the Weddell Sea Gyre (Fig. 4).

Eddies are likely to be important features increasing the retention of krill within the Scotia Sea, giving rise to the normally observed high abundance (Figs. 1 and 4). Breakdown of the eddy system would therefore release krill into the main current flow and these would be dispersed downstream. However, it is hard to envisage complete breakdown of eddy structure in the Scotia Sea, as the eddies extend to the sea-bed (Bryden 1983) and such a change would also produce corresponding destruction of the multicore system of the Antarctic Circumpolar Current. Such a major change is not considered feasible.

Ship-borne and remotely-sensed surface temperature distributions indicated that the integrity of the Polar Frontal Zone and Antarctic Circumpolar Current was main-

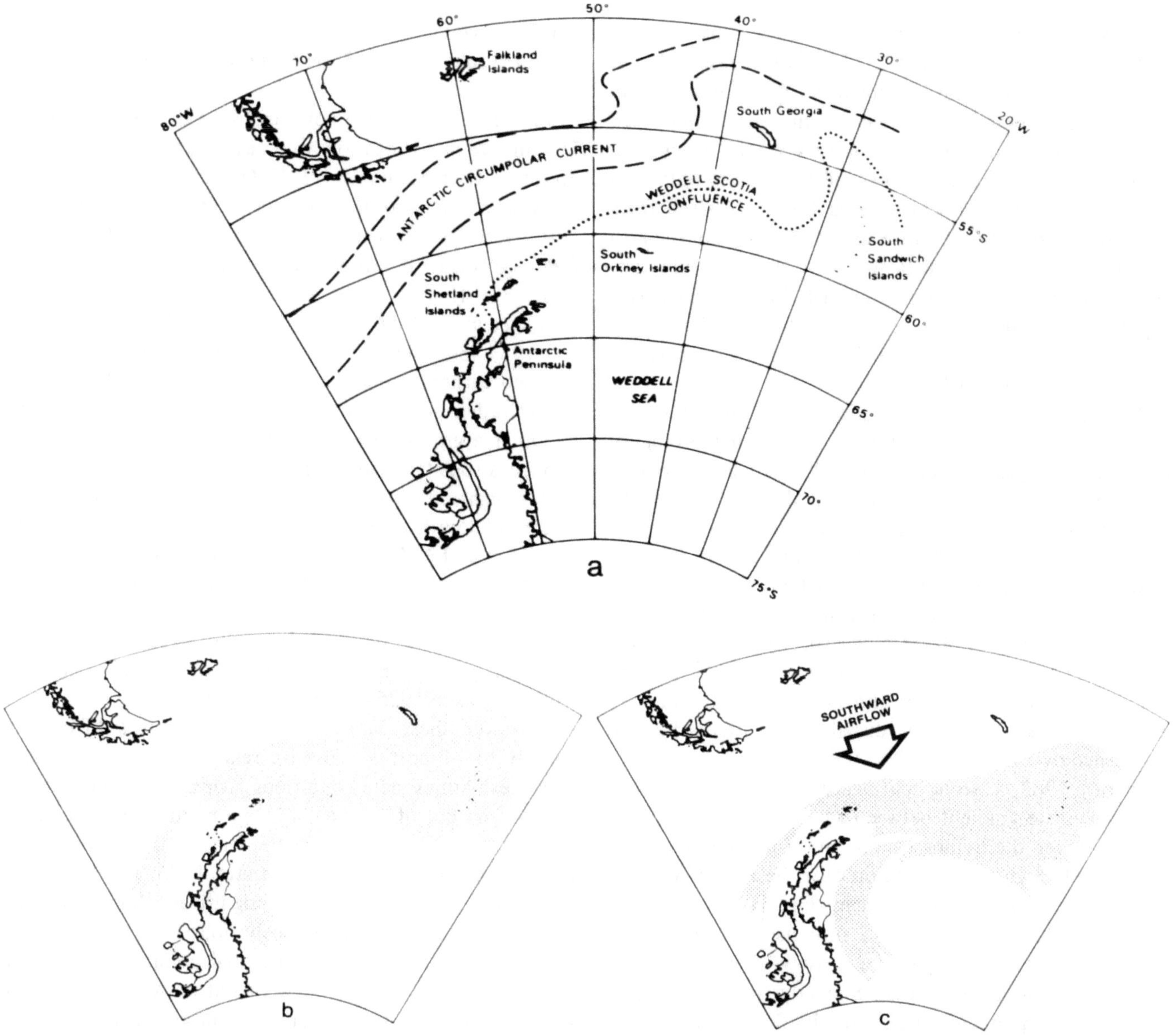

Fig. 4a–c. Hypothetical behaviour of the area of high krill abundance in the Scotia Sea in response to changes in mesoscale hydrography. **a** Mean position of fronts within the Scotia Sea. (After Gordon 1967). **b** Normal situation (cf. Fig. 1) in which krill derived from the Weddell Sea Gyre and from the Bransfield Strait are entrained within the Antarctic Circumpolar Current, reaching parts of the northern Scotia Sea including South Georgia. **c** Extreme conditions in which a southward airflow displaces frontal zones and breakdown of surface structure of eddies releases entrained krill into the main current flow. A conjected deflection of the influx of krill from the Weddell Sea is also indicated

tained during the 1983–84 season (Heywood et al. 1985), and another possible anomalous season in 1986–87 (Heywood unpublished data). However, prolonged periods of strong winds are sufficient to "smear" the structure of the upper 100 m or so of the water column. Sub-Antarctic Zone Water has been observed to overlie Polar Frontal Zone Water in the vicinity of South Georgia (Heywood unpublished data). Under these conditions, eddy structure will be weakened or broken down in surface waters although it persists below this. Krill inhabit the surface waters (Marr 1962) and they would no longer be retained within the eddies but released into the main current flow.

These mesoscale processes provide mechanisms by which krill may be redistributed in the Scotia Sea region, which is normally noted as an area of high abundance. However, either of these changes will produce reduction in krill biomass in the study area only over a comparatively short time and are indequate to explain the 6–9 month period of krill paucity noted in 1983–84 (see Sect. 2.4).

4.2 Postulated Disruption of Areas of High Krill Biomass

We envisage a combination of circumstances under which the changes outlined above could act together. If the surface eddy structure broke down when the polar frontal system was moved to the southern limit of its range, krill would be both displaced from the South Georgia region and then released to pass downcurrent and to the east of the Scotia Sea region. Return of the frontal zones to more northerly positions would fail to recapture these krill and they could not become entrained in eddies reforming in the surface layers. Under these circumstances, replenishment of the Scotia Sea krill biomass would depend on influx of animals carried by Weddell Sea surface water and water in the Antarctic Circumpolar Current derived from the South East Pacific Basin. Such "restocking" of the area would depend on several factors, but clearly would take longer than the original loss of animals from the area. This hypothesis is consistent with the reported duration of the period of krill paucity in the Scotia Sea region between July 1983 and March 1984.

This is a considerably oversimplified picture. However, there is circumstantial evidence from meteorological analyses for 2 years – 1983 and 1986 – which supports the hypothesis (there are insufficient data for 1977). We have examined monthly mean distributions of sea-level atmospheric pressure and height of the 500 hPa isobar (data from European Centre for Medium Range Weather Forecasts for 1983 and from National Climate Centre, Australian Bureau of Meteorology for 1986). These show that in the winters preceding summer seasons of krill paucity (see Sect. 4.3 for discussion of the 1986–87 season), anomalous pressure distribution patterns became established over the Weddell and Scotia Seas.

In July 1983, a strong southward airflow was produced by the pressure gradient between an anomalously low pressure area over the Bellingshausen and Amundsen Seas and an unusually high pressure region over the eastern Weddell Sea and the South Atlantic Ocean (Fig. 5). This persisted with little change during the 2 months until the end of August, during which time the airflow altered direction slightly towards the southeast. Similarly, in June 1986, a blocking high pressure pattern in the South East Pacific area and an intense centre of low pressure over Drake Passage (Fig. 6) also produced a strong southwards airflow into the Weddell Sea. Wind direction was towards the southwest over the southern Scotia Sea and southeastwards in the northern part (close to South Georgia).

In both years, an extraordinary wind-forced flow of surface waters was indicated by anomalous sea-ice distribution. The ice-edge was situated approximately 300 km south of the mean position for much of the winter (NOCD 1985) (Fig. 6c,d). Southerly displacement of sea surface isotherms in the Scotia Sea during 1986 appeared to confirm the suggested movement of surface waters (NOAA synoptic charts).

Therefore data for both winters suggest that southward-directed airflow in the Scotia Sea persisted for at least 1 month, producing a displacement of frontal zones. This could have given rise to release of krill through the breakdown of hydrographic conditions which normally contribute to the high krill abundance in the area. Influx from the Weddell Sea to the southern Scotia Sea might also have been affected (Fig. 4b).

4.3 Postscript – Possible Links to Large-Scale Climatic Phenomena

Croxall et al. (this Vol.) point out that the seasons 1977–78 and 1983–84 both followed years when strong El Niño/Southern Oscillation events (ENSO's) occurred. The 1986–87 season has also shown some anomalies in predator reproductive performance. It is too early to evaluate these fully, especially as this may involve assessment of data for 1987–88. If this does confirm that reproductive performance in the 1986–87 season resembled that in 1977–78 and 1983–84, then the latest event will have differed from the two earlier events because it coincided with a (weak) ENSO rather than lagging approximately one year behind it. Other events relating to changes in krill abundance can also be associated with ENSO's. The direct estimation of krill biomass during the period 1983–85 showed that a season of low krill biomass (1983–84) was similarly one year after an ENSO year, with the coincident poor breeding performance of land-based predators at Bird Island already noted. The older data set reviewed here, that relating to whale catches at South Georgia, also indicates a possible relationship with ENSO's. "Fin-Whale years" coincided with ENSO's in 1925–26 and 1929–30.

However, these data sets are short and it is difficult at present to support a causative relationship on this basis. The ocean-atmosphere conditions giving rise to the El Niño phenomena do, of course, only represent one extreme of the range of behaviour of the Pacific Ocean Southern Oscillation (Philander 1986; Ramage 1986; van Loon and Shea this Vol.). To date, no significant links between ENSO's and Southern Ocean behaviour have been demonstrated. However, Carleton (personal communication and 1988) notes that there is an apparent correlation between warm event years in the Pacific Ocean and increased open water in summer in the northern Weddell Sea. If further work indicates a link between ENSO's and ocean-atmosphere interactions in the Southern Ocean, then aspects of krill distribution and associated biological phenomena might be interpreted as results of larger-scale hydrographic changes than those described here.

5 Conclusion

Abundance of krill in parts of the Scotia Sea and Bransfield Strait may undergo significant interannual variation, as indicated by direct observation in 1983–85 and by somewhat longer data sets on the performance of krill predators. Such variation, where krill abundance is sometimes drastically reduced at localities normally recognized as areas of high biomass, cannot be attributed to intrinsic features of krill biology. Mass mortality is not implicated. Rather, redistribution of krill by physical forces is suggested. Normal mesoscale fluctuations in the positions of frontal zones and in eddy activity do not last long enough to cause

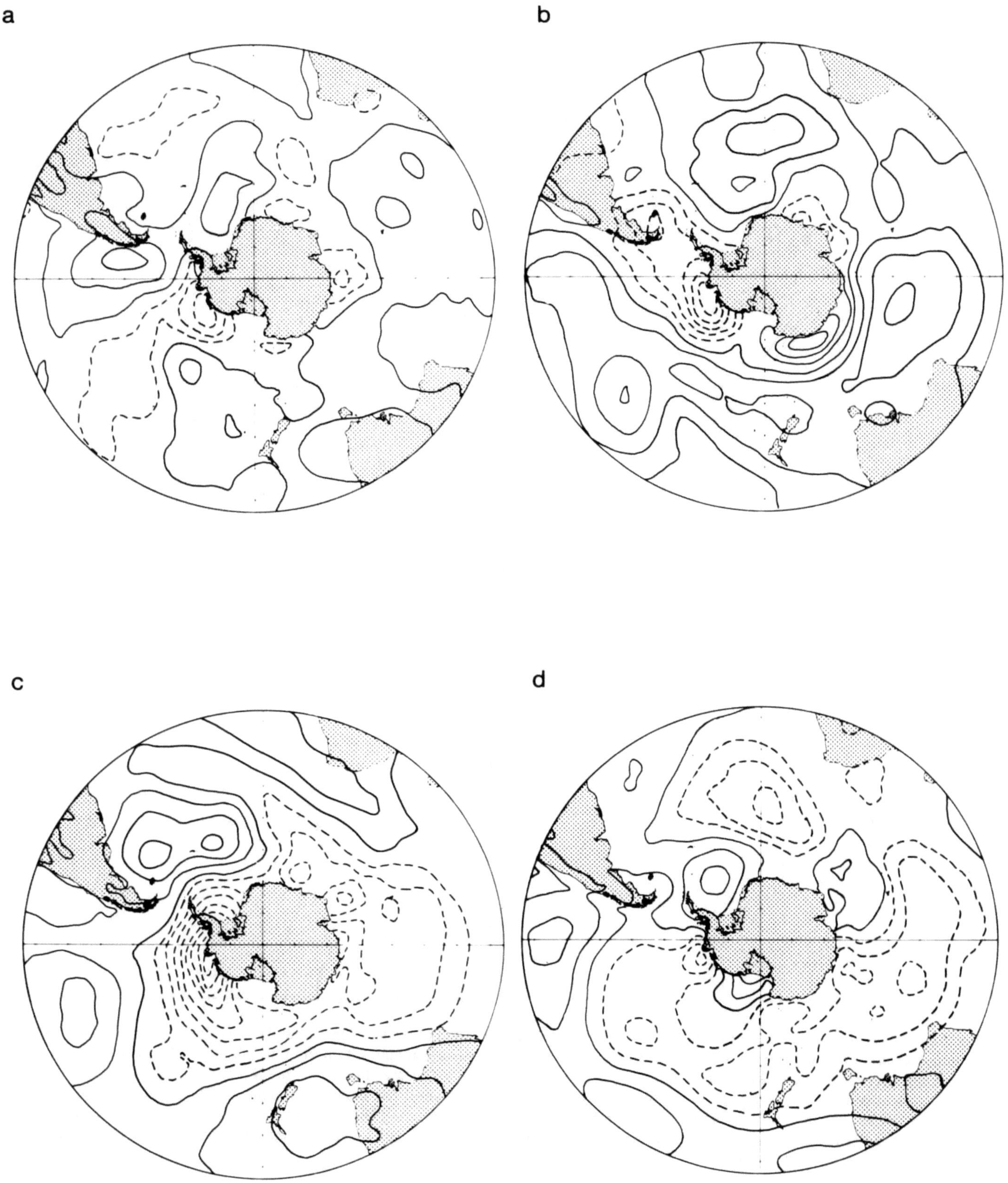

Fig. 5a–d. Patterns of monthly mean sea-level atmospheric pressure anomaly for winter 1983. **a–d** June to September respectively. Contour intervals are 2.5 hPa anomalies, with negative anomaly contours indicated by *broken lines.* (Redrawn from monthly summary charts supplied by the European Centre for Medium Range Weather Forecasts)

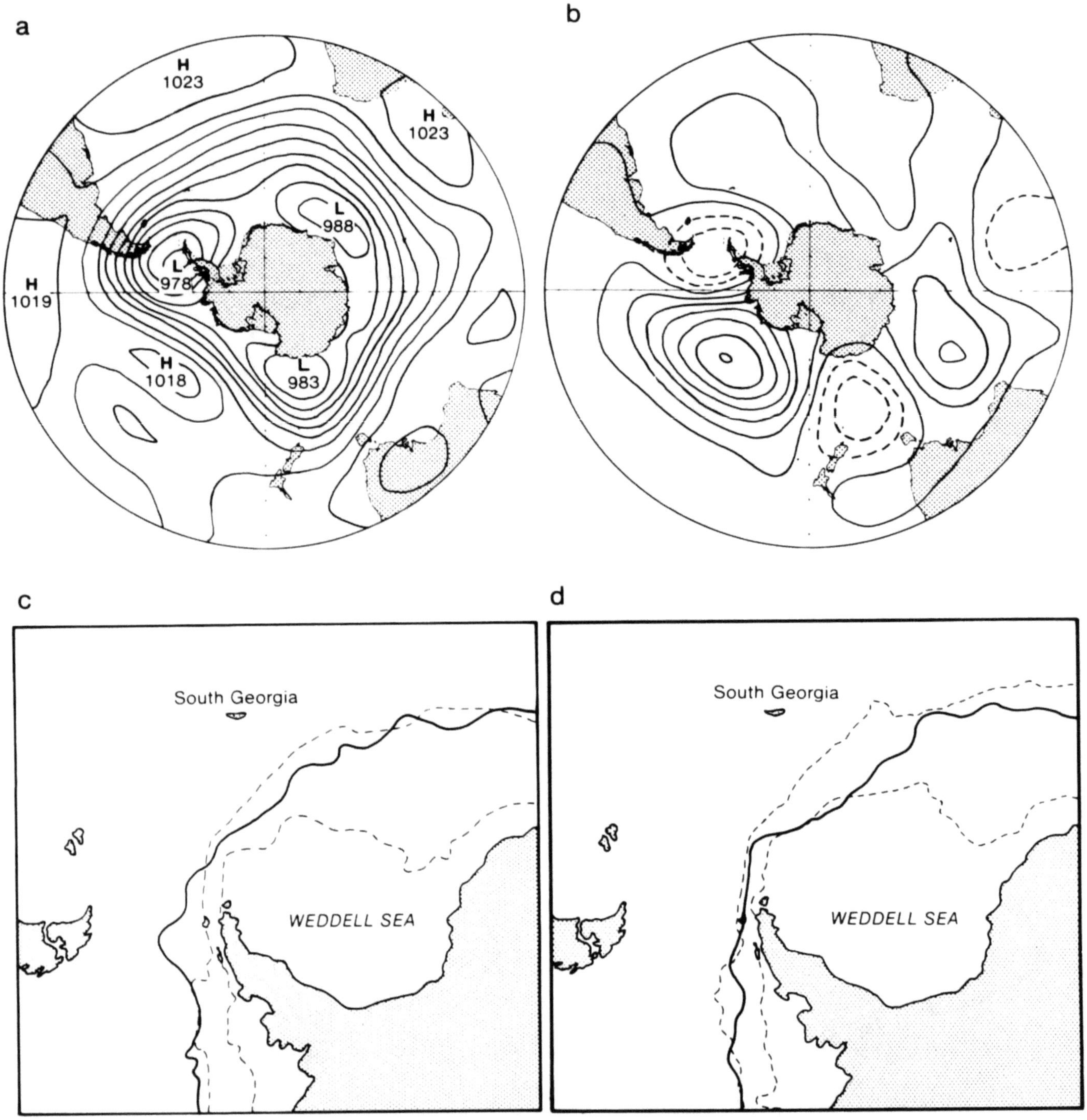

Fig. 6a–d. Pattern of atmospheric pressure at sea-level and change in position of sea-ice edge in June 1986. **a** Monthly mean sea-level atmospheric pressure. Contour intervals are 5 hPa. **b** Monthly mean sea-level pressure anomaly. Contour intervals are 5 hPa anomalies, with negative anomaly contours indicated by *broken lines*. (Redrawn from monthly Climate Bulletin of the National Climate Centre, Australian Bureau of Meteorology). **c, d** Position of the sea-ice edge in early and late June 1986 showing apparent southward movement during the month. *Solid line* indicates northern limit of sea-ice on 12 June 1986 (c) and 26 June 1986 (d). (Data from US Navy-NOAA Sea-ice Centre). *Broken lines* indicate corresponding mean and minimum limits of ice, c and d for mid- and end-June respectively, based on a 10-year climatology (1973–82: NOCD 1985)

the observed changes, which may persist for several months. However, we postulate that disruption of the normally observed mesoscale pattern of krill distribution would result not only in immediate dispersal of entrained krill but would be followed by a recovery phase during which high biomass was gradually restored. An ocean-atmosphere mechanism by which this initial disturbance could arise is suggested, in which southward airflow forces surface (top 100 m) water to the south in the Scotia and northern Weddell Seas and disrupts eddy activity. Testing of this hypothesis is dependent on long series of synoptic mesoscale observations and development of fine-resolution Antarctic ocean circulation models.

Acknowledgements. We are grateful to several colleagues for comments at various stages in the preparation of this paper, particularly David Limbert, Nigel Bonner and Pete Rothery (BAS), Max Rowe and Neil Wells (Southampton), Sidney Brown (SMRU) and Tim Smith (GST Ltd). We have also benefited from discussions with several participants in the CCAMLR-IOC Seminar on Antarctic Ocean Variability. David Warren and John Turner (BAS) provided assistance with remote-sensing data. Christine Thulborn typed the manuscript and Roger Missing prepared the figures.

References

Amos AF (1984) Distribution of krill (*Euphausia superba*) and the hydrography of the Southern Ocean: large-scale processes. J Crustacean Biol 4 (Spec No 1):306–329

Angel MV, Fasham MJR (1983) Eddies and biological processes. In: Robinson AR (ed) Eddies in marine science. Springer, Berlin Heidelberg New York, pp 492–524

Astheimer H (1986) A length class model of the population dynamics of the Antarctic krill *Euphausia superba* Dana. Polar Biol 6: 227–232

Baker DJ, Nowlin WD, Pillsbury RD, Bryden HL (1977) Antarctic Circumpolar Current: space and time fluctuations in the Drake Passage. Nature (London) 268:696–699

Beddington JR, May RM (1982) The harvesting of interacting species in a natural ecosystem. Sci Am 247(5):42–49

Beklemishev CW (1960) Southern atmospheric cyclones and the whale feeding grounds in the Antarctic. Nature (London) 187: 530–531

Brodie PF, Sameoto DD, Sheldon RW (1978) Population densities of euphausiids off Nova Scotia as indicated by net samples, whale stomach contents and sonar. Limnol Oceanogr 23:1264–1267

Bryden HL (1983) The Southern Ocean. In: Robinson AR (ed) Eddies in marine science. Springer, Berlin Heidelberg New York, pp 265–277

Carleton AM (1988) Antarctic sea ice–atmosphere signal of the Southern Oscillation in the Weddell Sea, Antarctica. J Climat 1:379–388

Carmack EC, Foster TD (1975) On the flow of water out of the Weddell Sea. Deep-Sea Res 22:722–724

Chernyy EI (1966) The influence of hydrologic conditions on the distribution of baleen whales in the Antarctic. TINRO 58:223–228 (in Russian – English translation 1973 by Naval Oceanogr Off, Washington DC, No 567)

Clarke A, Morris DJ (1983) Towards an energy budget for krill: the physiology and biochemistry of *Euphausia superba* Dana. Polar Biol 2:69–86

Croxall JP, Lishman GS (1987) The food and feeding ecology of penguins. In: Croxall JP (ed) Seabirds: feeding ecology and role in marine ecosystems. Cambridge Univ Press, Cambridge, pp 101–133

Croxall JP, Prince PA (1987) Seabirds as predators on marine resources, especially krill, at South Georgia. In: Croxall JP (ed) Seabirds: feeding ecology and role in marine ecosystems. Cambridge Univ Press, Cambridge, pp 347–368

Croxall JP, Everson I, Kooyman GL, Ricketts C, Davis RW (1985) Fur seal diving behaviour in relation to vertical distribution of krill. J Anim Ecol 54:1–8

Croxall JP, Davis RW, O'Connell MJ (1988) Diving patterns in relation to diet in gentoo and macaroni penguins, *Pygoscellis papua* and *Eudyptes chrysolophus*, at South Georgia. Condor 90:157–167

Everson I (1976) Antarctic krill: a reappraisal of its distribution. Polar Rec 18:15–23

Everson I (1981) Antarctic krill. In: El-Sayed SZ (ed) BIOMASS, vol 2. Selected Contrib Woods Hole Conf Living Resources of the Southern Ocean 1976. SCAR, Cambridge, pp 31–45

Everson I, Bone DG (1986) Detection of krill (*Euphausia superba*) near the sea surface: preliminary results using a towed upward-looking echo-sounder. Bull Br Antarct Surv 72:61–70

Fevolden SE (1984) Biotic and physical environmental impact on genetic variation of krill. J Crustacean Biol 4 (Spec No 1):206–223

Foster TD, Middleton JH (1984) The oceanographic structure of the eastern Scotia Sea. Deep-Sea Res 31:529–550

Foxton P (1956) The distribution of the standing crop of zooplankton in the Southern Ocean. Discovery Rep 28:191–236

Gambell R (1985) Birds and mammals – Antarctic whales. In: Bonner WN, Walton DWH (eds) Key environments – Antarctica. Pergamon, Oxford, pp 223–241

Georgi DT (1978) Fine structure in the Antarctic Polar Front Zone: its characteristics and possible relationships to internal waves. J Geophys Res 83:4579–4588

Gordon AL (1967) Structure of Antarctic waters between 20° W and 170° W. Antarctic Map Folio Ser 6, Am Geogr Soc, 10 pp

Gordon AL, Goldberg RD (1970) Circumpolar characteristics of Antarctic waters. Antarctic Map Folio Ser 13, Am Geogr Soc, 5 pp

Grant WS (1983) Population genetics of krill and comparison with other marine organisms. Ber Polarforsch Sonderh 4:246–266

Gulland JA (1983) Fish stock assessment: a manual of basic methods. FAO-Wiley series on food and agriculture, vol 1. Wiley, New York, 223 pp

Hancock DA (1973) The relationship between stock and recruitment in exploited invertebrates. Rapp PV Reun Cons Int Explor Mer 164:113–131

Hardy AC, Gunther ER (1935) The plankton of the South Georgia whaling grounds and adjacent waters, 1926–1932. Discovery Rep 11:1–456

Harmer SF (1931) Southern whaling. Proc Linn Soc London Session 142 (1929–30):85–163

Heywood RB, Everson I, Priddle J (1985) The absence of krill from the South Georgia zone, winter 1983. Deep-Sea Res 32: 369–378

Jarogov A (1969) On the physico-geographical conditions of the krill habitat. TINRO 66:85–101 (in Russian)

Kemp S, Bennett AG (1932) On the distribution and movements of whales on the South Georgia and South Shetland whaling grounds. Discovery Rep 6:165–190

Laws RM (1984) The ecology of the Southern Ocean. Am Sci 73: 26–40

Legeckis R (1977) Ocean polar front in the Drake Passage – satellite observations during 1976. Deep-Sea Res 24:701–704

Longhurst A (1984) Heterogeneity in the ocean – implications for fisheries. Rapp PV Reun Cons Int Explor Mer 185:268–282

Macaulay MC, English TS, Mathisen OA (1984) Acoustic characterisation of swarms of Antarctic krill (*Euphausia superba*) from Elephant Island and Bransfield Strait. J Crustacean Biol 4 (Spec No 1):16–44

MacDonald CM, Schneppenheim R (1983) Breeding structure and stock identity in the Antarctic krill *Euphausia superba* Dana. Ber Polarforsch Sonderh 4:240–245

MacDonald CM, Williams R, Adams M (1986) Genetic variation and population structure of krill (*Euphausia superba* Dana) from the Prydz Bay region of Antarctic waters. Polar Biol 6:233–236

Mackintosh NA (1972) Life cycle of Antarctic krill in relation to ice and water conditions. Discovery Rep 36:1–94

Marr JWS (1962) The natural history and geography of the Antarctic krill (*Euphausia superba* Dana). Discovery Rep 32:33–464

Maslennikov VV (1980) Modern concepts on the large-scale circulation of the Antarctic water and routes of mass drift of *Euphausia superba*. Trudy VNIRO, Biological resources of the Antarctic krill. ISSNO 372–2864:8–27 (in Russian)

Miller DGM (1985) The South African SIBEX 1 Cruise to the Prydz Bay region, 1984: IX. Krill (*Euphausia superba* Dana). S Afr J Antarct Res 15:33–41

Miller DGM (1986) Results from biological investigations of krill (*Euphausia superba*) in the Southern Ocean during SIBEX 1. Mem Natl Inst Polar Res Spec Issue 40:117–139

Miller DGM, Hampton I (ms) A review of the biology and status of the Antarctic krill (*Euphausia superba* Dana). Sea Fisheries Res Inst, Roggebaai, South Africa, 261 pp

Mizroch SA (1984) The development of balaenopterid whaling in the Antarctic. Cetus 5(2):6–10

Mizroch SA, Rice DW, Bengtson JL, Larson SW (1985) Preliminary atlas of balaenopterid whale distribution in the Southern Ocean based on pelagic catch data. Rep SC-CAMLR-IV/BG/21 to CCAMLR Sci Comm, 9 pp

Morris DJ, Priddle J (1984) Observations on the feeding and moulting of the Antarctic krill, *Euphausia superba* Dana, in winter. Bull Br Antarct Surv 65:57–63

Mujica A, Asencio V (1985) Fish larvae, euphausiids and community structure of zooplankton in the Bransfield Strait (SIBEX – Phase 1) 1984. Ser Cient INACH (Engl Ed) 33:131–154

NOCD (1985) Sea ice climate atlas, vol 1. Antarctic. Naval Oceanography Command Detachment, Asheville, 131 pp

Nowlin WD, Klinck JM (1986) The physics of the Antarctic Circumpolar Current. Rev Geophys 24:469–491

Philander SGH (1986) Predictability of El Niño. Nature (London) 321:810–811

Priddle J, Hawes I, Ellis-Evans JC, Smith TJ (1986a) Antarctic aquatic ecosystems as habitats for phytoplankton. Biol Rev 61: 199–238

Priddle J, Heywood RB, Theriot E (1986b) Some environmental factors influencing phytoplankton in the Southern Ocean around South Georgia. Polar Biol 5:65–79

Ramage CS (1986) El Niño. Sci Am 254(6):55–61

Rojas R (1985) Description of the thermal structure of the Bransfield Strait based on XBT observations. Ser Cient INACH (Engl Ed) 33:83–114

Rosenberg AA, Beddington JR, Basson M (1986) Growth and longevity of krill during the first decade of pelagic whaling. Nature (London) 324:152–154

Schneppenheim R, MacDonald CM (1983) Population genetics of krill (*Euphausia superba*). Ber Polarforsch Sonderh 4:239

Schneppenheim R, MacDonald CM (1984) Genetic variation and population structure of krill (*Euphausia superba*) in the Atlantic Sector of Antarctic waters off the Antarctic Peninsula. Polar Biol 3:19–28

Siegel V (1986) Untersuchungen zur Biologie des antarktischen Krill, *Euphausia superba,* im Bereich der Bransfield Straße und angrenzender Gebiete. Mitt Inst Seefisch Hamburg 38:1–244

Terazaki M, Wada M (1986) Euphausiids collected from the Australian Sector of the Southern Ocean during the BIOMASS SIBEX cruise (KH-83-4). Mem Natl Inst Polar Res Spec Issue 40:97–109

Witek Z, Kittel W, Czykieta H, Zmijewska MI, Presler E (1985) Macrozooplankton in the southern Drake Passage and in the Bransfield Strait during BIOMASS-SIBEX (December 1983–January 1984). Polish Polar Res (Engl Ed) 6:95–115

Zwally HJ, Comiso JC, Parkinson CL, Cambell WJ, Carsey FD, Gloersen P (1983) Antarctic sea-ice, 1973–76: satellite passive-microwave observations. Nat Aeronaut Space Administr, Washington DC, 206 pp

Hydrography, Krill and Fish and Their Possible Relationships Around Elephant Island

F. Nast, K.-H. Kock, D. Sahrhage, M. Stein, and J.E. Tiedtke[1]

Summary. Krill, fish, and hydrological conditions around Elephant Island were studied with a station grid program (Elephant Island box) between the 1975/76 and 1985/86 seasons on an almost annual basis. During the nine times of observation krill distribution and abundance was assessed by standard hauls with the RMT 1+8-nets, that of fish by commercial-sized bottom trawsl, both supplemented by vertical CTD profiles. The area north of Elephant Island was found to be characterized by the confluence of Scotia Sea and Weddell Sea waters. The dynamic topography elucidated the nature of a stationary meander in the vicinity of Elephant Island. Krill and fish abundances were related to the different water masses. Median krill abundances varied between 15 individuals/1000 m³ in November 1983 as a maximum to $\tilde{x}$ = 0 in May 1986, showing the normal seasonal cycle but no clear interannual trend within this time series, pointing to a more stable situation than around South Georgia. Also krill distribution appeared in a persistent spatial pattern along seasons and years. Total fish biomass was found to be generally higher (up to more than 100 times) in krill-rich areas than in krill-poor areas. Analysis is aggravated by the observed remarkable changes in biomass and population structure of the most dominant fish species between 1977/78 and 1980/81 due to heavy fishing activities.

1 Introduction

In 1976 the BIOMASS Program was established to stimulate investigations on the life history and dynamics of organisms at different trophic levels of the marine Antarctic ecosystems and their interactions with the environment. Emphasis should be given to those elements which are believed to contribute significantly to the energy flow through higher trophic levels and (or) are currently under exploitation or have potential for future larger-scale harvesting (BIOMASS 1977). Since the Convention for the Conservation of Antarctic Marine Living Resources (CCAMLR) came into force in 1982, its Scientific Committee has underlined the need for such studies as prerequisites for a sound management of Antarctic marine living resources.

Following the recommendations of the BIOMASS-Program, the Federal Republic of Germany established a routine sampling program with a 75 × 75 nautical miles grid ("box") around Elephant Island in 1977/78 (Hempel et al. 1979) to monitor the seasonal and year-to-year fluctuations in krill abundance and hydrographical conditions in a selected area of the seasonal pack ice zone. In 1983 the "box" was extended further to the north and investigations on demersal fish within the 50–500 m depth range around the island were included. The box was surveyed in 1977/78 (November–March), November 1983, 1984/85 (November–March) and May 1986. Additional but more limited information for the area was available from echosounder recordings (Klindt 1986), two exploratory fishing campaigns with chartered fishing vessels in 1975/76 and 1977/78 (Sahrhage et al. 1978; Hempel et al. 1979) and investigations during FIBEX in February–March 1981 (Anonymous 1982). In the following an attempt is made to synthesize the results of all surveys and to describe possible relationships between hydrographical conditions, krill abundance, and fish around Elephant Island.

2 Material and Methods

In 1977/78 the "box" consisted of four longitudinal transects from 60°45′S to 61°45′S at 54°15′W, 55°00′W, 56°00′W and 56°45′W. Routine stations were 15 nm in longitudinal and 20 nm in latitudinal direction apart from each other. In 1983 the transects were extended to 60 and 62°S and between each station an XBT probe was launched down to 750 m depth.

Routine station observations consisted of a vertical CTD profile recording temperature, conductivity and sound velocity (see Stein 1986 for detailed description), standard meteorological observations and an oblique haul with RMT 1+8. Maximum fishing depth was 0–140 m (in 1977/78) and otherwise always 200–0 m (see Wörner 1979; Nast 1986 for detailed description). Fifteen of the total of 25 hauls in May/June 1986 were fished from a depth of 400 m to the surface due to an erroneous recording of the net monitor. Values were corrected to 200 m maximum fishing depth by assuming that 90% of the krill is inhabiting the upper 200 m of the water column. The calculation of filtered volumes, catch and sample treatment, sorting, length measurements, length-weight relationships and the determination of sex and maturity stages are described by Pommeranz et al. (1983) and Nast (1986).

[1] Institut für Seefischerei, Bundesforschungsanstalt für Fischerei, Palmaille 9, 2000 Hamburg 50, FRG

Antarctic Ocean and Resources Variability
(ed. by D. Sahrhage)

Table 1. Number of stations and measurements performed each season

Period of investigation		Name of vessel	Type of survey	No. of stations CTD	XBT	RMT 1+8	Bottom trawl
1975/76	Jan.–Febr.	*Weser*	Exploratory	–	–	–	17
		Walther Herwig		–	–	15	–
1977/78	November	*Walther Herwig*	Grid	24	–	17	3
	January	*Walther Herwig*	Grid	20	–	18	4
	March	*Walther Herwig*	Grid	22	–	20	–
	Febr.–March	*Julius Fock*	Exploratory	–	–	–	24
1980/81	March	*Walther Herwig*	Random	12	–	11	13
1983/84	November	*Polarstern*	Grid	36	35	36	–
	November	*Polarstern*	Random	12	–	–	12
1984/85	February	*Walther Herwig*	Random	37	–	35	37
1985/86	May	*Polarstern*	Grid	36	35	25	–
	May–June	*Polarstern*	Random	31	–	–	37

Fish were sampled between 65 and 550 m depth by means of 140′ and 200′ bottom trawls. Fishing stations were chosen randomly but restricted mostly to the north and west of the island's shelf where fishing conditions were known to be moderate or good (see Kock 1986 for detailed description). In general a CTD profile and in 1981 and 1985 also an RMT 1+8 haul could be carried out concurrently to each fishing haul.

Exploratory fishing using only commercially sized trawls was primarily directed to dense krill aggregations with a perspective of several tons catch and to fish concentrations detected northwest of the island. A bottom trawl survey subsequently to the FIBEX participation based on a random survey design had to be given up after 13 hauls due to a machinery failure of FRV *Walther Herwig* (Kock 1982).

"Quality" of the data sets particularly for fish may thus vary considerably from year to year. A list of stations and measurements carried out is given in Table 1.

Due to the contagious distribution of krill and fish, resulting catches were not normally distributed ($\tilde{x}/\bar{x} < 0.3$; Sachs 1984). Preference was thus given to the median ($\tilde{x}$) to estimate average abundance, and nonparametric statistics were applied to the data.

3 Results

3.1 Hydrography

The hydrography of the Elephant Island region is described in a recent paper by Stein (1986). Accordingly, the area north of Elephant Island is characterized by the confluence of Scotia Sea and Weddell Sea waters, the Weddell-Scotia Confluence (WSC). The frontal zone separating these two water masses is a southern boundary for one of the major water masses which contribute to the general Θ,S-curve, the Warm Deep Water. This water mass obtains its characteristics from the North Atlantic Deep Water which flows southward from the North Atlantic near the bottom to rise sharply in the vicinity of the Antarctic Convergence. It is underlying the Antarctic Circumpolar Current and is found in the Antarctic Ocean at depths around 500 m. As noted during the Second Post-FIBEX Hydrographic Data Interpretation Workshop (BIOMASS 1983), there is indication of a deep flow of water from the Bransfield Strait passing on opposing courses east and west of Elephant Island. Both current branches meet and mix within a canyon off the north coast of the island. "One overall effect was to restrict the ingress of Scotia Sea water into the area" (BIOMASS 1983).

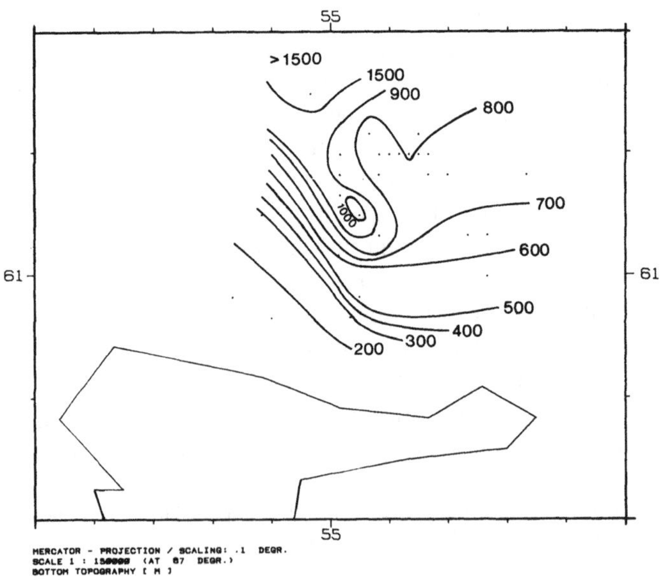

Fig. 1. The bottom topography north of Elephant Island

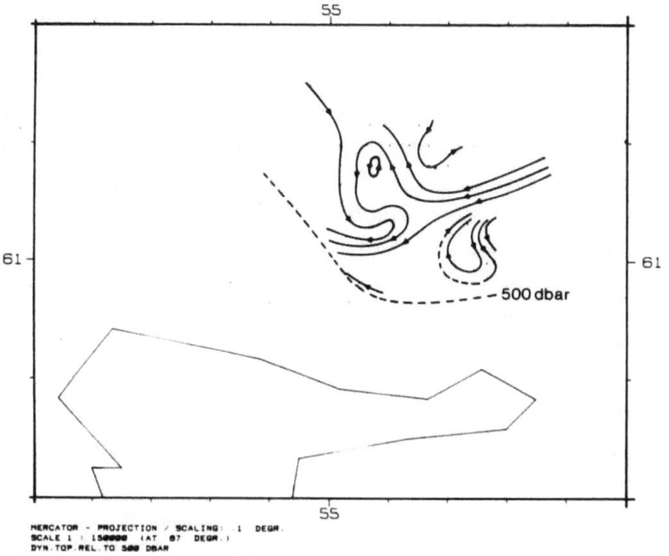

Fig. 2. The dynamic circulation at 500 dbar north of Elephant Island during FIBEX, 1981

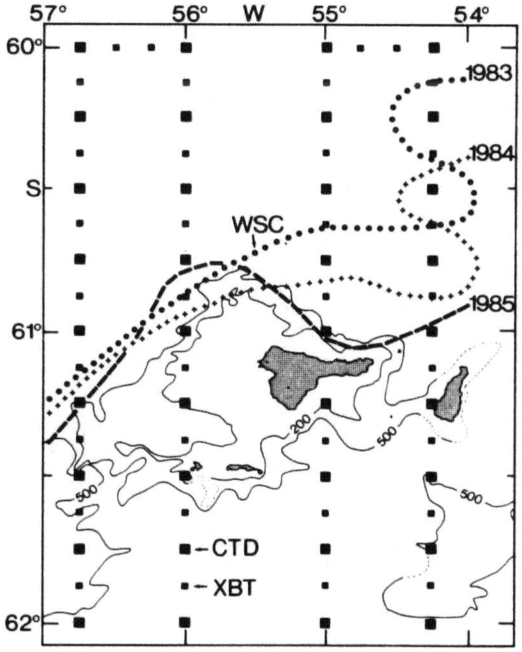

Fig. 4. The path of the Weddell-Scotia Confluence during spring 1983, late spring 1984 and autumn 1985

Figures 1, 2, and 3 indicate the bottom topography, the dynamic circulation and the percentage of Deep Water of the Bransfield Strait at 500 dbar as obtained during FIBEX 1981. Figure 4 outlines the path of the WSC during spring 1983, late spring 1984 and autumn 1985. The meander in the vicinity of 54°W denotes an area where water masses from the Weddell Sea flow into the Scotia Sea after passing the South Scotia Ridge east of the Elephant Island plateau.

Detailed studies on the variability of the baroclinic flow around Elephant Island are given by Stein (this Vol.). From this paper the 1983 and 1985 circulation pattern 0/500 dbar is depicted (Fig. 5a,b). Both parts of the figure reveal a bending of the stream-lines when approaching the shelf northwest of Elephant Island. The solid arrows in both parts of the figure denote the mean current measured in the southern Drake Passage (60°54.6′S, 57°06.0′W) at a depth of 439 m between December 2, 1980 and January 21, 1981 (Wittstock and Zenk 1983).

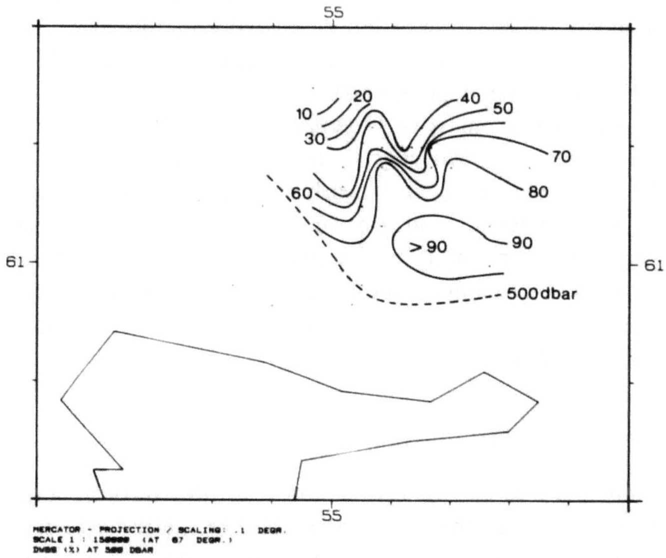

Fig. 3. Percentage of Deep Water of the Bransfield Strait at 500 dbar north of Elephant Island during FIBEX, 1981

There is a divergent flow pattern further to the north which reflects in part the impact of the submarine topography on the distribution of water masses (Stein this Vol.). Between King George Island and Elephant Island the Loper Channel (62°S, 57°W) is one of the major outlets for Weddell Sea water to the southern Drake Passage. This water mass, enriched by the steady flow from the northern Bransfield Strait, is seen to bend eastward to flow in a northeasterly direction between Elephant and Clarence Island (61°15′S, 54°30′W).

Figure 6 displays in different hatching the distribution of on-shelf water masses in February 1985, slope water masses and deep water masses. The southwestern and northwestern approaches of Elephant Island as well as the strait between Elephant and Clarence Island are influenced by water masses ranging from 34.4 to 34.6 in salinity and 0.65°C to 0.12°C in temperature at the seafloor (large dots). Confined to the shallower parts of the area are the water masses characterized by a lower salinity and warmer temperature. Contoured by the 200 m isobaths one finds salinities of the bottom water ranging from 34.25 to 34.4, and temperatures between 0.65°C and 1.16°C. As mentioned above, the water mass confined between the 200 m and 500 m isobaths reflects deep water flowing from the Bransfield Strait on opposing courses east and west of Elephant Island, to meet and mix north of the island, forming an anticyclonic gyre.

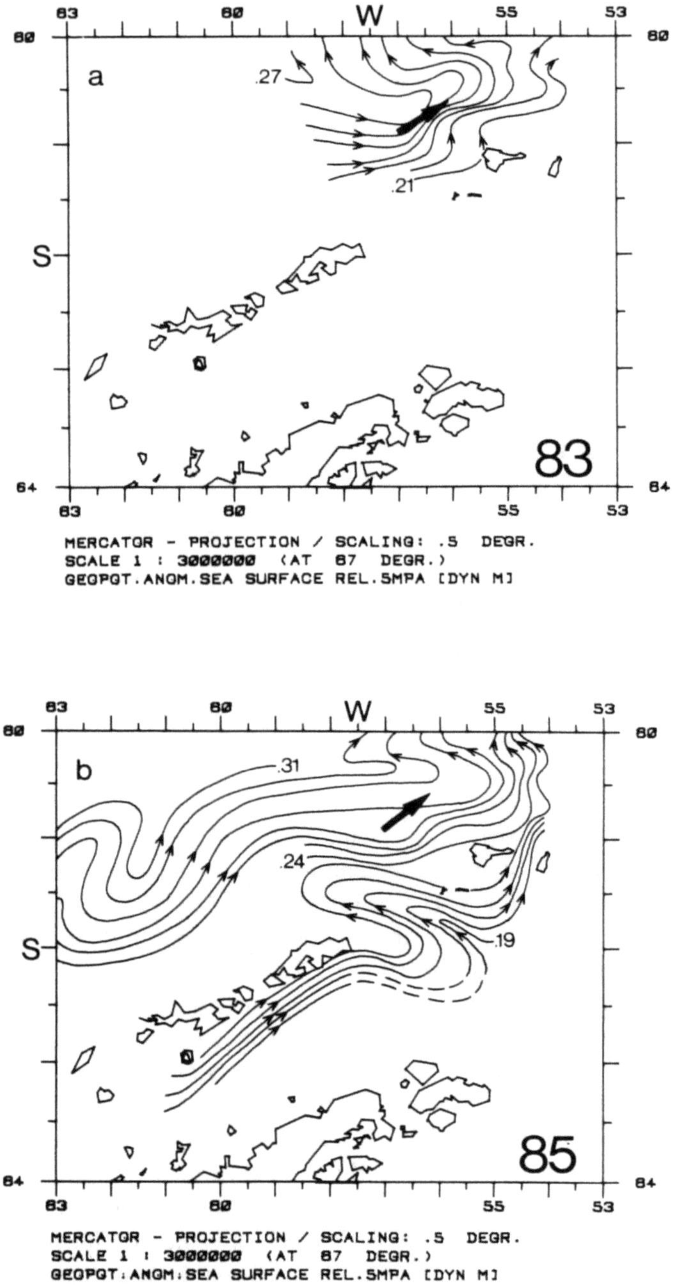

Fig. 5a,b. The baroclinic flow around Elephant Island in a 1983 and b 1985. *Solid arrows* denote the mean current

3.2 Fish

3.2.1 The Composition of the Fish Fauna

Eleven families comprising 44 species (excluding 7 pelagic and mesopelagic families with 13 species) have been encountered around the island. Sixteen (36.4%) of them have a Lesser Antarctic (partly including South Georgia) or circum-Antarctic type of range whereas 26 species (59.1%) are confined to circumpolar high Antarctic waters. Two species (4.5%) are Patagonian immigrants (e.g., Permitin 1977). Only 12 species are common (> 40 records), 10 of them have a Lesser Antarctic or circum-Antarctic type of distribution. Ten occur rarely but regularly (> 10 records), 21 are rare (< 10 records). Except for *Pseudochaenichthys georgianus* and *Pagothenia hansoni* all have a circumpolar high Antarctic type of range (Table 2).

The majority of species (52.5%) are members of the two families Nototheniidae (13 species) and Channichthyidae (10 species) (Table 2). The most dominant five of them, *Notothenia rossii marmorata, N. gibberifrons, N. neglecta, Champsocephalus gunnari,* and *Chaenocephalus aceratus* make up more than 98% of the fish biomass between 50 and 500 m depth. All have a Lesser Antarctic (4) or circum-Antactic (1) type of distribution.

Despite the uneven coverage of the area by trawling from year to year, there is evidence that fish concentrated each year on the northern and northwestern part of the shelf. Except for a single catch of 5.1 tons in November 1983 (Fig. 11) when a postspawning aggregation of *N. gibberifrons* was met, catch rates south of 61°10'S were in general 2–5 x less than north of it (Figs. 8–13).

In between 1977/78 and 1980/81 the composition of the fish fauna changed remarkably in terms of both biomass and population structure of the most abundant species. In 1976–78 *N. rossii marmorata* and *C. gunnari* made up a significant proportion of the biomass although its relative importance of 75% and 22% is overestimated to an unknown extent due to the nonrandom distribution of hauls. From 1981 onward the proportion of both species was considerably less: 7.1–13.0% and 1.4–2.8% depending on the year. The most dominant species was now *N. gibberifrons,* making up 65–80% of the total catch. The second most abundant species was *Ch. aceratus.*

Concentrations of *N. rossii marmorata* in 1976–1978 consisted predominantly of adult fish of age classes 7–13 (Freytag 1980). From 1981 onward, attempts to locate the adult stock failed although trawling was carried out in the area of its maximum biomass in previous seasons. Catches of *N. rossii marmorata* now comprised mostly 5–7-year-old fish which were about to recruit to the adult part of the population. The proportion of sexually mature fish, particularly that of females, was negligible compared to 1976–78 (Kock unpublished).

The high biomass of *C. gunnari* in 1977/78 was due to the presence of two very abundant year-classes (probably 1973, 1974). They had nearly disappeared in 1980/81. Since then length and age samples demonstrated a high year-to-year variability in the composition of the stock at a much lower biomass level but without any significant trend. Year-classes of the strength of 1973 and 1974 have not been observed since.

3.2.2 Vertical Distribution

Many species occurred over most of the vertical range covered by trawling. The fish fauna became more diverse with increasing depth. A comparison of the vertical range of maximum biomass (i.e., where at least 80% of the stock was present) revealed the presence of four groups of

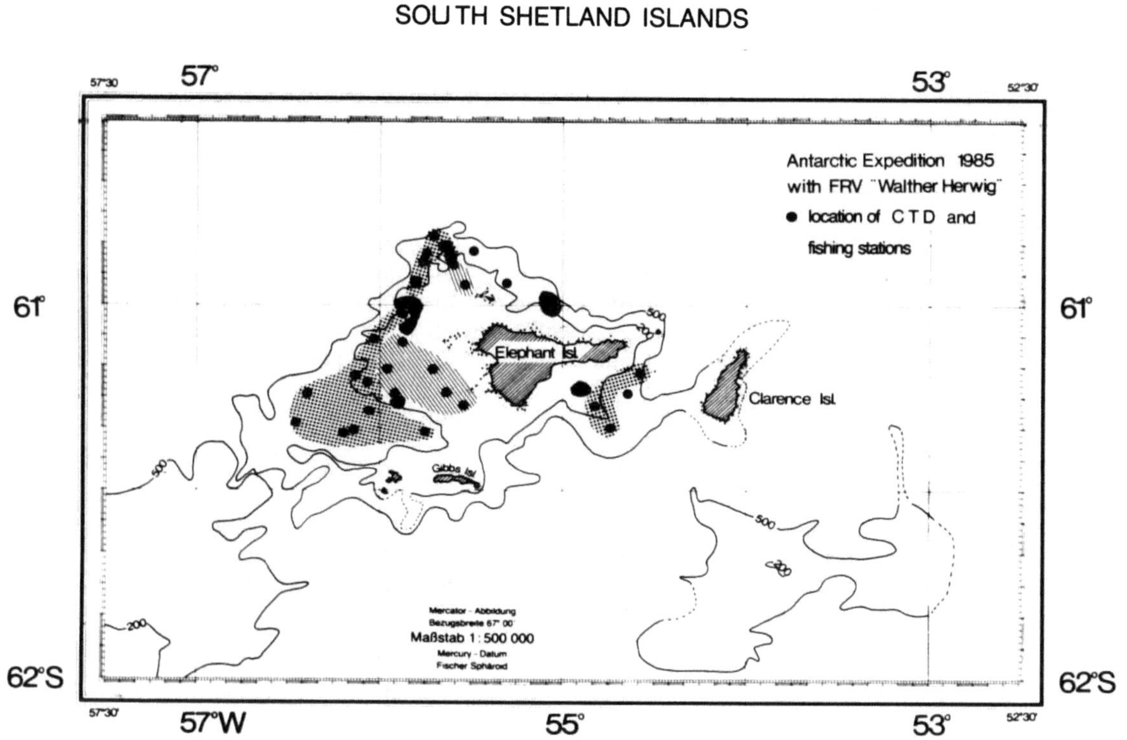

Fig. 6. The distribution of on-shelf water masses (*hatched*), slope water (*dots*), and deep water masses (*double hatched*) around Elephant Island in February 1985

species, altough the resolution of the depth zones by 50 m steps was rather coarse (Table 2):

1. *Shallow water species,* abundant only shallower than 200–250 m. This group comprises only three species: *Harpagifer antarcticus, Nototheniops nudifrons,* and *Parachaenichthys charcoti.*
2. *Eurybath species,* abundant over most of the depth range down to 350–400 m. Most of the species with a Lesser Antarctic type of distribution are eurybath (e.g., *N. rossii marmorata, N. gibberifrons,* and *C. gunnari*).
3. *Deeper water species,* in general occurring below 200–250 m or even deeper. Most of the species with a circumpolar high Antarctic type of distribution belong to that group.
4. *Continental slope species,* present only below 450 m. This group is represented by *Macrourus holotrachys* only.

Some of the species obviously prefer different vertical ranges during their annual or life cycle. Catches of *Pleuragramma antarcticum* in less than 200 m depth were in general juveniles of 7–10 cm (total) length (age class 2+, according to Kellermann 1986), whereas sexually mature fish usually occur only below 300 m. Vertical range of maximum biomass in *Notothenia neglecta* was much deeper in May/June due to the migration of the species to deeper waters for spawning.

3.2.3 Food Composition

A comprehensive study on the food composition is in preparation. So far stomach content analysis is still fragmentary and confined to some of the most abundant species (Table 3). It is, however, obvious that *Euphausia superba* is the staple food in *N. rossii marmorata* and *C. gunnari* and also an important food item for *N. larseni* and *Ch. aceratus.* Differences in the weight proportion of *E. superba* in the food composition of the latter two species between the years was mainly due to the different size of the fish investigated. In juvenile *Ch. aceratus* the proportion of krill in the diet decreased with size (Kock 1981; Gubsch 1982), whereas its proportion in *N. larseni* increased with size of the fish.

No information is so far available for *N. gibberifrons.* The species is primarily a benthos feeder. North and east of the South Shetland Islands, however, *E. superba* made up a significant proportion in its diet at least in some years (Tarverdiyeva and Pinskaya 1980; Takahashi 1983).

3.3 Krill

Variability in plankton is observed on different scales in time and space (e.g., Longhurst 1981; Mackas et al. 1985). Krill as a highly mobile micronekton organism (Hamner 1984) was demonstrated to be able to change its abundance

Table 2. Fish species (excluding mesopelagic species) collected around Elephant Island in the course of the Antarctic Expeditions 1975/76–1986 of the Federal Republic of Germany

Family *Species*	Geographical distribution Type of range	Vertical distribution depth range (m)	Of maximum abundance	Status No. of records	Occurrence	Remarks
Nototheniidae						
Notothenia rossii marmorata	Lesser Antarctic	50–550	300–550	101	Common	
N. gibberifrons	Lesser Antarctic	50–550	150–350	133	Common	
N. neglecta	Circum Antarctic	50–550	Summer: 50–250	74	Common	
			Winter: 150–450			
N. kempi	Circum Antarctic	100–550	350–550	52	Common	
Nototheniops larseni[a]	Lesser Antarctic	50–550	150–450	106	Common	
Nototheniops nudifrons	Lesser Antarctic	50–350	50–250	50	Common	
Pagothenia hansoni	Circum Antarctic	350–493	–	2	Rare	
P. bernacchii	Circumpolar high Antarctic	50–250	–	6	Rare	
P. brachysoma	Circumpolar high Antarctic	40– 90	–	1	Rare	Pelagic juvenile
Trematomus newnesi	Circumpolar high Antarctic	50–400	–	10	Rare	
T. eulepidotus	Circumpolar high Antarctic	100–550	350–550	23	Regular	
Dissostichus mawsoni[b]	Circumpolar high Antarctic	50–550	300–550	36	Regular	
Pleuragramma antarcticum	Circumpolar high Antarctic	50–550	100–200	21	Regular	
Harpagiferidae						
Harpagifer antarcticus[d]	Lesser Antarctic	50–100	<50	7	Common	
Artedidraconidae						
Artedidraco skottsbergi	Circumpolar high Antarctic	198–219	–	1	Rare	
Pogonophryne barsukovi	Circumpolar high Antarctic	200–220	–	1	Rare	
P. marmorata	Circumpolar high Antarctic	400	–	1	Rare	
P. phyllopogon[c]	Circumpolar high Antarctic	400	–	1	Rare	
P. permitini	Circumpolar high Antarctic	450–470	–	1	Rare	
P. scotti	Circumpolar high Antarctic	227–340	–	2	Rare	
Channichthyidae						
Champsocephalus gunnari	Lesser Antarctic, Kerguelen	50–550	100–300	114	Common	
Chaenocephalus aceratus	Lesser Antarctic	50–550	150–350	122	Common	
Pseudochaenichthys georgianus	Lesser Antarctic	100–550	150–350	23	Regular	
Chionodraco rastrospinosus	Lesser Antarctic (?)	100–550	250–550	80	Common	
Cryodraco antarcticus	Circumpolar high Antarctic	150–550	250–400	20	Regular	
Chaenodraco wilsoni	Circumpolar high Antarctic (100)	250–550	–	7	Rare	(100) Pelagic juvenile
Dacodraco hunteri	Circumpolar high Antarctic	392–406	–	1	Rare	
Pagetopsis macropterus	Circumpolar high Antarctic	100	–	1	Rare	Pelagic juvenile
Neopagetopsis ionah	Circumpolar high Antarctic	199–493	–	6	Rare	
Chionobathyscus dewitti	Circumpolar high Antarctic	147–185	–	1	Rare	
Bathydraconidae						
Gerlachea australis	Circumpolar high Antarctic	200–550	–	8	Rare	
Gymnodraco acuticeps	Circumpolar high Antarctic	100–550	–	7	Rare	
Parachaenichthys charcoti	Lesser Antarctic	50–400	50–200	16	Regular	
Prionodraco evansii	Circumpolar high Antarctic	250–400	–	3	Rare	
Racovitzia glacialis	Circumpolar high Antarctic	227–490	–	5	Rare	
Rajidae						
Bathyraja eatonii	Circumpolar high Antarctic Kerguelen	50–350	–	13	Regular	
Bathyraja maccaini	Circumpolar high Antarctic	50–550	150–450	49	Common	
Bathyraja sp. n. II	Circumpolar high Antarctic	100–550	100–300	35	Regular	
Zoarcidae						
Lycodichthys antarcticus	Circumpolar high Antarctic	50–400	–	13	Regular	
Ophthalmolycus amberensis	Lesser Antarctic (?)	279–336	–	5[e]	?	
Liparididae						
Paraliparis gracilis	Circum – Antarctic	392–406	–	1	Rare	

Table 2 (continued)

Family *Species*	Geographical distribution Type of range	Vertical distribution depth range (m)	 Of maximum abundance	Status No. of records	 Occurrence	Remarks
Muraenolepidae *Muraenolepis microps*	Patagonian-Antarctic	200–550	200–400	12	Regular	
Gadidae *Micromesistius australis*	Patagonian-Antarctic	292–405	–	5	Rare	
Macrourida *Macrourus holotrachys*[d]	Circum – Antarctic	445–552	> 500	2	Common	

[a] Identical with *Nothenia nybelini* Balushkin 1976.
[b] Two specimens of *D. mawsoni* were earlier misidentified as *D. eleginoides* (see Kock 1982).
[c] Misidentified as *P. dolichobranchiata* in Kock (1982).
[d] Vertical range not adequately covered by trawling.
[e] Preliminary, as Zoarcids are still under preparation.

Table 3. Proportion of *Euphausia superba* (in % of wet weight) in the food of four fish species around Elephand Island in the period 1976–1985

	February 1976		February 1978		March 1981		November 1983		February 1985	
	No. of stomachs investigated	% wet wt. of *E. superba*	No. of stomachs investigated	% wet wt. of *E. superba*	No. of stomachs investigated	% wet wt. of *E. superba*	No. of stomachs investigated	% wet wt. of *E. superba*	No. of stomachs investigated	% wet wt. of *E. superba*
Notothenia rossii marmorata	793	> 95[a]	745	> 98[a]	–	–	–	–	232	> 90[a]
Nototheniops larseni	–	–	–	–	–	–	61	25.0	66	93.9
Champsocephalus gunnari	–	–	224	99.2	265	97.8	231	90.8	284	96.9
Chaenocephalus aceratus	–	–	16	3.2	379	99.0	–	–	62	5.0

[a] Preliminary data.

within seconds, whereas Macaulay et al. (1984) showed a krill swarm area to persist for some weeks.

3.3.1 Vertical Distribution and Abundance

On a 4 days' time-station south of Elephant Island (Nast 1978/79), krill density could be demonstrated to be a function of time of day and depth. Maximum catch rates were decreasing by one order of magnitude per depth stratum:

Fishing depth interval (m)	20–0	50–20	200–50	400–200	600–400
Max.krill catch (n/1000 m^3)	8384	4005	238	79	1

Highest concentrations of the entire time-station were recorded at night in the 20–0 m stratum. When plotting the relative variability $(s_{\bar{x}}/\bar{x})$ and the depth in which an "average individual" (Pennak 1943) can be found against the time of day (Fig. 7) it becomes obvious that variability was highest among catches during the nighttime when dense concentrations were mainly met near the surface. During daylight the krill descended to $\sim$ 140 m depth and catch rates were much less variable. Higher nighttime values indicating a nighttime increase in the intensity of variability or patchiness have also been reported by Greenblatt et al. (1982) in patchiness studies off California.

3.3.2 Horizontal Distribution and Abundance

The horizontal distribution of krill is presented in the form of maps. To mark the krill-rich and krill-poor areas, isolines of krill abundances (in g, whenever possible) were calculated by linear interpolation between stations. As krill density changed considerably between sampling intervals, limits

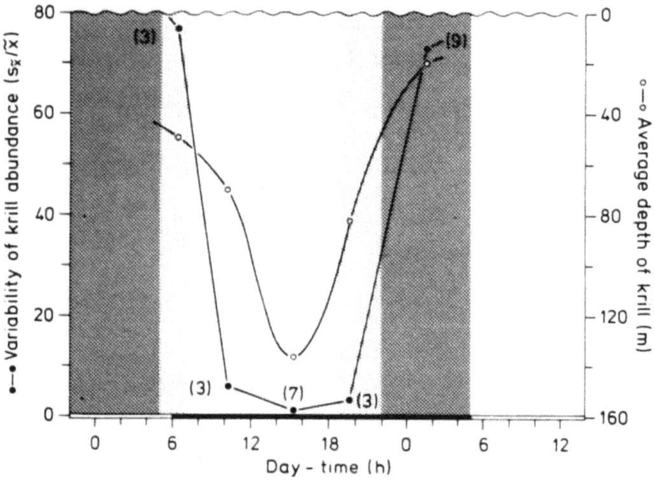

Fig. 7. The relative variability (standard deviation of the median/median) and the average depth of krill at different times of the day as observed during the time station south of Elephant Island, 6–10 February 1976

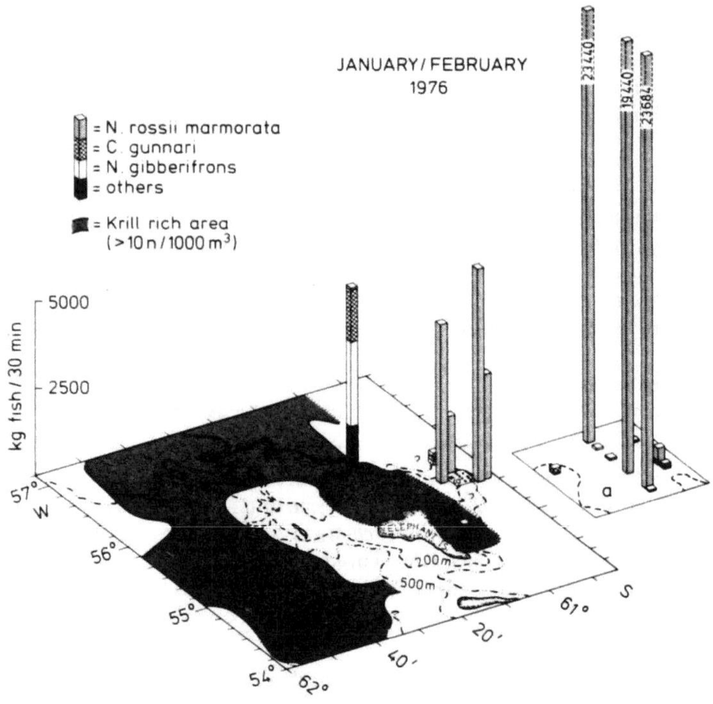

Fig. 8. Krill and fish abundance around Elephant Island in January/February 1976

for separation were set relative to the actual abundance and thus varied between 10–50 g/1000 m^3.

In January/February 1976 the highest krill abundance (n/1000 m^3) was recorded north of Elephant Island and the krill-rich area (> 10 individuals/1000 m^3, Fig. 8) covered the shelf west of the island, but not the southern shelf, whereas a band of high krill abundance extended south of 61°30'S from west to east. A relative high value of 33 animals/1000 m^3 (outside the 95% confidence limits of the median for this time of observation, Table 4) measured on the west plateau was supported by an aimed haul two days later recording a swarm density of more than 17 000 krill/1000 m^3.

Two years later, January 1978, the median krill abundance was two to three times higher, the variability expressed as the percentage standard error (Table 4) was found to be more than five times higher (327%). The krill-rich area was centered in a circle of 20–30 nautical miles around the island. In contrast to the observation two years before, krill now also occurred south of the island, even with the highest recorded value of this month.

In March 1978 krill abundance was centered in the north of the box (Fig. 9); south of 61°10'S krill occurred in very low numbers, dropping the biomass median by 20 times for the total box. However, the biomass of krill increased by a factor of 2.4 in the krill-rich area.

The March 1981 survey was restricted to the northern area of Elephant Island. Krill was abundant north of the island, from Seal Rocks to northwest of Cape Valentine with a narrow band extending to the west. Less krill was observed on the shelf west and northwest of Seal Rocks (Fig. 10).

At the earliest time of sampling in the season, November 1983, the krill-rich areas were not found as bound to the island as shown for seasonal later sampling intervals. Nevertheless, the krill-rich area northwest of the island (Fig. 11) covered the slope by its southern extension. A second krill-rich area was found southwest of Elephant Island as well as a small area south of Clarence Island. These areas of relative high krill biomass also showed a different maturity stage composition of krill in comparison to low-biomass areas, where juveniles dominated (Nast 1986).

Table 4. Number of RMT 8-stations (n), median abundance ($\tilde{x}$) of krill, 95% confidence limits and maximum abundance in individuals /1000 m^3, percentage standard error of the median. St.E. ($\tilde{x}$) and median/mean ratios ($\tilde{x}/\bar{x}$) for each sampling interval 1976–1986

	Jan./Feb. 1976	Jan. 1978	March 1978	March 1981	Nov. 1983	Feb. 1985	May 1986
n	15	18	20	11	36	35	25
$\tilde{x}$	3.9	9.0	0.85	2.1	15	2.0	0
95% C.L.	0–11.4	1–123	0.2–4	0.2–85	3–39	0.7–10	0–0.2
Max.	417.3	624	956	144	4321	2388	6
St.E. ($\tilde{x}$)	± 61%	± 327%	± 92%	± 915%	± 68%	± 84%	± 0.1%
$\tilde{x}/\bar{x}$	0.10	0.10	0.01	0.07	0.07	0.02	0

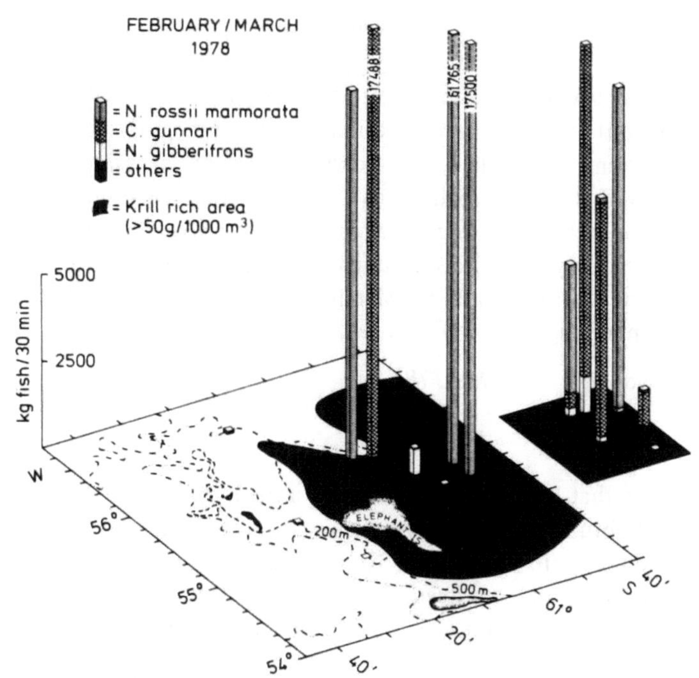

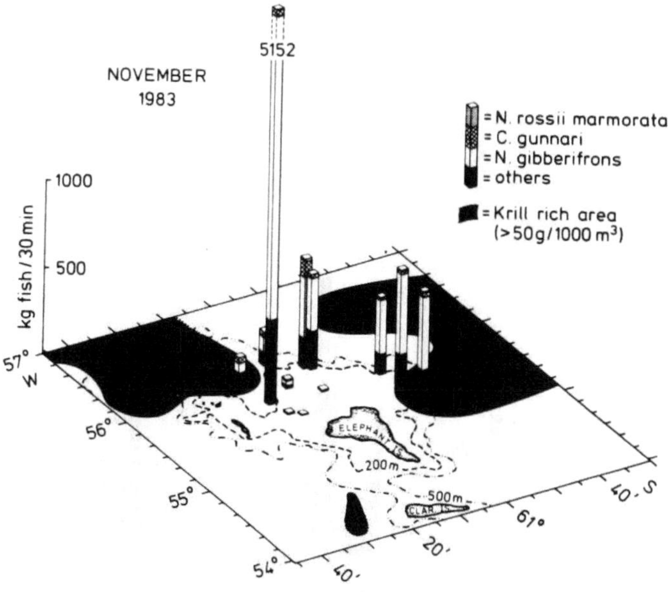

Fig. 9. Krill and fish abundance around Elephant Island in February/ March 1978

In February 1985, again three areas of high krill biomass were recognized; along the northwestern slope, on the west plateau and south of Elephant Island (Fig. 12) where the highest value of nearly 1500 g krill/1000 m^3 was located. The northwestern krill-rich area was extended to the east and to the southwest in recognition of dense krill concentration in that area, fished upon by a commercial krill trawl, yielding 7–10 tons of krill by 30–40-min tows.

The early-winter survey in May/June 1986 showed a marked reduced krill abundance ($\tilde{x}$ = 0 g/1000 m^3, maximum = 6 individuals/1000 m^3), whereas the spatial structure looked nearly the same as in other years and seasons (Fig. 13).

Therefore, one could summarize all observations of krill abundance in stating the north-northwest area of Elephant Island to be a preferred area for krill concentrations, as well seasonally as interannually persistent, even in winter. A second krill-rich area was detected on the western plateau southwest of the island from November to February, but not in March (Figs. 8–13), pointing to temporal variability in the abundance of krill.

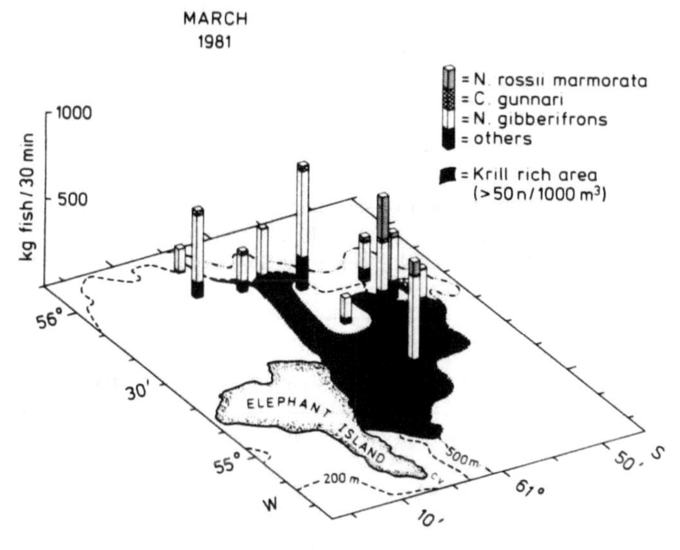

Fig. 10. Krill and fish abundance around Elephant Island in March 1981

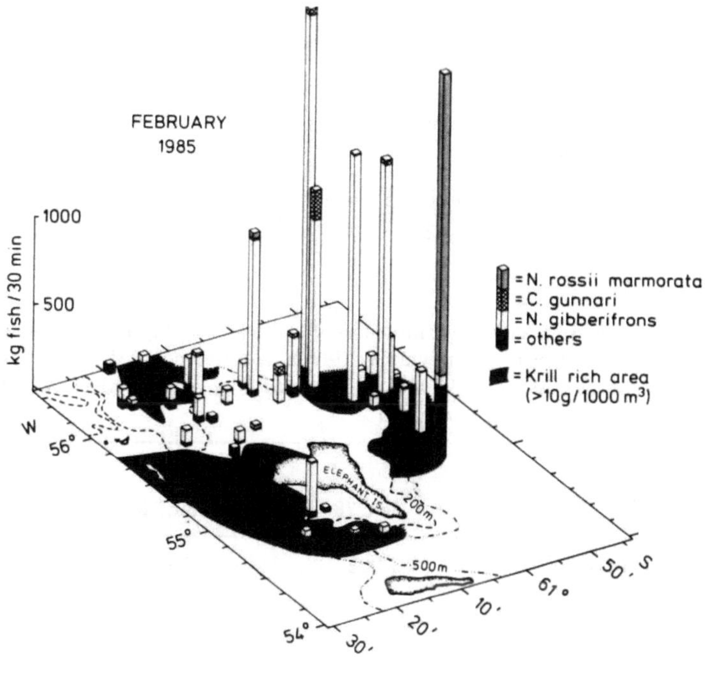

Fig. 12. Krill and fish abundance around Elephant Island in February 1985

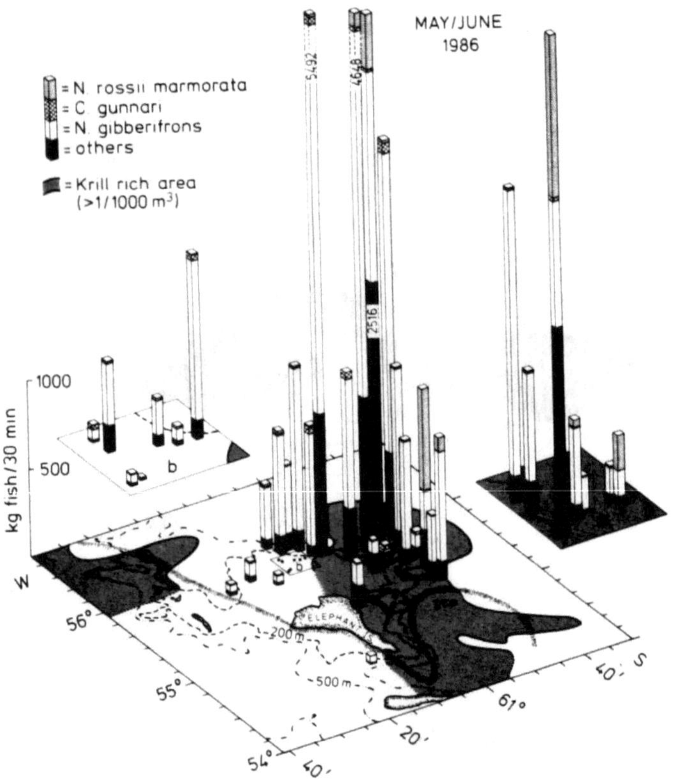

Fig. 13. Krill and fish abundance around Elephant Island in May/June 1986

3.3.3 Temporal Distribution and Abundance

Numerical abundance of krill was very variable in time (Table 4), values ranging from 0–4321 individuals/1000 m^3. This maximum value was recorded in November 1983 when median abundance was also highest. Omitting data for March 1981, which may have been biased due to a different survey design, median abundances were ordered in the following sequence.

	Nov.'83 >	Jan.'78 >	Jan./Feb.'76 >
$\tilde{x}$ (n/1000 m^3)	15	9	3.9
	Feb.'85 >	Mar.'78 >	May '86
	2.0	0.85	0

A Kruskal-Wallis analysis of variance revealed that differences between sampling periods were statistically significant ($\hat{H} = 39.13 > 20.52 = \chi^2_{5;0.001}$).

Mann-Whitney U-tests were then applied to the data to find out which sampling periods were significantly different from each other. As shown in Table 5, all sampling intervals differed significantly from the May 1986 survey and both first ranking observation times (November 1983 and January 1978) also from both times with the lowest observed krill abundance, March 1978 and February 1985. Summarizing the significant results in

Nov.'83		Mar.'78		
=	>	=	>	May '86
Jan.'78		Feb.'85		

shows that krill abundance was at a medium level in January/February 1976, only different from the very low abundance in May 1986. No annual trend is obvious in these data and to assess for a seasonal trend, more data from a single season (i.e., 1977/78) are still under preparation.

In krill biomass, the sequence of sampling periods was slightly different:

	Jan.'78 >	Jan./Febr.'76 >	Febr.'85 >
$\tilde{x}$ (g/1000 m^3)	6	(3.1)	2.1
	Nov.'83 >	Mar.'78 >	May '86
	1.2	0.3	(0).

Numbers in brackets should be considered with some reservation, as length-weight relationships from other sampling periods have been used to convert numerical abundance to biomass. They were thus not included in Mann-Whitney U-tests to reveal statistical differences between the other sampling periods.

Biomass was only significantly higher in January 1978 than in March of the same year but not statistically different between the years (Table 6). Temporal variability in biomass seemed to be less pronounced than in numerical abundance due to a seasonal reduction in numbers but an increase in weight.

Table 5. A survey of differences of krill's median ($\tilde{x}$) numerical abundance (n/1000 m^3) between two sampling intervals at a time. Given are the $\hat{z}$-values of Mann-Whitney U-tests. A difference is significant at $\hat{z} > 1.96 = z_{0.05}$ marked by an asterisk; $\hat{z} > 2.576 = z_{0.01}$ marked by two asterisks; $\hat{z} > 3.291 = z_{0.001}$ marked by three asterisks. An asterisk in brackets denotes a $\hat{z}$-value below the 5% margin, but higher than a z-value, if tested one-sided

$\tilde{x}$: 3.9	9 Jan.'78	0.85 Mar.'78	2.1 Mar.'81	15 Nov.'83	2.0 Feb.'85	0 May '86
Jan./Febr. 76	1.002	1.087	0.247	1.230	0.407	3.184**
Jan.'78		2.220*	0.899	0.055	1.832(*)	4.711***
Mar.'78			1.240	10.469***	0.797	3.136**
Mar.'81				0.892	0.670	3.592***
Nov.'83					4.551***	5.335***
Febr.'85						4.151***

Table 6. A survey of differences of krill's median ($\tilde{x}$) biomass (g/1000 m³) between two sampling intervals at a time. Given are the $\hat{z}$-values of Mann-Whitney U-tests. A difference is significant at $\hat{z} > 1.96 = z_{0.05}$. Marking by asterisks see legend of Table 5

$\tilde{x}$: 6	0.3 Mar.'78	1.2 Nov.'83	2.1 Febr.'85
Jan.'78	2.190*	1.799(*)	1.447
Mar.'78		0.868	1.206
Nov.'83			0.293

Table 7. Number of stations and differences in mean levels of numerical abundance and weight north ($\tilde{\mu}_N$) and south ($\tilde{\mu}_S$) of the Weddell-Scotia Confluence. A difference is statistically significant at P ($\hat{z}$) < 0.05 (Mann-Whitney U-test)

Survey		n/1000 m³ North		Krill South	North		g/1000 m³ South
	n	6		11	6		10
Nov. 1977	$\tilde{\mu}$	$\tilde{\mu}_N$	>	$\tilde{\mu}_S$	$\tilde{\mu}_N$	>	$\tilde{\mu}_S$
	ratio	87	:	1	324	:	1
	P ($\hat{z}$)		0.049			0.033	
	n	8		12	8		11
Mar. 1978	$\tilde{\mu}$	$\tilde{\mu}_N$	>	$\tilde{\mu}_S$	$\tilde{\mu}_N$	>	$\tilde{\mu}_S$
	ratio	2.7	:	1	3.3	:	1
	P ($\hat{z}$)		0.246			0.116	
	n	5		19	5		19
Nov. 1984	$\tilde{\mu}$	$\tilde{\mu}_N$	>	$\tilde{\mu}_S$	$\tilde{\mu}_N$	>	$\tilde{\mu}_S$
	ratio	9.5	:	1	15	:	1
	P ($\hat{z}$)		0.944			0.522	
	n	7		17	7		17
Mar. 1985	$\tilde{\mu}$	$\tilde{\mu}_N$	<	$\tilde{\mu}_S$	$\tilde{\mu}_N$	<	$\tilde{\mu}_S$
	ratio	1	:	6.7	1	:	6.3
	P ($\hat{z}$)		0.162			0.226	

3.4 Krill Abundance Relative to the Weddell-Scotia Confluence

The Weddell-Scotia Confluence (WSC) is the main hydrographic phenomenon in the area. Differences in numerical abundance and biomass between stations north and south of the WSC may indicate that the confluence acts as a biological boundary. Due to the variable position of the WSC, which lay outside the range of the box in some sampling intervals, and the uneven coverage of the area by stations from year to year, the WSC could not be located in each year. The analysis had thus to be limited to November and March both in 1977/78 and 1984/85. Data for the latter season were reduced to the extension of the 1977/78 survey to allow for direct comparison. Numerical abundance north of the WSC was 2.7 to 87 times higher than south of the WSC, except in March 1985, when slightly more krill was caught south of the WSC. This difference in abundance north and south of the WSC, however, was only statistically significant in November 1977 [Mann-Whitney U-test, P ($\hat{z}$) < 0.05; Table 7]. North to south ratios in biomass were even more distinct than in numerical abundance, but were only statistically significant in November 1977 (Table 7). This was mostly due to the lower proportion of juveniles encountered in the north (Kock et al. 1980; Nast 1986).

3.5 Fish Abundance in Relation to Krill Abundance

Prey distribution and abundance is one of the most important factors governing fish abundance. An attempt was thus made to correlate krill abundance with fish abundance both on a coarse (1–100 km) and a fine scale (1–1000 m).

3.5.1 Coarse Spatial Scale

Catch rates of fish (in kg/30 min), sectioned into the most abundant *Notothenia rossii marmorata, N. gibberifrons, C. gunnari,* and the remaining species (primarily *N. neglecta* and *Ch. aceratus*), in relation to krill abundance are given in Figs. 8–13. To allow for an easier comparison, differences in krill abundance have been condensed to krill-rich areas and areas of lower abundance.

Bottom trawling in January/February 1976 was primarily directed toward concentrations of *N. rossii marmorata* and *C. gunnari* previously detected by echo sounder northwest of the island. No direct observation on krill abundance was available for the trawling area. The extension of the krill-rich area north of the island and its vicinity to the trawling area may, however, indicate that krill was also abundant in the trawling area (Fig. 8).

As in 1976, fishing in January and February/March 1978 aimed predominantly at previously detected concentrations of *N. rossii marmorata* and *C. gunnari.* Large catches were restricted to areas of high krill abundance northwest of the island. Catches in areas of lower krill abundance, although confined to a few hauls, were negligible compared to krill-rich areas in both months (Fig. 9, January not shown here).

Patterns were less clear in 1981 and 1983 due to the small number of hauls. In March 1981 the survey had to be restricted to the northern and northwestern shelf, whereas in 1983 the western shelf could be surveyed additionally. *N. gibberifrons* was the most dominant species in the catches in both years. The most abundant species, particularly those of catch rates of *N. rossii marmorata* in 1981, were larger in areas of higher krill abundance. This difference, however, was not statistically significant (Table 8). The largest catch in 1981 was taken north of the island (Fig. 10), where slope water masses occurred (see Fig. 6). The best catch in 1983 consisted of a postspawning aggregation of *N. gibberifrons* which was incidentally met west of the island (Fig. 11).

In 1985 fish abundance was remarkably less south of 61°10'S in areas of lower krill abundance (Fig. 12). Catch rates in krill-rich areas, however, were only significantly

Table 8. Median and maximum abundance of Krill and fish, 95% confidence limits, median/mean ratios, median abundance in krill-rich and krill-poor areas and Mann-Whitney U-test for each sampling interval 1976–1986, significant results marked by asterisks as in legend of Table 5

Time of observation	Krill/fish	Median	Max	95% Conf.limits	$\tilde{x}/\bar{x}$	Median abundance in areas Krill-rich	Krill-poor	Mann-Whitney U-test [P ($\hat{z}$)]
	Krill (n/1000 m³)	3.9	417	0 – 11.4	0.10			
	tot.fish (kg/30 min)	480	23684	82 – 5143	0.09			
Jan./Febr.1976	*N. rossii* (kg/30 min)	255	23684	28 – 6081	0.05			
	C. gunnari (kg/30 min)	0	1481	0 – 0.45	0			
	N. gibb. (kg/30 min)	2	2459	0 – 30	0.01			
	Krill (n/1000 m³)	6	342	0.6 – 68	0.13			
	tot.fish (kg/30 min)	1441	8750	83 – 7031	0.60			
Jan. 1978	*N. rossii* (kg/30 min)	40	8750	0 – 7031	0.02			
	C. gunnari (kg/30 min)	10	3467	0 – 542	0.03			
	N. gibb. (kg/30 min)	7	87	0 – 83	0.28			
	Krill (n/1000 m³)	0.3	517	0.08– 3	0.006	240	0.25	
	tot.fish (kg/30 min)	5000	61765	70 –10633	0.53	8638	60	0.020*
Mar. 1978	*N. rossii* (kg/30 min)	65	61765	0 – 9380	0.01	158	0	0.057(*)
	C. gunnari (kg/30 min)	0	16765	0 – 960	0	0	0	0.589
	N. gibb. (kg/30 min)	10	1050	0 – 158	0.06	71	0.1	0.308
	Krill (n/1000 m³)	2.1	144	0.23– 85	0.07	85	1.5	
	tot.fish (kg/30 min)	255	808	139 – 631	0.75	469	233	0.165
Mar. 1981	*N. rossii* (kg/30 min)	1.5	248	0.2 – 26	0.05	40	0.9	0.121
	C. gunnari (kg/30 min)	4.8	34	0.3 – 21	0.16	4	3.5	1.000
	N. gibb. (kg/30 min)	161	496	117 – 390	0.72	269	147	0.124
	Krill (n/1000 m³)	1.2	718	0.2 – 6.5	0.03	330	0.55	
	tot.fish (kg/30 min)	321.0	5675	12 – 630	0.43	608	196	0.312
Nov. 1983	*N. rossii* (kg/30 min)	1.4	23	0 – 5	0.32	2	1.05	0.779
	C. gunnari (kg/30 min)	13.0	90	0.8 – 31	0.66	21	6.1	0.230
	N. gibb. (kg/30 min)	198.0	5152	1.9 – 365	0.33	365	90	0.230
	Krill (n/1000 m³)	2.1	1492	0.6 – 6.9	0.04	29.5	0.74	
	tot.fish (kg/30 min)	140	2206	82 – 258	0.37	320	91	0.011*
Febr. 1985	*N. rossii* (kg/30 min)	0	1736	0 – 1	0	0	0	0.582
	C. gunnari (kg/30 min)	3	169	1 – 4	0.27	1.7	3.2	0.110
	N. gibb. (kg/30 min)	91	2030	64 – 221	0.32	271	66	0.026*
	Krill (n/1000 m³)	0	6	0 – 0.2	0	2.5	0	
	tot.fish (kg/30 min)	490	6692	220 – 810	0.48	810	201	0.008**
May/June 1986	*N. rossii* (kg/30 min)	7	940	2 – 12	0.09	11	0.95	0.002**
	C. gunnari (kg/30 min)	10	58	5 – 14	0.70	9	11.25	0.509
	N. gibb. (kg/30 min)	302	5492	146 – 621	0.45	621	152	0.031*

higher in *N. gibberifrons* but not in *N. rossii marmorata* and *C. gunnari* (Table 8). The second-best catch, which consisted mostly of *N. rossii marmorata* and *N. neglecta,* was again taken north of the island (Fig. 12) in slope water masses (see Fig. 6) at the same position as in 1981.

In May/June 1986 trawling had to be confined to the north and west of the island as the northward-moving pack ice prevented a similarly even coverage of the area by stations as in 1985. Catch rates were again largest in areas of higher krill abundance north and west of the island (Fig. 13). Differences in catch rates between krill-rich and krill-poor areas were, however, only statistically significant in *N. gibberifrons* and in *N. rossii marmorata* but not in *C. gunnari* (Table 8). One of the best catches was again taken north of the island on the same position as in 1981 and 1985 (Fig. 13).

3.5.2 Fine Spatial Scale

In 1981 and 1985 RMT catches and bottom trawl hauls were carried out concurrently. This offered an opportunity to look for possible relationships between krill abundance and fish abundance on a fine scale. Spearman rank correlations were applied to the data.

No statistically significant correlations could be found between krill abundance and fish abundance on a fine

Table 9. Spearman rank correlation coefficients between krill abundance and fish abundance

Abundance	1981 (n = 11) r_{SB}	1981 (n = 11) $r^*_{S/0.05}$	1985 (n = 35) r_{SB}	1985 (n = 35) P ($\hat{z}$)
Krill – all fish species	0.0545	0.527	−0.089	0.604
Krill – *N. rossii marmorata*	0.391	0.527	−0.202	0.238
Krill – *C. gunnari*	−0.236	0.527	−0.590	0.0007*
Krill – *N. gibberifrons*	0.127	0.527	−0.097	0.569

spatial scale except for a negative relationship between the krill abundance and that of *C. gunnari* in 1985 (Table 9).

4 Discussion

Our survey results, and among them particularly abundance estimates, are subject to systematic and random errors (Nast 1982). In krill the rigorous standardization of the nets (RMT 1+8 only), survey design, and sample treatment from the outset in 1975/76 should have kept systematic errors to a minimum. Some influence might be expected from the lower maximum fishing depth of 140 m in 1977/78 compared to 200 m in all other sampling periods, but the bulk of krill was in general concentrated in the upper 100 m of the water column, at least in oceanic waters. The extension of the station grid to the north from 1983 onwards did not change the results (e.g., abundance and biomass estimates) significantly (Nast in prep.) but resulted in a better statistical validity. The box was also extended southward to the pack ice border. Siegel (1986) found low to medium krill concentrations south of the box in October 1983. Klindt (1986), however, observed a higher krill biomass south of 62°S along the easternmost transect by hydroacoustics. In March 1985 both authors recorded high krill abundance south of 62°S, supporting our findings of a higher krill abundance south of the WSC at that time.

The gear types and the station grid themselves, however, are filters which could have converted "real patterns" into "observed patterns" thus affecting particularly the validity of abundance estimates (Fig. 14). This was obvious in March 1981 when the area of high krill abundance north and northwest of the island ("super swarm area") was surveyed independently by our RMT 1+8 hauls, Bongo net catches (Brinton and Antezana 1984) and acoustics (Macaulay et al. 1984). A comparison of results revealed good agreement in the delineation of areas of high and low krill abundance but substantial differences in abundance estimates between random net hauls and those aimed at acoustic targets and abundance estimates derived from echointegration: Abundance estimates were 0 g/1000 m^3 and 0.7 g/1000 m^3 from random Bongo and RMT 8 catches but 203 g/1000 m^3 from aimed Bongo tows and 200–500 g/m^3 from acoustic measurements. The latter value led to an estimate of 2.1 million metric tons for the "super

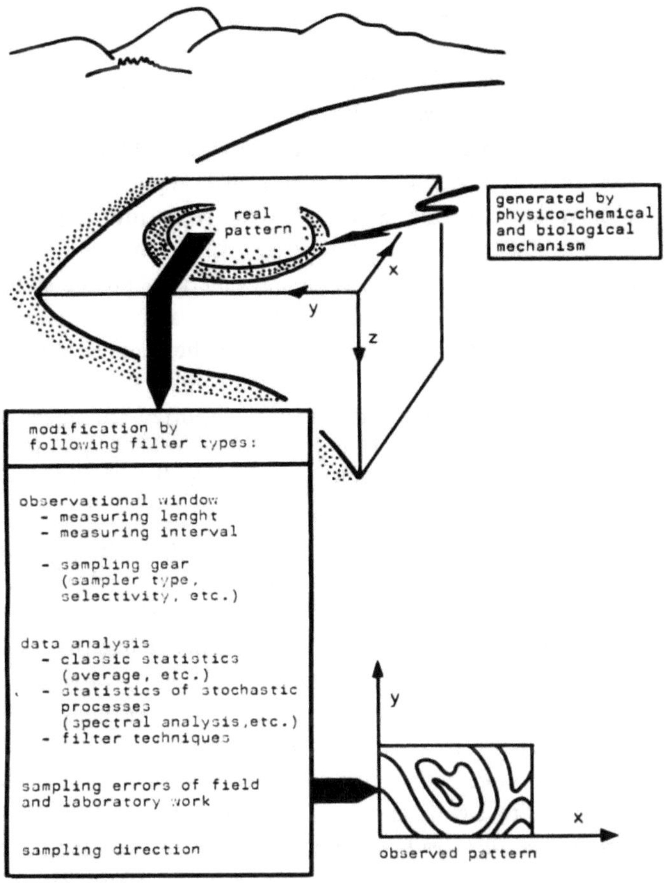

Fig. 14. Main kinds of filters which convert "real patterns" into "observed patterns". (Postel 1983, after Haury et al. 1978)

swarm" (Macaulay et al. 1984). This outstanding high biomass, however, was not evident from our data: neither median nor maximum numerical abundance or biomass were significantly higher than in other years. Both examples clearly underlined the filter effect by the gears used and the different spatial scales of the sampling grids. Hence, absolute numbers in our abundance estimates should be considered only relative to our surveys. They are most likely underestimates (of an unknown order of magnitude) of the "real abundance".

Net avoidance and patchiness, a term to describe variability, are probably most important among random errors. Net avoidance is likely to be not constant but dependent on the time of day, illumination, sex, developmental stage, and absolute density (Wiebe et al. 1982). Median abundance derived from ten replicate hauls on a krill swarm previously detected by echosounder (Nast in prep.), for example, was 3522 ± 33% (n/1000 m^3). This low standard error of the median compared to those estimated during our surveys (Table 4) may indicate that the efficiency of the RMT 8 to catch krill was increasing with increasing absolute density of krill.

Physicochemical and bioiogical mechanisms generating and maintaining patchiness in general are well documented (Fig. 14, Postel 1983), and may be partly applied to our data. The krill swarm area northwest of Elephant Island, observed at all times, coincides with the area of permanent water mass boundary (WSC). This region, from the topographic point of view, denotes one of the major outlets for Weddell Sea water which might add to the concentration of krill in the area. Everson (1984) reported a similar observation for the South Georgia area. Inferring krill distribution from whale catches, he found the major concentrations off the north coast and generally starting about 10–15 miles offshore in an area of water-mass mixing of different origin. These observed higher concentrations of zooplankton/micronekton at some distances from the coast are a common phenomenon, reported for instance by Hargreaves (1978), Pearcy (1976), Leis (1982) and for krill by Witek et al. (1981), Hampton (1983), Rakusa-Suszczewski (1983), Macaulay et al. (1984) and Shulenberger et al. (1984). The mechanisms generating these aggregations are up to now poorly understood and the controversial discussion ranges from active horizontal migration (Kanda et al. 1982) to enhancements due to local eddies (Everson 1984). Off Elephant Island hydrographic conditions are favored in producing the seasonally persistent krill-rich area to the north of Elephant Island.

In fish, systematic errors probably influenced our survey results more significantly than in krill. Due to the preponderance of tows north and northwest of the island and the nonrandom distribution of hauls in 1976–78 the dominance of *N. rossii marmorata* and *C. gunnari* in these years is most likely overestimated, that of *N. gibberifrons* underestimated.

Net avoidance may be less important, as Antarctic fish seem to be capable of high maximum swimming speeds for only a short period of time (Montgomery and Macdonald 1984; Dunn and Johnston 1986). The efficiency of the trawl (the proportion of fish in the area sampled which is actually caught), however, was probably higher in the more bottom-dwelling "slurp" and "ambush" feeders *N. gibberifrons* and *Ch. aceratus* than in the more actively hunting *N. rossii marmorata* and the bentho-pelagic *C. gunnari*, which exhibit regular vertical migrations (Kock 1985).

Low median/mean ratios (Table 8) demonstrated the patchy distribution of the most abundant species. In *N. rossii marmorata* aggregations were confined to very small areas on the northwestern shelf but were stable there over two years (1976–78) (Figs. 8 and 9). *C. gunnari* shoaled along the western and northwestern shelf in 1977/78 (Fig. 9). Concentrations were stable over at least two months. The effect of the uneven distribution of fish on our results is difficult to assess. It must have decreased in recent years due to the more even coverage of the shelf by random tows and the larger number of hauls.

Despite these limitations a number of conclusions can be drawn:

The species inventory around Elephant Island is similar to that reported for the other South Shetland Islands (Permitin 1977). It is formed by two ichthyofaunistic elements: Lesser Antarctic and circumpolar high Antarctic species. High Antarctic species were scarce on the upper shelf above 200 m. Their proportion in the number of species was increasing below 200 m, where slightly colder and more saline waters prevailed. Due to the weak gradients between water masses on the shelf, however, it seems more likely that depth and other factors (food, habitat) rather than hydrographical conditions are governing species diversity.

Although dominant in the number of species, the biomass of circumpolar high Antarctic species is negligible (< 1%) compared to the Lesser Antarctic species. This underlines the fact that Elephant Island is part of the Lesser Antarctic (Western Antarctic) zoogeographical province (Permitin 1977), where high Antarctic species are at their northernmost limit of distribution.

The ichthyofauna is (was) dominated by only five species, which made up 98% of the biomass: *N. rossii marmorata, N. gibberifrons, N. neglecta, C. gunnari,* and *Ch. aceratus.* Although their importance may have been overestimated, *N. rossii marmorata* and *C. gunnari* were probably the most abundant species until 1978. Between 1978 and 1981 their abundance dropped substantially and their length and age compositions changed remarkably (Kock 1986; Kock unpublished). This was most likely due to large-scale harvesting in the Peninsula region in 1978/79 and 1979/80 (Table 10). Elephant Island was the most important fishing ground, particularly for *N. rossii marmorata* and *C. gunnari.* Catches were in the order of the rough biomass estimate for the area (Kock et al. 1985) and must have affected species composition and population structure significantly. Since 1981 *N. gibberifrons* predominated, making up 65–80% of the biomass between 50 and 500 m depth. Proportions of the most abundant species in the catches and length compositions (except in *C. gunnari*) have remained relatively stable since then.

Euphausia superba was the staple food in the diet of the most abundant species except probably in *N. gibberifrons.* It is most likely that a correlation exists on a coarse spatial scale between the permanent high abundance of krill northwest and north of Elephant Island and the much higher fish biomass compared to the south of the island's shelf where krill was only temporarily abundant. Due to the patchiness of krill and fish which is generated and maintained by different mechanisms (mostly passive in krill, active local movements in fish) we failed to demonstrate a similar relationship on a fine scale. The large biomass of primarily benthos-feeding *N. gibberifrons* in the area of high krill abundance, however, may indicate that factors other than krill abundance (e.g., habitat, benthos), which we were unable to investigate, govern fish abundance.

Annual krill consumption was estimated to be 250–320 000 metric tons before the onset of commercial exploitation (Kock 1985), but must have dropped by one order of magnitude since then. It seems highly unlikely

Table 10. Nominal annual catches (t) in the Peninsula area (Anonymous 1986)

Species	1978/79	1979/80	1980/81	1981/82	1982/83	1983/84
Notothenia rossii marmorata	470	18763	0	0	0	0
Notothenia gibberifrons	3280	765	50	0	0	0
Champsocephalus gunnari	35917	1087	1700	0	2604	0
Chaenocephalus aceratus	1393	153	0	0	0	0

that predation by fish has had any significant effect on krill in recent years.

Our analysis should be considered as a first attempt to illuminate the variability and interrelationships of organisms at different trophic levels and their interactions with the environment in an ecosystem of the seasonal pack ice zone. It had been confined so far, however, to a coarse spatial scale. Due to the lack of appropriate statistical methods (time-series analysis, spectral analysis) which were not available to us, we were unable to separate between seasonal and interannual components in temporal variability of krill (Slutsky-Yule effect; Kendall 1973). Analysis of variability of krill in time is thus still pending.

Further improvement of our data base can be expected from a reduction of systematic errors, particularly in bottom trawl surveys in the near future. A resolution higher than that during our surveys, however, may be obtained only by a combination of nets and acoustic methods (fine spatial scale) and the additional use of underwater television to record individual krill not detected by echosounding (micro spatial scale) (Schröder 1961) and by an increased number of bottom trawl tows. Although our knowledge on the life history of krill and fish around Elephant Island has increased considerably, it is most obvious that we are still far from modeling species and environmental interactions around the island.

Acknowledgements. We are very much indebted to Mrs. J. Willführ and Mrs. S. Beecken for their technical assistance and the preparation of the graphs.

References

Anonymous (1982) Antarktis-Expedition 1981 der Bundesrepublik Deutschland mit FFS *Walther Herwig*. Arch Fischereiwiss (Beih 1) 33:1–176

Anonymous (1986) Draft summary of catch and effort statistics. SC-CAMLR V/BG/8:1–64

BIOMASS (1977) Biological investigations of marine antarctic systems and stocks, vol 1. Research proposals. Cambridge, 79 pp

BIOMASS (1983) 2nd Post-FIBEX Hydrographic Data Interpretation Workshop Hamburg, FRG, 16–20 May 1983. BIOMASS Rep Ser 31

Brinton E, Antezana T (1984) Structures of swarming and dispersed populations of krill (*Euphausia superba*) in Scotia Sea and South Shetland Waters during January–March 1981, determined by Bongo nets. J Crust Biol 4 (Spec No 1):45–66

Dunn JF, Johnston IA (1986) Metabolic constraints on burst-swimming in the Antarctic teleost *Notothenia neglecta*. Mar Biol 91 (4):433–440

Everson I (1984) Zooplankton. In: Laws RM (ed) Antarctic ecology, vol 2. Academic Press, London New York, pp 463–490

Freytag G (1980) Length, age and growth of *Notothenia rossii marmorata* Fischer 1885 in the West Antarctic waters. Arch Fischereiwiss 30 (1):39–66

Greenblatt PR, Shulenberger R, Wormuth JH (1982) Small-scale distribution of zooplankton biomass. Deep-Sea Res 29 (4A): 443–457

Gubsch G (1982) Zur Verbreitung und Biologie der Eisfische (Chaenichthyidae) im atlantischen Sektor der Antarktis. Fischereiforschung 20 (2):39–47

Hamner WM (1984) Aspects of schooling in *Euphausia superba*. J Crust Biol 4 (Spec No 1):67–74

Hampton I (1983) Preliminary report on the FIBEX acoustic workshop to estimate the abundance of *Euphausia superba*. Proc Biomass Colloq 1982. Mem Natl Inst Polar Res Spec Issue 27: 165–175

Hargreaves PM (1978) Relative abundance of zooplankton groups in the northwest African upwelling region during 1968 and 1972. In: Boje R, Tomczak M (ed) Upwelling ecosystems. Springer, Berlin Heidelberg New York, pp 62–72

Hempel G, Sahrhage D, Schreiber W, Steinberg R (1979) Antarktis-Expedition 1977/78 der Bundesrepublik Deutschland. Arch Fischereiwiss (Beih 1) 30:1–119

Kanda K, Takagi K, Seki Y (1982) Movement of the larger swarms of Antarctic krill *Euphausia superba* population off Enderby Land during 1976–1977 season. J Tokyo Univ Fish 68 (1/2): 25–42

Kellermann A (1986) Zur Biologie der Jugendstadien der Notothenioidei (Pisces) an der Antarktischen Halbinsel. Ber Polarforsch 31:1–155

Kendall MG (1973) Time series. Griffin, London, 197 pp

Klindt H (1986) Acoustic estimates of the distribution and stock size of krill around Elephant Island during SIBEX I+II in 1983, 1984 and 1985. Arch Fischereiwiss (Beih 1) 37:107–127

Kock K-H (1976) Untersuchungen über mögliche Nutzfischbestände im atlantischen Sektor der Antarktis: *Dissostichus eleginoides* und *D. mawsoni*. Inf Fischwirtsch 23 (6):160–165

Kock K-H (1981) Fischereibiologische Untersuchungen an drei antarktischen Fischarten: *Champsocephalus gunnari* Lönnberg, 1905, *Chaenocephalus aceratus* (Lönnberg, 1906) und *Pseudochaenichthys georgianus* Norman, 1937 (Notothenioidei, Channichthyidae). Mitt Inst Seefisch Hamb 32:1–226

Kock K-H (1982) Fischereibiologische Untersuchungen bei Elephant Island im März 1981. Arch Fischereiwiss (Beih 1) 33:127–142

Kock K-H (1985) Krill consumption by Antarctic notothenioid fish. In: Siegfried WR, Condy PR, Laws RA (eds) Antarctic nutrient cycles and food webs. Springer, Berlin Heidelberg New York Tokyo, pp 437–444

Kock K-H (1986) The state of exploited Antarctic fish stocks in the Scotia Arc region during SIBEX (1983–1985). Arch Fischereiwiss (Beih 1) 37:129–186

Kock K-H, Nast F, Stein M (1980) Some interactions between hydrography, krill and fish near Elephant Island in 1977/78. Coun Meet ICES, Biol Oceanogr Comm L3:1–20

Kock K-H, Duhamel G, Hureau JC (1985) Biology and status of exploited Antarctic fish stocks: a review. BIOMASS Sci Ser 6: 1–143

Leis JM (1982) Nearshore distribution gradients of larval fish (15 taxa) and planktonic crustaceans (6 taxa) in Hawaii. Mar Biol 72:89–97

Longhurst AR (1981) Significance of spatial variability. In: Longhurst AR (ed) Analysis of marine ecosystems. Academic Press, London New York, pp 415–441

Macaulay MC, Saunders English T, Mathisen OA (1984) Acoustic characterization of swarms of Antarctic krill (*Euphausia superba*) from Elephant Island and Bransfield Strait. J Crust Biol 4 (Spec No 1):16–44

Mackas DL, Denman KL, Abbott MR (1985) Plankton patchiness: Biology in the physical vernacular. Bull Mar Sci 37 (2):652–674

Montgomery JC, Macdonald JA (1984) Performance of motor systems in Antarctic fishes. J Comp Physiol A 154:241–248

Nast F (1978/79) The vertical distribution of larval and adult krill (*Euphausia superba* Dana) on a time station south of Elephant Island, South Shetlands. Meeresforschung 27:103–118

Nast F (1982) The assessment of krill (*Euphausia superba* Dana) biomass from a net sampling programme. Meeresforschung 29: 154–165

Nast F (1986) Changes in krill abundance and in other zooplankton relative to the Weddell-Scotia Confluence around Elephant Island in November 1983, November 1984 and March 1985. Arch Fischereiwiss (Beih 1) 37:73–94

Pearcy WG (1976) Seasonal and inshore-offshore variations in the standing stocks of micronekton and macrozooplankton off Oregon. Fish Bull US 74:70–80

Pennak RW (1943) An effective method of diagramming diurnal movements of zooplankton organisms. Ecology 24:405–407

Permitin YuYe (1977) Species composition and zoogeographical analysis of the bottom fish fauna of the Scotia Sea. J Ichthyol 17 (5):710–726

Pommeranz T, Herrmann C, Kühn A (1983) Mouth angles of the Rectangular Midwater Trawl (RMT 1+8) during paying out and hauling. Meeresforschung 29:267–274

Postel L (1983) Problems in identifying distribution patterns of oceanological parameters. Medd Havsfiskelab, Lysekil 293: 1–12

Rakusa-Suszczewski S (1983) The relationship between the distribution of plankton biomass and plankton communities in the Drake Passage and the Bransfield Strait (BIOMASS-FIBEX, February–March 1981). Proc Biomass Colloq 1982. Mem Natl Inst Polar Res Spec Issue 27:77–83

Sachs L (1984) Angewandte Statistik: Anwendung statistischer Methoden. Springer, Berlin Heidelberg New York Tokyo, 552 pp

Sahrhage D, Schreiber W, Steinberg R, Hempel G (1978) Antarktis-Expedition 1975/76 der Bundesrepublik Deutschland. Arch Fischereiwiss (Beih 1) 29:1–96

Schröder R (1961) Untersuchungen über die Planktonverteilung mit Hilfe der Unterwasser-Fernsehanlage und des Echographen. Arch Hydrobiol Suppl Falkan Schr 25:228–241

Shulenberger E, Wormuth JH, Loeb VJ (1984) A large swarm of *Euphausia superba:* overview of patch structure and composition. J Crust Biol 4:75–95

Siegel V (1986) Structure composition of the Antarctic krill stock in the Bransfield Strait (Antarctic Peninsula) during the 2nd Int BIOMASS Exp (SIBEX). Arch Fischereiwiss (Beih 1) 37: 51–72

Simard Y, Ladurantaye de R, Therriault J-C (1986) Aggregation of euphausiids along a coastal shelf in an upwelling environment. Mar Ecol Progr Ser 32:203–215

Stein M (1986) Variability of water masses and currents off the Antarctic Peninsula during SIBEX. Arch Fischereiwiss (Beih 1) 37:25–50

Takahashi M (1983) Trophic ecology of demersal fish community north of the South Shetland Islands with notes on the ecological role of the krill. Mem Natl Inst Polar Res Spec Issue 27:183–192

Tarverdiyeva MJ, Pinskaya IA (1980) The feeding of fishes of the families Nototheniidae and Chaenichthyidae on the shelves of the Antarctic Peninsula and the South Shetlands. J Ichthyol 20 (4):50–60

Wiebe PH, Boyd SH, Davis BM, Cox JL (1982) Avoidance of towed nets by the euphausiid *Nematoscelis megalops.* Fish Bull 80 (1): 75–91

Witek Z, Kalinowski J, Grelowski A, Wolnomiejski N (1981) Studies of aggregation of krill (*Euphausia superba*). Meeresforschung 28: 228–243

Wittstock R-R, Zenk W (1983) Some current observations and surface T/S distribution from the Scotia Sea and the Bransfield Strait during early austral summer 1980/81. *Meteor* Forschungsergeb Reihe A/B 24:77–86

Wörner FG (1979) Zooplankton und Mikronektonfänge. In: Hempel G, Sahrhage D, Schreiber W, Steinberg R (eds) Antarktis-Expedition 1977/78 der Bundesrepublik Deutschland. Arch Fischereiwiss (Beih 1) 30:40–61

Can We Satisfactorily Estimate Variation in Krill Abundance?

I. EVERSON[1]

Summary. A knowledge of abundance and rates of change of abundance are basic requirements for fisheries models. Catch per unit of effort (CPUE) has traditionally been used in demersal fisheries assessments as an estimator of abundance. It is less satisfactory for pelagic fisheries. Evidence is given that CPUE estimated from data reported to the Commission for the Conservation of Antarctic Marine Living Resources (CCAMLR) does not provide a realistic index of krill abundance. Recognizing that there is a large natural year-to-year variation in krill abundance, it is important that this be quantified and separated from fishery-induced variation. The time-scale for such information is discussed in the light of the needs of fishery management.

1 Introduction

In studies of the variability of krill (*Euphausia superba*) populations, assessment of the impact of fisheries on the resource is of crucial importance. In this chapter I propose to look at information on variability in krill abundance from net haul and acoustic surveys and compare this with data from the fishery.

The krill fishery began in the late 1960's and expanded during the 1970's to a maximum annual catch of just over half a million tonnes (Fig. 1). Relative to estimated predator consumption rates the current total catch is small. However, if this catch were concentrated within a region where krill predators were restricted by, for example, foraging range from breeding sites, such a modest annual take could have serious implications. In this light it is essential to be able to distinguish between variations in krill abundance resulting from the effect of the fishery and those resulting from either natural variation in the krill population dynamics or variation induced by environmental factors.

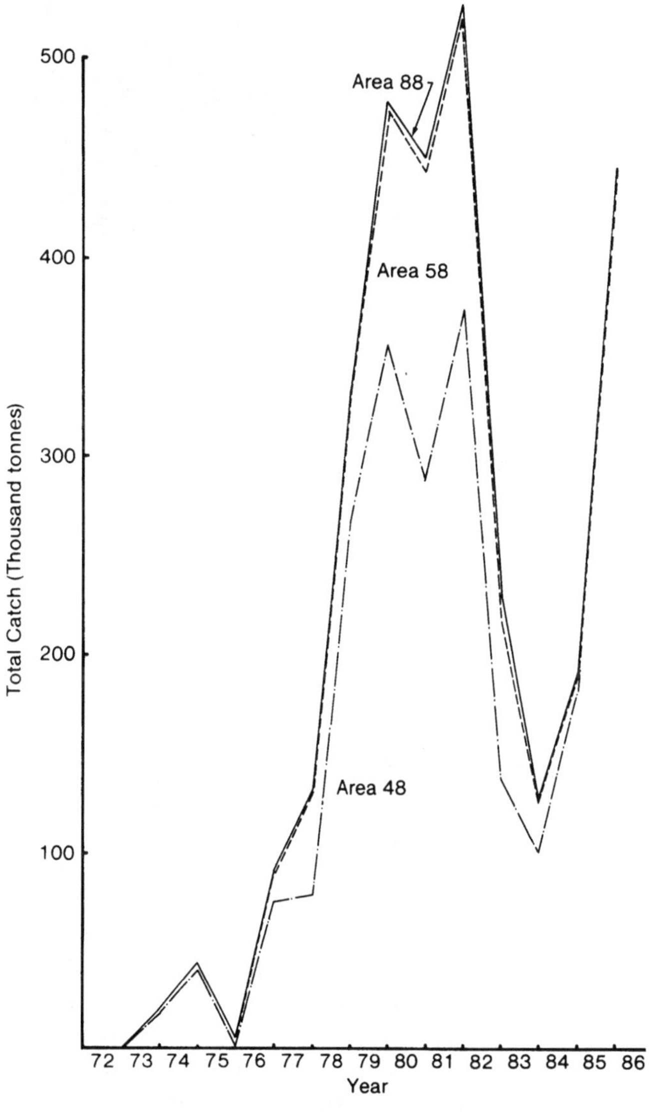

Fig. 1. Annual catch of the krill fishery by major area

2 Fishery Data for Use as an Estimator of Abundance

Traditionally Catch Per Unit of Effort (CPUE) has been used in fishery stock assessment as an estimator of abundance. The underlying assumption is that for a given unit of effort, measured in terms of hours of fishing, number of tows or some similar index, the catch will be proportional to the abundance of the resource. While this assumption holds reasonably well for many demersal resources, assessment of pelagic resources poses preater problems because of their greater tendency to aggregate. This means that a

1 British Antarctic Survey, Natural Environment Research Council, High Cross, Madingley Road, Cambridge CB3 OET, United Kingdom

Antarctic Ocean and Resources Variability
(ed. by D. Sahrhage)

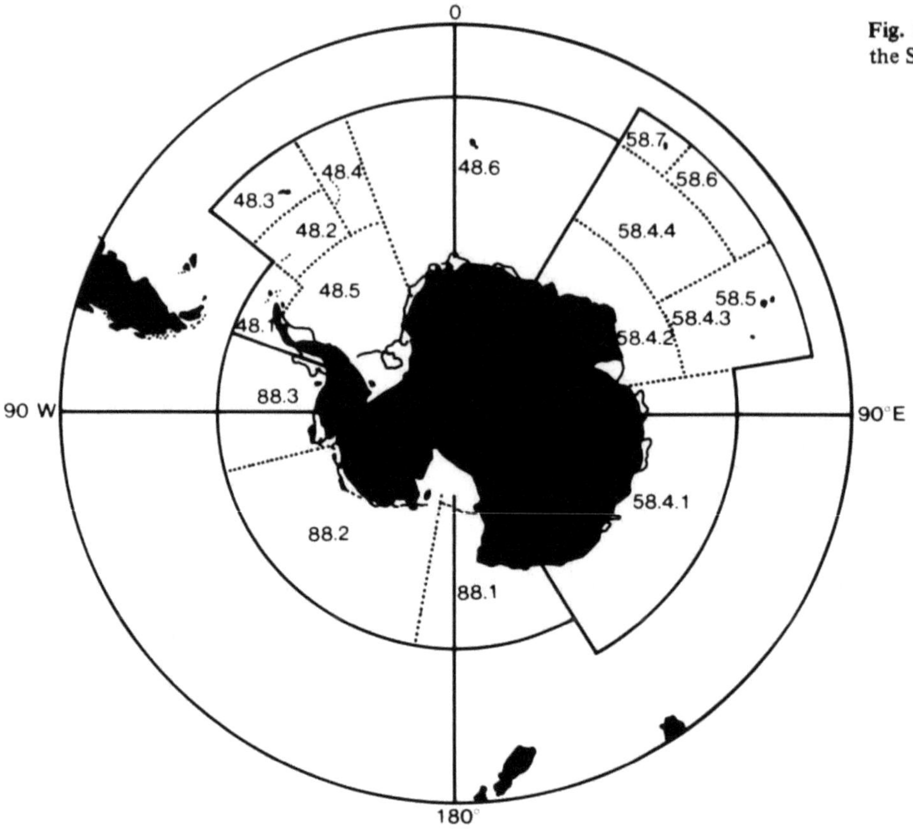

Fig. 2. Boundaries of statistical reporting areas in the Southern Ocean

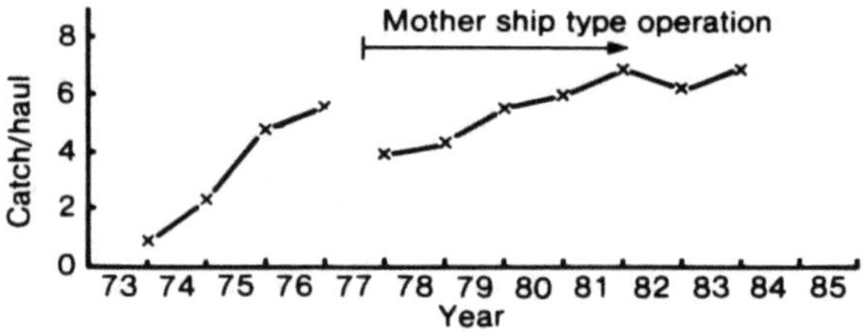

Fig. 3. Mean catch (t) per haul for the Japanese krill fishery

significant proportion of the effort is diverted towards searching for fishable swarms, thus adding another factor that needs to be incorporated into the equation.

Summary catch and effort data are supplied annually to the Commission for the Conservation of Antarctic Marine Living Resources (CCAMLR) in the STATLANT A and STATLANT B formats and additional information is available in the scientific literature. The STATLANT data contain information on catch, identified by species, for each year, month, major gear type, vessel type, geographical subarea (Fig. 2) and target species. Unfortunately it has not been the universal practice to apportion the fishing effort between the various species either in terms of actual catch or intended catch. Where more than one species has been reported it is impossible to provide valid estimates of CPUE in mixed fishery situations unless the target species is identified. The present analysis has been restricted to situations where krill was the only species reported. The Japanese have provided STATLANT data over the full period of the fishery and have concentrated their fishing effort almost exclusively on krill. Their data have, therefore, been used as an example in this analysis.

3 The Japanese Krill Fishery

Combining data from all subareas within each season (Shimadzu 1985), there is a general increase in the mean annual catch per haul (Fig. 3). The only season when there was a decline of greater than 10% was in the 1977/78 season, when the pattern of the operation changed. Prior to that time small numbers of vessels had been operating more or less independently while afterwards a mothership type operation using smaller nets was in progress until 1981/82. The mothership operation, by restricting the geographical range over which the catching could operate at any time, may have been the cause of the reduction. Subsequent to the 1977/78 season there was a steady increase in mean catch per haul, which may be a result of increased effectiveness of the vessels as in earlier years. The alternative explanation, that the low value in 1977/78 was a result purely of generally low krill abundance, cannot, however, be discounted.

A better CPUE index is likely to be derived from using time actually spent fishing as the unit of effort (Shimadzu 1985; Shimadzu and Ichii 1985). In terms of catch per hour's fishing, using the STATLANT B data, there is a wider year-to-year fluctuation leading to a major increase for the last three seasons (Fig. 4). A comparison of catch

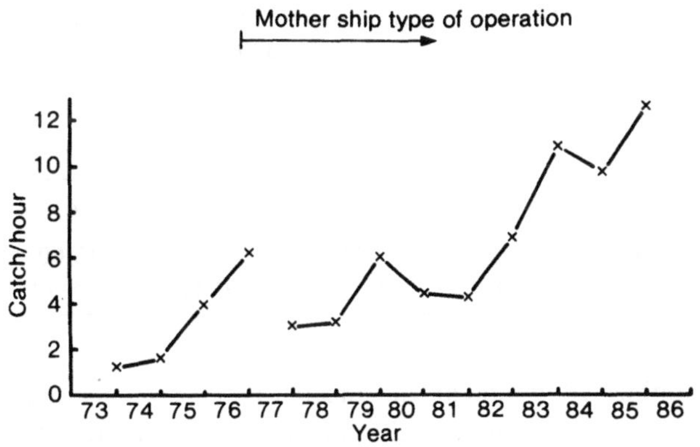

Fig. 4. Mean catch (t) per hour fishing for the Japanese krill fishery

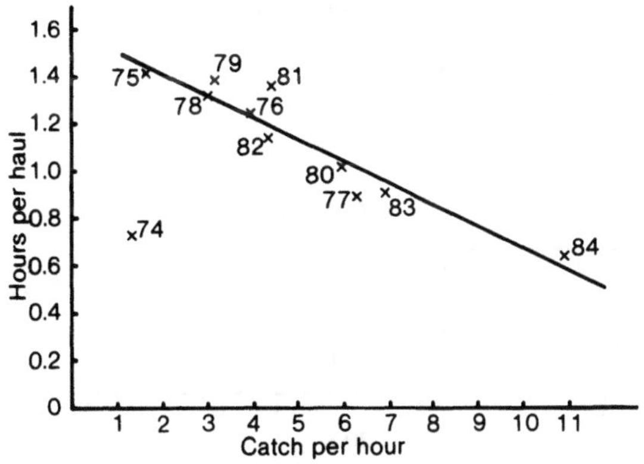

Fig. 6. Mean catch per hour plotted against mean haul duration (h). The line has been fitted to all data points excluding 1974, when only one ship was fishing. The equation of the line is:
Haul duration = 1.60–0.093 catch per hour
r = 0.946, n = 10

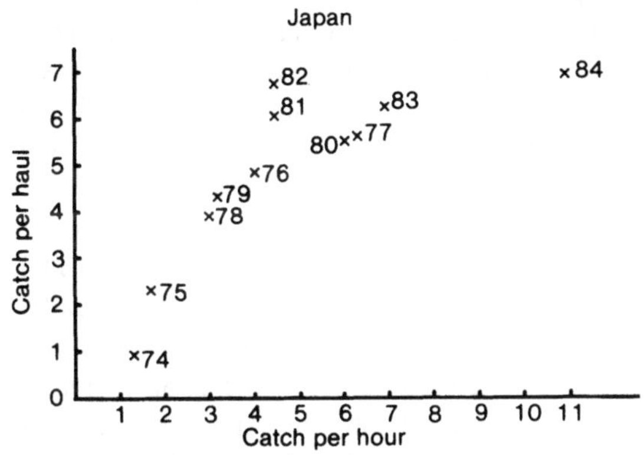

Fig. 5. Mean catch per hour plotted against mean catch per haul for the Japanese krill fishery. The *numbers* are the second years of the split year season

per hour with catch per haul (Fig. 5) indicates that there is a more or less proportionate increase in the two estimators. Combining data from STATLANT B and Shimadzu (1985), a mean haul duration has been calculated. Excluding data from the first season (1973/74), when only one vessel was operational, there is a strong negative correlation (r = –0.946, df = 8, P 0.01) between the catch per hour and the duration of the haul (Fig. 6). Considered in conjunction with the data on catch per haul, the results indicate a strong tendency to organize the fishery so as to give a more or less standard catch of krill per tow. This would make available a steady supply of fresh krill to the processing plant. Such a strategy has been described (Gulland 1985) and is to be expected, given the rapid spoilage rate of bulk krill (Grantham 1977).

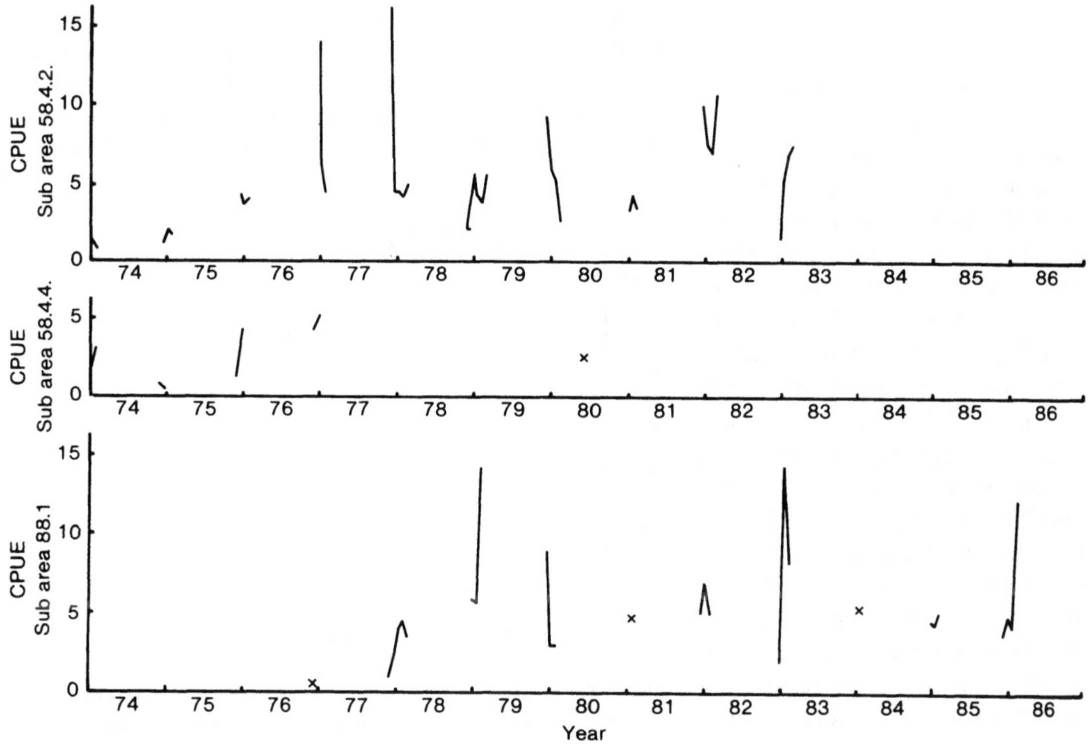

Fig. 7. Monthly CPUE values for the Japanese fishery in the Indian and Pacific Ocean sectors

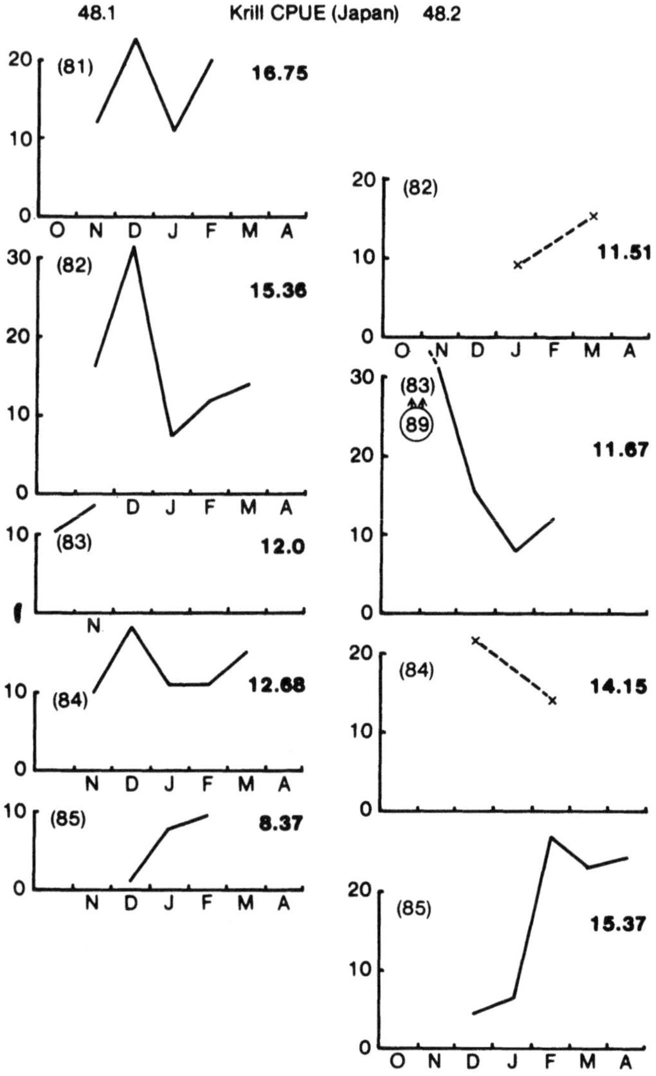

Fig. 8. Monthly CPUE values for the Japanese fishery in the Atlantic Ocean sector

Catch per hour's fishing has therefore been used as the CPUE index for the remainder of this paper.

Dividing the data between subarea and months of the year provides a large number of CPUE values and these are plotted in Figs. 7 and 8. In the Indian Ocean sector [subareas 58.4.2 (coastal) and 58.4.4 (offshore)] the fishery moves southwards as the season progresses, following the retreating ice edge into the East Wind drift zone. The monthly values fluctuate widely and no dominant trend is apparent. Published information is available for some parts of a few of the seasons. Sugimoto (1977) describes fishing on a super-swarm in the 1976/77 season. The high catch rates achieved under these conditions for a few weeks may explain the unusually high CPUE value (30.97) for December 1976 in subarea 58.4.2. in the following season (the first year of Japanese mothership operations) Nasu (1983) and Shimadzu and Ichii (1985) describe the location of fishing operations in the Southeastern Indian Ocean. During 14–16 December fishing was restricted to the region very close to the ice edge. Catch rates for that period have not

Table 1. Mean CPUE values (t/h) fishing) for all seasons by subareas. Seasons refers to the year in which the split year ended (i.e. 75 refers to the 1974/75 season)

	Subareas			
	58.4.2	88.1	48.1	48.2
Mean CPUE	4.79	5.76	11.19	13.60
Seasons	75–83	79–85	84–85	83–85

been published although the mean CPUE for the month was 4.69 t/h. By January the ice edge had retreated two degrees further south. Fishing was then concentrated between latitudes 64° to 64°30′S and longitudes 113° and 117°E. Initially the focus of activity moved westwards, then slightly northwards and finally eastwards. This would suggest that the same patch of krill was being fished and also that it might have been held within a gyre. CPUE for the month (4.63) was similar to that for the preceding December. The highest CPUE (16.15) for that season was in November 1977, although there is no published information to indicate whether such a large value coincided with any other phenomenon, such as a super-swarm.

Japanese krill fishing operations extended eastwards around the Southern Ocean into area 88 from 1977. CPUE values throughout the period are similar to those of area 58, suggesting that the type of operation and the catchability of krill were more or less the same. In the 1980/81 season the fishing operation moved into the Atlantic sector and the resulting CPUE was very much higher than for the other subareas (Table 1 and Fig. 8).

The presence of fishing activity in two adjacent subareas of the Atlantic sector (48.1 and 48.2) allows a comparison to be made between them both within and between seasons. Comparing the mean values for the seasons either in terms of all fishing months or just the mid-season months December, January and February indicates that there is little in common between the two subareas (Fig. 9). If, however, the monthly values are compared, there is a tendency for

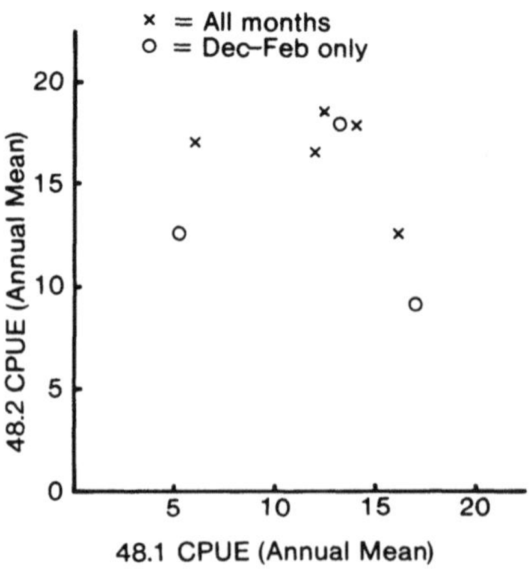

Fig. 9. Seasonal mean CPUE values for the Antarctic Peninsula (48.1) subarea compared with those from the South Orkney subarea (48.2)

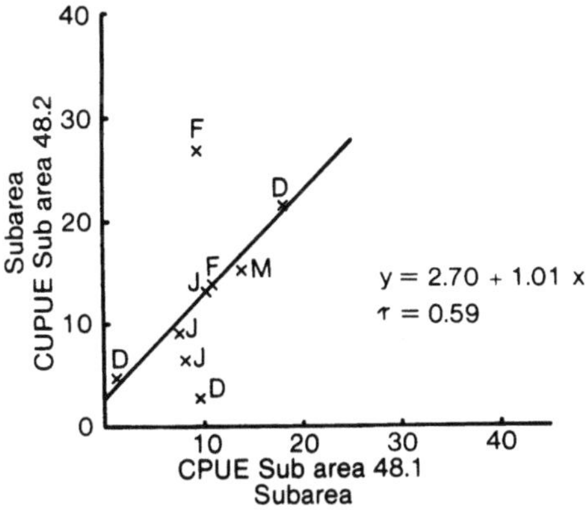

Fig. 10. Comparison of monthly mean CPUE values for the mid-season period (December to March) in subareas 48.1 and 48.2. (*D* December; *J* January; *F* February; *M* March)

Table 2. Monthly CPUE (t/h) for the 1983/84 season in a period during and immediately following a period of low krill abundance

Subarea	CPUE for the month				
	Nov.	Dec.	Jan.	Feb.	Mar.
48.1	9.89	18.15	10.70	10.71	15.15
48.2	–	21.63	–	14.07	–

high CPUE in subarea 48.1 to be associated with a high CPUE in 48.2, suggesting a link between krill catchability in the two subareas at the same time (Fig. 10). I have found no information to indicate why this relationship should be present, whether it represents a "real" effect or is purely a chance occurrence. Such a similarity would not be expected because subarea 48.1 receives krill largely from the Bellingshausen Sea whereas 48.2 is largely within the Weddell Sea. In any case, since 48.2 is effectively downstream of 48.1, one would have expected high values in 48.1 to be followed by high values in 48.1 sometime, perhaps a month, later. Such a relationship is not present.

The high CPUE values in area 48 are interesting for another reason: Evidence from research vessels working in the region from mid-1983 until 1984 clearly indicated that krill abundance in the area was extremely low (Heywood et al. 1985). The CPUE values for the months when fishing was in progress are shown in Table 2 and these indicate a high catchability for krill even though abundance in the preceding months was apparently low.

4 The Fisheries of Other Nations

Seven countries have, in addition to Japan, supplied STATLANT data to CCAMLR. Bulgaria and the German Democratic Republic reported small bycatches of krill between 1978 and 1985, their total catch amounting to about 300 t. France reported a catch of 6 t in November 1979 from area 58.4. Chile has had a small fleet operating intermittently in the Atlantic sector, giving a total catch of a little over 8000 t from 1976 to the present. The only season when the fishery continued over a series of months was in 1982/83 and these data are plotted in Fig. 11. Poland reported catches totalling 6966 t in 1976/77 but, since these amounted to less than 50% of the total catch from a mixed fishery and no effort data were provided, analysis is not possible. Korea has reported krill catches from a directed fishery in the Indian Ocean sector: CPUE values are plotted in Fig. 12.

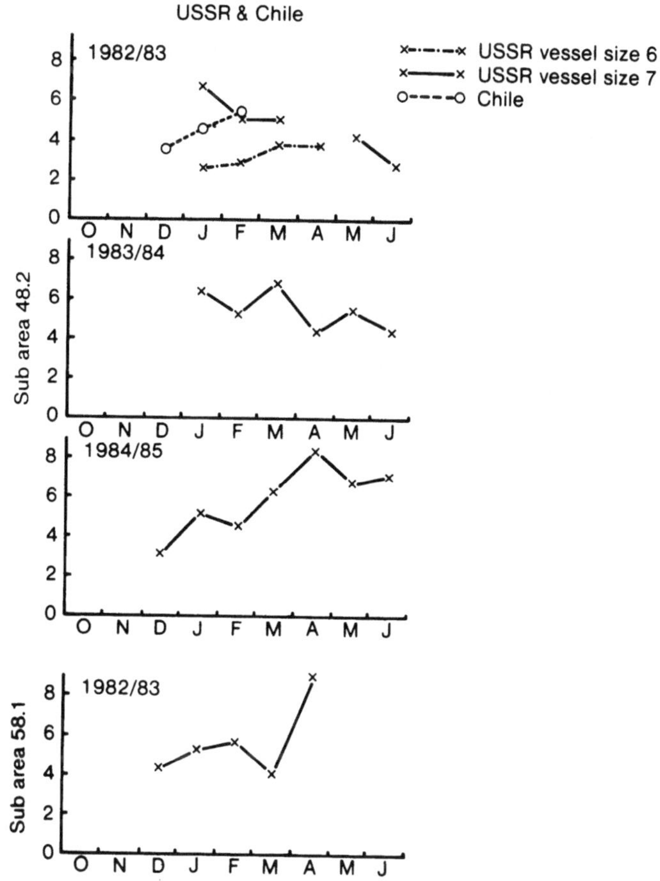

Fig. 11. Monthly CPUE values for USSR and Chile in subarea 48.2 and 58.1

The USSR has been responsible for the largest catches of krill. STATLANT reports have been provided from 1974 but regrettably without any effort data prior to the 1982/83 season. Furthermore, since krill are caught by the same gear type (as identified in the STATLANT classification) as fin-fish, it is often impossible to determine CPUE due to the mix of species reported. This serves to highlight one of the inadequacies of the reporting system. Data from months when only krill catches were reported are shown in Fig. 11. In subarea 48.2, although substantially lower than the Japanese data from the same period, they give no indication that krill were scarce during the 1983/84 season, as indicated by research vessel catches noted above. A further peculiarity is that CPUE from bottom trawls in subarea 58.4 was higher than those from the more typical krill fishing gear (midwater trawls) in the 1984/85 season. This

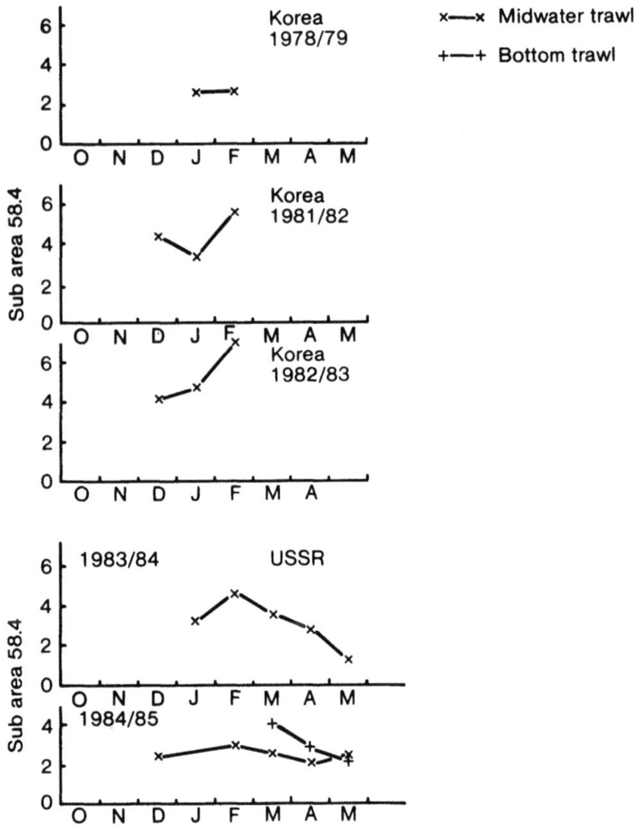

Fig. 12. Monthly CPUE values for USSR and Korea in subarea 58.4

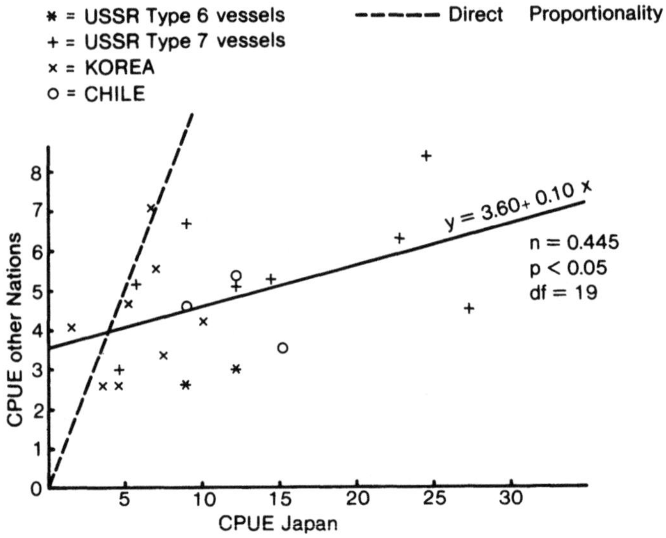

Fig. 13. Comparison of CPUE values from Japan with values from other nations in the same subarea during the same month

may indicate a different behavioral pattern by the krill either geographically or at a specific time.

5 Between-Nation Comparison of CPUE

Even though the data sets extend for over 10 years, there are relatively few months in which comparisons can be made between nations fishing in the same subarea. Data from months where the comparison can be made have been plotted in Fig. 13. With only two exceptions, the Japanese CPUE is greater than that of other fishing nations, often by a very large amount.

6 Comparison of CPUE and Independent Abundance Estimates

Krill abundance surveys that are independent of commercial fishing activity have been described for the Bransfield Strait and Prydz Bay regions. The BIOMASS 1979, 1980, 1982) provided the initiative for much of this work. The two methods employed are acoustic, using calibrated echosounders and echointegrators, and systematic net sampling, using scientific nets. The results of these surveys are set out in Table 3. The data are grouped according to the sampling method used and also by the research team which undertook the analysis so as to minimize bias due to the analytical methods.

The surveys reported by Klindt (1986) and Nast (1986) were undertaken on the same ship and at the same time. Even so, the estimates of abundance differ widely. Part of this variation is due to intrinsic errors which are associated with the respective methods. In the case of nets, net avoidance is a problem with the sampling method as a contagious distribution introduces problems in sampling design. In the case of acoustics, vertical migration to the surface (out of the transducer range) and the inability of the echosounder to detect krill at very low densities are both unquantified errors that lead to underestimation of total abundance. Analytical methods can lead to widely differing results from the same data set (Table 3). In addition, the poor information on krill target strength introduces a constant error to the absolute although not to the relative abundance estimates. Even though the results differ greatly, it is impossible to determine which method provides the better estimate of abundance. Modelling the distribution of krill could provide a means of combining information from the two methods to provide more precise estimates. For the present it must be accepted that the two methods are equally valid, and since the biomass estimates are well in excess of the krill catch for the months in question they are equally believable.

Given the poor degree of congruence between these independent estimates of abundance, it is not surprising that CPUE, an additional index of abundance, does not apparently follow any trend relative to the other two estimates. The fishing operation is known to be concentrated in relatively small localities within the general area and CPUE is a reflection of abundance in the vicinity of the fleet. Abundance is obviously high in these small localities, but it is not known how extensive these are or whether they might represent the only high abundance localities within the area.

Table 3. Comparison of CPUE and independent survey estimates of abundance during the same months. Pairs of lines that are bracketed together as for different analyses on the same data sets. * assumes 1 t per nautical square mile = 0.292 g/m^2. All values assume a depth range of 100 m

Date	Location or study	CCAMLR subarea	Biomass (t)	Density (published)	Density estimated (g/m^2)	CPUE (Japan)	Reference for survey
Acoustic surveys (FRG)							
Oct./Nov. 1983	SIBEX I	48.1	51 680	7.2 g/1000 m^3	0.72	10.25, 13.75	Klindt (1986)
Nov./Dec. 1984	SIBEX II	48.1	379 750	54.8 g/1000 m^3	5.48	– 1.06	Klindt (1986)
Mar./Apr. 1985	SIBEX II	48.1	16 490	2.59 g/1000 m^3	0.26	(9.46 Feb.)	Klindt (1986)
Acoustic surveys (Poland)							
Feb./Mar. 1981	FIBEX (Drake Passage)	48.1	(1 195 572	28.9 t/nm^2	8.4 g/m^2 *)	20.08, –	Kalinowski (1982)
			(62 000		)	20.08, –	BIOMASS (1986)
	FIBEX (Drake Passage)	48.1	(70 827	34 t/nm^2	9.93 *)	20.08, –	Lillo and Guzman (1982)
			(182 000	7.15/m^3	715)	20.08, –	BIOMASS (1986)
	FIBEX (Bransfield Strait)	48.1	(2 271 360	346 t/nm^2	100 g/m^2 *)	20.08, –	Kalinowski (1982)
			(136 000	4.68 g/m^3	468)	20.08, –	BIOMASS (1986)
	FIBEX (Bransfield Strait)	48.1	(448 795	76.2 t/nm^3	22.26 *)	20.08, –	Lillo and Guzman (1982)
			(800 000	32.25 g/m^3	3225)	20.08, –	BIOMASS (1986)
Dec./Jan. 1983/84	SIBEX I (Drake Passage)	48.1	122 470	4.01 t/nm^2	1.17 *	18.15, 10.70	Kalinowski et al. (1985)
	SIBEX I (Bransfield Strait)	48.1	70 593	3.03 t/nm^2	0.88 *	18.15, 10.70	Kalinowski et al. (1985)
Acoustic survey (Australia, France, Japan, Republic of South Africa)							
Feb./Mar. 1981	FIBEX	58.4.2	4 512 000		1.97 g/m^2	4.24, 3.60	BIOMASS (1986)
Net haul surveys (FRG)							
Oct./Nov. 1983	SIBEX I	48.1	723 000	103.18 g/1000 m^3	10.32	10.25, 13.75	Nast (1986)
Nov./Dec. 1984	SIBEX II	48.1	252 000	36.01 g/1000 m^3	3.60	– , 1.06	Nast (1986)
Mar./Apr. 1985	SIBEX II	48.1	164 000	23.39 g/1000 m^3	2.34	(9.46 Feb.)	Nast (1986)

7 Discussion

The above consideration of CPUE calculation, net sampling and acoustic survey as methods of estimating abundance demonstrates the magnitude of the variation between these estimates and highlights the disparity between some of their implications. No mention has been made here of using predators to provide abundance indices derived from predator biology as this topic is dealt with elsewhere (Croxall et al. this Vol.). In order to identify how to improve the quality of abundance estimates it is necessary to understand clearly what are the important sources of variation in krill abundance we need to estimate. In this chapter I consider the situation purely in terms of fishery management; i.e. the kind of information that is required to make management decisions.

The Convention for the Conservation of Antarctic Marine Living Resources (The Convention) specifies two interrelated management strategies. Firstly, that resources should be managed on a sustainable basis, and secondly that harvesting of one species should not reduce the stock size of dependent species below their level of greatest net annual increment. In this context increment is synonymous with production.

The first strategy considers krill in isolation. If annual production balances mortality due to fishing, predation and other natural causes over the long term, then the stock will be sustained. Of these factors, the amount extracted by the fishery can be determined by a reporting procedure such as is currently in operation, but the other factors are more difficult to determine. Annual production can be derived from knowledge of initial standing stock, growth and mortality rates. Information on these is currently very imprecise.

The second strategy requires a careful consideration of the effects of predators on krill population dynamics. While some major predators such as whales and Crabeater seals may, for simplicity, be considered to operate over the whole range of the krill distribution, others, specifically those restricted in foraging range by their need to return to breeding sites, clearly have only a local impact at certain times of year. Thus a broad Southern Ocean framework is adequate for krill on its own, but krill in the context of The Convention should be considered in terms of a series of subunits, many of which would be defined in local predator dependence on krill.

Provided the sum of such subunits were to cover the full distributional range of krill, then, because a self-sustaining krill population is implicitly necessary to maintain predators, it follows that the resource management needs of krill could be met within the framework of a system that took account of key predator requirements.

These management subunits need to be designated on ecological grounds based on those requirements of the species that are most sensitive to changes in krill abundance. The following is a list of characteristics that should be considered:

1. Predator factors
Foraging range
Period in life cycle most sensitive to krill availability
Ecophysiological indicator to use as krill availability index
The above three would give an index of krill abundances which, with predator abundance, would indicate predator impact.

2. Krill factors
Immigration/emigration rates
Standing stock
Productivity
Natural mortality from all causes
Fishing mortality

The present chapter is concerned with only one aspect, that of abundance or standing stock, although it is accepted that variability in physical oceanographic factors will affect the other parameters.

Three different methods are available for estimating absolute krill abundance or indices of abundance: direct surveys, CPUE and indirect indices from predator-prey interactions. Direct surveys, whether using hydroacoustics or net-hauls, demonstrate that there can be great variations in krill density within an area. Differences between results from net-hauls and acoustics are a reflection of the sampling method as well as sampling design and analytical procedure. Net-hauls on a grid provide point estimates, whereas hydroacoustic surveys sample along narrow line transects. An improved abundance estimate would clearly be obtained by amalgamating these two sets of results. Currently net-haul surveys can stand alone, whereas acoustic surveys must make use of net-haul data to provide confirmation of acoustic targets and krill size frequency distributions to derive target strength values. Even the results from more refined analytical models are still likely to indicate a high degree of variability in abundance that most probably reflects the real situation.

CPUE, as demonstrated by the analysis presented here, gives the impression that abundance is not subject to great variation. This does not reflect the true situation over a large area because the fishing fleets move to, and remain in, favourable fishing localities. Consequently, CPUE will provide a resonable index in the vicinity of the fleet but not for the subarea as a whole (Gulland 1985). It is essential to determine how this index relates to the rest of the subarea where little or no fishing takes place; this is one of the main objects of a simulation study sponsored by CCAMLR. In the long term it should therefore be possible to estimate krill abundance using a combination of CPUE, acoustic and scientific net-haul surveys.

Although seasonal mean CPUE and, to a lesser extent, the monthly mean CPUE tend to be reasonably stable, there are several anomalously high monthly values. One of these is known to be when the fleet was fishing a super-swarm (Sugimoto 1977). I have not seen any published information to indicate whether this was the case for the other high CPUE months. Super-swarms are characterized by having a high krill density over a large area and a wide depth range. From the fishing point of view, this means that the depth at which the net is fished is important but not critical to its success and also that the net is continuously towed through water containing krill rather than through distinct swarms with clear water between. Under the circumstances, consistently high catch rates are to be expected. Data for those months where CPUE was greater than 10 t/h are summarized in Table 4. The high values come predominantly from the Atlantic sector and this may indicate a greater tendency for the formation of dense aggregations in that area. If these high values are excluded there is still a higher CPUE in the Atlantic sector than elsewhere. This greater "catchability" may be a result of differing behaviour patterns of krill in terms of size, density or detectability of swarms. CPUE therefore provides an index of local abundance, but at this stage we cannot relate this to wider areas.

Indications of variability in krill abundance derived from monitoring programmes on warm-blooded predators are discussed fully by Croxall (this Vol.). No programmes are in progress that are monitoring predator-prey interactions of fish. It has been suggested (SC-CAMLR-1986a) that the distribution of krill has been in the past, and is currently, influencing the stocks of *Champsocephalus gunnari* in the South Orkney region. These fish, it is suggested, rely on the presence of stable concentrations of krill. Coincident with the reduction in density of *C. gunnari* has been the estab-

Table 4. Months in which the Japanese krill fishery achieved high CPUE

Season	Months	Subarea	Catch (t)	Hours fished	CPUE (t/h)	Comments
80/81	Dec.	48.1	1 847	81	22.80	
	Feb.		783	39	20.08	
81/82	Dec.		2 030	65	31.23	Super swarm reported in region during FIBEX although not at this time
82/83	Oct.	48.2	89	1	89.0	
	Nov.		812	26	31.23	
83/84	Dec.		173	8	21.63	
84/85	Feb.		6 229	232	26.96	
	Mar.		11 317	493	22.96	
	Apr.		266	11	24.18	
85/86	Mar.		12 597	619	20.35	
76/77	Dec.	58.4.2	2 106	68	30.97	Fishing on super swarm (Sugimoto 1977)

Table 5. Total catches of *Champsocephalus gunnari* and *Euphausia superba* (krill) in subarea 48.2

Species	Season								
	77/78	78/79	79/80	80/81	81/82	82/83	83/84	84/85	85/86
Champsocephalus	138 895	21 439	5 231	1 861	557	5 948	4 499	2 361	2 682
Krill	–	18	173 765	60 540	257 695	125 605	64 112	123 830	224 744

lishment of a significant krill fishery (Table 5). This krill fishery would have been based on large-scale stable krill concentrations which in view of the reported catches must have been present. Heavy fishing on krill would reduce the amount potentially available to the fish and this in turn might be manifested by a change in distribution or spawning and hence ultimately recruitment of the fish stock. Since shore-based predators have been increasing over the same period (Croxall et al. this Vol.), the alternative explanation that low catches are purely a result of overfishing *C. gunnari* (SC-CAMLR 1986b) seems far more likely.

Returning to the initial question posed in the title of this paper "Can we satisfactorily estimate variation in krill abundance?" the answer must be "NO" because although we can estimate the variation within each method we have no information to indicate how these estimates relate to total krill abundance over the Southern Ocean. Even within subareas the information is little better. This is serious in terms of understanding the dynamics of the krill population and therefore our ability to provide the information necessary for the long term management of the resource. The requirements in this context are year-to-year estimates of abundance. Such information could come from direct surveys (which due to the long time required to undertake them can only be considered quasi-synoptic) or else from CPUE data (provided some valid model becomes available to permit realistic extrapolation to cover the whole area). Of more immediate importance is the situation for species which depend on krill, that is the population subunits outlined above.

From the fishery management point of view, it is important to be able to restrict fishing activity when fishing is currently causing or is likely to cause a decline in the predator stocks below their level of greatest net annual increment. Our understanding of the ecosystem is such that we are currently unable to determine predator population dynamics with this accuracy. What may be possible is to provide indications as to when the Krill abundance is reduced, whether by fishing activity or natural variation, to a level likely to adversely affect dependent predators. Such a system would need to operate at two spatial scales: the restricted range of shore-based predators such as penguins, and the open-ocean scale of entirely pelagic predators such as whales and Crabeater seals. Some form of monitoring and control is clearly necessary now and the following generalized scheme is proposed as a system that might be practical in the long term for this purpose.

1. Shore-based predators

1.1 Initial requirements

a) Identification of species that are sensitive to local changes in krill abundance.
b) Identification of predator parameters that can be used as "quantitative real-time" indices of krill abundance, e.g. predator foraging trip duration.
c) Determination of the foraging locality of the predator (Fig. 14).

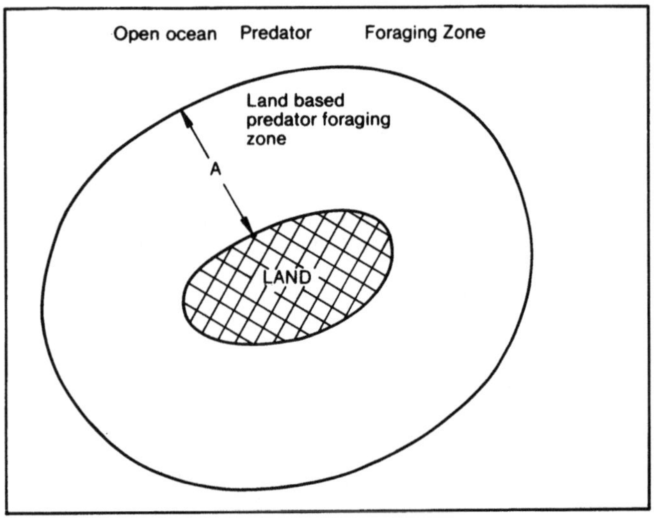

Fig. 14. Diagrammatic representation of a fishery management area. The land based predator foraging zone would be determined by the key predator species and its estimated foraging range (*A*)

1.2 Monitoring requirements

a) Determine variability in predator parameters.
b) Determine values of predator index that indicate low krill availability.

1.3 Suggested management procedure

a) Obtain CCAMLR agreement on size and locality of foraging zones.
b) Establish monitoring programmes at selected sites, including reporting of data at short time intervals (e.g. 10 days).
c) Obtain CCAMLR agreement on predator index levels that indicate when krill availability is low. (Two possible levels might be considered, a warning level and a danger level at which the local fishery would be temperarily closed until the index went above the warning level).

Although theoretically feasible, a monitoring programme of this type is currently well beyond the scope of any current monitoring programmes.

2. Open ocean pelagic predators

2.1 Initial requirement

a) The "best" species is likely to be Crabeater seal.
b) Suitable indicator parameters need to be identified.
c) The localities would be subdivisions of the Southern Ocean that are not included in (1.3b) above.

The monitoring requirement and management procedure could be undertaken in the same way as for the shore-based predators.

The scheme outlined above, if approved, would be a step towards fulfilling the requirements of Article II of the Convention. It is not, however, management on an eco-sytem basis within the meaning of the Convention. To achieve this it is essential to gain a greater understanding of ecosystem interactions and their natural variability.

Acknowledgement. I am grateful to the CCAMLR Secretariat for supplying the STATLANT data used in this paper. Graham Medley (Imperial College London) and Alistair Murray (BAS) very kindly reformatted the data and the latter undertook the statistical analyses. Thanks are also due to John Gulland FRS and my colleagues At BAS, Dr. R.M. Laws FRS, Nigel Bonner, Alistair Murray, Drs. John Croxall and Barry Heywood for constructive comments on the draft manuscript. Finally my thanks are due to Christine Thulborn and Gemma Hurrell for typing the manuscript.

References

BIOMASS (1979) Meeting of the Group of Specialists on Living Resources of the Southern Oceans. BIOMASS Rep Ser No 7
BIOMASS (1980) FIBEX implementation and coordination. BIOMASS Rep Ser No 13
BIOMASS (1982) Technical Group on Programme Implementation and Coordination. 3rd Meet. BIOMASS Rep Ser No 29
BIOMASS (1986) Report on the Post-FIBEX Acoustic Workshop. BIOMASS Rep Ser No 40
Grantham GJ (1977) The utilization of krill. FAO South Ocean Fish Surv Programme, GLO/SO/77/3. FAO, Rome, 61 pp
Gulland JA (1985) Krill – catch per unit effort Commission for the Conservation of Antarctic Marine Living Resources Krill. Working Group. Krill WG/1985/Doc 2, 4 pp
Heywood RB, Everson I, Priddle J (1985) The absence of krill from the South Georgia zone, winter 1983. Deep-Sea Res 32:369–378
Kalinowski J (1982) Distribution and stocks of krill in the Drake Passage and the Bransfield Strait, during the BIOMASS-FIBEX expedition 1981. Pol Polar Res 3:243–251
Kalinowski J, Godlowska M, Klusek Z (1985) Distribution and stock of krill in the Bransfield Strait and the Drake. Passage during December 1983–January 1984, BIOMASS-SIBEX. Pol Polar Res 6:151–158
Klindt H (1986) Acoustic estimates of the distribution and stock size of krill around Elephant Island during SIBEX I + II in 1983, 1984 and 1985. Arch Fischereiwiss (Beiheft 1) 37:107–127
Lillo S, Guzman O (1982) Study of the abundance, distribution and behaviour of krill at the Bransfield Strait and Drake Passage by means of hydroacoustic techniques. Ser Sci Inst Antarct Chileno No 28:17–46
Nast F (1986) Changes in krill abundance and in other zooplankton relative to the Weddell-Scotia Confluence around Elephant Island in November 1983, November 1984 and March 1985. Arch Fischereiwiss (Beiheft 1) 37:73–94
Nasu K (1983) On the geographic boundary of Antarctic krill distribution. Ber Polarforsch Sonderh 4:216–222
SC-CAMLR (1986a) Rep Sci Comm Paragr 4.62
SC-CAMLR (1986b) Rep Ad-hoc Working Group on Fish Stock Assessment. SC-CAMLR 1986, Annex 4
Shimadzu Y (1985) A brief summary of Japanese fishing activity relating to Antarctic Krill, 1972/73–1982/83. Commission for the Conservation of Antarctic Marine Living Resources. Select Pap Present Sci Comm CCAMLR 1982–1984, part 1, pp 439–452
Shimadzu Y, Ichii T (1985) Some considerations on the usefulness of CPUE data from Japanese krill fishery in the Antarctic. Pap SC-CAMLR Working Group on Krill CPUE Ref Krill. WG/1885/ Doc 4, 10 pp
Sugimoto S (1977) (On the status of the Antarctic krill fishery in 1976/77 season). (Rep Meet Fish Top, Jpn Soc Sci Fish, No 11 in Japanese)

Patterns of Fluctuations in the Hydrological Conditions of the Antarctic and Their Effect on the Distribution of Antarctic Krill

V. V. MASLENNIKOV and E. V. SOLYANKIN[1]

Summary. On the basis of published reports, long-term trends in the hydrometeorological conditions in the Atlantic sector of the Southern Ocean are described with the appearance of "cold" and "warm" epochs. It is shown that during these periods there are major differences in the hydrological conditions of the Scotia Sea and around South Georgia. Fluctuations in the position of the secondary frontal zone (Weddell-Scotia Confluence) are of great importance for the accumulation or lack of concentrations of krill.

1 Introduction

Our studies have shown that the density of krill stocks and features of their distribution vary significantly with time. Areas that have been known to be rich in krill in some years are practically devoid of them in others. The reason for this, in our view, lies in fluctuations of oceanological conditions. Comprehensive investigations, including meteorological studies, of the environment and the distribution of krill indicate that sharp changes in the density of krill occur when there is significant reconstruction of hydrophysical fields often accompanied by anomalously warm air temperature over the ocean. Less intensive environmental changes also affect the occurrence and density of the krill concentrations. These conclusions are supported by investigations which show that in Antarctic waters physical and biological phenomena undergo interannual changes. It should be noted, however, that these relatively short-lived fluctuations take place against the background of long-term changes with unidirectional trends.

In the present study an attempt is made to analyze hydrometeorological fluctuations and to evaluate their effects on the quantitative parameters characterizing the distribution of krill in the Atlantic sector of the Southern Ocean. We are fully aware of the immense scope of this problem, and hope that even its partial solution will be of practical value.

Our study is based on data reported in literature, including results of oceanological and biological research in the Atlantic sector of the Southern Ocean. We also used air temperature data and atmospheric measurements made at sea level at several hydrometeorological stations. The meteorological data were analyzed by plotting integral curves of anomalous readings. This made it possible to determine long-term trends in climatic fluctuations and changes.

1 All-Union Research Institute of Marine Fisheries and Oceanography (VNIRO), Moscow, USSR

2 Long-Term Variability

2.1 Fluctuations in Hydrometeorological Conditions

To elucidate the climatic conditions that prevailed over a period of many years in the Atlantic sector of the Southern Ocean, we analyzed data beginning from 1903. Time series of air temperature and near-surface atmospheric pressure during the warm season in the Southern Hemisphere (December–March inclusive) at Grytviken station (South Georgia) and Lauri Island (South Orkney Islands) were used.

On the basis of the inflection of the trends on the integral curves of anomalous air temperature (Fig. 1), we can determine when cold periods replaced warm ones. These are relatively long periods that are characterized by only positive or negative anomalies of air temperature (Maslennikov 1979a). Changes from one epoch to another in different places differ somewhat with respect to the time of their occurrence. Thus, according to data from the southernmost station (South Orkney Islands), a cold epoch began in 1910 and was replaced by a warm period in 1950. In the northern part of the Scotia Sea (South Georgia) the onset of a cold and a warm climatic trend occurred in 1919 and 1959, respectively. The time of the occurrence of a cold and a warm period in the southern Scotia Sea was more definite than in the north, where anomalous temperatures of the opposite sign were often observed.

These conclusions are confirmed by data on the fluctuations in the mean annual air temperature in the 1904–1972 period at Orcada station (South Orkney Islands), from which it follows that a stable transition from negative anomalous air temperatures to positive ones occurred in 1951 (Petrov and Lyubersky 1980).

Long periods of negative or positive anomalies of air temperatures in the western Atlantic sector could have had

Antarctic Ocean and Resources Variability
(ed. by D. Sahrhage)

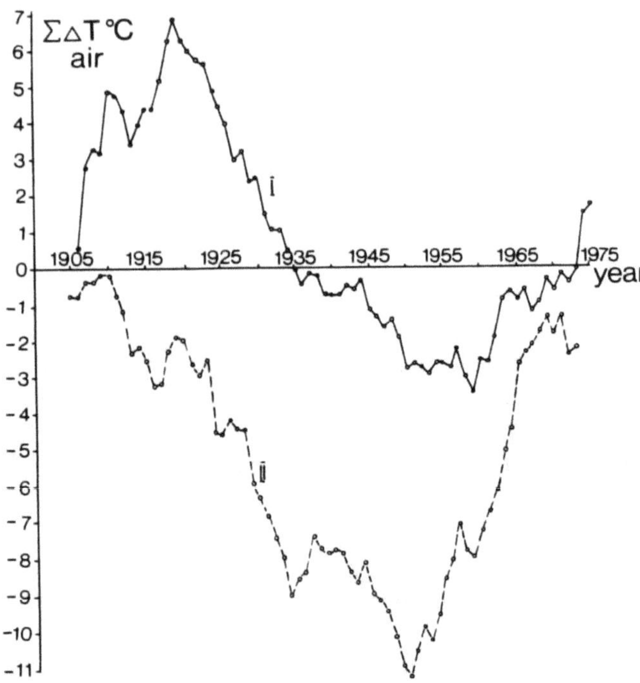

Fig. 1. Integral plots ($\Sigma\Delta T_{air}$°C) of anomalies in mean air temperature during summer (December–March). *I* Data from station at Grytviken (South Georgia); *II* data from station at Lauri Island (South Orkney Islands. (Maslennikov 1979a)

an effect, in our opinion, on the hydrological conditions in the Scotia Sea. This consequently affected the biological processes taking place there. The existence of different climatic epochs was clearly manifested first of all in the thermic regime of the surface waters. This is clearly indicated in Fig. 2A,B showing the distribution of the mean water temperature of the layer of 0–50 m in the Scotia Sea for December during the cold and warm epochs. The difference of the temperature field is sufficiently large for the given latitudes. This difference is evident not only in the absolute values of the water temperatures but also in the configuration of the isotherms themselves. This configuration indicates the increase of expansion of waters from the Weddell Sea during the cold epoch, and demonstrates that during the warm epoch the influence of the Antarctic Circumpolar Current prevailed. The differences in the distribution of the intermediate minimum of water temperature around South Georgia also support this conclusion. During the warm epoch T_{min} were higher than during the cold epoch (Fig. 3). This indicates a decrease in the advection of water from the Weddell Sea in the first case and a corresponding increase in the influence of the Antarctic Circumpolar Current.

This raises the question of whether the climatic epochs established for the western part of the Atlantic sector of

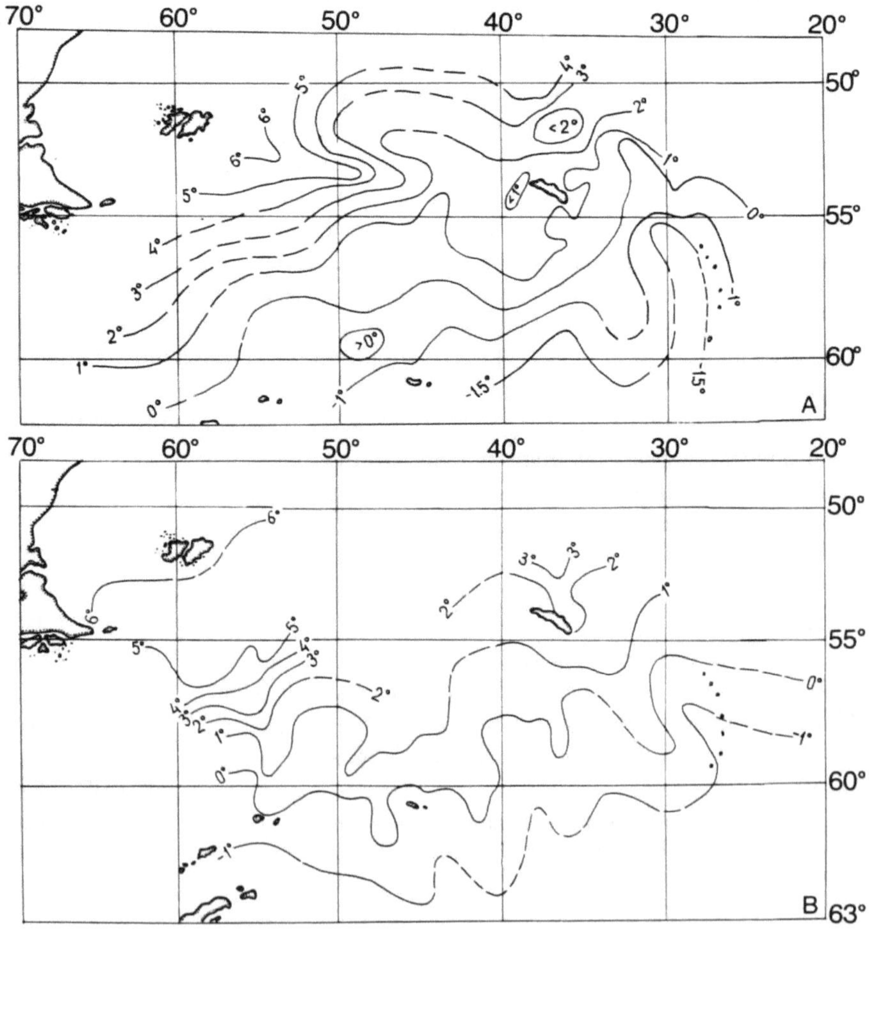

Fig. 2A,B. Distribution of mean water temperature in the 0–50 m layer during December. (Maslennikov and Solyankin 1979). **A** Cold period (1911–1937). **B** Warm period (1955–1975). (Maslennikov 1979a)

the Southern Ocean are global in nature. Are they related to a general trend of a warming up or fall in temperature for the entire region of the Southern Ocean or do they characterize merely the local climatic processes?

An analysis of the mean seasonal air temperatures (Voskresensky et al. 1986) at various continental and insular stations located at different latitudes indicates that the climatic fluctuations observed in the years 1955–1980 have different sign. Thus, while in the western part of the continent the warm period lasted from the end of the 1950's to the early 1970's (gradually replaced by the cold period), in the eastern part of the continent the warm period began 5–7 years later and lasted until the end of the 1970's.

The presence of an "inflection point" in the ice-edge boundary (fast ice) in the Atlantic sector (Mackintosh 1972) can serve as indirect confirmation of the diverse nature of climatic fluctuations in the adjacent regions of the Southern Ocean. A southward shift of the ice edge (lying west of the zero meridian) causes the adjacent ice edge (lying to the east of the zero meridian) to shift northward.

Contemporary measurements (Ryzhakov et al. 1980) show that there are three main forms of macrocirculation of atmospheric processes in the Southern Hemisphere: a zonal and two meridional (the latter are characterized by the opposite position of the baric crests and troughs). These results can also be considered as confirmation of large-scale regional differences in the dynamics of long-range changes of thermobaric indices.

In spite of what has been said above, we should bear in mind that there are climatic changes of longer duration which encompass a large territory of the Southern Ocean and which affect the climate of the continent as a whole. It is known, for example, that in the 1947–1970 period the iceberg boundary moved far to the south, as compared with the boundary for the period from 1773 to 1947 (Bunitsky 1975).

2.2 Effect of Long-Term Hydrometeorological Fluctuations on the Distribution of Krill

The biomass of krill in the Scotia Sea consists of specimens of different origin. They are carried to these waters by currents from the Weddell Sea, the Bransfield Strait and from the eastern part of the Pacific sector (the Bellingshausen Sea and possibly from farther west). However, most of the krill come from the Weddell Sea. Large aggregations of them are formed in the south of the Scotia Sea within the boundaries of the so-called secondary frontal zone (Weddell-Scotia Confluence). Long-term climatic variations appear to have affected the position of this zone and in general the expansion of waters from the Weddell Sea in the Scotia Sea and around South Georgia. Naturally this will affect the distribution of krill.

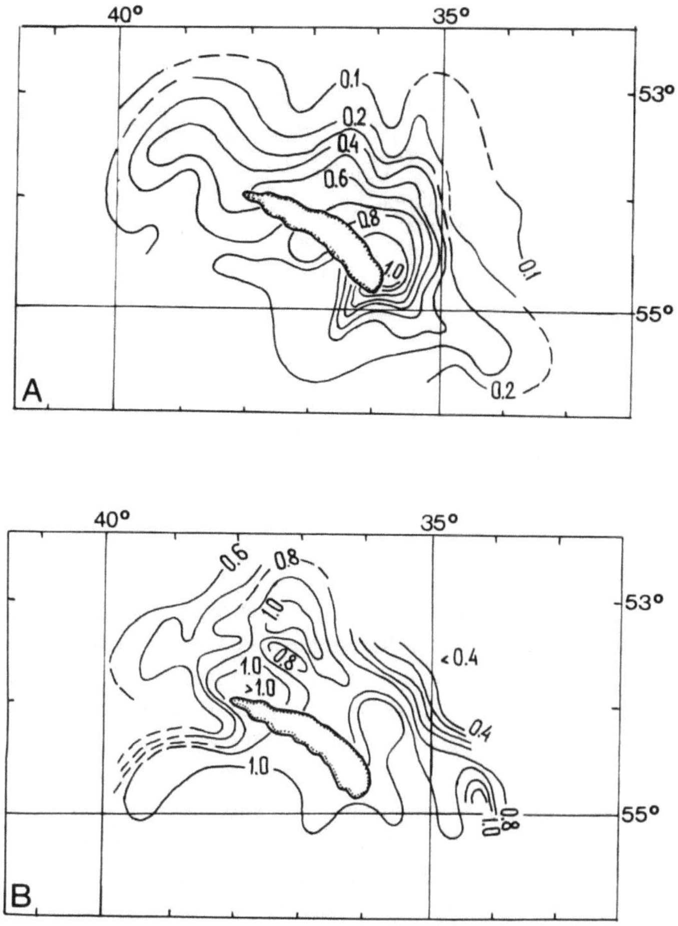

Fig. 3A,B. Distribution of the intermediate minimum of water temperature (T_{min}) in the vicinity of South Georgia. **A** Cold period (1911–1937). **B** Warm period (1955–1975). (Maslennikov 1979a)

Beginning from 1975, waters of high-latitude modification constantly increased, with occasional interruptions, their influence on the waters of the southern Scotia Sea (Fig. 4). One may assume that a gradual change from a warm to a cold epoch took place. Of interest in this connection is the interpretation of data on the location of the mixing zone between Weddell Sea and Bellingshausen Sea waters (Weddell-Scotia Confluence) during the culmination of the cold epoch, i.e. the 1920's–1930's (Fig. 4, Deacon 1933). Since a cooling of the atmosphere near the surface is accompanied by an increase in the distribution of waters of high-latitude modification (Maslennikov 1976), it is possible that during the cold epoch waters from the Weddell Sea covered a larger area. However, it seems unlikely that the location of the secondary frontal zone would undergo a radical change. There may, however, be separate intrusions of waters from the Weddell Sea that could reach low latitudes. This phenomenon could lead to an increase in the mass transfer of krill.

From the point of view of variability of hydrophysical fields and quantity and distribution of krill, the natural conditions near South Georgia (adjacent to the Scotia Sea)

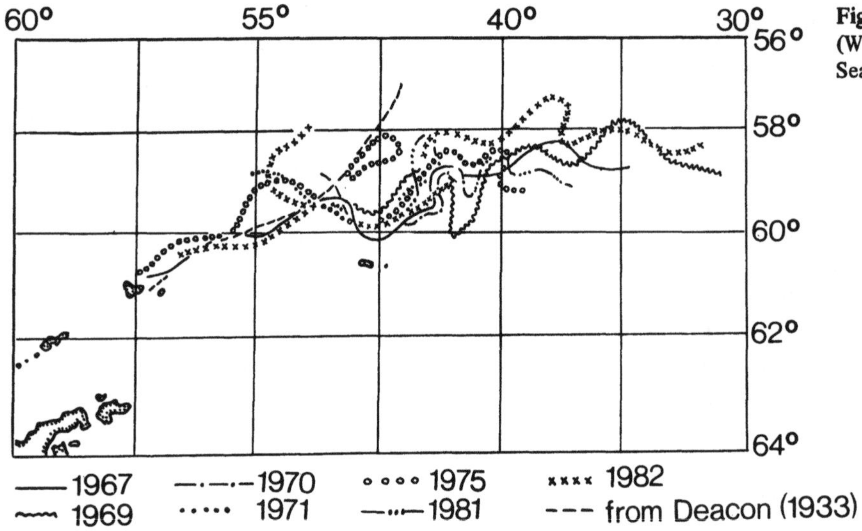

Fig. 4. Location of the secondary frontal zone (Weddell-Scotia Confluence) in the southern Scotia Sea

are more contrasting. During the current (warm) epoch frontal waters close to the insular slope and spreading into a northwestern direction carry krill together with waters from high-latitude modification, predominantly from the Weddell Sea. In part these waters penetrate via depressions the shelf area (Maslennikov 1979b). This is evident from the high gradient values of physical and chemical properties (Maslennikov 1979c). Favorable dynamic conditions for the formation of krill concentrations are created here with the inflow of waters from the Weddell Sea which carry krill. However, during some years the situation can change drastically. In this case, for example in 1969, the vortex zones characteristic of the dynamic situation are absent. Over the shelf and its declivity and beyond, a single current flows in northwestern direction. As a result, no krill accumulate in this area (Makarov et al. 1980).

During the cold epoch, as noticed above, transformed high-latidue waters spread over a large area, engulfing part of the insular shelf. This is illustrated by the water distribution characteristic around South Georgia in March 1981 during research under the FIBEX program (Maslennikov et al. 1983). The hydrological conditions during this season were very similar to those during the cold epoch. This again shows that in spite of definite climatic changes during some years an anomalous deviation of opposite sign can take place. It may be assumed that with a possible increase in the krill stock over the insular shelf (owing to a wide and relatively free advection of high-latitude waters) the distribution of krill under the conditions of diminishing gradients of physical and chemical characteristics should on the whole be more regular. This has indeed been observed in March 1981. The examples cited show that a wider distribution of high-latitude waters and an expansion of the area of krill drift during the cold epoch can lead to various effects. Under conditions of latitudinal displacement of the secondary frontal zone in a northern direction (for example, in 1982), the dynamic conditions on a mesoscale necessary for the accumulation of krill are far from being optimal.

References

Bunitsky VKh (1975) Ice flows and icebergs in Antarctica. Leningrad State Univ, pp 1–254

Deacon GER (1933) The general account of the hydrology of the South Atlantic Ocean. Discovery Rep 7:171–238

Limbert DWS (1974) Variations in the mean annual temperature for the Antarctic Peninsula, 1904–1972. Polar Rec 17 (108):303–306

Mackintosh NA (1972) Life cycle of Antarctic krill in relation to ice and water conditions. Discovery Rep 36:1–94

Makarov RR, Maslennikov VV, Solyankin EV, Shevtsov VV (1980) Quantitative distribution and conditions for formation of swarms of *Euphausia superba* Dana with reference to some areas in the Atlantic and Pacific Southern Ocean. In: Lubimova TG (ed) Biological resources of the Antarctic krill. VNIRO, Moscow, pp 114–145

Maslennikov VV (1976) Horizontal circulation and distribution of macrozooplankton in near-Antarctic waters of the Atlantic Ocean. Trudy VNIRO 112;50–56

Maslennikov VV (1979a) Long-term variability in hydrometeorologic characteristics in the Southwest Antarctic Atlantic Ocean. Trudy VNIRO 136:50–56

Maslennikov VV (1979b) Horizontal circulation of waters in the vicinity of the South Georgia island. In: Antarctica, vol 18. Nauka, Moscow, pp 140–143

Maslennikov VV (1979c) Regional characteristics of the Antarctic surface waters in the southwestern part of the Atlantic sector. In: Antarctica, vol 18. Nauka, Moscow, pp 134–139

Maslennikov VV, Solyankin EV (1979) Inter-annual shift of zones where the waters of the Weddell Sea and the Antarctic Circumpolar Current interact. In: Antarctica, vol 18. Nauka, Moscow, pp 118–122

Maslennikov VV, Solyankin EV, Spiridonov VA, Sysoeva MV (1983) A contribution to the study of the origin of *Euphausia superba* Dana occurring off South Georgia. In: Lubimova TG (ed) Antarctic krill. Distribution pattern and environment. Nauka Tr VNIRO, Moscow, pp 74–85

Petrov LS, Lyubersky AN (1980) Variations in thermobaric and ice conditions over a period of many years in the area of South Orkney Islands. In: Research into the climate of Antarctica. Gidrometeoizdat, Leningrad, pp 59–65

Ryzhakov LYU, Rabtsevich SV, Savitsky GB (1980) Some new characteristics of the forms of circulation in the southern hemisphere and types of synoptic processes in Antarctica. In: Research into the climate of Antarctica. Gidrometeoizdat, Leningrad, pp 178–184

Voskresensky AI, Lyubersky AN, Subbotin VV (1986) Main trends of changes in thermic conditions over a period of many years in Antarctica. In: Meteorological research in Antarctica. Gidrometeoizdat, Leningrad, pp 33–39

Differences in the Hydrology, Biomass, and Species Distribution of Plankton, Fishes, and Birds in the Bransfield Strait and the Drake Passage During FIBEX 1981 and SIBEX 1983/84

S. RAKUSA-SUSZCZEWSKI[1]

Summary. The areas of the Bransfield Strait and the southern Drake Passage were investigated by Polish expeditions in February–March 1981 and in December–January 1983/84. This chapter presents a comparison of the results from the cruises in both periods as regards the hydrological situation, phyto- and zooplankton, juvenile fish occurring simultaneously with krill, and observations on birds.

1 Introduction

In the years 1981 and 1983/84 complex investigations were carried out within the BIOMASS Program in the Drake Passage and the Bransfield Strait (Rakusa-Suszczewski 1982a; Rakusa-Suszczewski and Lipski 1985). This research was a continuation of the investigations conducted in the same area in 1976 and 1977 (Rakusa-Suszczewski 1978; Witek et al. 1980).

2 Results 1981 Versus 1983/84

2.1 Hydrological Conditions

During surveys it was found that the geostrophic currents in relation to the 500 dbar surfaces always had similar directions. In summer the water masses around the South Shetlands move from SW toward NE, and in the Bransfield Strait at the Antarctic Peninsula they flow from NE to SW. The research area is a meeting place of water masses incoming from the Bellingshausen Sea and the Weddell Sea, with the local shelf waters of the Antarctic Peninsula and of the Bransfield Strait. The T/S values undergo seasonal changes, and their distribution may vary from year to year (Anonymus 1983; Grelowski and Tokarczyk 1985). For example, in the western part of the research area the water temperatures in December 1983 and January 1984 were higher than in February–March 1981, in spite of the fact that the measurements were done much earlier in the season. Salinity values were distinctly higher in 1983/84 than in 1981. It is probable that there was an inflow of waters from the north in the period preceding the investigations. The hydrological situation prevailing in this area in winter and early spring seems to be responsible for the situation observed in summer. The hindrance of transport of the water masses from the SW area of the Bellingshausen Sea, caused by an inflow of waters from the north, may have partly contributed to the observed lowering of the biomass of krill (*E. superba*) in 1983/84 (Kalinowski 1982; Kalinowski et al. 1985).

2.2 Phytoplankton

In 1981 the average net phytoplankton biomass in the Bransfield Strait equalled 0.4 g/m^2 dry weight (Kopczyńska and Ligowski 1982) while in 1983/84 it was 3.5 g/m^2 (Kopczyńska and Ligowski 1985). In the Drake Passage the respective values were 5.3 g/m^2 and 1.6 g/m^2. In 1983/84 the highest phytoplankton biomass occurred NW of Anvers Island and in the northern and eastern parts of the Bransfield Strait (average 5.2 g/m^2 dry weight; Kopczyńska and Ligowski 1985). In February and March 1981 maximal biomass was found north of Anvers Island and in the Drake Passage. The area of Anvers Island is particularly interesting since high phytoplankton biomass occurred there in both seasons. The causes of this phenomenon are not known, but it might be that a greater hydrological stability of waters in this area could provide an adequate explanation.

Among phytoplankton species, *Phaeocystis* spp. occurred in 1981 west of Anvers Island and in the Bransfield Strait near the Antarctic Peninsula. In 1983/84 it was abundant northwest of Joinville Island. Other common species such as *Chaetoceros neglectus* and *C. tortissimum* were confined to the near-shore waters around the South Shetland Islands in late summer of 1981, whereas in early summer 1983/84 they were widely distributed in the entire research area. In 1981 *Corethron criophilum* occurred from SW to NE through the central part of the Bransfield Strait, while in 1983/84 it was mainly found in the SW part of the Strait. In both seasons of 1981 and 1983/84 *Nitzschia* spp. were observed in the Drake Passage.

The greatest similarities in the species compositions of net phytoplankton in both seasons are observed along the

1 Department of Polar Research, Institute of Ecology, Polish Academy of Sciences, Dziekanow Lesny, 05-092 Lomianki, Poland

Antarctic Ocean and Resources Variability
(ed. by D. Sahrhage)

SW-NE directions, which corresponds to the direction of currents. Greatest differences are noted from SE toward NW.

The distribution of chlorophyll a was generally similar in both seasons. Large values were recorded NW of Anvers Island and in the neighborhood of Elephant Island. In 1983/84 very large amounts of chlorophyll a were found near Joinville Island, in contrast to very small quantities noted there in 1981. During FIBEX 1981 maximum values of chlorophyll a were 4.4 mg/m^3 and 332 mg/m^2 in a 150 m water column and were observed west of Elephant Island (Lipski 1982). During SIBEX 1983/84 the maximum values of chlorophyll a were 11.2 mg/m^3 and 635 mg/m^2 and were found NW of Anvers (Lipski 1985). The results of El-Sayed (1967) showed that high primary productivity is often associated with the continental shelf.

2.3 Zooplankton

In 1981 variations in the zooplankton distribution were evident. In a region of large krill concentration there were few Copepoda (Rakusa-Suszczewski 1982b). The contribution to the plankton by such species as *Clausocalanus* sp., *Calanus propinquus, C. acutus* and *Thincalanus gigas* decreased from NW to SE. In the Bransfield Strait where there was much *E. superba*, the dominant species were *Oncaea curvata* and *Metridia gerlachei,* numerous in deeper water (100–300 m) below the layer of krill occurrence (Jażdżewski et al. 1982). The mutual exlcusion of large krill concentrations and of Copepoda may result not only from food competition but probably primarily from the fact that copepods are being chased away by krill. The smallest predacious *Cithona* and *Oncaea* do not need much space and are not influenced by the proximity of krill in the way that Calanidae or *Rhincalanus* probably are.

In 1983/84 the distribution of zooplankton did not vary greatly. Krill abundance was low, and *E. superba* and Calanoidea occurred together (Witek et al. 1985; Żmijewska 1985). Salps were very numerous, mainly near Elephant Island and in the Drake Passage. Salps were absent from the waters characterized by T/S values typical of the Weddell Sea water. Indicator species for Weddell Sea water are the copepods *Paralabidocera antarctica* (Żmijewska 1985), the pelagic polychaete *Rhynchonerella bongraini* and the siphonophore *Dimophyes antarctica* (Witek et al. 1985).

The greatest abundance of zooplankton in 1983/84 was noted in the vicinity of the Palmer Archipelago and in the Weddell Sea, while the lowest quantities were observed in the southern part of the Bransfield Strait. There were substantial quantitative differences between the distributions of *E. superba* larvae in the summers of 1981 and 1983/84, while the spatial distribution showed certain similarities (Rakusa-Suszczewski 1984; Kittel et al. 1985). In 1981 larvae of *E. superba* occurred in high numbers in the eastern part of the Bransfield Strait and NE from the Weddell Scotia Confluence beyond the area of main krill concentrations (Rakusa-Suszczewski 1984). The picture thus obtained of krill larvae distribution supports the hypothesis of Marr (1962) on the development and vertical distribution of the early life stages of krill. In 1983/84 the numbers of larvae were small and they occurred in the eastern and northeastern parts of the Bransfield Strait (Kittel et al. 1985). In both seasons, in spite of substantial differences in the average sizes of *E. superba* – in 1981 krill were much larger than in 1983/84 –, the characteristic feature of the populations was a diminution of krill sizes from SW to NE and the occurrence of smaller krill in the Bransfield Strait and on the shelf of the Antarctic Peninsula (Wolnomiejski et al. 1982; Czykieta et al. 1986).

Among the remaining species of krill, *Euphausia frigida* in 1981 was found in the Drake Passage and in the eastern part of the Bransfield Strait (Kittel and Stępnik 1983). In 1983/84 this species was more numerous and occurred mainly in the Drake Passage and near Elephant Island. It was also found north of Anvers Island and in the northern part of the Bransfield Strait (Kittel et al. 1985). *Euphausia triacantha* in 1981 was present exclusively in the northern part of the Drake Passage area investigated by us, while in 1983/84 it was found much farther south in the western Bransfield Strait and in the vicinity of Anvers Island (Kittel et al. 1985). Adult *E. crystallorophias* were found in 1981 north of Anvers and north of Livingstone Island and also in the Bransfield Strait. In 1983/84 larvae of *E. crystallorophias* were mainly distributed in the surface stratum of the southeastern part of the area, while adult forms were present in the central Bransfield Strait. *Thysanoessa macrura* was in 1981 more numerous in the Drake Passage than in the Bransfield Strait, while in 1983/84 it was more or less evenly distributed in the whole area of research (Kittel et al. 1985). The distribution of *Euphausia* species conforms to a scheme (John 1937), but in the given area it may change from year to year.

Energy flow through the populations of *E. superba, Calanus acutus, Calanus propinquus,* and *Rhincalanus gigas* was calculated on the basis of oxygen consumption and on theoretical assumptions pertaining to the parameters of energy balance (Rakusa-Suszczewski and Godlewska 1984; Godlewska 1988). The energy flow through 1 g wet weight was for the respective species: 28.0 cal/day/m^2; 18.2 cal/day/m^2; 16.3 cal/day/m^2; 21.0 cal/day/m^2. In the case of *E. superba,* energy flow is proportional to the biomass and it may be expressed by the equation

$$A = 18.685 \times B^{1.01},$$

where A = Assimilation (cal/day/m^2), B = Biomass (g/m^2). The values obtained for Copepoda and *Euphausia superba* are similar, which supports the idea of space competition and the observed mutual exclusion of large quantities of biomass. Energy flow between primary productivity and the first consumers such as krill and Copepoda depends first of all on the size of consumers' biomass and the age structure of the crustacea (body sizes).

In the investigated area, the biomass and size and age structure of crustacea populations undergo, as observed, tremendous changes from year to year. The calculated values of food assimilation by krill are very small in relation to primary productivity and the biomass of phytoplankton. Krill consumption equals about 3% of the available primary production (Miller et al. 1985), and food is not a decisive factor in the formation of swarms and fluctuations of biomass.

2.4 Krill Abundance

Analysis of the Polish hydroacoustic data of krill biomass indicated that in 1981 there were 2.3 mio. tons of krill (100 g/m^2) in the Bransfield Strait, and 1.2 mio. tons (8.4 g/m^2) in the Drake Passage (Kalinowski 1982). In the coastal zone of the Antarctic Peninsula and of the South Shetland Islands, the biomass concentration exceeded by more than one order of magnitude the biomass in the Drake Passage. Krill biomass in 1983/84 in the Bransfield Strait was estimated at 70 000 t (1.07 g/m^2) and in the Drake Passage at 122 000 t (1.42 g/m^2) (Kalinowski et al. 1985). The difference in krill biomass in the Bransfield Strait between 1981 and 1983/84 amounted to about two orders of magnitude (100 times) while in the Drake Passage it equalled about six times. In 1983/84 a large part of the zooplankton biomass was contributed by the salps. In spite of the differences in the values of krill biomass in both seasons, the areas of large krill concentrations were similar. Krill occurred in the area of the Palmer Archipelago, between King George and Elephant Islands on the islands' shelf and slope, over valleys and bottom elevations. This fact supports the theory of the hydrodynamic causes of krill concentrations and swarms (Rakusa-Suszczewski 1978, 1981).

2.5 Fishes

In 1981, the presence of 23 species of juvenile forms of Notothenidae occurring simultaneously with krill was observed above the shelves of the Antarctic Peninsula, in the Bransfield Strait and on the north side of the South Shetland Islands. Most numerous were *Chionodraco rastrospinosus, Chaenodraco wilsoni,* and *Cryodraco antarcticus,* which feed on krill (Ślósarczyk and Rembiszewski 1982). In the same area in 1983/84 (December, January) 13 species of juvenile fishes were found with *Pleuragramma antarcticum* as the most numerous form (Ślósarczyk and Cielniaszek 1985).

2.6 Birds

Observations of birds in the season 1983/84 showed that their numbers were twice as high as in 1981 (Starck and Wyrzykowski 1982; Starck 1985). This pertains especially to species feeding on krill such as *Pygoscelis* spp., *Daption capense, Diomedea melanophris, Fulmarus glacialoides, Pachyptila* spp., and *Fregata tropica.* Greater numbers of birds in December–January 1983/84 as compared to February–March 1981 seem to be a result of the increased number of feeding trips either due to the commencement of hatching or due to the lower krill abundance in 1983/84. In spite of the trend generally accepted for the Antarctic toward an increase in both numbers and biomass of krill consumers, such as birds and seals, the energy flow and circulation of matter may undergo locally large seasonal or annual fluctuations and even take on the appearance of an ecological disaster (Laws 1985). Variability is one of the basic features of the functioning of the Antarctic marine ecosystem, which is characterized by a short food chain.

3 Conclusions

1. Observations of the distribution of water masses and geostrophic currents repeatedly conducted in the region of the Bransfield Strait and in the southern part of Drake Passage in the summer season revealed a high similarity in the direction of water movements over the season from spring to autumn. It seems that the hydrological situation in winter and early spring, caused by changes in cyclonic activity in the Pacific and Atlantic regions, is responsible for the hydrological situation observed in the region of our study in summer.
2. The hydrological situation in the region of the Bransfield Strait and the southern part of the Drake Passage near South Shetlands and the Antarctic Peninsula is modified by ice conditions, freshwater runoff from the land, and topography of the bottom. The layers of the water masses with similar hydrochemical characteristics and similar origin are stratified, and this stratification varies in the study area with depth. Locally, processes of up- and downwellings are observed, along with whirling movements.
3. Large dynamics of water masses in the study region enhances retention and formation of aggregations of phyto- and zooplankton, especially krill, on the leeward side of islands, in bays of islands, above the shelf slope, bottom mounds, slopes of shelf troughs, sometimes on the bottom, and also at the junction of water masses of different origin in the zone of whirling movements. Temperature, salinity, and nutrient content do not have a direct effect on the formation of krill aggregations, but they can be indicators of hydrodynamic processes enhancing the formation of such aggregations.
4. Variation in krill biomass in the study region is related with seasonal changes in the life cycle (multiannual) of *E. superba*, but this variation is smaller by two orders of magnitude than the extreme changes in biomass observed in this region and caused by immigration of the krill. Krill seem to immigrate mostly from the Bellingshausen Sea. Under suitable hydrodynamic condi-

Table 1. Mean air temperature on King George Island and krill biomass evaluated in the Bransfield Strait and south Drake Passage during Polish expeditions

	1975	1976	1977	1978	1979	1980	1981	1982	1983	1984	1985	1986	1987
T°C Mean year Arctowski				−1.9	−1.0	−3.6	−2.0	−1.1	−1.1	−1.2	−0.6		
T°C Mean winter 06, 07, 08, 09				−6.5	−4.0	−8.3	−5.1	−4.9	−3.8	−2.8	−2.4	−6.9	
T°C Mean year Bellingshausen			−2.8	−2.6	−1.8	−4.0	−2.6	−1.8	−1.8	−1.8	−1.1		
T°C Mean winter 06, 07, 08, 09			−7.1	−6.8	−4.7	−9.4	−5.9	−5.3	−4.4	−3.5	−3.2		
Evaluation of krill biomass	Moderate		High		High		High			Low			Moderate

tions, krill aggregations can stay in one geographical area for a rather long time, which can account for formation of distinct cohorts. Changes in the krill biomass in the study region tend to take more than 1 year. That is, the summer season with a very high biomass is not followed by a season with a very low biomass, and conversely (Table 1). In the second half of the 1970's until 1981, krill was more abundant, and this was a period of cooler years as compared with the years which followed, which were warmer, and when krill occurred in low numbers. In 1986, temperatures were low again in the study area and krill biomass was higher than in earlier years.

5. The species composition of phyto- and zooplankton varied as a result of seasonal succession, but also due to changes caused by the inflow of waters from the Pacific, Bellingshausen Sea, and Weddell Sea into the study region, carrying different species known as indicators of these waters. Not only species but also the age structure and population sizes of *E. superba* and dominant Copepoda can be indicators of the origin and similarity of water masses. Krill individuals occurring above the shelf were typically younger than outside the shelf.
6. Large aggregations of krill exclude phytophageous Copepoda, which seems to be a result of competition for food and a mechanical effect of large, dense krill aggregations on Copepoda. Salps, if present, typically occur deeper, under krill aggregations, or in the period of its relative scarcity.
7. In the region of the Bransfield Strait under study and in the southern part of the Drake Passage, the estimate of krill biomass consumed by penguins and flying birds in the summer of 1983/84 was several times higher than the krill biomass estimated by hydroacoustic methods, which would also imply that krill was permanently inflowing from adjacent areas.
8. A marked interannual inconstancy of the biomass of phytoplankton, zooplankton, fish, and birds was recorded in the study region, but this inconstancy should not be confused with the instability characteristic of this fragment of the marine ecosystem of the Antarctic. Multiannual trends locally observed in this region can be related with global trends of changes in production and biomass.

References

Anonymus (1983) First Post-FIBEX Hydrographic Data Interpretation Workshop Hamburg, F.R.G. 20–26 September 1982. Pol Polar Res 4 (1–4):155–162

Czykieta H, Kittel W, Witek Z, Wolnomiejski N (1986) Biological characteristics of the population of *Euphausia superba* Dana from the southern part of the Drake Passage and in the Bransfield Strait carried out during the BIOMASS-SIBEX expedition (December 1983–January 1984). Pol Polar Res 7 (3):261–275

El-Sayed SZ (1967) On the productivity of the southwest Atlantic Ocean and the waters of the Antarctic Peninsula. In: Biology of the Antarctic Seas, vol 3. Antarct Res Ser 11:15–47

Godlewska M (1988) Energy flow through the main phytophagous species in the Bransfield Strait and Drake Passage during SIBEX. IEMBA Conf, in press

Grelowski A, Tokarczyk R (1985) Hydrological conditions in the region of Bransfield Strait and southern part of Drake Passage in the period from December 10, 1983 and January 8, 1984 (BIOMASS-SIBEX). Pol Polar Res 6 (1–2):31–41

Jażdżewski K, Kittel W, Łotocki K (1982) Zooplankton studies in the southern Drake Passage and in the Bransfield Strait during the austral summer (BIOMASS-SIBEX, February–March 1981). Pol Polar Res 3 (3–4):203–242

John DD (1937) The Southern species of the genus *Euphausia*. Discovery Rep 14:195–324

Kalinowski J (1982) Distribution and stocks of krill in the Drake Passage and the Bransfield Strait, during the BIOMASS-FIBEX expedition 1981. Pol Polar Res 3 (3–4):243–251

Kalinowski J, Godlewska M, Klusek Z (1985) Distribution and stock of krill in the Bransfield Strait and the Drake Passage during December 1983–January 1984 (BIOMASS-SIBEX). Pol Polar Res 6 (1–2):151–158

Kittel W, Stępnik R (1983) Distribution of *Euphausia crystallorophias, E. frigida, E. triacantha* and *Thysanoessa macrura* (Crustacea, Euphausiacea) in the southern Drake Passage and Bransfield Strait. Pol Polar Res 4 (1–4):7–19

Kittel W, Witek Z, Czykieta H (1985) Distribution of *Euphausia frigida, Euphausia crystallorophias, Euphausia triacantha* and *Thysanoessa macrura* in the southern part of Drake Passage and in the Bransfield Strait during the 1983–1984 austral summer (BIOMASS-SIBEX). Pol Polar Res 6 (1–2):133–149

Kopczyńska E, Ligowski R (1982) Phytoplankton abundance and distribution in the southern Drake Passage and the Bransfield Strait in February–March 1981 (BIOMASS-FIBEX). Pol Polar Res 3 (3–4):193–202

Kopczyńska E, Ligowski R (1985) Phytoplankton composition and biomass distribution in the southern Drake Passage, the Bransfield Strait and the adjacent waters of the Weddell Sea in December 1983–January 1984 (BIOMASS-SIBEX). Pol Polar Res 6 (1–2):65–77

Laws RM (1985) Introduction: Antarctic biology, summary and conclusions, 1974–1983. In: Siegfried WR, Condy PR, Laws RM (eds) Antarctic nutrient cycles and food webs, vols 1–3. Springer, Berlin Heidelberg New York Tokyo

Lipski M (1982) The distribution of chlorophyll a in relation to the water masses in the southern Drake Passage and the Bransfield Strait (BIOMASS-FIBEX) February–March 1981. Pol Polar Res 3 (3–4):143–152

Lipski M (1985) Chlorophyll a in the Bransfield Strait and the southern part of Drake Passage during BIOMASS-SIBEX (December 1983–January 1984). Pol Polar Res 6 (1–2):21–30

Marr JWS (1962) The natural history and geography of the Antarctic krill (*Euphausia superba*). Discovery Rep 32:33–463

Miller DGM, Hampton I, Henry J (1985) The relationship between krill food requirements and phytoplankton production in a sector of the southern Indian Ocean. In: Siegfried WR, Candy PR, Laws RM (eds) Antarctic nutrient cycles and food webs. Springer, Berlin Heidelberg New York Tokyo, pp 362–371

Rakusa-Suszczewski S (1978) 1st Polish Antarctic marine research expedition on R/V *Profesor Siedlecki* and M/T *Tazar* in 1975/76. Pol Arch Hydrobiol 25 (3):505–510

Rakusa-Suszczewski S (1981) Polish Antarctic expedition 1975/76. BIOMASS, vol II. Selected contributions to the Woods Hole Conference in Living Resources of the Southern Ocean 1976. SCAR/SCOR, Cambridge, England

Rakusa-Suszczewski S (1982a) Report on the R/V *Profesor Siedlecki* expedition to the Antarctic in 1981 during the international BIOMASS-FIBEX programme. Pol Polar Res 3 (3–4): 137–141

Rakusa-Suszczewski S (1982b) The relationship between the distribution of plankton biomass and plankton communities in the Drake Passage and the Bransfield Strait (BIOMASS-FIBEX, February–March 1981). Mem Natl Inst Polar Res Spec Issue No 27. Proc BIOMASS Colloq 1982:77–83

Rakusa-Suszczewski S (1984) Krill larvae in the Atlantic sector of the Southern Ocean during FIBEX 1981. Polar Biol 3:141–147

Rakusa-Suszcuewski S, Godlewska M (1984) Energy flow through krill aggregations in Drake Passage and Bransfield Strait. J Crust Biol 4:198–205

Rakusa-Suszczewski S, Lipski M (1985) Report on the R/V *Profesor Siedlecki* expedition to the Antarctic during the BIOMASS-SIBEX in 1983-1984. Pol Polar Res 6 (1–2):7–19

Ślósarczyk W, Cielniaszek Z (1985) Postlarval and juvenile fish (Pisces, Perciformes and Myctophiformes) in the Antarctic Peninsula region during BIOMASS-SIBEX 1983/84. Pol Polar Res 6 (1–2):159–165

Ślósarczyk W, Rembiszewski JM (1982) The occurrence of juvenile Notothenioidei (Pisces) within krill concentrations in the region of the Bransfield Strait and the southern Drake Passage. Pol Polar Res 3 (3–4):299–312

Starck W (1985) Seabird observations in the region of the South Shetland Islands and South Orkney Islands during BIOMASS-SIBEX (December 1983–January 1984). Pol Polar Res 6 (1–2): 167–173

Starck W, Wyrzykowski R (1982) Seabird observations in the southern Drake Passage and the Bransfield Strait (BIOMASS-FIBEX programme) in February–March 1981. Pol Polar Res 3 (3–4): 313–332

Witek Z, Kalinowski J, Grelowski A, Wolnomiejski N (1980) Rozmieszczenie i charakterystyka skupień kryla w rejonie Półwyspu Antarktycznego w sezonie letnim 1977 r. Studia i Materiały. Wydawnictwo Morskiego Inst Rybackiego Ser A No 25:5–91

Witek Z, Kittel W, Czykieta H, Żmijewska M, Presler E (1985) Macrozooplankton in the southern Drake Passage and in the Bransfield Strait during BIOMASS-SIBEX (December 1983–January 1984). Pol Polar Res 6 (1–2):95–115

Wolnomiejski N, Czykieta H, Stępnik R, Jackowska H (1982) Biological characteristics of *Euphausia superba* Dana in the Southern Drake Passage and the Bransfield Strait in February–March 1981 (BIOMASS-FIBEX). Pol Polar Res 3 (3–4):259–271

Żmijewska M (1985) Copepoda in the southern part of Drake Passage and in Bransfield Strait during early summer 1983/84 (BIOMASS-SIBEX, December 1983–January 1984). Pol Polar Res 6 (1–2):79–93

A Concept of Seasonal Variation of Krill (Euphausia superba) Distribution and Abundance West of the Antarctic Peninsula

V. SIEGEL[1]

Summary. Data from six expeditions to the Antarctic Peninsula (from 1977 to 1986, in the months of October through June) were analyzed to obtain a general picture of the seasonal variability of krill distribution and abundance. In summer, during the spawning season the krill stock shows a spatial succession of developmental stages along the Peninsula from inner shelf (juveniles) to oceanic waters (adults). A temporal succession of abundance is observed from early spring (mostly juveniles) to summer (all stages) in occurrence of developmental stages and absolute krill abundance in the area. Within a given year average krill abundance differs by more than an order of magnitude, depending on the month of sampling, thus masking a possible interannual variability in the Peninsula region. Seasonal fluctuation in abundance and spatial distribution patterns are discussed as possible effects of seasonal changes in the location of the atmospheric pressure zones (influencing currents and drift of krill) as well as migration of krill.

1 Introduction

The coastal shelf region of the Antarctic Peninsula is known to be rich in primary production and is one of the principal spawning and feeding grounds of the Antarctic krill, *Euphausia superba* (Marr 1962; Hempel et al. 1979; Makarov 1979; Witek et al. 1980; Siegel 1985, 1986a,b). The structure of the marine ecosystem in this part of the Antarctic Ocean has been intensively studied by many national and international programs, beginning with the *Discovery* expeditions in the 1920's.

The physical oceanography of the Bransfield Strait was first described by Clowes (1934) and reanalyzed during FIBEX (BIOMASS 1982, 1983). Studies further south of the Peninsula are scarce (e.g., Kock and Stein 1978; Witek et al. 1981; Makarov et al. 1982) and contribute little to the knowlege of dynamic processes.

Although biological studies have been conducted on a large scale, the sampling strategies and data treatment have often only permitted descriptive results. A further problem is manifest by the relatively short periods during which studies have been conducted. In addition, results are very often incompatible due to different sampling strategies and equipment. Therefore almost nothing is known about seasonal or interannual variability of even the most dominant components of this system, and more specifically krill.

During the last decade, six expeditions of the Federal Republic of Germany took place in the vicinity of the Antarctic Peninsula. This presents an opportunity to analyze a great amount of data, collected during almost every period of the annual seasonal cycle. Before any opinion on the extent of interannual variability of the krill-related system can be formed, it is necessary to consider possible seasonal fluctuations in krill abundance and distribution. The major objective of the present analysis is therefore

a) to ascertain if spatial mesoscale distribution patterns of the krill stock west of the Peninsula occur regularly,
b) to elucidate any temporal changes in this pattern, and
c) to discuss how environmental factors might influence or even cause variability in the system.

2 Material and Methods

During the 1977/78 and 1980/81 seasons the area under investigation was surveyed several times between late November and March. During other years single surveys were conducted each season, concentrating research in short periods in February 1982, October/November 1983, November 1984, March/April 1985 and May/June 1986 (Hempel et al. 1979; Anonymous 1982, 1986; Hempel and Heywood 1982; Fütterer 1984; Hempel 1985). Most important advantage of the complete data set is that standard fishing gear and standardized survey methods were used throughout.

For all krill sampling the Rectangular Midwater Trawl (RMT 1+8) (Baker et al. 1973; Pommeranz et al. 1982) was fished randomly or on standard transects (i.e., target or aimed hauls on krill concentrations detected acoustically were excluded from the analysis). Only random oblique RMT tows were considered for quantitative analysis to avoid different catchability factors of various nets or procedures. Sampling depths ranged from 200 m to the surface, except in the 1977/78 season, when maximum sampling depth was 140 m. In February 1982 the multiple RMT 1+8

1 Institut für Seefischerei, Bundesforschungsanstalt für Fischerei, Palmaille 9, 2000 Hamburg 50, FRG

Antarctic Ocean and Resources Variability
(ed. by D. Sahrhage)

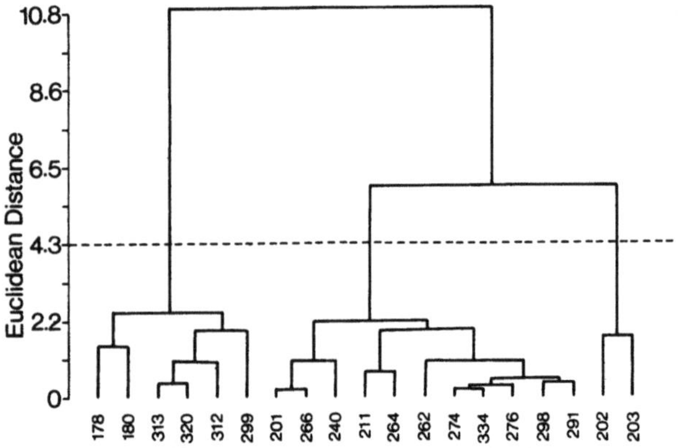

Fig. 1. Dendrogram of the cluster analysis computing similarity of length class and maturity stage frequencies between krill hauls in March/April 1985

(Roe and Shale 1979; Roe et al. 1980) was fished in three vertical layers down to 200 m. In shallow areas the net was opened close to the bottom. Normal towing speed of the net was between 2.5 and 3.0 kn so as to optimize the mouth opening of the net.

Total lengths of sampled krill were measured to the mm below from the anterior margin of the eye to the tip of the telson (Mauchline 1980; Siegel 1982a). In small samples up to 300 specimens were measured in total, larger catches were divided into subsamples of this size.

Maturity stages of individual krill were identified using Makarov and Denys' (1982) classification. Mature or spawning animals refer to stage IIID for females and IIIB for males; the term adult includes all stages III.

For comparative purposes and prior to computing abundance values for several hauls, krill catch frequencies were adjusted to a standard filtered volume of 1000 m^3. Statistical analyses were carried out using the CLUSTAN and SPSS 9 program packages.

3 Results

3.1 Spatial Distribution

The spatial distribution of the krill population was investigated using a cluster analysis to compare similarity between stations. The data set was divided on a monthly basis to allow an analysis of seasonal variability in the geographical distribution pattern. In this manner, eight groupings of stations were selected for various periods of the year. A similarity matrix was computed based on the frequency of each krill length class and maturity stage as station parameters. The hierarchical fusion of clusters was preferred using Ward's method to link homogeneous clusters and the Euclidean distance coefficient was applied for the diversity analysis. Since this method is sensitive to high dominant values, the size and maturity stage abundances were transformed to relative percentage frequency distributions before considering these for the computations of the similarity matrix. Results of the cluster analysis were plotted as dendrograms; examples are given in Figs. 1 and 2. A minimum similarity of at least 60% among clustered groups of stations is interpreted to be biologically meaningful.

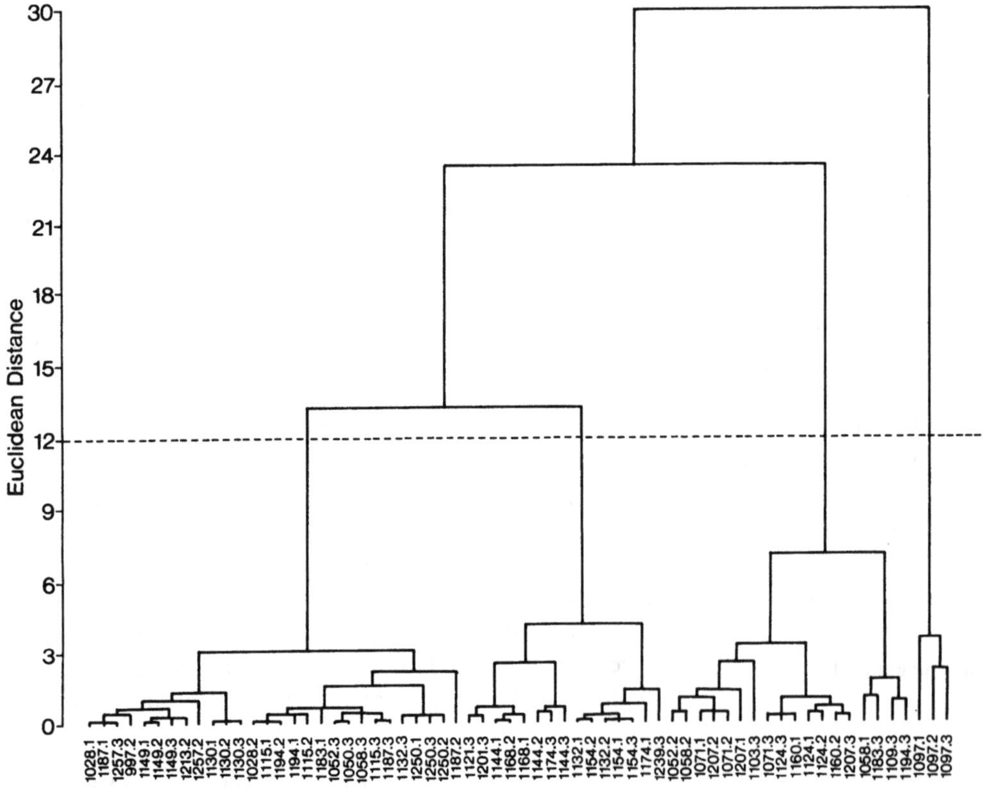

Fig. 2. Dendrogram of the cluster analysis computing similarity of length class and maturity stage frequencies between krill hauls in February 1982

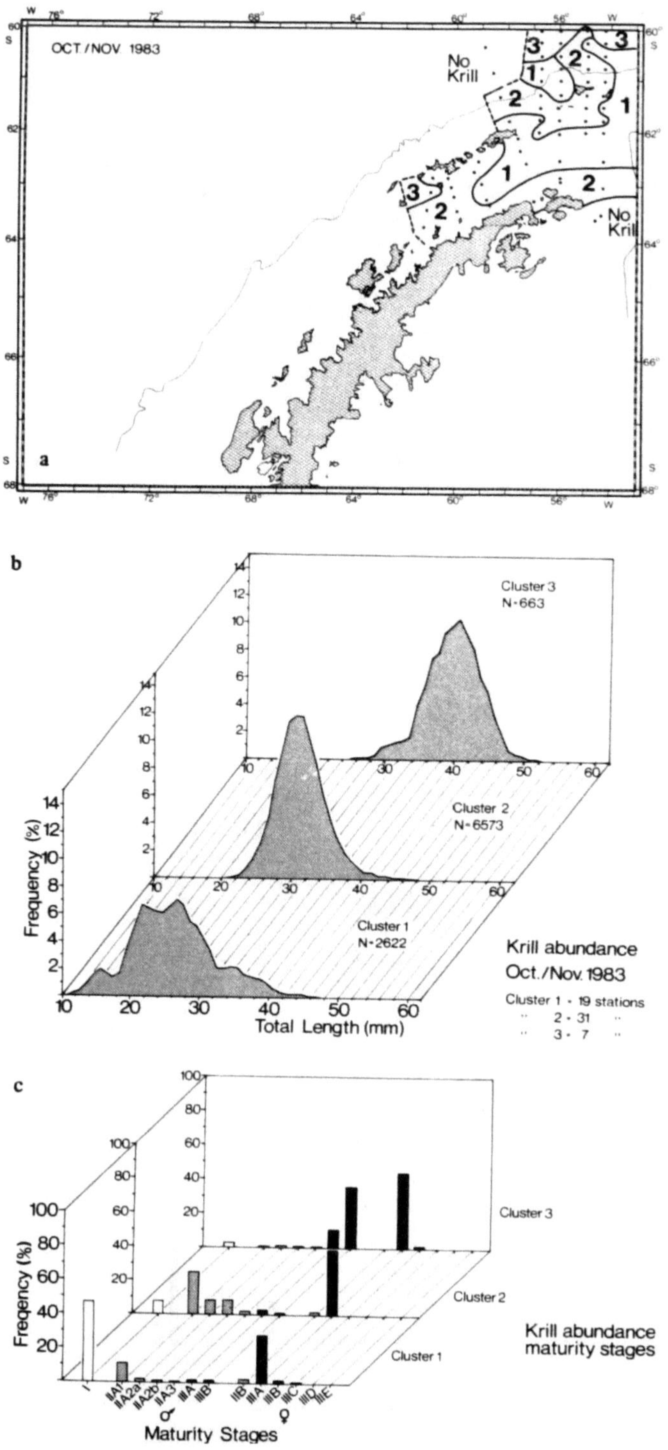

Fig. 3a–c. Similarity of stations along the Antarctic Peninsula derived from cluster analysis for October/November 1983. **a** Spatial distribution of station clusters. **b** Accumulated krill length frequency for each cluster. **c** Krill maturity stage composition for each cluster

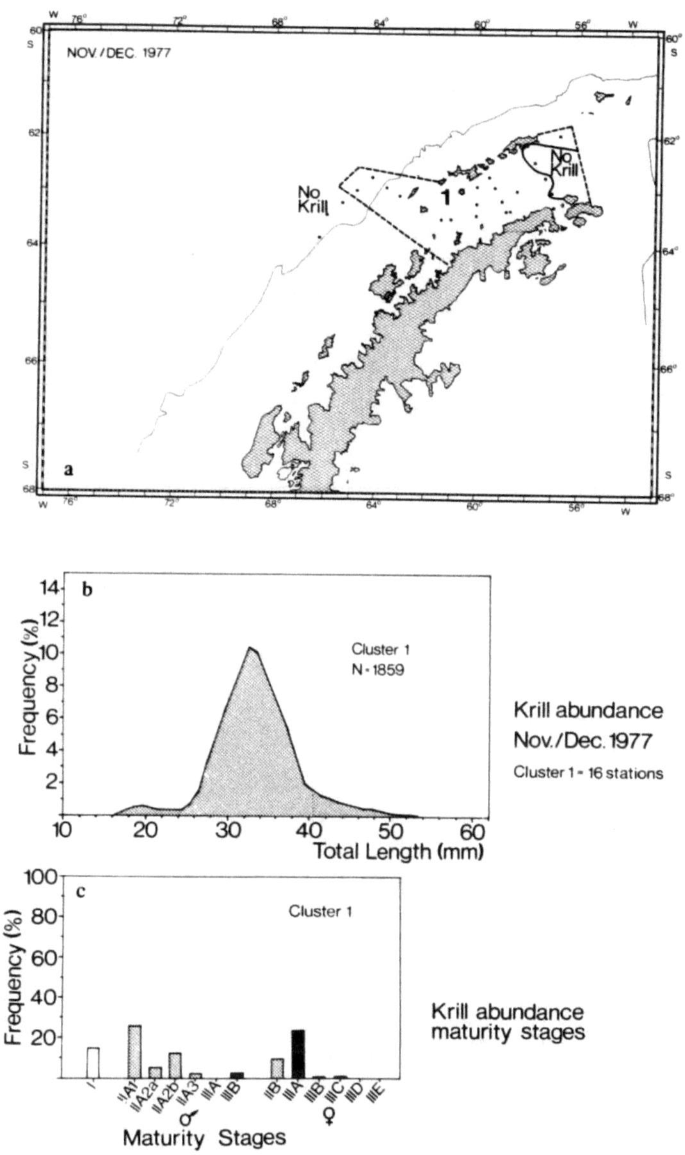

Fig. 4a–c. Similarity of stations in November/December 1977

Figures 3 to 10 reflect the interpretation of the dendrograms and the linkage of similar stations to clusters of higher order. The above figure shows the horizontal distribution of size and maturity stage station affinity groupings obtained from the cluster analysis. The respective figure in the middle gives the overall size composition of krill, and the lower figure of the mean maturity stage frequency distribution in each of the geographical cluster groupings.

In October/November 1983 three distinct groups of krill occurred in the area under investigation (Fig. 3). Cluster 1 represented stations of small krill, consisting mainly (47%) of juveniles. This cluster covered most of the central Bransfield Strait. Large areas of the western and northern Bransfield Strait, as well as the waters around Elephant Island, were dominated by subadults and pre-mature krill of 30 mm mean size. The westernmost and northern stations beyond the continental slope were characterized by adults of about 40 mm length. This group, however, occurred only in low numbers. Mature krill and size classes larger than 50 mm were absent throughout the entire area. No krill was found in the oceanic waters of the Drake Passage and in the Weddell Sea water.

In November/December 1977 the distribution pattern of krill was rather homogeneous (Fig. 4). No separation into

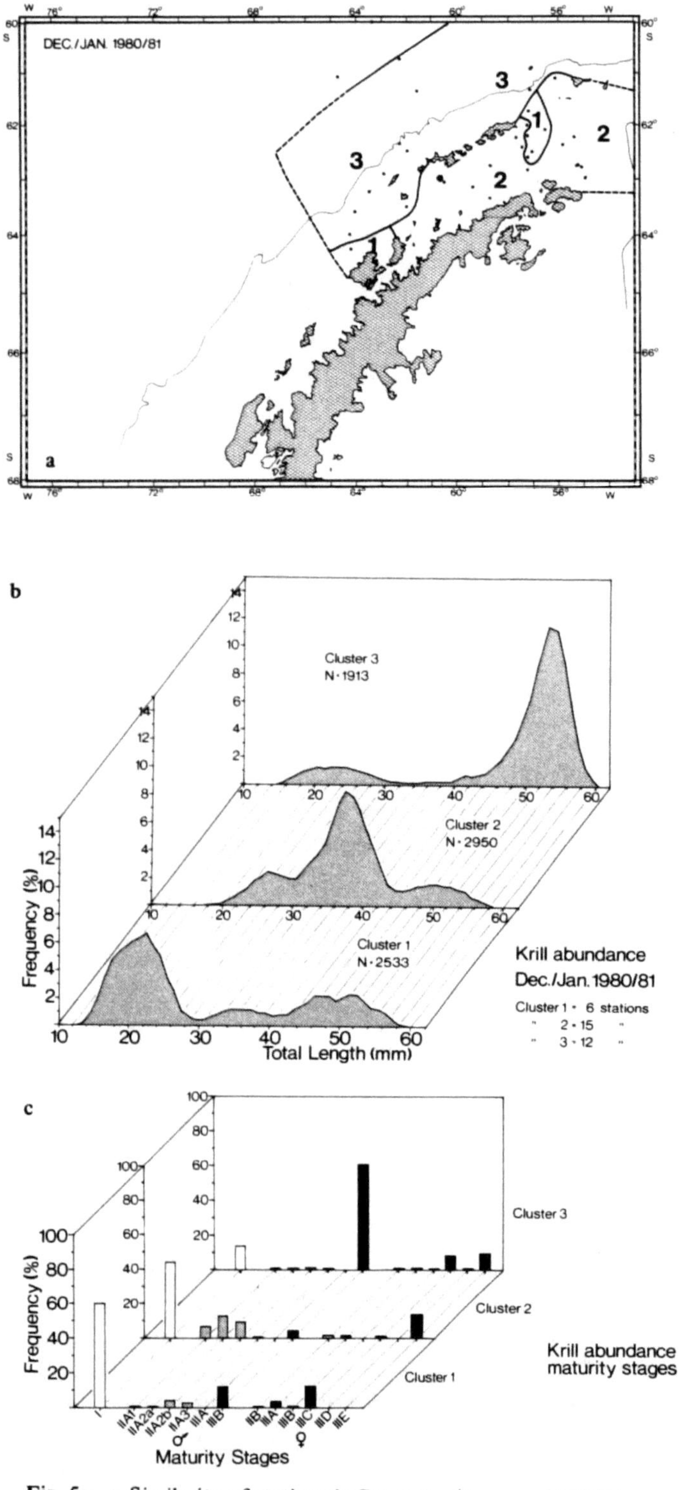

Fig. 5a–c. Similarity of stations in December/January 1980/81

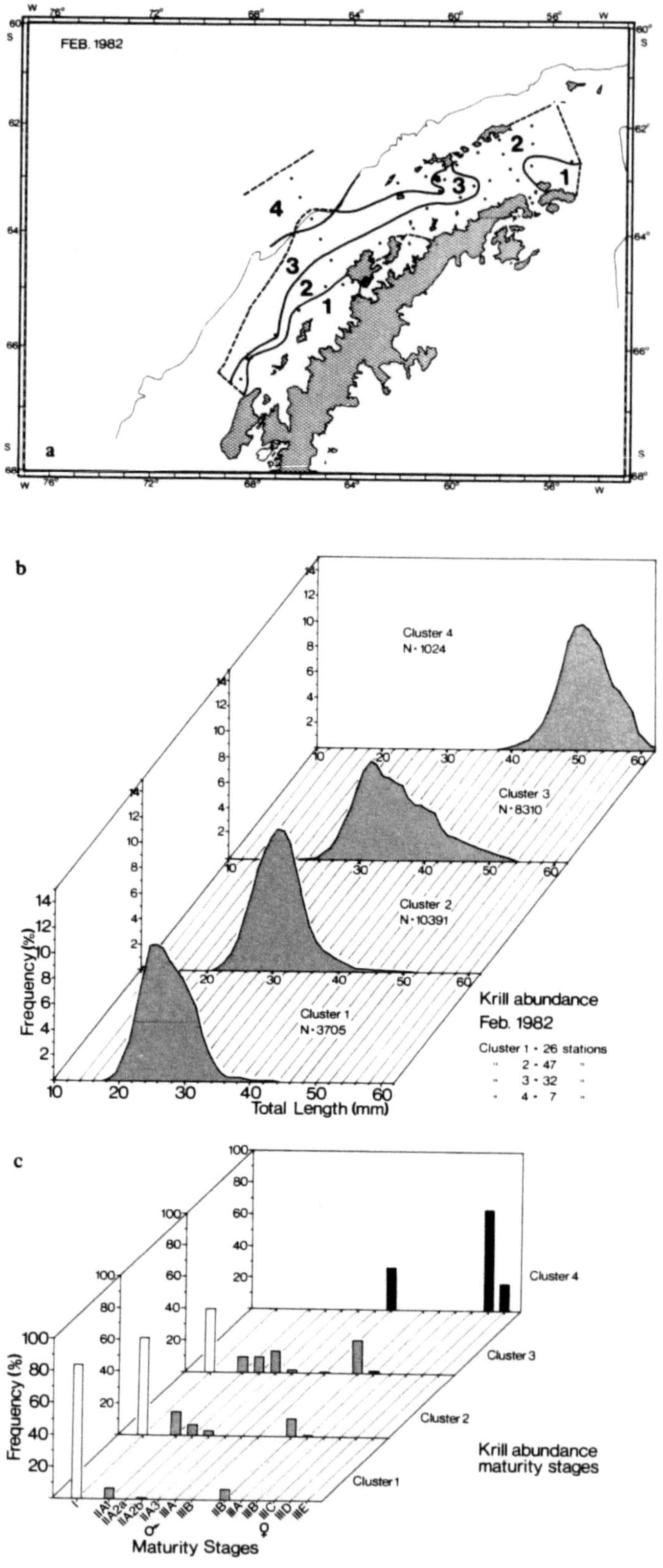

Fig. 6a–c. Similarity of stations in February 1982

different clusters was observed in the Bransfield Strait. Total krill abundance was very low and stations under the influence of Weddell Sea water yielded no krill. The size frequency distribution and the maturity stage composition of November/December closely resemble cluster 2 in October 1983 of medium size krill, which at that early time of the season were by far the most abundant group.

In December/January 1980/81 again three large station clusters were evident (Fig. 5). In the eastern part of the Bransfield Strait juvenile krill made up 60% of the krill stock in cluster 1, which, however, included only very few stations. In most areas of the Bransfield Strait medium-sized krill at all stages of maturity dominated the samples. Cluster 3 was situated in the Drake Passage and on the outer shelf areas of the Palmer Archipelago. Here, the krill population consisted almost exclusively of adult mature (December) and already postspawning (January) specimens.

A horizontal separation into four clusters was observed in February 1982 during the spawning season (Fig. 6). A continuous decrease of small, juvenile krill abundance was evident from nearshore to offshore waters. In the shelf waters of the Palmer Archipelago and the Bransfield Strait the proportion of juveniles was as high as 60% or reached even more than 80% in cluster 1, while the area along and beyond the continental slope was inhabited by mature, spawning krill (cluster 4).

In February/March 1978 the situation was similar to February 1982 with slight differences. Most juveniles occurred in coastal waters and in the Bransfield Strait (Fig. 7), but they did not exceed 40% and the stock was already mixed with great numbers of subadults. The older development stages dominated further offshore and in the northern approaches of the Weddell Sea (cluster 2). As in February 1982, large spawning krill were distributed near or beyond the continental slope. In cluster 3 more than 80% were mature animals.

In February/March 1981 the spawning season of krill was already completed, no gravid females were encountered, and most of the adults had reached the spent phase (Fig. 8). Almost the entire area was characterized by cluster 2 which represented medium- and large-sized animals. During that time adult krill also occurred in nearshore waters (cluster 3) where they had not been caught during the spawning period. This station cluster in coastal waters which was dominated by juvenile krill was no longer found during the late season. After the spawning season postlarval krill were almost absent from the oceanic waters of the Drake Passage.

A similar tendency in krill distribution was detected in March/April 1985 (Fig. 9). Krill were very rare in offshore areas. In the Bransfield Strait and south of Elephant Island all developmental stages occurred in cluster 2, while in the southernmost areas off the Peninsula and north of Elephant Island the proportion of large and adult krill increased in cluster 3. This reflected the former clustering of the spawning stock. The low number of juvenile krill in the entire area was quite obvious and these were mainly restricted to the eastern stations near the limits of the Weddell Sea (cluster 1).

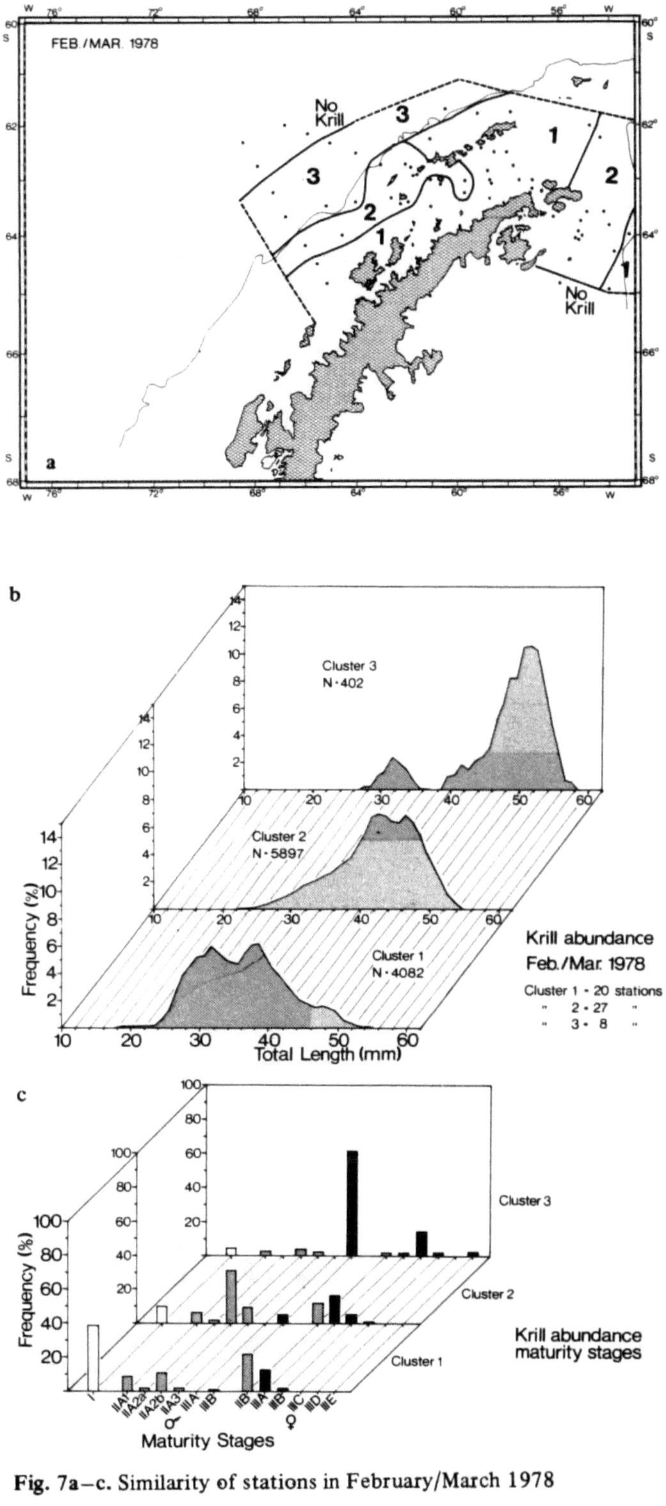

Fig. 7a–c. Similarity of stations in February/March 1978

In winter (May/June) 1986 only two clusters were evident from the computational analysis (Fig. 10). Cluster 1 in the Bransfield Strait, Gerlache Strait, and along the coast of the northwestern Peninsula was dominated by small and medium-sized juveniles and subadults. Cluster 2 covered the area north of Elephant Island and the outer shelf areas. The stock composition was only slightly different from cluster 1: the mode of the length frequency distribution shifted to larger classes, which indicated a higher percentage of adult krill at these stations. Krill were absent in most parts of the Drake Passage as they had been in autumn 1984/85.

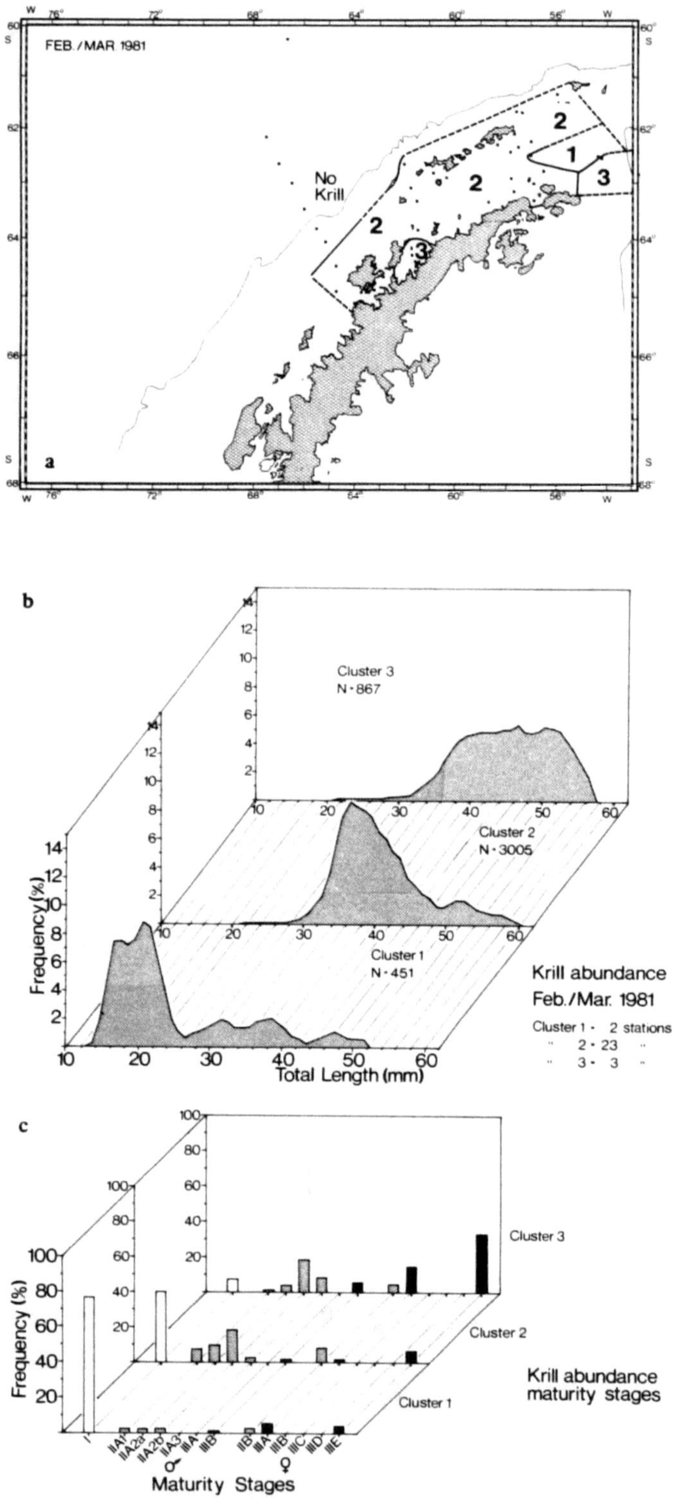

Fig. 8a–c. Similarity of station in February/March 1981

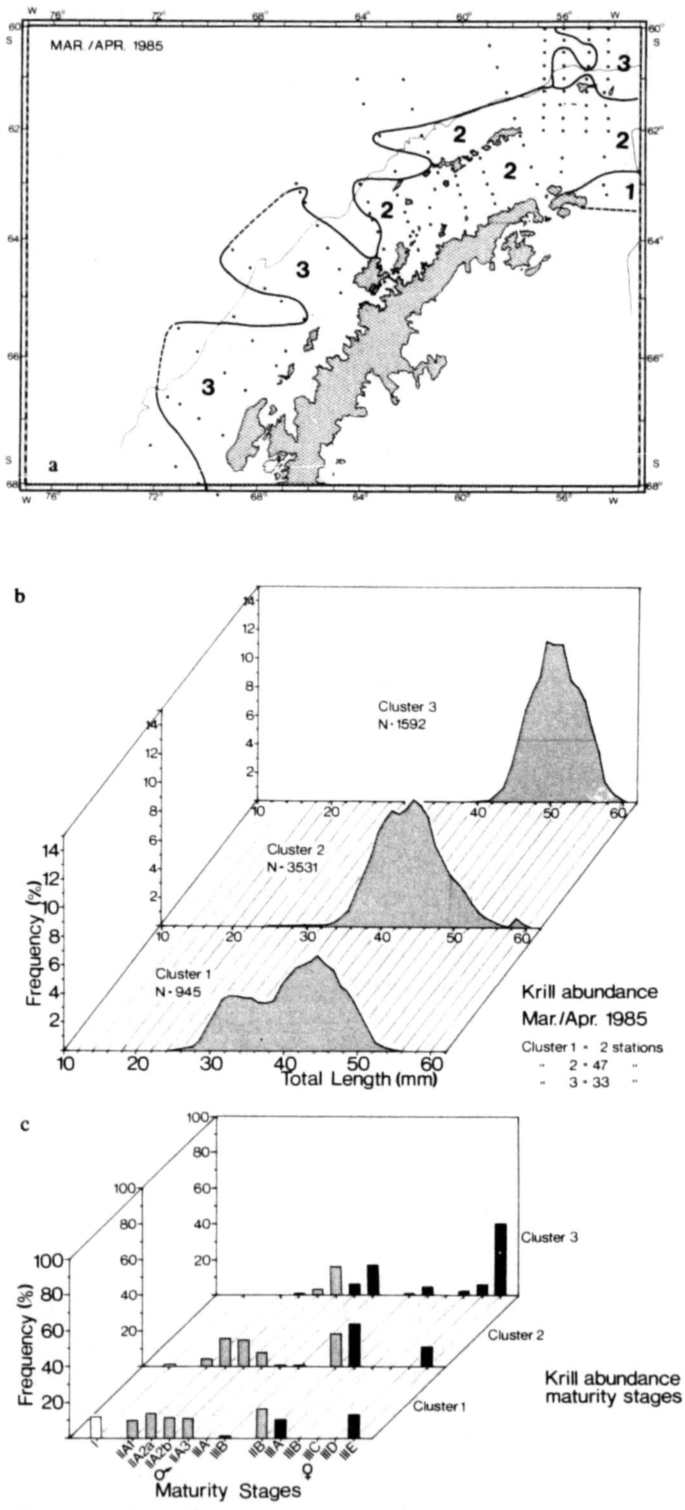

Fig. 9a–c. Similarity of stations in March/April 1985

3.2 Abundance

So far only relative frequencies of stock composition parameters have been analyzed to describe krill spatial distribution patterns. Changes in the occurrence of krill in the Drake Passage have been mentioned above and this gave a first impression of possible variation in absolute krill abundance values. To study the seasonal fluctuation of krill abundance all data which were randomly collected in a limited survey area were gathered and examined on a monthly time basis. The area chosen enclosed waters between 60 and 64°S and extended from 53 to 64°W, which includes Elephant Island and the Bransfield Strait. This region was surveyed regularly, whereas the southwestern shelf of the Antarctic Peninsula was excluded because of irregular sampling performance resulting from the variable ice cover. The average abundance within each time interval was calculated and plotted against the season (Fig. 11).

Although some of the results were derived from different years, the annual tendency of increasing and decreasing krill abundance with season was obviously the same in different years. Until early December mean krill abundance was low. Krill then occurred in greater number and from January to early March abundance was at its highest. From the beginning of March abundance decreased rapidly. The lowest level of abundance was observed in early winter, that is May/June. There was only one exception from this general trend.

In October/November 1983 the highest mean abundance was calculated for the whole data set. The mean of 62 stations exceeded 330 krill/1000 m^3. This does not fit the trend described above and will be discussed in more detail below. Another small set of data, however, was available to test the seasonal variability. During several years a standard transect was worked south of the survey area off Anvers Island into the Drake Passage. At least during the post-spawning autumn season krill abundance exhibited a similar decline during March (Fig. 12). Although the value for March 1978 lies apart from the other means, it is already much lower than that for the preceding February. This late decrease of krill abundance in 1978 might be due to a delayed spawning season until March, whereas, for example, in 1981 the spawning had already terminated before February. In 1981 krill attained minimum concentration by the middle of March.

4 Conclusion and Discussion

In the Bransfield Strait and adjacent areas mesoscale krill distribution has been intensively studied during the past years (e.g., Jazdzewski et al. 1978; Kock and Stein 1978; Makarov 1979; Witek et al. 1981; Quetin and Ross 1984; Siegel 1985, 1986a). Further investigations have contributed additional data on a more local scale (e.g., Rakusa-Suszczewski and Stepnik 1980; Kittel 1980; Jackowska 1980; Stepnik 1982; Nast 1986). All these studies, however, considered only a short period of the annual cycle, namely the high summer season. In connection with the results obtained during the present analysis, it seems reasonable to describe a generalized dynamic model of krill abundance and distribution off the Antarctic Peninsula for almost the entire annual cycle.

Few data are available from early spring (October/November), since most of the area southwest of the Bransfield Strait is covered by pack ice at that time. Krill abundance is low and the stock consists mostly of small juvenile and subadult animals. The waters north of the South Shetland Islands are almost devoid of krill.

A further indirect indication of low numbers of large krill during spring, as described above, is given by the obser-

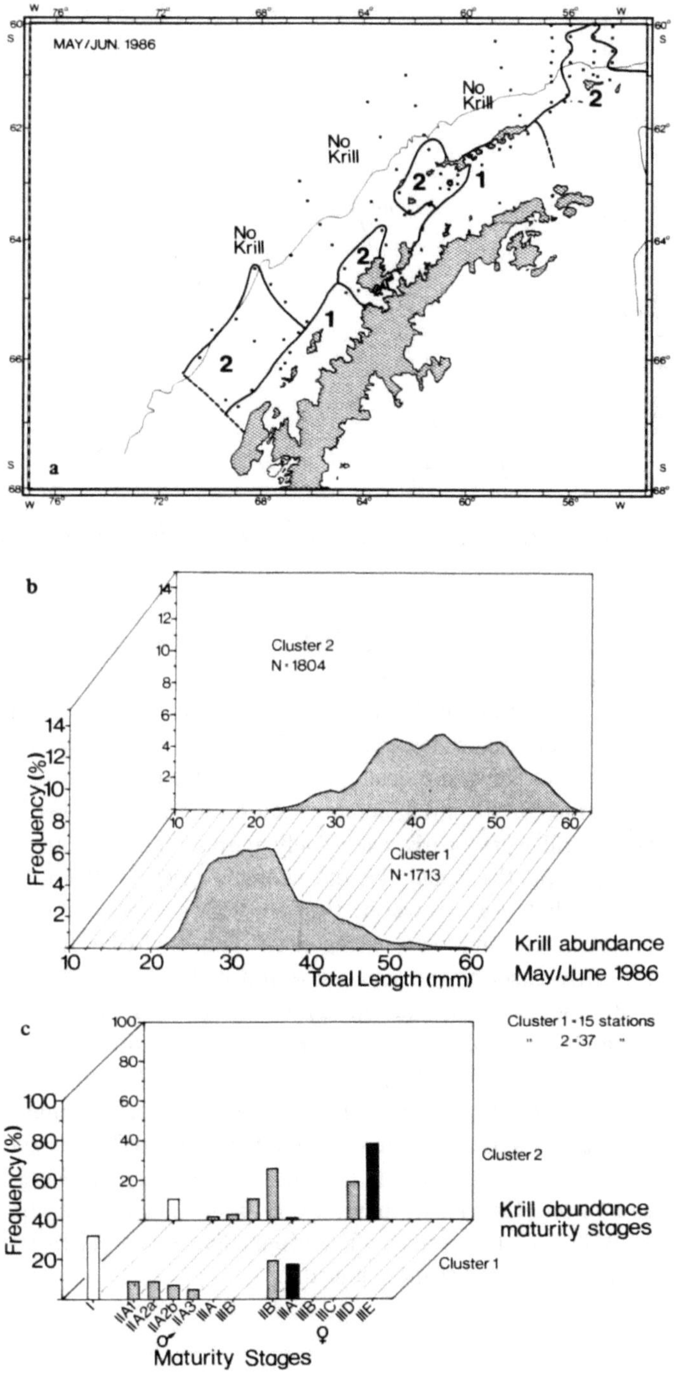

Fig. 10a–c. Similarity of stations in May/June 1986

vations of whalers. They draw a distinction between small blue whale and large fin whale krill. According to Peters (1955) catches during November consisted almost exclusively of blue whales, while fin whales did not occur before the end of December in the whaling grounds. The temporal succession of krill size groups therefore may have led to the impression of different prey sizes for blue and fin whales. Peters (1955), however, was able to demonstrate that in the later summer season the same krill sizes are found in the stomachs of both whale species. Until

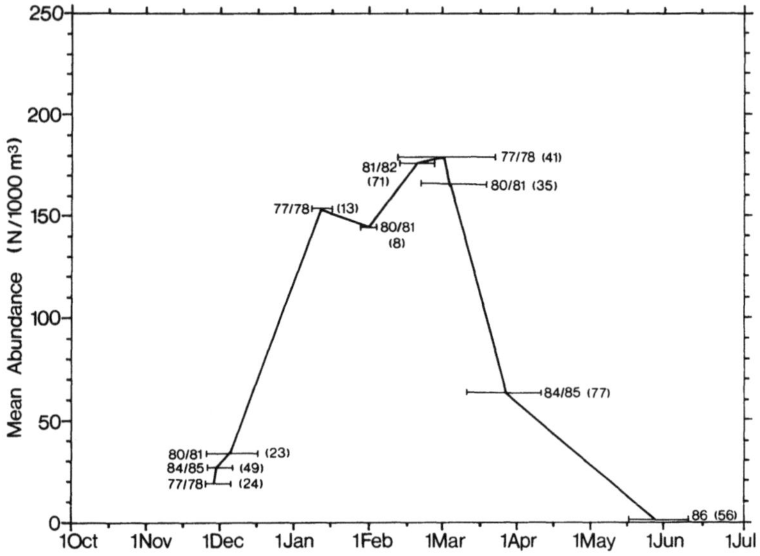

Fig. 11. Mean monthly abundance (N/1000 m³) versus time of the season for the Bransfield Strait and adjacent areas

the end of December the krill stock also consists of all size and age groups in the area.

The situation in November still reflects the biological winter conditions reported by Stepnik (1982), Guzman (1983) and Heywood et al. (1985). This would be particularly true for the stock composition during October 1983, when krill were nearly exclusively represented by a single age group, the small 1+ old animals (Siegel 1986a,b). During that year, however, the very high numbers of krill found in the area were exceptional. This might be due to the early onset of the summer season in 1983. In October 1983 the pack ice had already retreated from the whole research area and adjacent waters. Possibly the drift of krill into the area started much earlier than usual. Following this early onset, krill abundance again decreased dramatically and Rakusa-Suszczewski and Lipski (1985) reported rather low concentrations of small krill in summer. Witek and Kittel (1985) also found a very poor spawning success of krill in that season. Therefore the season 1983/84 could be regarded as more unusual.

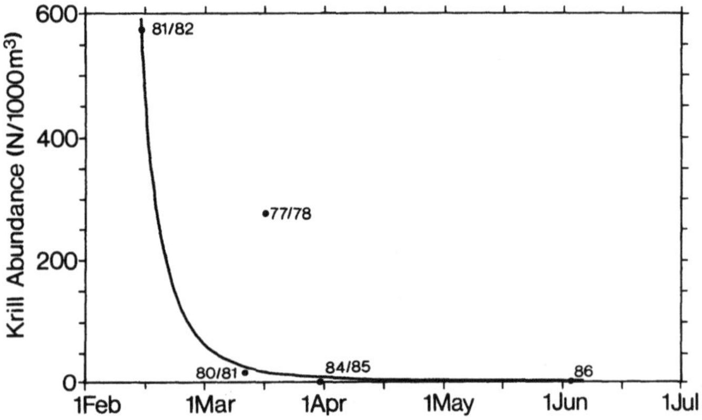

Fig. 12. Autumn decline in krill abundance derived from a standard transect off Anvers Island into the Drake Passage

Under usual conditions krill abundance increases after the retreat of the pack ice. The krill obviously originate from the Bellingshausen Sea, since waters influenced by the Weddell Sea are abandoned by krill in spring. The increase in krill cannot be attributed to increased growth and production as Mackintosh (1973) supposed, since not only does the total biomass of krill increase, but all developmental stages, juveniles to adults, also increase in number. In late December to early January the stock reaches its highest levels of abundance which last until the beginning of March. A generalized picture of the krill distribution is given in Fig. 13.

The krill stock at the onset of the spawning season then shows a distinct spatial separation in the maturity stages. Gravid and spawning adults occur along the continental slope and in oceanic waters (Jazdzewski et al. 1978; Makarov 1979; Witek et al. 1981; Wolnomiejski et al. 1982; Quetin and Ross 1984; Siegel 1985, 1986b). Nearer to the coast and in the Bransfield Strait subadult krill dominate, while the juvenile stages are confined to coastal shelf waters (Fig. 13). A small number of adult krill seem to drift into the Bransfield Strait and spawn along its deep central basin. The total amount, however, remains negligible (Kittel 1980; Jackowska 1980; Fevolden and George 1984) compared to the oceanic area north of the South Shetland Islands (Quetin and Ross 1984) or Elephant Island (Nast 1982). Despite high abundances and the largest distributional range during summer, krill's northward distributional limit rarely reaches the Antarctic Convergence in the Drake Passage, as it does, for example, in the Scotia Sea off South Georgia (Marr 1962).

The spatial horizontal succession of size and maturity stages along the Peninsula must be at least partly a result of active migration and not only induced by advective drift. Otherwise it could be expected that smaller and weaker animals would drift faster and be more easily carried away into offshore waters. Larval krill, for example,

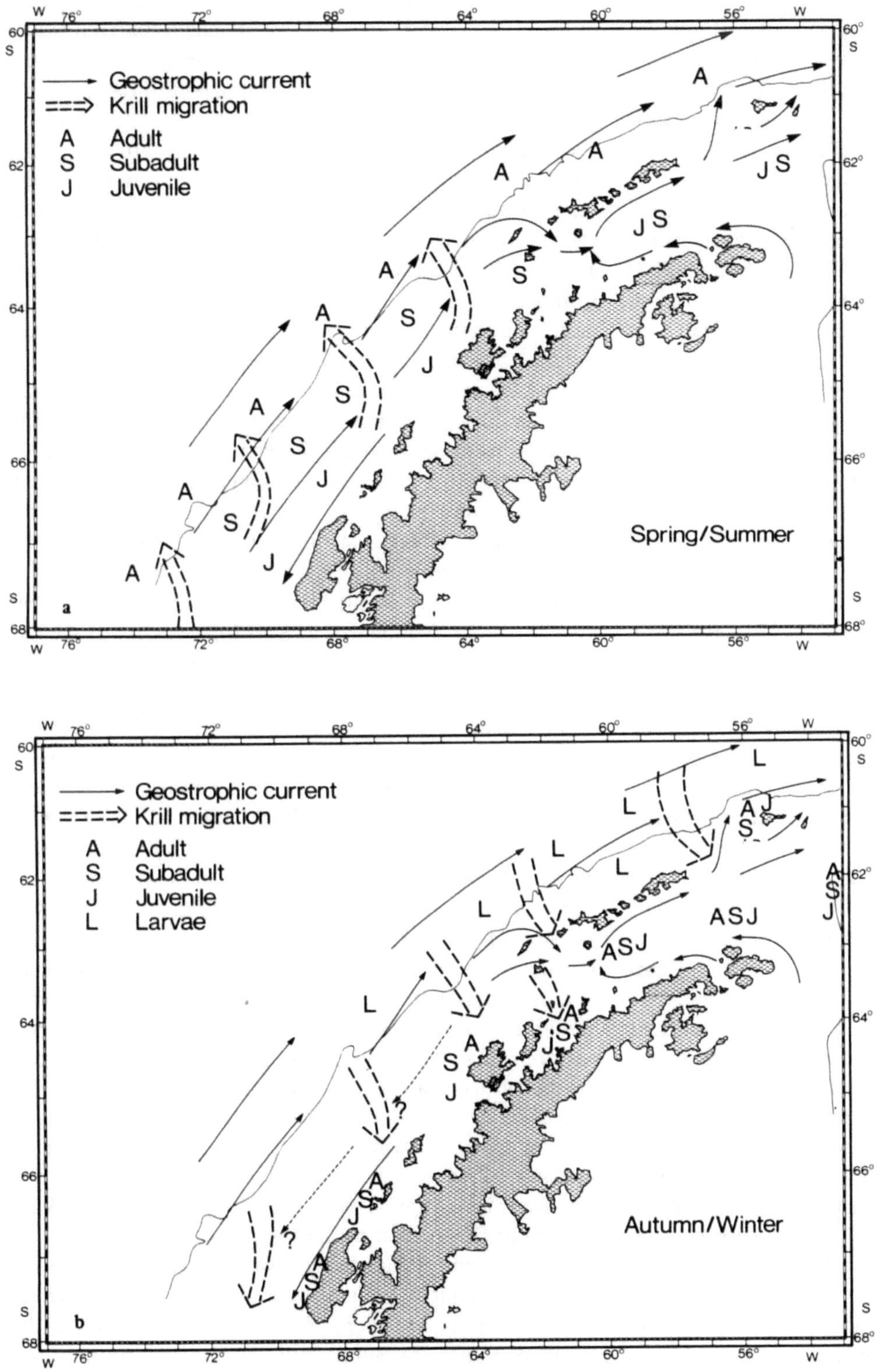

Fig. 13a,b. Seasonal distribution pattern of juvenile, subadult and adult krill developmental stages off the Antarctic Peninsula. **a** Krill migration into oceanic waters during spring/summer spawning season. **b** Postspawning movement into neritic areas (for detailed explanation see text)

drift passively from the Peninsula region into the Scotia Sea (Rakusa-Suszczewski 1984; Brinton and Townsend 1984; Brinton 1985) while concentrations of postlarval krill at least move more slowly away (Siegel 1986b). In this connection, Kanda et al. (1982) have shown that adult krill swarms are able to migrate considerable distances against the current.

After spawning, adults leave the oceanic regions where larvae occur at that time in surface waters, and remigrate into neritic areas (Fig. 13b). The presence of migration can be supported also by the fact that spent females are found more frequently in coastal shelf waters and in the Bransfield and Gerlache Straits than gravid-stage animals observed in the preceding spawning season.

In autumn and winter the number of various distribution clusters is reduced, because all developmental stages are concentrated and mixed in nearshore waters, causing a more homogeneous distribution pattern for the stock. Although the distribution clusters during these late months often extend as far as the continental slope, absolute krill abundance is extremely low in the outer shelf areas.

From March onwards abundance usually decreases. Again the same tendency has been reported from whaling records. Mackintosh and Wheeler (1929) and Peters (1955) have reported that there is a clear increase in number of empty whale stomachs from the beginning of March as compared to the preceding months. This decrease cannot be attributed only to a higher mortality rate of the spawned krill stock, since the dramatic decline in abundance is detectable for all size classes and developmental stages.

The largest concentrations of krill are only found in zones protected from wind and currents, e.g , the Gerlache Strait (Siegel 1982b, 1986a). The shortage of krill found in May/June 1986 and reported from the Bransfield Strait and Elephant Island during the winter season (Stepnik 1982; Guzman 1983; Heywood et al. 1985) therefore reflect normal winter conditions and cannot be considered as extraordinary. According to the study of Stepnik (1982), low winter abundances persist from March to the following October, which corresponds with the results of the present analysis.

The discussion on the biological aspects of this phenomenon must be speculative at this stage, but some ideas should be raised briefly. The proximate factors of the annual rhythmicity for the starting of the migration might be the photoperiod in October and March, the breakup of the pack ice in spring and the ice growing in March. Commonly gonadal stages are inferred to act as stimuli via neurohormonal pathways (Herrnkind 1983). In autumn, the spawning and resulting physiological changes might act as stimuli for the southward migration. The development of ovars and testes are possibly stimuli in spring, although it is not clear by which factor gonad development is stimulated to start. This would explain, however, why adults are migrating further to the north (oceanic areas) in summer than juveniles do. Movements may be orientated by perception of guideposts such as polarized light (Denys 1982).

The ultimate factors for the backward migration of adults to the south might be seen in the reduction of intraspecific food competition between adults and larvae. In oceanic areas food supply is lower than in the neritic zone and larger krill probably have a greater ability to move and to switch to a cryopelagic or benthopelagic life during winter and graze on ice algae or sea-bed detritus in shallow water (Spiridonov et al. 1985; Kawaguchi et al. 1986). The interspecific competition is also reduced in oceanic areas during winter, because copepods, for example, are undergoing seasonal vertical migration and are leaving the surface zone, which gives the krill larvae a better chance to survive.

This strategy would indicate that *Euphausia superba* utilizes its habitat up to the carrying capacity, which besides other parameters (longevity, large size, rematuration) characterizes the krill as apparently relatively K-selected.

On the one hand, the movement of adults after spawning into coastal waters can be explained by migration. On the other hand, it seems doubtful whether this argument is valid for obvious autumn decreases in total abundance and also for the dramatic increase of all stages during spring. In this case seasonal environmental changes seem to have a more important impact on the occurrence of krill. Indications of seasonal variability of prevailing currents may be derived from fluctuations of wind and atmospheric pressure over the Drake Passage. These parameters influence the monthly mean sea level off the Peninsula (van Loon 1972) which evidently results in a biannual oscillation. Such a seasonal change in the sea level slope towards Antarctica is thought to effect a stronger eastward surface current during the extreme seasons, i.e., summer and winter. The transport of water is significantly higher during summer than in winter (Whitworth 1980). First studies of relative dynamic height indicate changes in geostrophic flow in the northern Bellingshausen Sea (Makarov et al. 1982).

Two main current systems occur west of the Peninsula, the West Wind Drift and the East Wind Drift. The mean position of the transition zone, the Divergence, is found in the area around Adelaide Island (Deacon 1982). (Not far south of this area the "Belgica" drift started to the west). It must be appreciated, however, that measurements are scarce in the region southwest of the Bransfield Strait. The location of the Antarctic Divergence is influenced by high and low pressure areas and this influence varies seasonally as during winter areas of low pressure extend farther north. Therefore it seems reasonable to assume that the respective positions of East and West Wind Drift vary seasonally along the Peninsula coast. It would follow from this that the impact of the East Wind Drift is greater during winter and that it extends farther offshore and to the north of Adelaide Island – possibly as far as the western entrance of the Bransfield Strait – whereas during summer the West Wind Drift flows closer to the Peninsula.

Under these circumstances two different but simultaneous events might explain the variability of krill distribution and abundance along the Peninsula. The first is the active migration of adult krill during summer into offshore waters which are usually influenced by the West Wind Drift. The second is the passive drift out of the Bellingshausen Sea to the north as a result of environmental influences. A strong summer flow of the West Wind Drift affects even coastal waters and would carry juvenile and subadult krill to the NE, resulting in high concentrations of the species all over the area considered. In autumn both procedures are reversed. The krill stock from offshore areas migrates into shelf waters. Simultaneously the shelf areas are affected by a stronger winter East Wind Drift flow and krill would be carried south. In those areas which are not directly influenced by the East Wind Drift (i.e., the more northern Bransfield Strait) krill would dirft to the east into the open Scotia Sea. As long as the winter situation persists – and this might coincide with the period of the pack-ice coverage – the coastal East Wind Drift flow would prevent krill from drifting northwards. The total abundance in the Bransfield Strait and adjacent areas would therefore remain at low winter levels, thereby maintaining a krill winter "vacuum" off the Antarctic Peninsula.

The described interaction of migration and drift and the resulting seasonal fluctuations and geographical successions of the krill stock must be considered as a baseline before coming up with any interpretation and explanation of good or bad years in the krill stock off the Antarctic Peninsula and possibly other areas.

The model might help to explain year to year fluctuations of krill abundance: a strong East Wind Drift influence during summer might restrict a large proportion of the krill population in the south thereby inhibiting successful spawning.

Acknowledgments. The author is very grateful to D.G.M. Miller (Cape Town) for comments and language editing of the English manuscript. The research was supported by funds of the Federal Ministry of Research and Technology (MFE 534.3).

References

Anonymous (1982) Antarktis-Expedition 1981 der Bundesrepublik Deutschland mit FFS *Walther Herwig*. Arch Fischereiwiss 33: 1–176

Anonymous (1986) Antarctic expedition of the Federal Republic of Germany with RV *Polarstern* (Legs II/2, 1983 and III/2, 1984) and FRV *Walther Herwig* in 1985. Arch Fischereiwiss (Beih 1) 37:1–234

Baker A de C, Clarke MR, Harris MJ (1973) The N.I.O. Combination Net (RMT 1+8) and further developments of rectangular midwater trawls. J Mar Biol Assoc UK 53:167–184

BIOMASS (1982) 1st Post-FIBEX Hydrographic Data Interpretation Workshop, Hamburg FRG, 20–26 September 1982. BIOMASS Rep Ser 30

BIOMASS (1983) 2nd Post-FIBEX Hydrographic Data Interpretation Workshop, Hamburg FRG, 16–20 May 1983. BIOMASS Rep Ser 31

Brinton E (1985) The oceanographic structure of the eastern Scotia Sea III. Distributions of euphausiid species and their developmental stages in 1981 in relation to hydrography. Deep-Sea Res 32 (10):1153–1180

Brinton E, Townsends AW (1984) Regional relationships between development and growth in larvae of Antarctic krill, *Euphausia superba* from field samples. J Crust Biol 4 (Spec No 1):224–246

Clowes AJ (1934) Hydrology of the Bransfield Strait. Discovery Rep 9:1–64

Deacon GER (1982) Physical and biological zonation in the Southern Ocean. Deep-Sea Res 29 (1A):1–15

Denys CJ (1982) Ommochrome pigments in the eyes of *Euphausia superba* (Crustacea, Euphausiacea). Polar Biol 1:69–76

Fevolden SE, George RY (1984) Size frequency pattern of *Euphausia superba* in the Antarctic Peninsula waters in the austral summer of 1983. J Crust Biol 4 (Spec No 1):107–122

Fütterer D (1984) Die Expedition Antarktis-II mit FS *Polarstern* 1983/84. Ber Polarforsch 18:1–92

Guzman FO (1983) Distribution and abundance of Antarctic krill (*Euphausia superba*) in the Bransfield Strait. In: Schnack SB (ed) On the biology of krill *Euphausia superba*. Ber Polarforsch Spec 4:169–190

Hempel G (1985) Die Expedition ANTARKTIS III mit FS *Polarstern* 1984/85. Ber Polarforsch 25:1–209

Hempel G, Heywood RB (1982) Joint Biological Expedition on RRS *John Biscoe*, February 1982. Ber Polarforsch 5:1–38

Hempel I, Hempel G, Baker A de C (1979) Early life history stages of krill (*Euphausia superba*) in Bransfield Strait and Weddell Sea. Meeresforschung 27:267–281

Herrnkind WF (1983) Movement patterns and orientation. In: Bliss DE (ed) The biology of crustacea, vol 7. Behavior and ecology. Academic Press, New York, pp 41–105

Heywood RB, Everson I, Priddle J (1985) The absence of krill from the South Georgia Zone, winter 1983. Deep-Sea Res 32 (1):369–378

Jackowska H (1980) Krill monitoring in Admiralty Bay (King-) George Island, South Shetland Islands) in summer 1979/1980. Pol Polar Res 1 (4):117–125

Jazdzewski K, Dzik J, Porebski J, Rakusa-Suszczewski S, Witek Z, Wolnomiejski N (1978) Biological and populational studies on krill near South Shetland Islands, Scotia Sea and South Georgia in the summer 1976. Pol Arch Hydrobiol 25 (3):607–631

Kanda K, Takagi K, Seki Y (1982) Movement of the larger swarms of Antarctic krill *Euphausia superba* population off Enderby Land during 1976–1977 season. J Tokyo Univ Fish 68 (1/2): 25–42

Kawaguchi K, Ishikawa S, Matsuda O (1986) The overwintering strategy of Antarctic krill (*Euphausia superba* Dana) under the coastal fast ice off the Ongul Island in Lützow-Holm Bay, Antarctica. Mem Natl Inst Polar Res Spec Issue 44:67–85

Kittel W (1980) Population studies on *Euphausia superba* Dana, 1852 (Euphausiacea, Crustacea) in waters of the Admiralty Bay during Antarctic summer of 1978. Pol Arch Hydrobiol 27 (2): 267–272

Kock K-H, Stein M (1978) Krill and hydrographic conditions off the Antarctic Peninsula. Meeresforschung 26 (1/2):79–95

Loon H van (1972) Half-yearly oscillations in the Drake Passage. Deep-Sea Res 19:525–527

Mackintosh NA (1973) Distribution of post-larval krill in the Antarctic. Discovery Rep 36:95–156

Mackintosh NA, Wheeler JFG (1929) Southern blue- and finwhales. Discovery Rep 1:259–540

Makarov RR (1979) Size composition and conditions of existence of *Euphausia superba* Dana (Crustacea, Euphausiacea) in the eastern part of the pacific sector of the Southern Ocean. Oceanology 19 (5):582–585

Makarov RR, Denys CJI (1982) Stages of sexual maturity of *Euphausia superba*. BIOMASS Handb 11:13 pp

Makarov RR, Maslennikov VV, Movchan OA, Solyankin EV (1982) Oceanographical conditions and regional peculiarities of several successions in plankton off Antarctic Peninsula (Russ.). In: The Antarctic, Sov Comm Antarct Res; Rep 21, Nauka, Moscow, pp 101–117

Marr JWS (1962) The natural history and geography of the Antarctic krill (*Euphausia superba* Dana). Discovery Rep 32:33–464

Mauchline J (1980) Measurement of body length of *Euphausia superba* Dana. BIOMASS Handb 4:9 pp

Nast F (1982) The assessment of krill biomass from a net sampling programme. Meeresforschung 29 (3):154–165

Nast F (1986) Changes in krill abundance and in other zooplankton relative to the Weddell-Scotia Confluence around Elephant Island in November 1983, November 1984 and March 1985. Arch Fischereiwiss (Beih 1) 37:73–94

Peters H (1955) Über das Vorkommen des Walkrebschens *Euphausia superba* Dana und seine Bedeutung für die Ernährung der südlichen Bartenwale. Arch Fischereiwiss 6 (5–6):288–304

Pommeranz T, Herrmann C, Kühn A (1982) Mouth angles of the Rectangular Midwater Trawl (RMT 1+8) during paying out and hauling. Meeresforschung 29:267–274

Quetin LB, Ross RM (1984) School composition of the Antarctic krill *Euphausia superba* in the water west of the Antarctic Peninsula in the austral summer of 1982. J Crust Biol 4 (Spec No 1):96–106

Rakusa-Suszczewski S (1984) Krill larvae in the Atlantic sector of the Southern Ocean during FIBEX 1981. Polar Biol 3 (3):141–147

Rakusa-Susczewski S, Lipski M (1985) Report on the R/V *Profesor Siedlecki* expedition to the Antarctic during the BIOMASS-SIBEX in 1983/1984. Pol Polar Res 6 (1/2):7–19

Rakusa-Suszczewski S, Stepnik R (1980) Three species of krill from Admiralty Bay (King George, South Shetlands), in summer 1978/79. Pol Arch Hydrobiol 27 (2):273–284

Roe HSJ, Shale DM (1979) A new multiple rectangular midwater trawl (RMT 1+8) and some modifications to the Institute of Oceanographic Science's RMT 1+8. Mar Biol 50:283–288

Roe HSJ, Baker A de C, Carson RM, Wild R, Shale DM (1980) Behaviour of the Institute of Oceanographic Science's Rectangular Midwater Trawl: Theoretical aspects and experimental observations. Mar Biol 56:247–259

Siegel V (1982a) Relationship of various length measurements of *Euphausia superba* Dana. Meeresforschung 29:114–117

Siegel V (1982b) Untersuchungen an Nachlaichkonzentrationen des antarktischen Krills *Euphausia superba.* Arch Fischereiwiss (Beih 1) 33:113–125

Siegel V (1985) The distribution pattern of krill, *Euphausia superba,* west of the Antarctic Peninsula in February 1982. Meeresforschung 30:292–305

Siegel V (1986a) Structure and composition of the Antarctic krill stock in the Bransfield Strait (Antarctic Peninsula) during the Second International BIOMASS Experiment (SIBEX). Arch Fischereiwiss (Beih 1) 37:51–72

Siegel V (1986b) Untersuchungen zur Biologie des antarktischen Krill, *Euphausia superba,* im Bereich der Bransfield Straße und angrenzender Gebiete. Mitt Inst Seefisch 38:1–244

Spiridonov VA, Gruzov EN, Pushkin AE (1985) Observations on schools of the antarctic *Euphausia superba* (Crustacea, Euphausiacea) under ice. Zool Zh 64 (11):1655–1660

Stepnik R (1982) All-year populational studies of Euphausiacea (Crustacea) in the Admiralty Bay (King George Island, South Shetland Islands, Antarctic). Pol Polar Res 3 (1/2):49–68

Whitworth T (1980) Zonation and geostrophic flow of the Antarctic Circumpolar Current at Drake Passage. Deep-Sea Res 27A: 497–507

Witek Z, Kittel W (1985) Larvae of the species of the genus *Euphausia* (Euphausiacea, Crustacea) in southern part of the Drake Passage and Bransfield Strait during the BIOMASS-SIBEX (December 1983–January 1984). Pol Polar Res 6 (1/2):117–132

Witek Z, Koronkiewicz A, Soszka GJ (1980) Some aspects of the early life history of krill *Euphausia superba* Dana (Crustacea). ICES Biol Oceanogr Comm CM 1980/L:48, 7 pp

Witek Z, Kalinowski J, Grelowski A, Wolnomiejski N (1981) Studies of aggregations of krill (*Euphausia superba*). Meeresforschung 28:228–243

Wolnomiejski N, Czykieta H, Stepnik R, Jackowska H (1982) Biological characteristics of *Euphausia superba* Dana in the Southern Drake Passage and the Bransfield Strait in February-March, 1981 (BIOMASS-FIBEX). Pol Polar Res 3 (3–4):259–271

Variability in Population Density of Antarctic Krill in the Western Scotia Sea in Relation to Hydrological Conditions

R. R. MAKAROV, V. V. MASLENNIKOV, E. V. SOLYANKIN, V. A. SPIRIDONOV[1], and V. N. YAKOVLEV[2]

Summary. Based on large data sets from the period 1964–1985, this paper shows that the genesis and persistence of krill aggregations in the area of Elephant Island is mainly determined by hydrodynamic conditions. Oceanographic observations cover a wide area from the Scotia Sea to the Pacific sector of the Southern Ocean. Interannual and short-term variations in the distribution of different water masses are described in relation to the distribution and density of krill accumulations. Correlations are so evident that a prognosis of krill concentrations seems feasible.

1 Introduction

Irregular distribution of the Antarctic krill (*Euphausia superba*) is a well-known feature of the biology of this species. Its distribution is characterized by spatial fluctuations and apparently also by fluctuations in time. These variations are rather permanent and appear first of all on a regional macroscale. The population density of species inhabiting high latitudes is always greater than in the waters of the Antarctic Circumpolar Current (Marr 1962; Mackintosh 1973; Maslennikov 1980). When the scale of spatial nonuniform distribution of krill decreases, it is affected to a larger degree by fluctuations in time. Naturally, an elucidation of the factors responsible for these variations is of great scientific interest.

In our opinion this variability is due to a complex of biological and physical factors (Makarov et al. 1980). The dynamics of the krill population and the well-balanced growth of successive generations apparently play an important role here. To analyze such phenomena, we need a thorough knowledge of the spatial structure of the euphausiid population (Latogursky 1979). Presently there is still a dearth of knowledge on this subject. It is hardly likely that the dynamics of the population of the species in subpopulations is of a synchronous nature. Therefore, we need to know the nature of the specimens and particularly what subpopulation they belong to, when they gather at the time of observation in a certain region of the ocean. The degree of the relative local abundance of krill depends on the phases of the biological season.

In the open Scotia Sea, within the boundaries of the secondary frontal zone (Weddell-Scotia Confluence) a definite correlation has been observed between the phases of the krill reproduction cycle and the density of their population (Makarov et al. 1980). Some authors believe that krill is an actively migrating species (Kanda et al. 1982). Such behavior would have an effect on the distribution of krill but until now there is no proof of their migratory activity. Of course, in analyzing the variability of the population density of krill in a certain area we should also take into account the biological factors. Yet the diversity of the effects of these factors (often of an ambivalent nature) makes an evaluation of the causal relationships difficult.

Presently we have already a fairly good idea of the role of oceanological factors on the distribution of krill. We know how the circulation and distribution of various modifications of water masses affect the drift of planktonic crustaceans, and the possibility of their accumulation and aggregation.

2 Effects of Variability in Abiotic Factors on the Distribution of Krill

To elucidate the effects of variability in abiotic factors on the distribution of krill we have selected the northeastern part of the Antarctic Peninsula and the western Scotia Sea. Research efforts concentrated on the area around Elephant Island (Mordvinov Island), where extensive fluctuations in the krill population have been observed. The data available for this area make it possible to evaluate the variability of environmental conditions as well as fluctuations in the krill biomass, and to establish the areas of krill aggregations.

Since the area covered by our study lies between the Atlantic and Pacific sectors of the Southern Ocean, we should expect that the physicochemical properties of its waters may vary, and so may the density of the krill population. One can assume that changes in the distribution of different water masses will affect the distribution and abundance of krill.

1 All-Union Research Institute, of Marine Fisheries and Oceanography (VNIRO), Moscow, USSR
2 Atlantic Research Institute of Marine Fisheries and Oceanography (AtlantNIRO), Kaliningrad, USSR

Antarctic Ocean and Resources Variability
(ed. by D. Sahrhage)

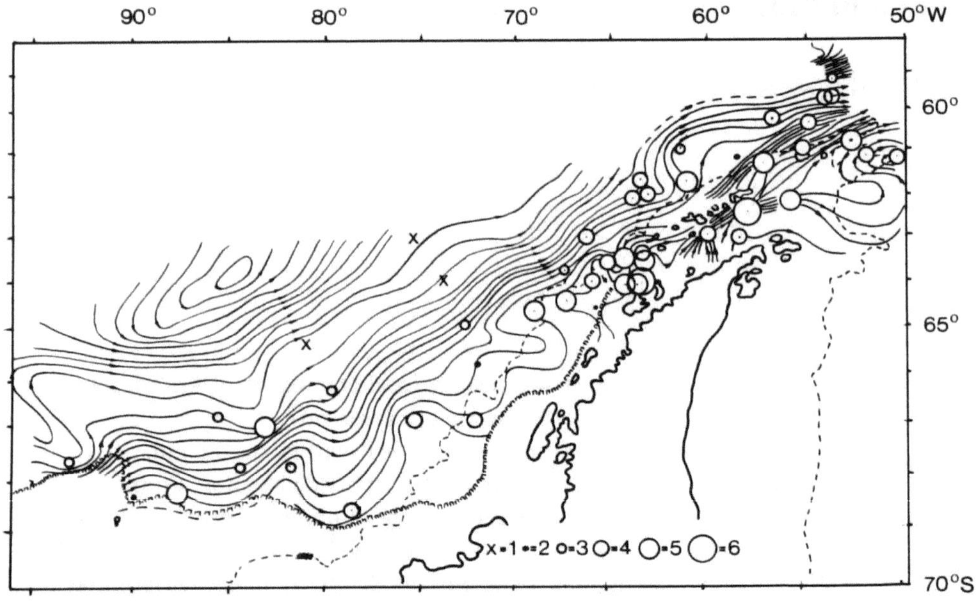

Fig. 1. Surface geostrophic currents (relative to 1000 dbar) and distribution of the krill population during January–February 1978 in the region around the Antarctic Peninsula and the South Shetland Islands. *1* no krill catch; *2* few krill catches; *3* 10–100 kg/h of krill; *4* 100–1000 kg/h of krill; *5* 1000–10000 kg/h of krill; *6* > 10000 kg/h of krill. 0–1000 dbar dynamic topography in dynamic mm (for every 10 dyn/mm), also in Figs. 4 and 5. ⊓ ⊓ ⊓ denotes ice edge (also in Figs. 2 and 5)

Observations made in the waters of Elephant Island show that the formation and stability of krill aggregations depend largely on the hydrodynamic conditions, in particular on the vortices of the currents. Waters of different characteristics are populated by krill of different sizes (Makarov 1979). This enhances the spatial contrast in the biological characteristics of macroplankton in areas where different water masses meet (Makarov 1980).

2.1 The Main Water Masses

We have found that in the vicinity of Elephant Island the composition of waters is very complex and can be characterized as follows:

a) The Weddell Sea water (originally rich in krill).
b) The waters of the Bransfield Strait, belonging to the high-latitude modification with a specific vertical structure (also rich in krill).
c) The waters over the shelf and continental slope along the western coast of the Antarctic Peninsula, belonging to the high-latitude type and partially mixed within the secondary frontal zone, near the continental slope, with waters from the southern periphery of the Antarctic Circumpolar Current (amount of krill here can fluctuate depending on the dynamic conditions within the frontal boundaries).
d) The waters at the southern periphery of the Antarctic Circumpolar Current (characterized by a sharp drop in the concentrations of krill, with occasional transfers by the current of individual krill aggregations – sometimes even large ones – into the frontal zone (Bogdanov et al. 1980a; Makarov and Maslennikov 1980).

A correct analysis of the composition of the waters in the area covered by our study depends also on our knowledge of the environmental conditions outside its boundaries. This is especially true for the waters around Elephant Island, where certain situations in the water structure and krill distribution could be ascertained only by including wide regions to the west and south of the island. It is from there that water transfer takes place and with it transfer of the krill. Therefore we have included a much wider area that lies outside the immediate area of our study (Figs. 1 and 2).

The experience gained can be useful in planning further large-scale research aimed at establishing how the variability of the major dynamic systems affect the local water circulation and thus the krill distribution (Makarov et al. 1982).

2.2 Mixing of Water Masses

On the basis of results obtained mostly in the summers of 1977/78, 1978/79, 1983/84, and 1984/85, we evaluated the spatial fluctuations in the distribution of the various water masses. In an earlier work (Bogdanov et al. 1980a) we examined waters of different modifications in their "pure" state. Here we also analyze waters formed as a result of the mixing of the above-mentioned main water modifications. Such an approach involving a relatively detailed analysis of water structures has recently found increasing application (BIOMASS 1981; Grelowski and Tokarczyk 1985; Heywood 1985; Kelly et al. 1985). In these investigations use is made of the regional characteristics of T-S curves in accordance with the field of geostrophic currents.

In Figs. 3–5 the distribution of different modifications of the Antarctic waters in the study area and peculiarities of the current field as well as physical and chemical parameters are shown.

The vertical structure of the waters around the west coast of the Antarctic Peninsula (frequently referred to in the literature as the waters of the Bellingshausen Sea) is typical of the waters of the region studied. A well-defined

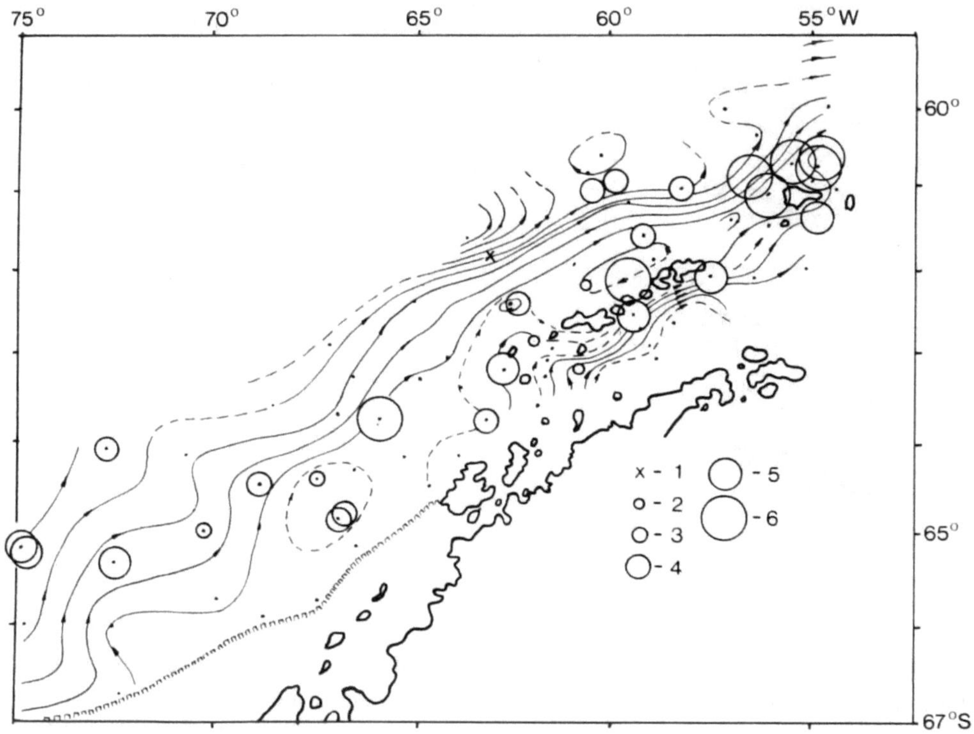

Fig. 2. Surface geostrophic currents (relative to 1000 dbar) and distribution of the krill population during December 1978–January 1979 in the region around the Antarctic Peninsula and the South Shetland Islands (notations are the same as in Fig. 1). 0–1000 dbar dynamic topography in dynamic mm (for every 20 dyn/mm)

intermediate temperature minimum corresponding to the cold residual water layer from the winter period changes to a well-defined maximum temperature characteristic of the upper deep-water layer. The salinity of the surface water is lower than that of other types of high-latitude waters in the area.

The Weddell Sea water, on the other hand, has the highest surface salinity. The vertical structure of these waters is characterized by slight temperature extremes.

Waters of the Bransfield Strait show an unusual vertical structure characterized by the absence of the temperature maximum throughout the depth of the Strait. The salinity of the surface waters here is quite high with a level intermediate between corresponding values for the Weddell Sea water and the waters of the Antarctic Peninsula region.

2.3 Fluctuations in the Water Masses Around Elephant Island

In the vicinity of Elephant Island, on its shelf and continental slope, waters of high-latitude modifications in various proportions predominate [i.e., types (a), (b) and (c) according to the classification given in the previous section]. Apparently these waters are not completely displaced by the Antarctic Circumpolar Current. However, there were times when water of the Antarctic Circumpolar Current predominated.

This occurred, for instance, during the summer of 1983/84 (Fig. 3C). Observations show that the Weddell Sea water was driven off in an east-southeastern direction. Only a small portion of other types of high-latitude waters from the Bransfield Strait and the Antarctic Peninsula region (west coast) washed the northern shore of Elephant Island. On the northern shelf and slope of Elephant Island, waters of the secondary frontal zone prevailed with a large portion of water of the Antarctic Circumpolar Current. This combination accounts for the unusual hydrophysical and hydrochemical properties of the waters in the area during that summer. The deep-water temperature maximum (one of the best parameters for locating the secondary frontal zone) showed values of 1.3–1.8°C instead of the 0.0–1.2°C usually encountered here. Correspondingly, the surface salinity was less than 34‰ instead of the usual levels of 34.0–34.3‰. Silicate concentrations near the surface were 30–40 μgat/l instead of the usually observed 40–75 μgat/l (Figs. 4 and 5).

Apparently a similar situation occurred also in the summer of 1964/65 with an unusually large expansion of the waters of the Antarctic Circumpolar Current.

2.4 Effects on the Krill Distribution

Such conditions, which are probably rare, result in a noticeable decrease in the krill population in this area. This may be explained by the fact that the greater proportion of water in the area came from the Antarctic Circumpolar Current, which is not rich in krill. It is also possible that the through-carrying nature of currents over the shelf and the slope of Elephant Island contribute to this situation. These are very unfavorable conditions for the accumulation of krill in this region.

During other periods, different types of water of high-latitude modifications spread over the shelf and slope of Elephant Island (Fig. 3A,B,D). These waters carry different

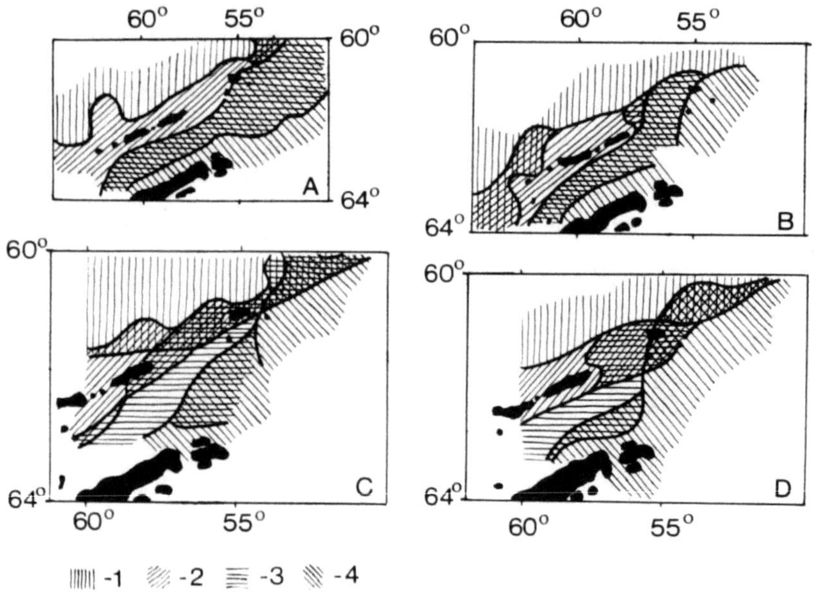

Fig. 3A–D. Distribution of different modification of the Antarctic waters in the area of the South Shetland Islands during January–February 1978 (A), December 1978–January 1979 (B), March 1984 (C), and December 1984 (D). *1* Antarctic Circumpolar Current waters; *2* waters of the Antarctic Peninsula region; *3* waters of the Bransfield Strait; *4* waters of the Weddell Sea

amounts of krill into the area. The fraction of each type of water present at a certain time could provide the basis for the quantitative distribution of krill.

Figure 3 illustrates the distribution of water of high-latitude modifications over the shelf and slope of Elephant Island in summer 1984 and during the summer seasons 1977/78, 1978/79, and 1984/85 (Makarov et al. 1980, 1982; Bogdanov et al. 1980a). It is known that during each of these seasons large and stable concentrations of krill were not observed close to the island.

In many respects an increase in the probability for the formation of krill concentrations appears to depend on the structure of the horizontal circulation on a mesoscale. The dynamic situation in summer 1979 (Bogdanov et al. 1980b) may serve as an example of favorable conditions for the accumulation of krill (Fig. 2). At that time, when

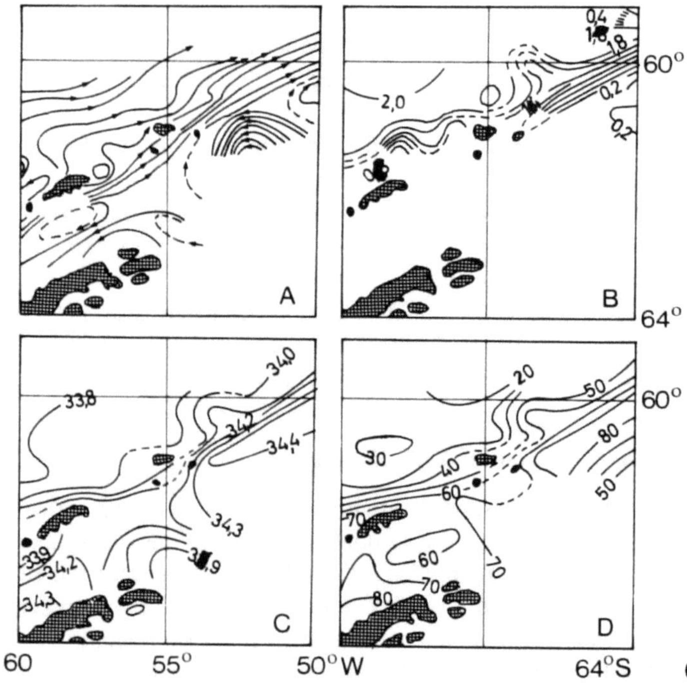

Fig. 4A–D. Surface geostrophic currents, relative to 1000 dbar (A); distribution of the Deep Water temperature maximum (B); surface salinity (C); silicate concentration (μgat/l) in surface waters (D), in an area around the South Shetland Islands during March 1984

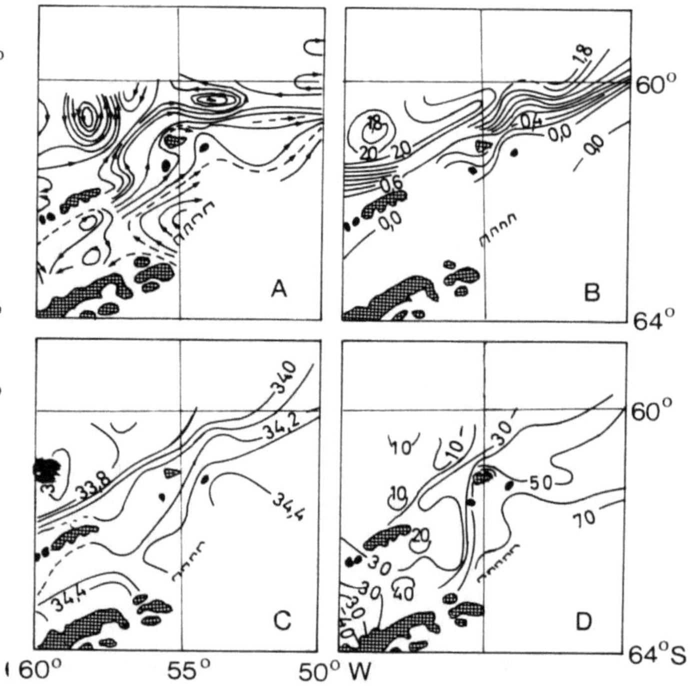

Fig. 5A–D. Surface geostrophic currents, relative to 1000 dbar (A); distribution of the Deep Water temperature maximum (B); surface salinity (C); silicate concentration (μgat/l) in surface waters (D) in an area around the South Shetland Island during December 1984

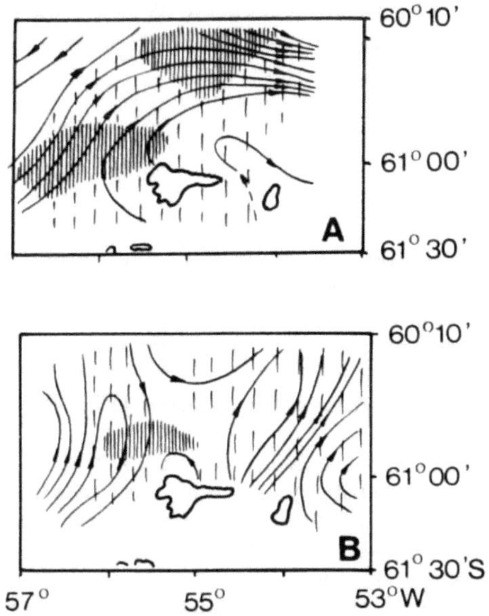

Fig. 6A,B. Surface geostrophic currents, relative to 1000 dbar, and distribution of krill in the area of Elephant Island. **A** Early December 1984; **B** end of December 1984. 0–1000 dbar topography in dynamic mm (for every 20 dyn/mm). The *density of shading* is proportional to krill density

owing to the topographic effect the current deviated to the north on its approaches to Elephant Island, a flow of Weddell Sea water (most likely of a compensational nature) to the north of the island and the formation of a cyclonic mesoscale meander were observed.

Over the northern shelf of Elephant Island there is nearly always a certain dynamically weak zone due to some "shadow" effect (Fig. 6). Under conditions of a slight water exchange with surrounding waters, krill transported into this zone could accumulate, thus increasing the density of the krill population. Fluctuations in the number of krill in this area are apparently determined by the extent to which water from the Weddell Sea (rich in krill) participate in the formation of the hydrological conditions of the area. These waters, which surround the island from the west and east, can play a major role in increasing the probability of locating krill formations (Fig. 3B). If these waters are driven off to the east (Fig. 3A,C,D), they practically do not reach the island shelf. At the exit into the Scotia Sea east of Elephant Island, the waters flowing to the east carry with them most of the krill and increase the krill population in the southern parts of the Scotia Sea. As may be seen from Fig. 3C, Weddell Sea water may mix with water of the Antarctic Circumpolar Current. This results in a drop in the number of krill in the region.

2.5 Short-Term Variation

The variants discussed above are concerned with the interannual variability. As an example of short-period variability, some of the data for December 1984 can be used (Fig. 6). This figure shows that there is a clear difference in the relative dynamic topography 0–1000 dbar in the area to the north of Elephant Island according to the data from two successive surveys at the beginning of December and at the end of December 1984. During these surveys 16 and 21 bathymetric stations were made, respectively. The stations were evenly distributed throughout the area. As may be seen from Fig. 6, no accumulation of krill was observed in the waters over the island shelf, in spite of favorable dynamic conditions prevailing there. Our analysis showed that the accumulations occurred at the frontal zones where different waters meet. These zones are marked by greater horizontal gradients of the hydrophysical and hydrochemical characteristics. Frontal zones hindered the expansion of krill beyond their boundaries.

In early December 1984 these accumulations were vaster and were observed to the west-northwest and north of Elephant Island (Fig. 6A). The aggregations consisted of krill carried apparently with waters from the Weddell Sea and the Bransfield Strait. One of the aggregations occurred in the zone where an interaction between the waters from the Weddell Sea, the Bransfield Strait, and the Antarctic Peninsula region takes place. The second aggregation was located where Weddell Sea water (mixed with other waters of high-latitude types) interacts with water of the southern periphery of the Antarctic Circumpolar Current.

At the end of December 1984 (Fig. 6B), in another dynamic situation, there were fewer krill, and the area of their distribution decreased. Krill was mainly found at the eastern periphery of the local cyclonic gyre of Bransfield Strait water which has its onset west of Elephant Island. The flow of Weddell Sea water is clearly seen east of Elephant Island and Clarence (Shishkov) Island.

In this connection it was of interest to analyze the effect of synoptic changes in the environment on the distribution of separate aggregations of krill. Over the northern shelf of Elephant Island the distribution, as expected, was influenced by synoptic changes. In March 1981 an attempt was made to trace these changes in an area of 10 × 8 nautical miles north of Elephant Island. The distance between stations was not more than 2–3 miles. During the first stage of the survey 12 stations were made (March 17–18), and the same stations were repeated on March 19–20. Prior to the first survey and between the two research phases, hydroacoustic measurements were made with tacks inside the testing ground, though within a somewhat smaller area. Salinity proved to be the most reliable indicator for identifying different types of water. The distribution of salinity at 50 m depth (34.104–34.244‰ during the first survey, and 34.073–34.228‰ during the second survey) shows that in the east-northeast and west-northwest parts of the testing area the waters were of different types. Quite significant changes in local water composition can occur within 24 h. These changes are apparently due to the reconstruction of the current field (Fig. 7A,B) which had a most pronounced effect on the distribution of salinity (Fig. 7C,D).

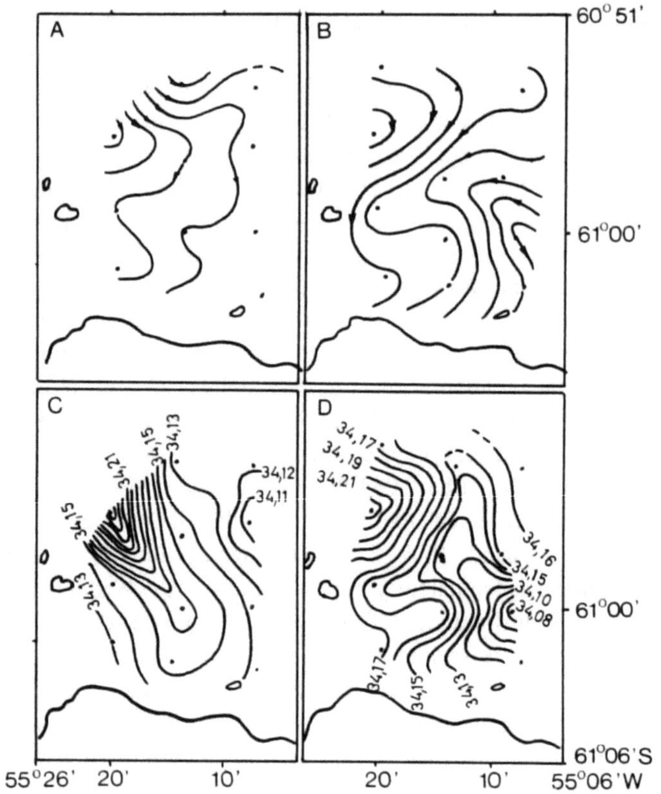

Fig. 7A–D. Surface geostrophic currents, relative to 200 dbar, during the first survey (**A**) and the second survey (**B**), and the distribution of salinity at 50 m during the first survey (**C**) and the second survey (**D**) in March 1981

Changes in the configuration of krill accumulations and a decrease in krill density can result from waters penetrating into the shelf area from the east. These waters are relatively less saline owing to increase of the portion from the Antarctic Circumpolar Current. Water from the latter are less rich in krill, and krill disappeared almost completely in the area where the penetration of this water was observed during the second survey. One may question whether we made a correct identification of the water masses, but the fact that krill definitely reacted to the changing situation confirmed that our approach to solving the problem is a correct and promising one for further research in this field.

3 Conclusion

Exploration of the "migration" routes of krill into the Elephant Island region is important for an understanding of the causes for the variability in the size and density of the krill population at different time scales. It is necessary to establish the relationship between the degree of abundance of krill and age-composition of their aggregations (for the latter see Makarov 1979) on the one hand, and the types of water modifications on the other. This problem is especially important for the regions with a complex composition of waters of different modifications. In practical terms, research on this problem will enable us to distinguish between natural fluctuations of krill populations and those due to anthropogenic influences.

References

BIOMASS (1981) Post-FIBEX Data Interpretation Workshop. BIOMASS Rep Ser No 20:1–38

Bogdanov MA, Solyankin EV, Rodionov SN (1980a) The distribution of mixed waters of the secondary frontal zone in the Sea of Scotia and the distribution of krill swarms. In: Lubimova TG (ed) Biological resources of the Antarctic krill. VNIRO, Moscow, pp 23–41

Bogdanov MA, Solyankin EV, Maslennikov VV, Popkov VV, Rodionov SN (1980b) Variability of oceanographic conditions and regularities in the distribution of krill off the West Antarctic Peninsula. In: Lubimova TG (ed) Biological resources of the Antarctic krill. VNIRO, Moscow, pp 55–72

Grelowski A, Tokarczyk R (1985) Hydrological conditions in the region of Bransfield Strait and southern part of Drake Passage in the period from December 10, 1983 and January 8, 1984 (BIOMASS-SIBEX). Pol Polar Res 6 (1–2):31–41

Heywood RB (1985) Environmental conditions in the Antarctic Peninsula area of the Southern Ocean during the Anglo-German Joint Biological Expedition, February 1982. Meeresforschung 30:220–239

Kanda K, Takagi K, Seki Y (1982) Movement of the larger swarms of Antarctic krill *Euphausia superba* population off Enderby Land during 1976–77 season. J Tokyo Univ Fish 68 (1/2):25–42

Kelly RJ, Blanco JL, Diaz MV (1985) Hydrography of the Bransfield Strait during 1984 southern summer. SIBEX-Phase I. Ser Cient INACH 33:15–45

Latogursky VI (1979) Interaction of independent populations of the Antarctic krill. Rybn Khoz Moscow 10:12–14

Mackintosh NA (1973) Distribution of post-larval krill in the Antarctic. Discovery Rep 36:95–156

Makarov RR (1979) Size composition and conditions of existence of *Euphausia superba* Dana (Crustacea; Euphausiacea) in the eastern part of the Pacific sector of the Southern Ocean. Oceanology 19 (5):582–585

Makarov RR (1980) The study of the composition of the population of *Euphausia superba* Dana, 1852 (Crustacea, Eucarida, Euphausiacea). In: Lubimova TG (ed) Biological resources of the Antarctic krill. VNIRO, Moscow, pp 89–113

Makarov RR, Maslennikov VV (1980) Distribution and age-composition of the larvae of Euphausiacea *Thysanoessa macrura* under hydrological conditions in the Pacific sector of Antarctica. Biol Morya 4:8–17

Makarov RR, Maslennikov VV, Solyankin EV, Shevtsov VV (1980) Quantitative distribution and conditions for formation of swarms of *Euphausia superba* Dana with reference to some areas in the Atlantic and Pacific Southern Ocean. In: Lubimova TG (ed) Biological resources of the Antarctic krill. VNIRO, Moscow, pp 114–145

Makarov RR, Maslennikov VV, Movchan OA, Solyankin EV (1982) Oceanographic conditions and regional peculiarities of seasonal successions of plankton in the coastal waters of the Antarctic Peninsula. In: Antarctica, vol 21. Nauka, Moscow, pp 101–117

Marr JWS (1962) The natural history and geography of the Antarctic krill (*Euphausia superba* Dana). Discovery Rep 32:33–464

Maslennikov VV (1980) Modern concepts on the large-scale circulation of the Antarctic waters and the routes of mass drift of the Antarctic krill. In: Lubimova TG (ed) Biological resources of the Antarctic krill. VNIRO, Moscow, pp 8–27

Formation of Antarctic Krill Concentrations in Relation to Hydrodynamic Processes and Social Behaviour

Z. WITEK, J. KALINOWSKI, and A. GRELOWSKI[1]

Summary. Based on results obtained during Polish marine Antarctic expeditions to the Atlantic sector, an attempt is made to present a conception concerning a mechanism of formation of krill concentrations in areas of meanders and current eddies. According to this conception, the mechanical factors (currents) cause the permanent inflow of animals to the regions of concentration and at the same time the behavioral factors (specific reactions to the turbulent water flow and the instinct of shoal formation) make krill stay and congregate in such regions. It was observed that larger quantities of krill occur in the areas of hydrological fronts, meanders, and current eddies, especially in regions characterized by high gradients of current velocity. It was found, on the basis of hydroacoustic measurements, that krill patches occurring in their concentration regions, are of larger size and are characterized by greater krill quantities than those occurring beyond the limits of such regions.

1 Introduction

It follows from the analyses of factors causing uneven spatial distribution of marine organisms made by Stavn (1971) and Parsons et al. (1977) that the following factors play an essential role here:

- environmental boundaries,
- advective effects and turbulence,
- reproduction rates,
- social behavior,
- intraspecific relations.

With respect to krill, two main types of aggregations can be distinguished (Kalinowski and Witek 1985):

- Patches (shoals, swarms, schools) – elementary forms, occurring locally and accessible to direct visual or hydroacoustic observations.
- Concentrations – aggregations in certain regions on a geographical scale, consisting of many patches.

Formation of swarms, the most frequent type of patches, is caused primarily by social behavior, while concentration formation must be connected with factors of a more global nature.

It seems that within the areas of krill occurrence, i.e., between the Antarctic continent (or minimum range of ice cover) and Antarctic Convergence (or maximum range of ice cover) there are no such factors gradients which could be taken to constitute environmental boundaries. Krill is a pelagic species of circumpolar distribution, both in neritic zones and in open waters. The research of Rakusa-Suszczewski (1978) and Witek et al. (1981) showed that no elements in the observed values such as salinity, temperature, oxygen conditions, nutrient content, and abundance of phytoplankton directly influence krill horizontal distribution.

The assumption that krill concentrations in certain regions are the result of their active migration should also be rejected. Active migration extending over several hundred nautical miles, which separate krill concentration regions appears unlikely.

On the other hand, there are numerous publications which link the formation of large krill concentrations with the circulation of water masses. Gathering of krill off South Georgia was explained by Marr (1962) as resulting from krill inflow on one of the current branches from the Weddell Sea. Makarov et al. (1970) pointed to the fact that krill concentrations are often located on the "lee side" of islands. Maslennikov (1972) associated the occurrence of krill concentrations off South Georgia with the presence of a system of currents flowing in opposite directions on the southeastern side of the island. Elizarov (1971) pointed to the fact that concentrations are formed in areas of sinking water, adjacent to the zones of intensive upwelling, separated from the rest of the water mass by a kind of "partitions" of either dynamic or orographic nature. The connection between the occurrence of commercial krill concentrations and eddies was pointed out by Wolnomiejski et al. (1978). Mauchline (1980), in his analysis of the formation of krill aggregations, included behavioral factors besides environmental parameters.

This chapter presents the distribution of krill aggregations in relation to the hydrodynamic situation (currents and fronts) and makes a comparison of swarm sizes occurring in concentration regions and those outside them. These data enabled the authors to propose a conception for a mechanism of krill concentration formation.

1 Sea Fisheries Institute, Al. Zjednoczenia 1, 81-345 Gdynia, Poland

Antarctic Ocean and Resources Variability
(ed. by D. Sahrhage)

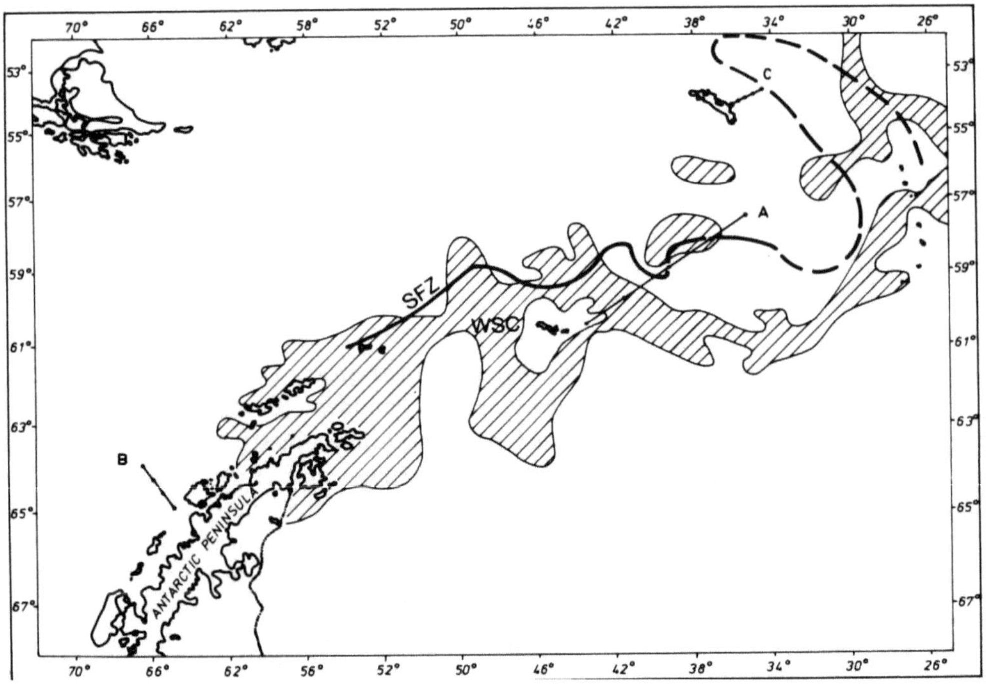

Fig. 1. Position of Weddell-Scotia Confluence (WSC), Secondary Frontal Zone (SFZ) and transects shown in Figs. 2, 3, 4. WSC defined on the basis of salinity distribution (higher than 34‰) at the depth of 20 m (after Patterson and Sievers 1980). Position of SFZ after Maslennikov (1979). A, B, C transects. *Dots* indicate stations

2 Material and Methods

The authors used materials collected on RV *Professor Siedlecki* in the Western Antarctic in the II Polish Marine Antarctic Expedition in 1976/77 and in the IV Polish Marine Antarctic Expedition in 1978/79.

Oceanographic sampling was conducted down to the depth of 2000 m with the help of both the Bissett-Berman STDO probe and Nansen bottles. Hydrological parameters were measured at standard levels: 0, 10, 20, 30, 50, 75, 100, 150, 200, 250, 300, 400, 500, 600, 800, 1000, 1250, 1500, 2000 m. Water temperature was measured with a probe and reversible thermometers with an accuracy of ± 0.01°C. Salinity was measured with the Autolab 601 salinometer with an accuracy of ± 0.003‰ or with the Bissett-Berman probe with an accuracy of ± 0.01‰. Temperature and salinity data from sampling stations served to determine the direction of flow of geostrophic currents by a dynamic method given by Zubov and Mamaiev (1956).

The following appliances were utilized in the hydroacoustical investigations:

- EK 120 vertical Simrad echosounder (frequency 120 kHz), used to measure the geometric and acoustic parameters of krill patches. The observation range was 10–130 m,
- QM MK II analog echointegrator, operated together with the EK-120 echosounder to measure the density of swarms.

To determine the regions of krill concentration the mean biomass of krill on the distance crossed during 1 h was calculated as follows:

$$\overline{B} = \frac{1852}{D} \sum_{i=1}^{n} \frac{W_i}{d_{ci}} ,$$

where: $\overline{B}$ = the mean biomass of krill (t/nm^2), D = distance crossed during 1 h (nm), n = number of krill swarms recorded during 1 h, W_i = the total weight of krill in swarm i (t), and d_{ci} = the length of swarm i (m).

The regions of biomass above 100 t per nautical square mile, noted on the distance not shorter than 10 nautical miles, were regarded as regions of krill concentrations.

The physical parameters of swarms (length, density, and total weight) were measured according to the method described by Kalinowski and Witek (1981).

3 Results

3.1 Krill Distribution with Respect to Hydrological Conditions

There was a significant relation between krill distribution and water circulation. Markedly larger quantities of krill occurred within frontal zones. At transects across the frontal zones (Fig. 1), larger biomass of krill was found at the sections of weakened and disturbed thermal stratification. This was shown on a transect made in March 1977 from the South Orkney Islands to the northeast through the Scotia Sea and across the Weddell-Scotia Confluence (Fig. 2), on the transect across a local frontal zone between oceanic and coastal waters of the Palmer Archipelago in February 1977 (Fig. 3), and on the transect across such a zone off South Georgia in April 1977 (Fig. 4). Such

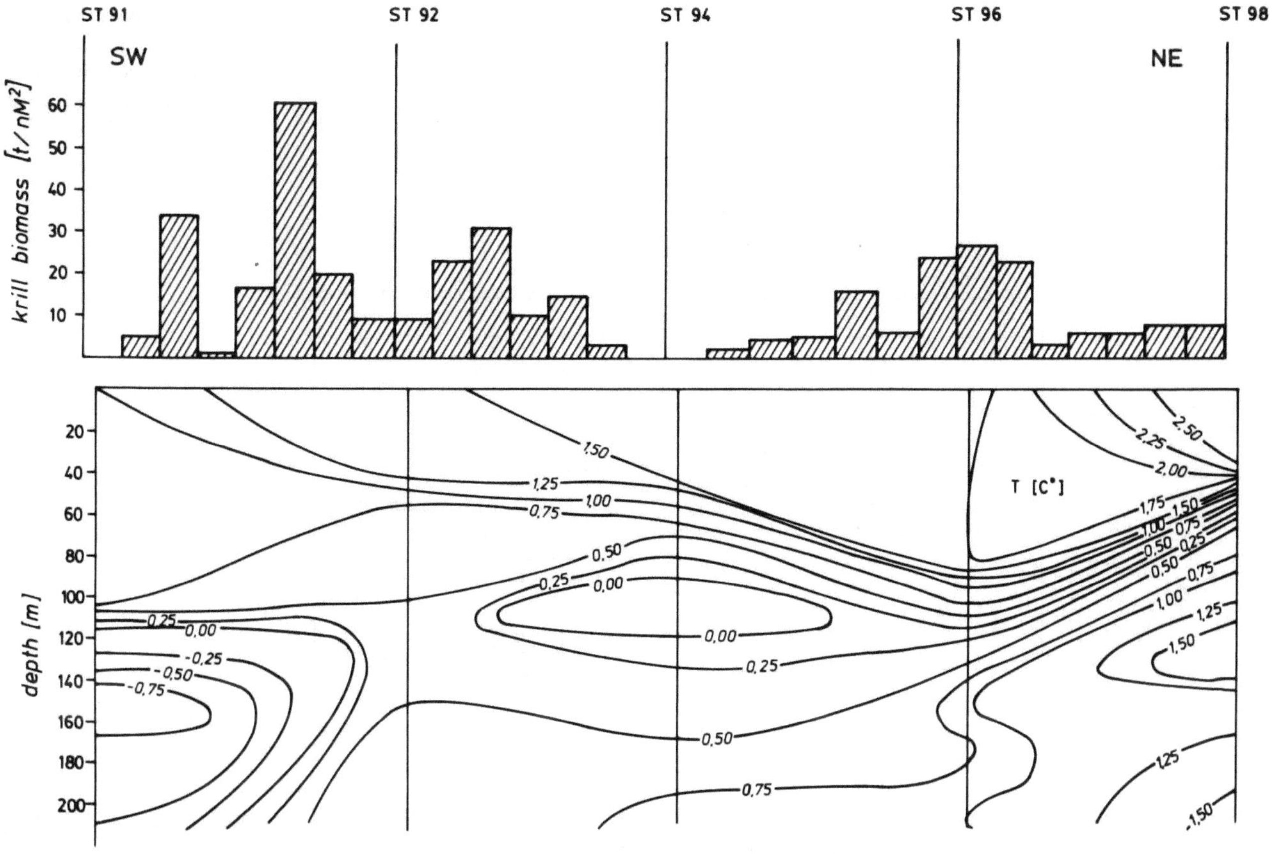

Fig. 2. Distribution of krill biomass and temperature (*T*) at transect A in the Scotia Sea (Fig. 1), March 4–6, 1977

Fig. 2. Distribution of krill biomass and temperature (*T*) at transect A in the Scotia Sea (Fig. 1), March 4–6, 1977

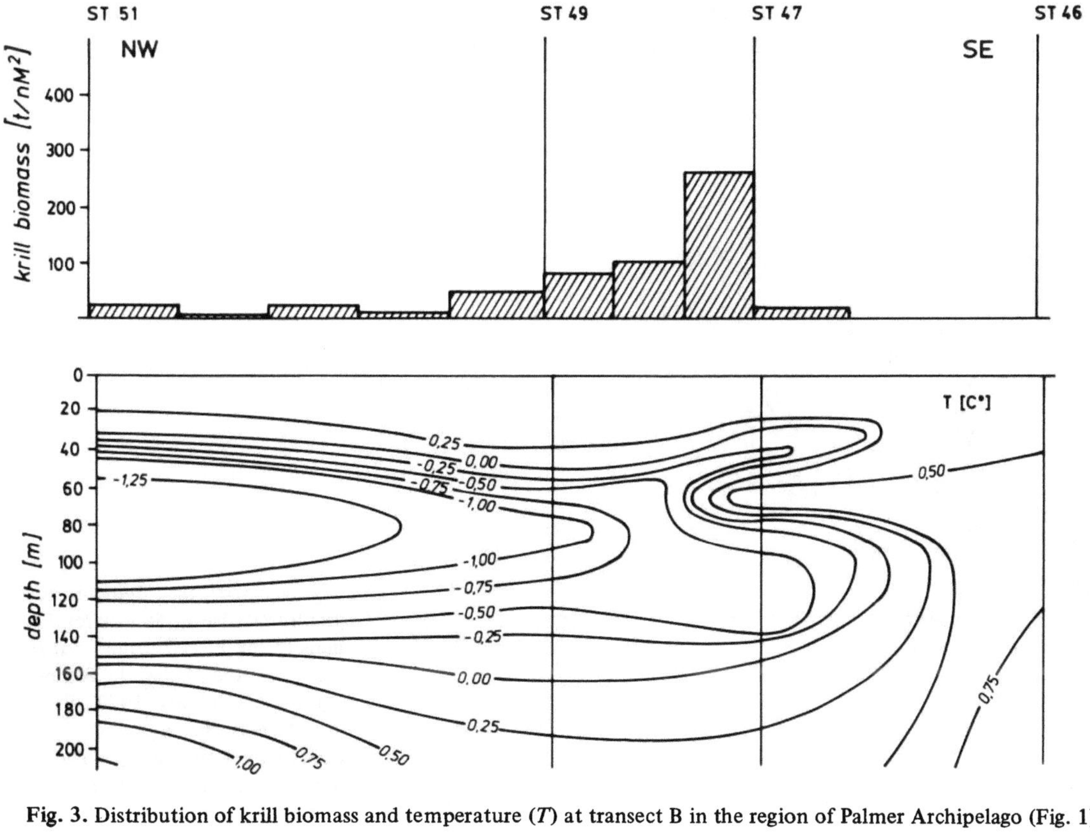

Fig. 3. Distribution of krill biomass and temperature (*T*) at transect B in the region of Palmer Archipelago (Fig. 1), February 21–22, 1977

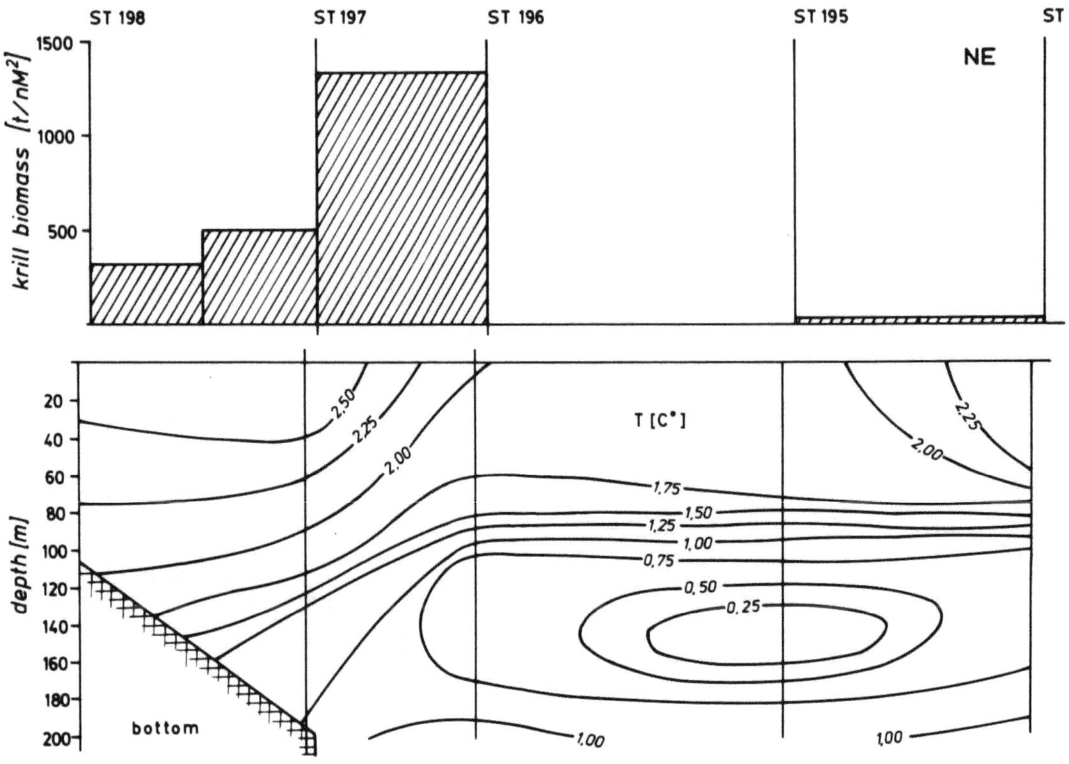

Fig. 4. Distribution of krill biomass and temperature (*T*) at transect C east of South Georgia (Fig. 1), April 14–15, 1977

deformations of thermal stratification were evidence of confluence of water and its uneven flow in a given area.

Maps of dynamic topography of the sea surface, reflecting the flow of currents, illustrate even better the relation between the location of krill aggregations and water circulation. They show that krill congregated in the area of meanders and eddies, especially in those places with high gradients of current velocity, i.e., where streams with intensive flow (high density of isolines) were next to the areas of slow drift, inside current meanders (Figs. 5–9).

3.2 Size of Krill Swarms

Figures 10 and 11 present the most frequent range of main krill parameters (length, density, and total weight of krill in a swarm) in particular regions. Since these para-

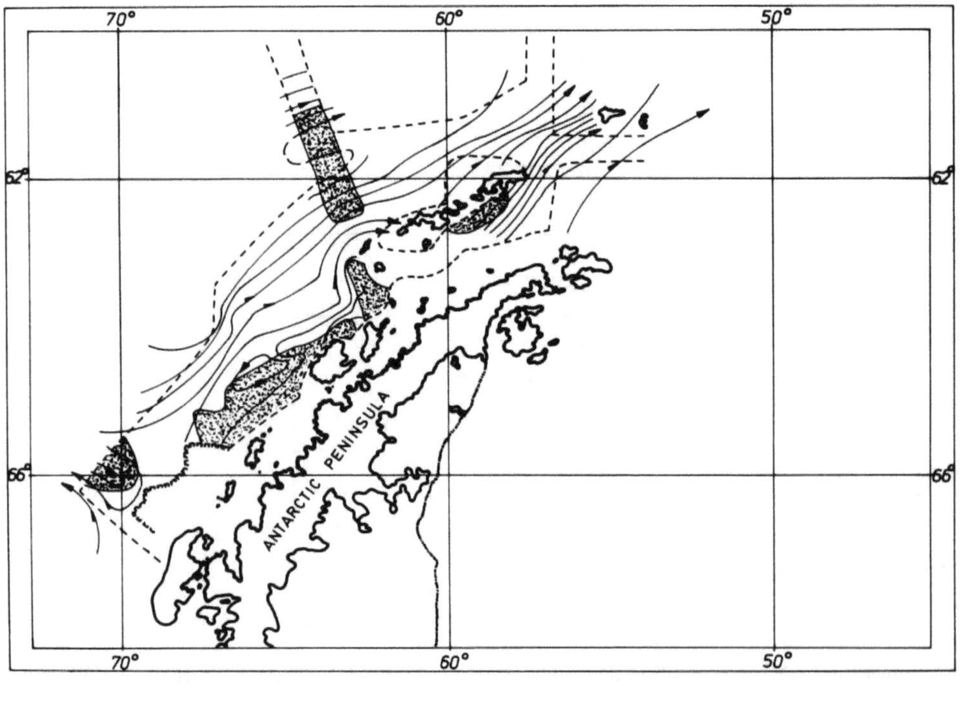

Fig. 5. Sea surface dynamic topography relative to 500 db level and density over 100 t/nm² (*dotted areas*) in the Antarctic Peninsula region in February 1977. *Dashed line* frames the region of hydroacoustic echosoundings. The isolines condensation testifies the higher speed of current

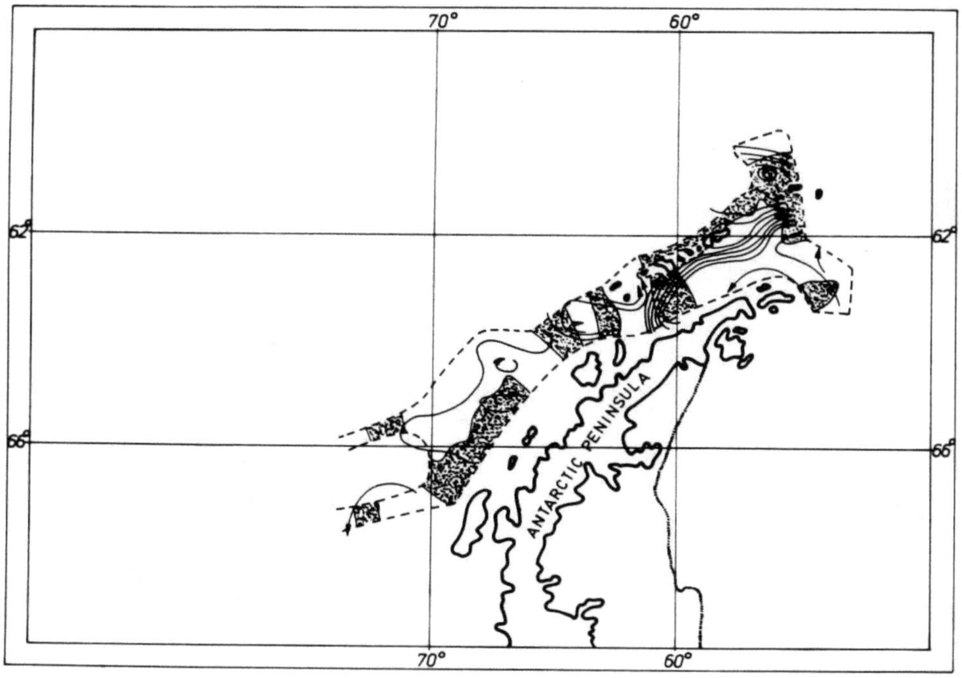

Fig. 6. Sea surface dynamic topography relative to 500 dbar level and areas of krill concentrations with biomass over 100 t/nm² in the Antarctic Peninsula region in January and February 1979. Comments as in Fig. 5

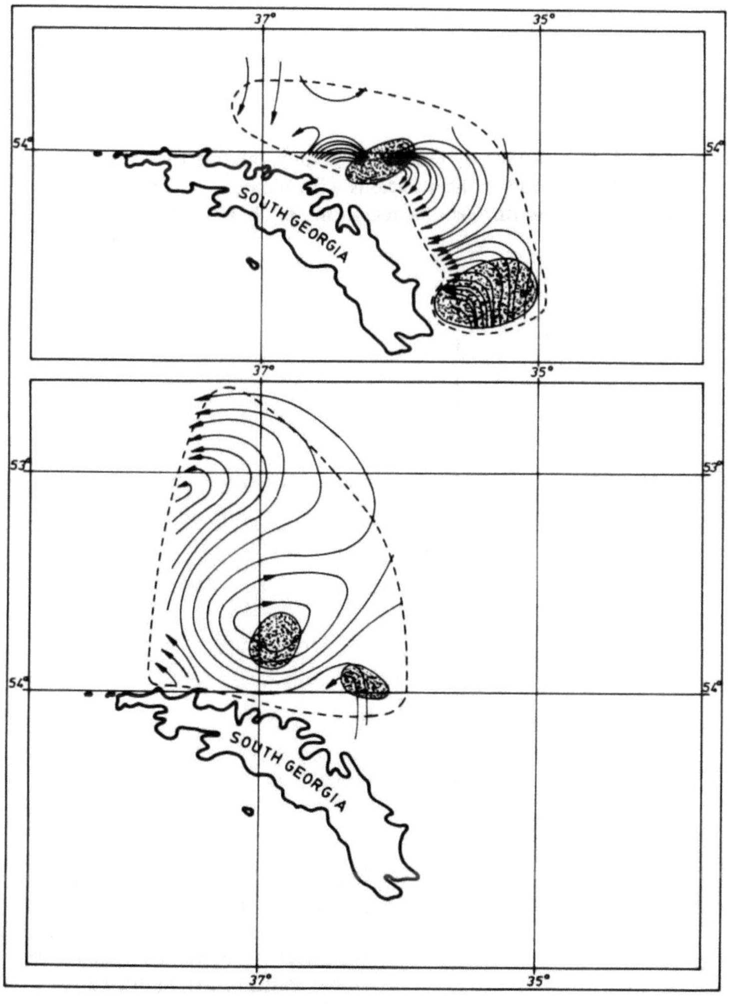

Fig. 7. Sea surface dynamic topography relative to 500 dbar level and areas of krill concentrations with biomass density over 100 t/nm² in the South Georgia region in 1976. *Above* February, *below* March. Comments as in Fig. 5

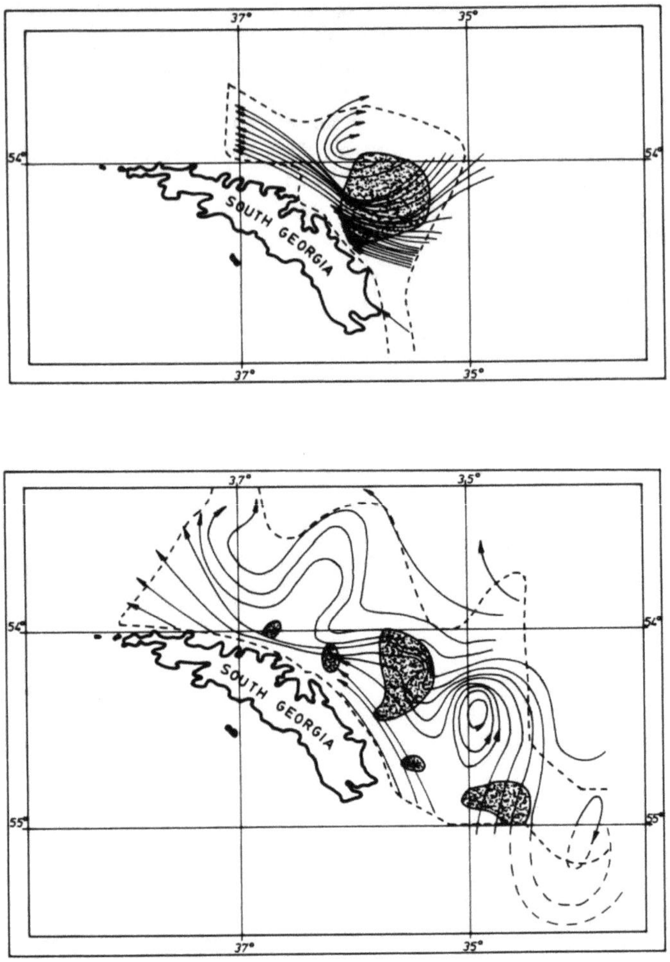

Fig. 8. Sea surface dynamic topography relative to 500 dbar level and areas of krill concentrations with biomass density over 100 t/nm² in the South Georgia region in 1977. *Above* March, *below* April. Comments as in Fig. 5

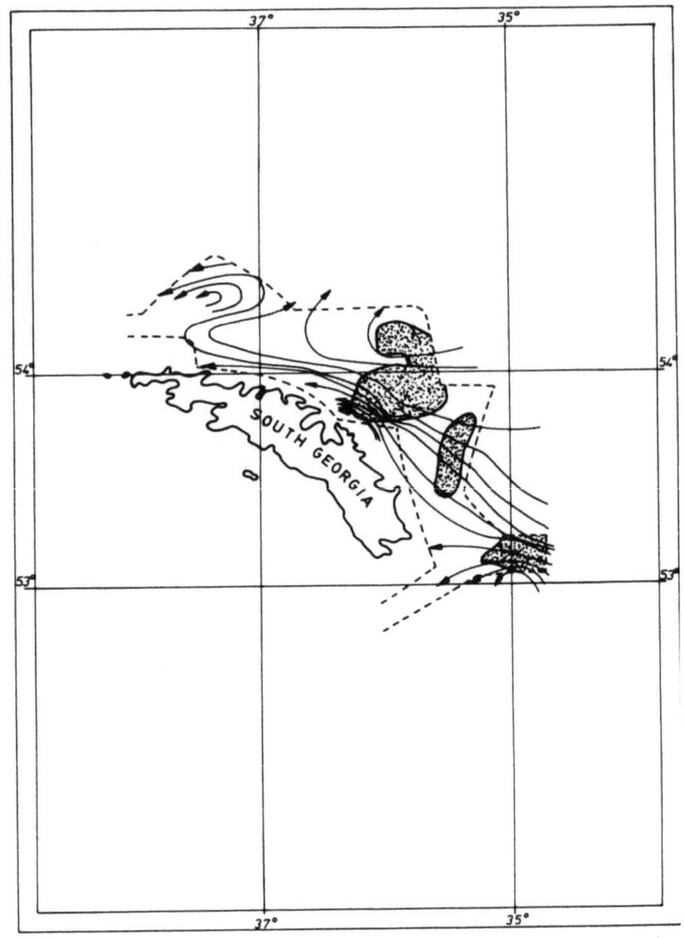

Fig. 9. Sea surface dynamic topography relative to 500 dbar level and areas of krill concentrations with biomass density over 100 t/nm² in the South Georgia region in March 1979. Comments as in Fig. 5

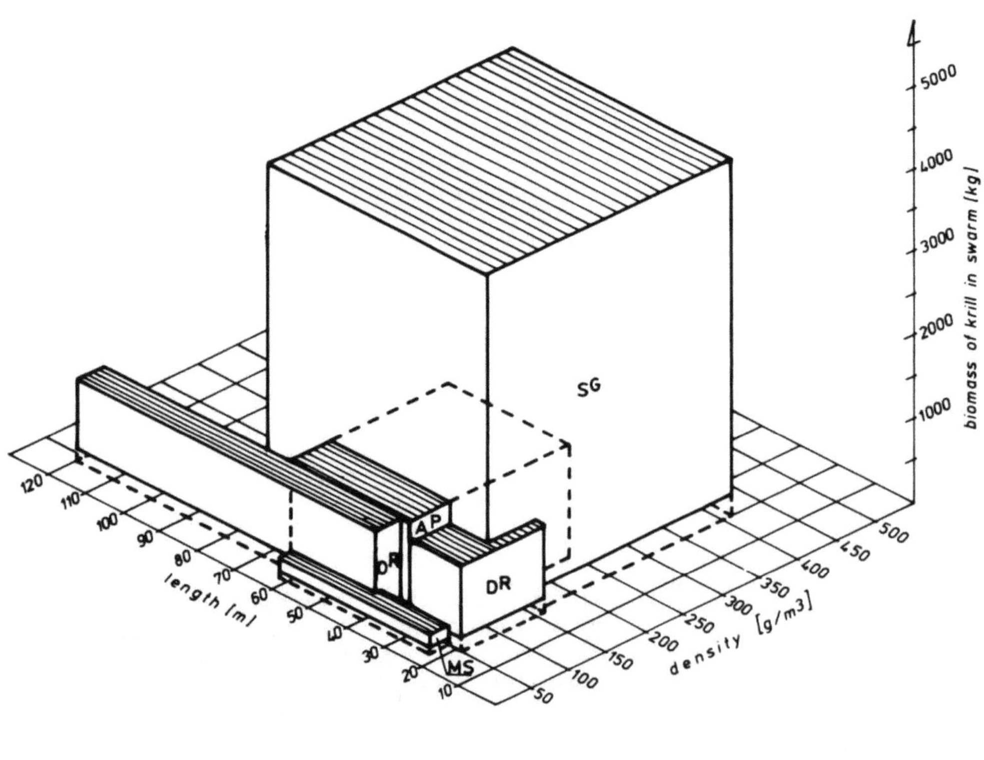

Fig. 10. The values of krill swarms parameters in different regions in summer 1976/77. *OR* South Orkneys; *AP* Antarctic Peninsula; *DR* Drake Passage; *SG* South Georgia; *MS* Scotia Sea

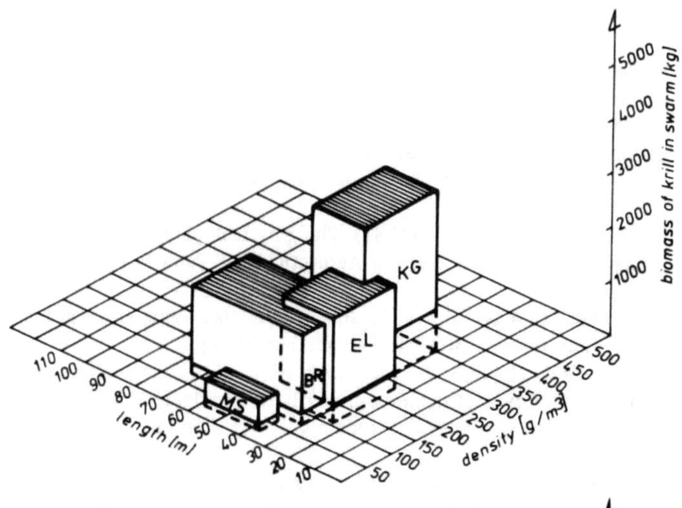

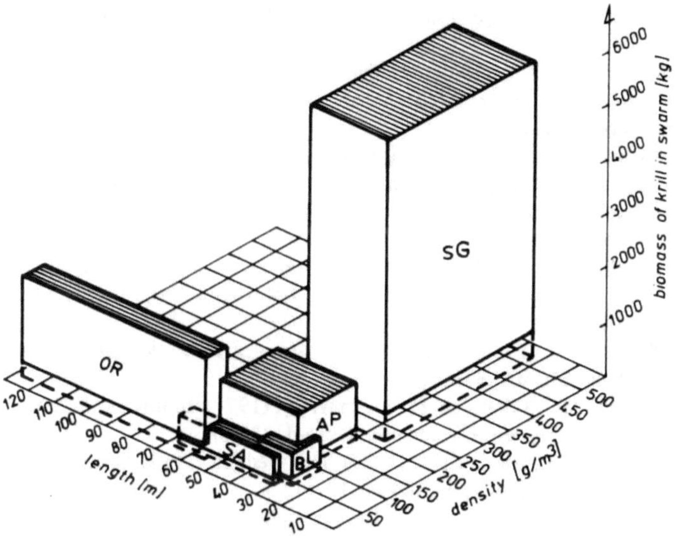

Fig. 11. The values of krill swarms parameters in different regions in summer 1978/79. *MS* Scotia Sea; *BR* Bransfield Strait; *EL* Elephant Island; *KG* King George Island; *OR* South Orkneys; *SA* South Sandwich Island; *AP* Antarctic Peninsula; *BL* Bellingshausen Sea; *SG* South Georgia

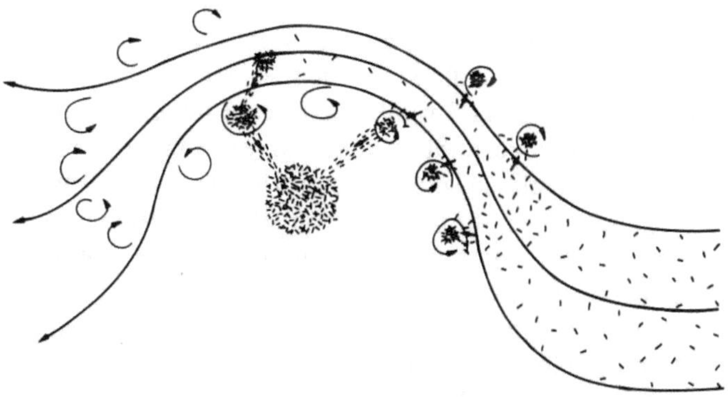

Fig. 12. Krill accumulation in current meanders. *Short lines* indicate krill individuals, *long arrows* indicate the direction of water flow, *short arrows* indicate the direction of active krill moving

meters are characterized by a log-normal frequency distribution, their ranges were determined by finding values corresponding to standard deviation from the arithmetic mean of logarithms.

The highest quantities of krill were recorded in swarms from the areas of the largest concentrations, especially from the shelf waters off South Georgia. The geometric mean of swarm length in these areas was 34 and 40 m, that of density 327 and 223 g/m^3, and that of total weight of krill in a swarm was 1596 and 1502 t in the 1976/77 and 1978/79 seasons, respectively.

In the open waters of the Scotia Sea, Bellingshausen Sea, and Drake Passage, swarms were generally composed of smaller quantities of krill (geometric mean of length of swarm 27–46 m, that of density 28–80 g/m^3, that of total weight 100–388 kg).

3.3 The Problem Concerning the Manner of Krill Concentration Formation

Although we can explain how currents cause the inflow of krill to a given area, it is difficult to find any hydrological mechanisms which would cause krill to stay in this area and form aggregations in spite of the constant outflow of water, balancing its inflow. Thus, it seems that besides hydrological factors, behavioral factors must play some role here: a specific reaction to turbulent water flow and a shoal-forming instinct. We propose the following reasoning.

In large stabilized currents, in the vast open spaces, the features of water flow resemble laminar flow. Krill carried by such waters are dispersed. The probability that individuals or swarms will meet is low; that is why swarms are composed of not too many individuals and large quantities of krill may be dispersed outside the swarms. Not having a reference point, and not feeling the turbulence, krill do not resist the current.

The situation is different in the areas of meanders and eddies. In the conditions of increased turbulence and high gradients of current velocity, current perceptibility for krill, especially those in swarms, may be higher. In the vicinity of land and submarine elevations which influence light conditions in the water, krill may find some permanent reference points for spatial orientation. In such conditions krill may show reflexes to find the stillest places. These reactions, reinforced by a shoal-forming instinct, would cause krill to stop and congregate in the centers of local eddies and current meanders, yet close to streams with intensive water flow (Fig. 12). Constant inflow of new krill would cause swarms to attain large dimensions. If such a situation existed over a sufficiently long period, a concentration would form.

Using simple calculations helps one to obtain an impression of the scale of the phenomenon of krill aggregation formation according to the mechanism proposed above. If the area of meanders and eddies were a square with a 30-nm side and the water inflow took place only through one side of the square at an average speed of 0.5 knots, the water would be replaced in this area after 60 h. If all the flowing krill were stopped in such a square, after 60 h the number of krill would double and after 1 month mean density of krill would increase 12 times. The smaller the area of concentration formation, the faster the water would be replaced and the more intensively krill would congregate – in our example inversely proportional to the length of the square side. For instance, if the side of the square were three times smaller (10 nm) and other conditions remained unchanged, after 1 month the number of krill would increase 36 times.

The most favorable conditions for concentration formation would exist near islands and submarine elevations, where meanders and eddies may be of a stationary nature. The best season would be summer, when light conditions enable individuals to maintain visual contact.

4 Discussion

The attempts made so far to explain the manner of krill aggregation dealing with concentrations, mentioned in the Introduction, generally described only the "mechanical" side of the problem. However, if mechanical factors were the only cause of planktonic animals' aggregation, then, besides krill concentrations, concentration of other planktonic organisms would have to form in the same areas. Although the distribution of other species is by no means even, none of them forms such concentrations as krill does. Therefore, there must be additional behavioral factors, characteristic of *Euphausia superba,* which stimulate concentration formation – most likely specific reactions to turbulent water flow and a shoal-forming instinct. There is enough evidence to support the existence of a shoal-forming instinct in krill (Mauchline 1980), while there is little information as regards reactions of krill to the current. One can suppose, for instance by analogy with mysids, some species which form swarms (Clutter 1969; Mauchline 1971) similar to those of *E. superba,* that these animals are quite sensitive to currents. Well-developed antennae of krill on which mechanoreceptors are concentrated also support such a supposition. It seems, however, that the problem may be solved only by direct observation in the environment and experimental investigations.

References

Clutter RI (1969) The microdistribution and social behaviour of some pelagic mysid shrimps. J Exp Mar Biol Ecol 3 (2):125–155

Elizarov AA (1971) Osobennosti dinamiki wod w mestach massovych skoplenij krila (*Euphausia superba* Dana). Trudy VNIRO 79:31–40

Kalinowski J, Witek Z (1981) The physical parameters of krill aggregations in the Western Antarctic. ICES Pap CM 1981/L:19, 4 pp

Kalinowski J, Witek Z (1985) Scheme for classifying aggregations of antarctic krill. BIOMASS Handb No 27, 9 pp

Makarov RR, Naumov AG, Shevtsov VV (1970) The biology and the distribution of the antarctic krill. In: Holdgate MW (ed) Antarctic ecology, vol 1. Academic Press, London New York, pp 173–176

Marr JWS (1962) The natural history and geography of the Antactic krill (*Euphausia superba* Dana). Discovery Rep 32:33–464

Maslennikov VV (1972) O vlijanii dinamiki vod na raspredelenije *Euphausia superba* Dana w rajonie o-va Južnaja Georgija. Trudy VNIRO 112:50–56

Maslennikov VV (1979) Regionalnye osobennosti Antarkticzeskoj poverchnostnoj vodnoj massy w jugo-zapadnoj czasti atlanticzeskogo sektora Antarktiki. In: Antarktika, vol 18. Izdat. Nauka, Moskwa, pp 118–122

Mauchline J (1971) Seasonal occurrence of Mysids (Crustacea) and evidence of social behaviour. J Mar Biol Assoc UK 51 (4): 809–825

Mauchline J (1980) Studies on patches of krill *Euphausia superba* Dana. BIOMASS Handb 6, 36 pp

Parsons TR, Takahashi M, Hargrave B (1977) Biological oceanographic processes. Pergamon, Oxford, 332 pp

Patterson SL, Sievers HA (1980) The Weddell Scotia Confluence. J Phys Oceanogr 10:1584–1610

Rakusa-Suszcewski S (1978) Environmental conditions within krill swarms. Pol Arch Hydrobiol 25 (3):585–588

Stavn RH (1971) The horizontal-vertical distribution hypothesis: Langmuir circulation and Daphnia distributions. Limnol Oceanogr 16:453–466

Witek Z, Kalinowski J, Grelowski A, Wolnomiejski N (1981) Studies of aggregations of krill (*Euphausia superba*). Meeresforschung 28 (4):228–243

Wolnomiejski N, Chłapowski K, Porębski J, Garbacik-Wesołowska A (1978) Obserwacje nad ekologią antarktycznego kryla (*Euphausia superba* Dana). Stud Mater MIR, Gdynia, A 23, 49 pp

Zubov NN, Mamaev OI (1956) Dinamitcheskij metod vytchislenija elementov morskich tetchenij. Gidrometeoizdat, Leningrad

Variability in the Physical and Biotic Environment of the Antarctic Krill (Euphausia superba Dana), South of Africa: Some Results and a Conceptual Appraisal of Important Interactions

D. G. M. MILLER[1] and P. M. S.. MONTEIRO[2]

Summary. Data from recent research (particularly the First International BIOMASS Experiment) south of Africa are used to describe the scale of variability in the distribution of hydrographic conservative properties and primary production. These are examined in relation to quantifiable variations in krill (*Euphausia superba* Dana) density obtained from hydroacoustic records. The result so obtained are integrated with currently available information in order to qualify understanding of mechanistic interactions between physical processes (especially turbulence in near-surface waters), primary production and krill. A simple conceptual model is presented to describe interactive elements and suggestions for future studies are made.

1 Introduction

Marine systems comprise dynamic, nonhomogeneous, fluid mediums with an enormous capacity for fixing heat, matter, and momentum (Bakun 1986). It is not difficult to demonstrate that the biological and environmental elements of such systems are inherently variable (Longhurst 1981). However, adequate quantification of this variability and improvement of our empirical understanding of important mechanistic linkages constitutes a persistent challenge (Steele 1978; Bakun 1986).

In the Southern Ocean, much interest has been focused on the relationships between physical and biological variability. To date, as much as a result of an ease of definition as of heuristic appeal, studies have tended to concentrate on hydrographic variability at frontal discontinuities and on any concomitant effects on biological productivity nearby (Lutjeharms et al. 1985). Such studies have demonstrated a close affinity between biomass and productivity on the one hand (Plancke 1977; Jacques and Minas 1981; Allanson et al. 1981) and major frontal features (i.e., the Sub-Tropical Convergence, the Antarctic Polar Front and the continental water boundary) on the other. Results have also indicated that not only do sea-surface contiguous zones play an important role in promoting localized nutrient enhancement in the Southern Ocean (cf. Lutjeharms et al. 1985), their subsequent effects on biological productivity are in turn intermittent, highly variable, and geographically restricted.

Since the limitation of essential nutrients is not a universal feature of Antarctic waters (El-Sayed 1984), it is generally accepted that primary production is affected predominantly by variable irradiance levels and temperature (Holm-Hansen et al. 1977; Tranter 1982; Tilzer and Dubinsky 1987). Of several somewhat related factors determining light available for phytoplankton growth, day length is the most obvious (Tranter 1982). A marked seasonal trend has been demonstrated in Antarctic phytoplankton productivity (Foxton 1964; El-Sayed et al. 1980) and, during summer, photosynthesis is greatest in near-surface waters (shallower than about 20 m depth). The depth of maximum phytoplankton activity is summarily related to the light absorption spectra of the phytoplankton species concerned (Holm-Hansen et al. 1977) and levels of primary productivity are as much a function of light availability as of light quality.

As light attentuates with depth (Z), primary production is also limited by the extent of vertical mixing in surface waters (Sverdrup 1953). Because of the persistently inclement weather (particularly high wind stress), evidence shows that primary production in the Antarctic summer is a direct function of water column stability (Fogg 1977; Tranter 1982; Heywood and Whitaker 1984). In these terms, a physical process (water column stability) serves as a critical link between an abiotic energy source (light) and biological productivity. The scale of this effect is in turn dependent on meteorological phenomena (e.g., wind and barometric pressure) most affecting local or regional water dynamics at a variety of temporal scales (Hellmer et al. 1985). For this reason, it is important that any investigation of the interactions between physical variability and biological productivity should consider stability of the water column.

With respect to higher-order biotic interactions, many authors conclude that since the highly asymmetrical distribution of Antarctic phytoplankton cannot otherwise be fully explained, zooplankton grazing constitutes an important limiting factor of local primary production and phytoplankton distribution (Weber et al. 1986; von Bodun-

1 Sea Fisheries Research Institute, P/Bag X2, Roggebaai, 8012, South Africa

2 Department of Oceanography, University of Cape Town, P/Bag, Rondebosch, 7700, South Africa

Antarctic Ocean and Resources Variability
(ed. by D. Sahrhage)

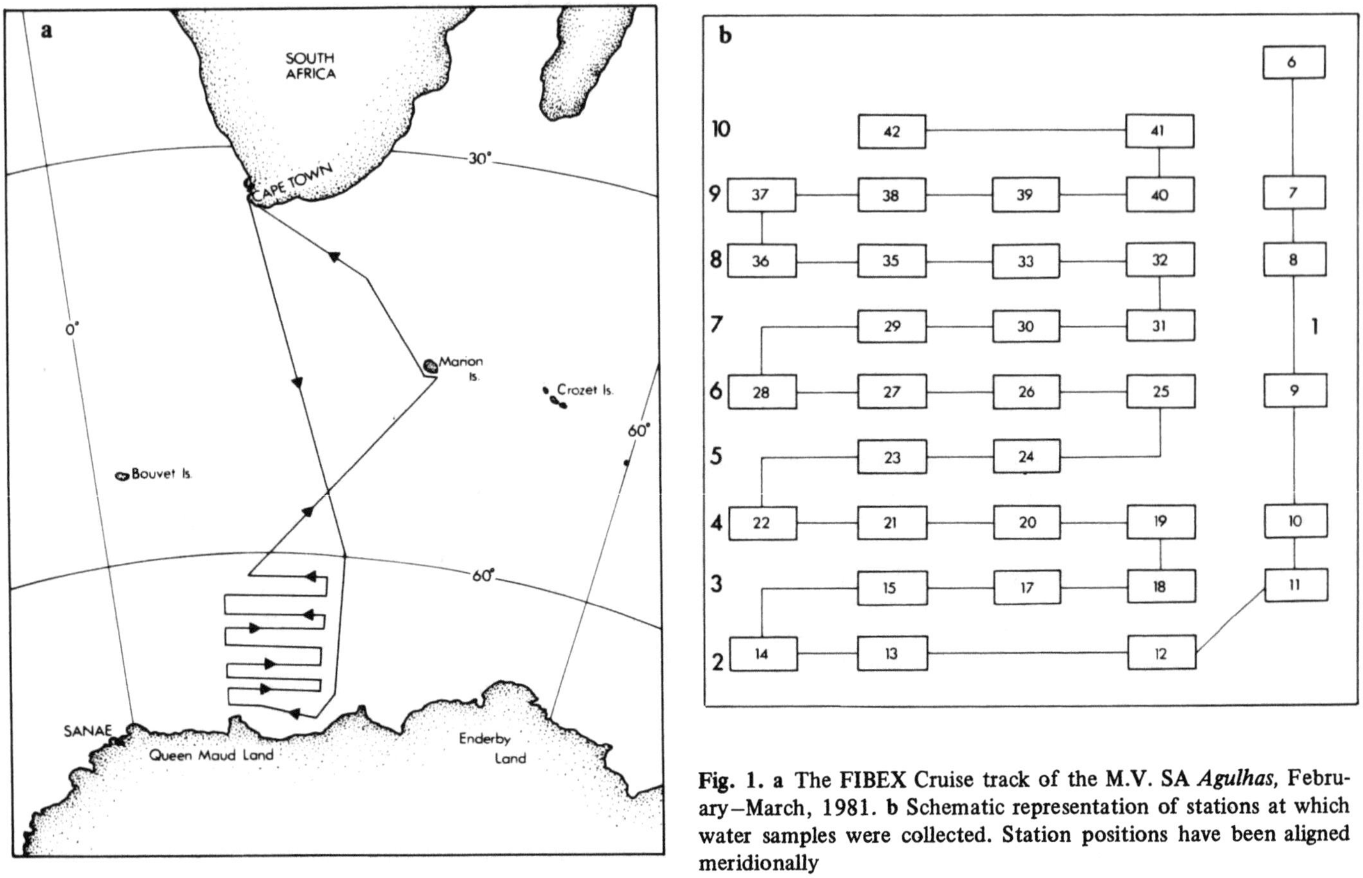

Fig. 1. a The FIBEX Cruise track of the M.V. SA *Agulhas*, February–March, 1981. **b** Schematic representation of stations at which water samples were collected. Station positions have been aligned meridionally

gen 1986). In these terms, krill (*Euphausia superba* Dana) is considered a key link between primary production and higher trophic levels (Knox 1984). Despite the fact that limited, localized (> 1–10 km) coherence between krill and phytoplankton has been demonstrated (Morris et al. 1984; Weber 1984; Weber and El-Sayed 1985), apparent incongruencies between krill grazing and phytoplankton on a more broad scale (± 100–1000's km) (Holm-Hansen and Huntley 1984; Miller et al. 1985) strongly emphasize a need to consider inter-relationships between these two parameters. As discussed by Haury et al. (1978) and shown by Murphy et al. (this Vol.), this requires the development of a suitable conceptual framework to explain both temporal and spatial variability in critical parameters. It also entails the proper spatial and temporal definition of the environmental matrix in which interactive linkages between physical parameters, primary production, and krill occur.

The main objective of this chapter, therefore, is to investigate how physical processes may affect biological productivity in an open-ocean area of the south-west Indian Ocean. Although substantial data on the ocean dynamics of this region exist (e.g., Lutjeharms 1985), data on linkages between physical processes and variability/distribution of biotic parameters, principally primary productivity and krill abundance, are limited. Important variable interactions are therefore considered in the light of previous statistical analyses of interactive data (Weber 1984; Weber and El-Sayed 1985; Weber et al. 1986), other available information (Hampton 1985; Miller et al. 1985) and in accordance with possible mechanistic associations between physical processes and krill.

2 Materials and Methods

Data were collected during FIBEX (First International BIOMASS Experiment) (16 February to 10 March, 1981) by the M.V. SA *Agulhas* in the southwest Indian Ocean (62–69°S; 15–30°E). The survey consisted of nine zonal transects, extending from the pack-ice edge northward (Fig. 1a and b). Water samples were collected from seven discrete depths at 32 stations, at 21 of which phytoplankton production was measured. Eight-liter Nisken samples were taken from depths corresponding to the penetration of 100, 40, 18, 8, 5, 1.25, and 0.25% of surface irradiance determined by use of a Secchi disk (Weber and El-Sayed 1985). All samples were analyzed according to prescribed methods for chlorophyll measurement (Evans and O'Reilly 1983). Integrated primary productivity in the euphotic zone (above the 1% light level) was determined by simulated in situ incubation on board ship (Strickland and Parsons 1972; O'Reilly and Thomas 1983). It was assumed that the primary productivity measured was net production.

Mean krill surface density per integration interval, P_i, was estimated from echosounder data following the procedures outlined by Hampton (1985). Given a strong periodicity between day and nighttime density estimates, mean values were calculated for day and night separately. When taking various distributional trends into consideration, nighttime density values were corrected by an amount proportionate to the mean daytime value. The following expression was used to derive mean density, P_d, at a variety of spatial scales:

$$P_d = \frac{\sum_{i=1}^{N} P_i L_i}{L_d} , \qquad (1)$$

where L_i is the distance steamed during the ith echo-integration interval, L_d is the total distance steamed for the interval(s) being considered and N is the total number of intervals concerned. The L_i/L_d values act as weighting factors and account for intervals of unequal distance.

3 Results

3.1 Physical Environment

The physics of the upper 200 m of the water column plays a vital role in biological productivity through its control of the environmental quality of both primary and secondary producers (Parsons and Takahashi 1973). Conventionally, mechanisms considered most important in the Southern Ocean are the depth of the surface-water layer, the 1% light absorption depth and the susceptibility of the water column to wind-induced turbulence (Heywood and Whitaker 1984).

Two scales of variability in the physical environment of the survey area were evident. These were the large-scale, accounting for spatial variability > 10 km and a seasonal time scale, and the small scale, which acounted for variability at a vertical resolution of 1–100 m and for hours to days. In terms of the categorization offered by Murphy et al. (this Vol.), these scales would correspond to interactions involving krill "concentrations" and individuals/"swarms" respectively.

3.1.1 Large-Scale Variability

Meridional Variability. Applying criteria originally proposed by Deacon (1937) and modified by Carmack (1977), four water masses were identified and these are schematically illustrated in a composite t-S section (Fig. 2). This section also demonstrates the t-S range for each water mass and the relative strength of their respective boundary zones. Typical meridional distributions of conservative properties (temperature t, salinity S and density σt) to a depth of 200 m are shown in Fig. 3a–c along a line nominally fixed at 28°E.

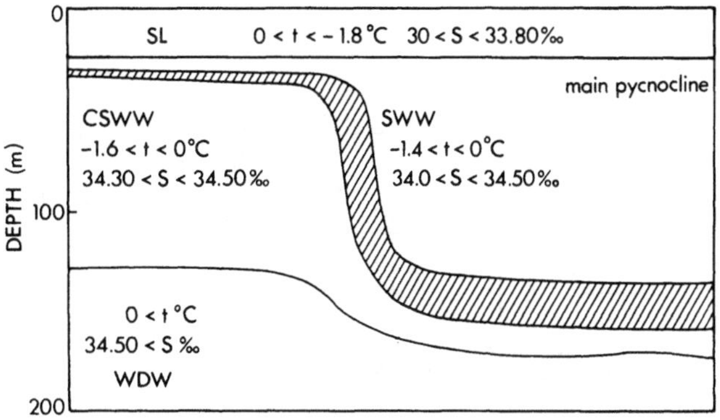

Fig. 2. A schematic representation of water masses present during the survey

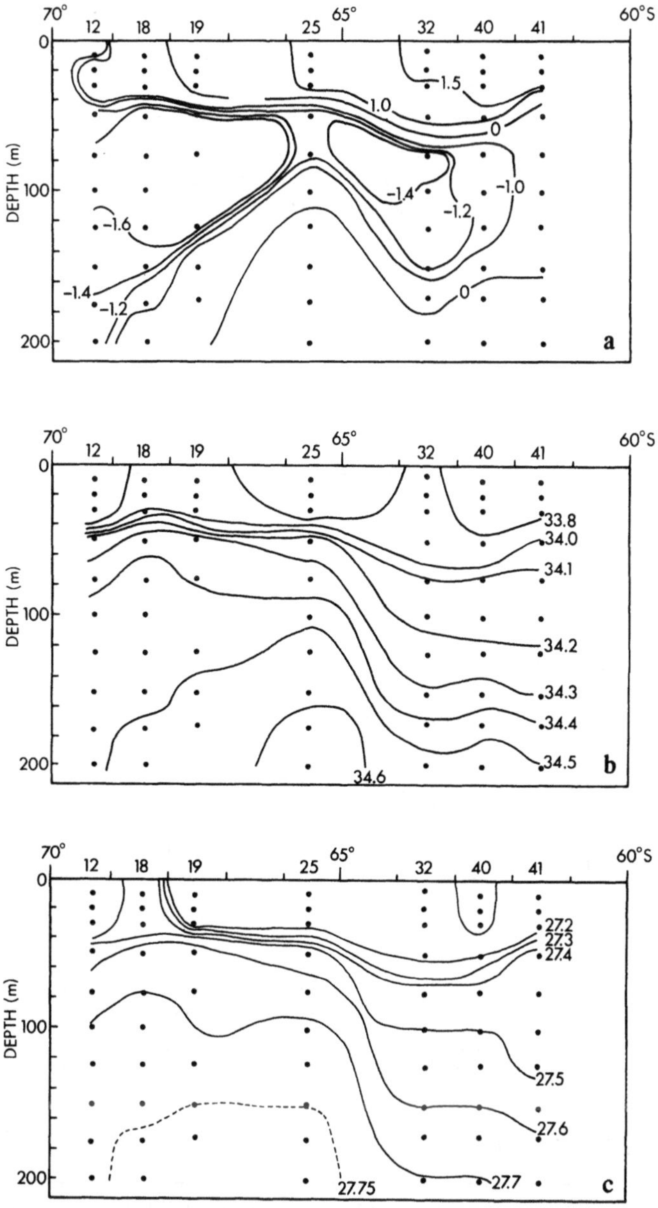

Fig. 3a–c. Meridional (along 28°E) **a** temperature, **b** salinity, and **c** density (σt) sections

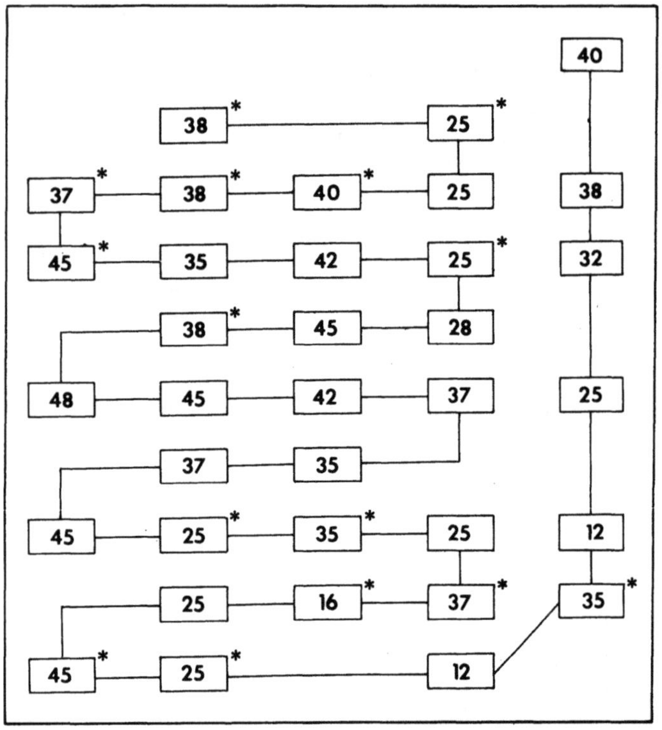

Fig. 4. Distribution and variation in depth (m) of the near-surface water (SL) layer above the pycnocline. (* Stations at which some degree of thermal stratification was evident. See also Fig. 12)

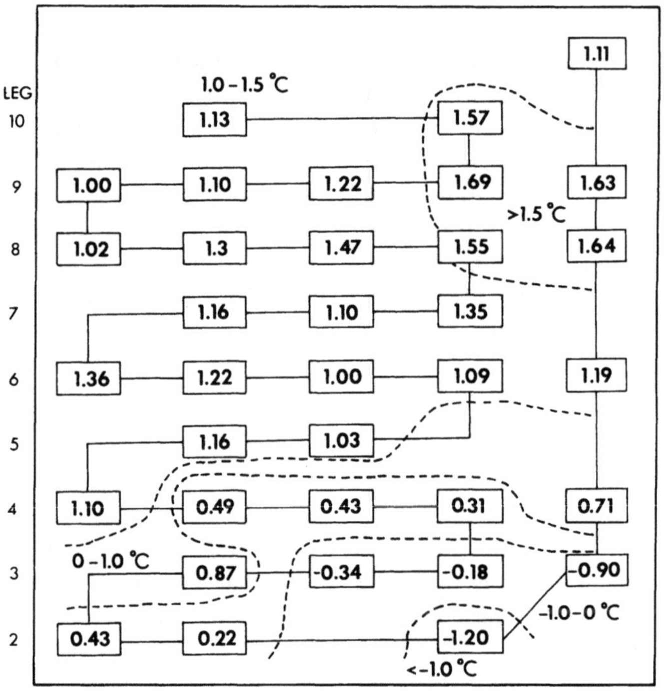

Fig. 6. Near-surface water (SL) salinity (‰) distribution and variation

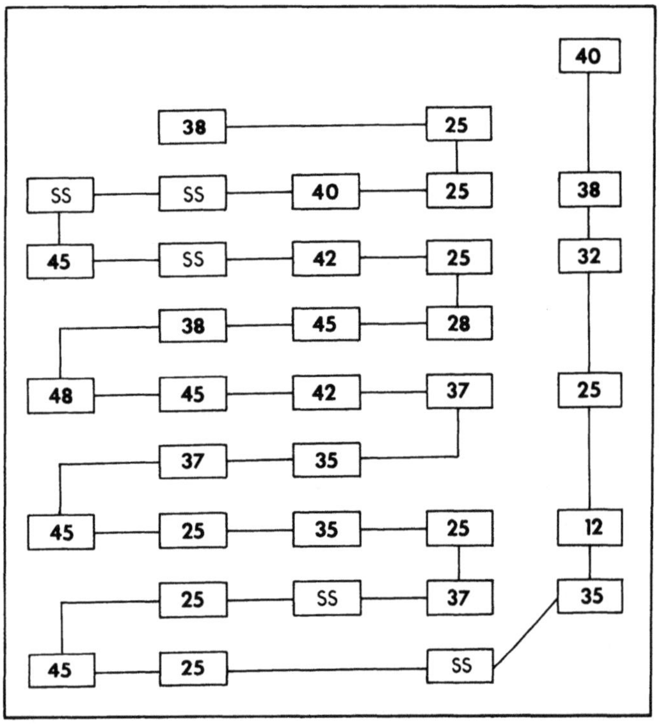

Fig. 5. Near-surface water (SL) temperature (°C) distribution and variation

The survey was thought to span the Antarctic Divergence (ADZ) (for definition cf. Deacon 1982; Lutjeharms and Valentine 1984 and others). Although no surface expression of divergent motion (i.e., upwelling) was evident, isopleths (Fig. 3a–c) showed a steep upturn in the weakly stratified deep waters below the pycnocline, between about 50 and 150 m depth. This suggested weak frontal expression in the vicinity of 65°S. The pycnocline appeared to be more strongly stratified in the south, thereby suggesting greater upwelling activity and steeper salinity gradients (Fig. 3b and c).

The near-surface layer (SL) was separated from the underlying winter (SWW) and continental shelf water (CSWW) by a strongly developed pycnocline at 30–50 m (Fig. 3c). In open Antarctic waters, the near-surface pycnocline is a seasonal feature which develops through the austral summer as a result of reduced wind stress, solar heating, and the concomittant surface-ice thaw (Tchernia 1980; Tranter 1982; Tilzer and Dubinsky 1987). The SL is thus generally warmer and less saline than the underlying SWW (± 0°C, 33.80‰ vs. ± –1.0°C, 34.00‰ respectively). Its depth varies according to the opposing effects of past wind activity (which tends to deepen it) (Oakey and Elliott 1982) and the intensity of any vertical advection (Tranter 1982). It can be seen from Fig. 4 that the depth of the SL (Mean = 34.63 m; S.D. = ± 8.78; Range = 12–48 m) and hence the pycnocline, varied throughout the survey area although it was marginally deeper in the northwest.

Interstation Variability. Interstation variability in SL temperature and salinity are shown in Figs. 5 and 6, where values

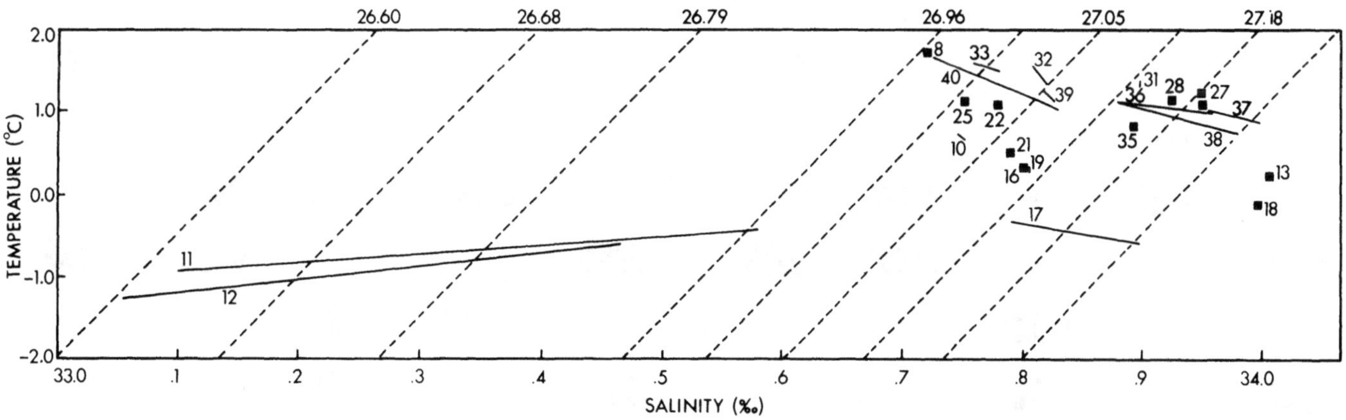

Fig. 7. Temperature-salinity (t-S) distribution in the SL (scaled according to the relative contributions of temperature (t) and salinity (S) to density (σ_t) (1:15 after Ruddick 1983)

represent a mean over the depth range of the SL. Surface water temperatures were between −1.2 and 1.69°C (Mean = 0.88°C; S.D. = ± 0.69) and salinity ranged between 33.06 and 34.07‰ (Mean = 33.82‰; S.D. = ± 0.21). This variability will be discussed later. The individual station variations of these parameters above the pycnocline were ± 0.0–0.62°C (Mean = 0.16°C) and ± 0.0–0.65‰ (Mean = 0.06‰) respectively.

The depth of the SL, which, as already stressed, is a function of the immediate wind history (cf. also Pollard et al. 1973; Brundrit 1985), varied between 12 and 50 m with most stations falling between 34 and 45 m (Fig. 4). The spatial variability of temperature and salinity were different, with the former increasing to the north, while the latter increased to the west (Figs. 5 and 6). This was most probably the result of the seasonal evolution of t-S properties under the two complementary forcing mechanisms of solar heating for temperature– and wind-induced turbulence entraining more saline SWW into the SL for salinity.

The greatest thermohaline stability above the pycnocline was observed on the two most southerly legs (Legs 2 and 3 in Fig. 1b) and in the extreme northwest (the western ends of Legs 8 and 9 – Fig. 1b). The t-S characteristics of these two areas are quite different (see Stations 11, 12, 17, 16 vs. Stations 36, 37, 38 in Fig. 7), which implies that stratification was induced by different processes operating at various stages of a seasonal cycle. The development of this cycle is beyond the scope of this chapter, although similar seasonal ocean kinetics have been demonstrated elsewhere in the Southern Ocean (e.g., Martinson et al. 1981).

3.1.2 Small-Scale Variability

Wind-induced turbulence in the SL was assumed to be the principal process determining variability in the physical and biotic environments at both the small temporal and spatial scales defined above. As such, wind-induced turbulence would have two effects; (a) to deepen the SL relative to the 1% irradiance level, and (b) to cause vertical displacement of phytoplankton at a time scale less than the photosynthetic light adaptation rate (cf. discussion in Tranter 1982; Heywood and Whitaker 1984; Sakshaug and Holm-Hansen 1986). The extent of wind-induced turbulence thus affects irradiance levels, thereby controlling photosynthesis (Lewis et al. 1984a).

As turbulence was not measured directly, t-S data from the SL were used to obtain a qualitative index of turbulence based on the degree of stratification at each station. The stations at which primary productivity was measured were categorized into either (a) a stable, thermally stratified (SS) condition where turbulence was at a minimum, and (b) a well-mixed (WM) condition with maximal turbulence. An SS condition would arise in the absence of significant wind forcing (cf. Oakey and Elliott 1982) with solar heat input inducing a weak but measurable thermal gradient near the surface (Martinson et al. 1981). The turbulence categories derived above were integrated with biological variables (see below) and differ from the stability quotient (which was based on the 0.25% light absorption depth and on limited near-surface density information) used by Weber and El-Sayed (1985).

3.2 Biological Variability

3.2.1 Phytoplankton Productivity and Distribution

The principal scales of variability considered were; (a) the horizontal scale, which is governed by large-scale physical variability of physical processes such as currents, weather, and light on a seasonal time scale, and (b) the vertical scale, which is governed by synoptic weather conditions, turbulence, diel light variation, and is superimposed on the large-scale seasonal signal.

Horizontal Variability. Integrated primary productivity and chlorophyll a over the 0–40 m depth range are shown in Figs. 8 and 9. For both parameters, maxima were found at the meridional extremes of the grid where primary production exceeded 15 mg C/m^2/h and chlorophyll a biomass

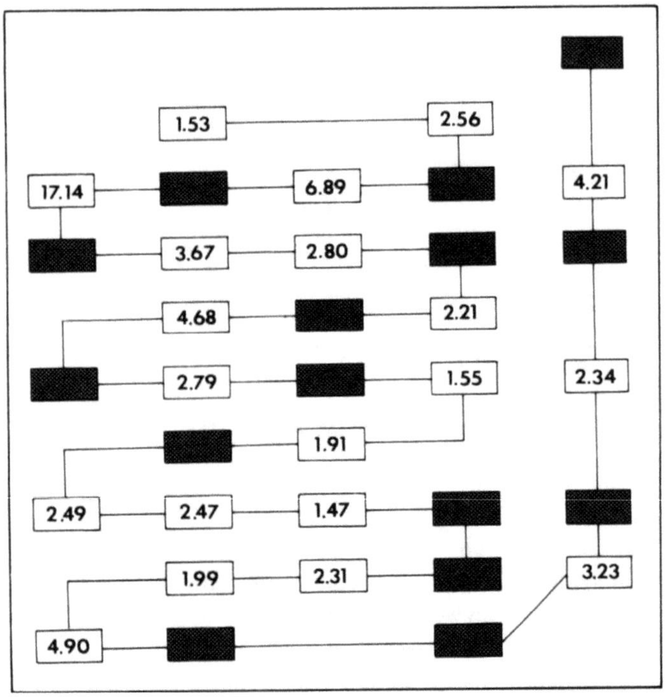

Fig. 8. Distribution and variation of integrated chlorophyll a (mg Chl-a/m²) over the 0–40 m depth range

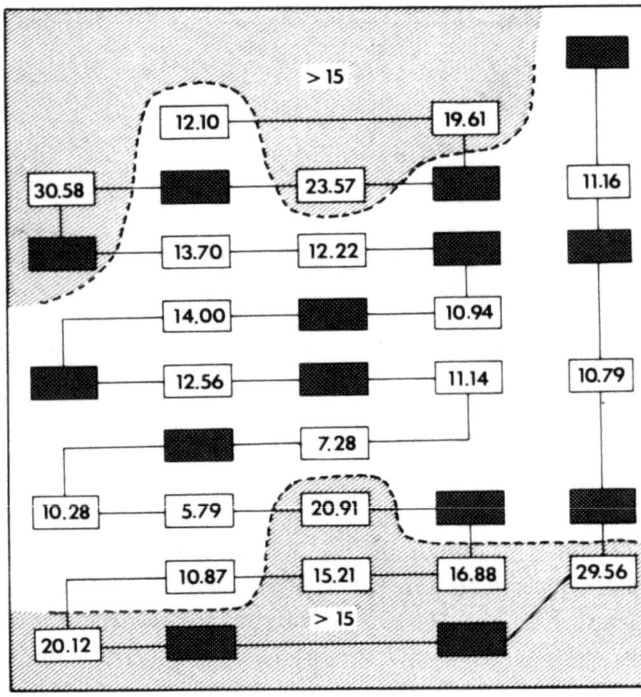

Fig. 9. Distribution and variation of integrated primary production (mg C/m² /h) over the 0–40 m depth range

3 mg/m². The overall ranges were 5.79–30.58 mg C/m²/h and 1.47–17.14 mg Chl-a/m² respectively. These ranges fell within limits reported from other Antarctic areas (Heywood and Whitaker 1984). The cut-off depth for primary production was taken as 40 m, since minimal production was measured deeper.

Primary production maxima were associated with areas of marked stratification in the SL in the northwestern and southern areas of the grid (Figs. 4 and 9). Although the current methodology used to estimate ^{14}C uptake does not realistically duplicate the turbulence regime, it is suggested that increased stratification with minimal turbulence would have given rise to conditions where higher photosynthetic rates were sustained (cf. Marra 1980; El-Sayed 1984; El-Sayed and Taguchi 1981). A study undertaken in the same area (Bidigare et al. 1986) also obtained a similar horizontal distribution of chlorophyll with two meridional maxima being separated by a region of lower biomass. Bidigare et al. concluded that this was the result of physical and biological conditions consistent with the late austral summer.

Vertical Variability. As discussed above, variability of primary production in the SL was a function of irradiance. A cumulative frequency distribution of primary production maxima with depth for all the stations (Fig. 10) indicated that maximum production occurred between 5 and 15 m with no production being measured deeper than 40 m. The production cut-off depth was above the 1% irradiance level (± 50 m), suggesting that the accepted convention of using this level in the estimation of areal production was inappropriate for the present survey. In contrast, the chlorophyll maxima (Fig. 11) were predominantly deeper than 40 m, suggesting that maxima were only associated with growing phytoplankton at a few stations. The stations at which primary production and chlorophyll maxima coincided were located at the meridional extremes of the

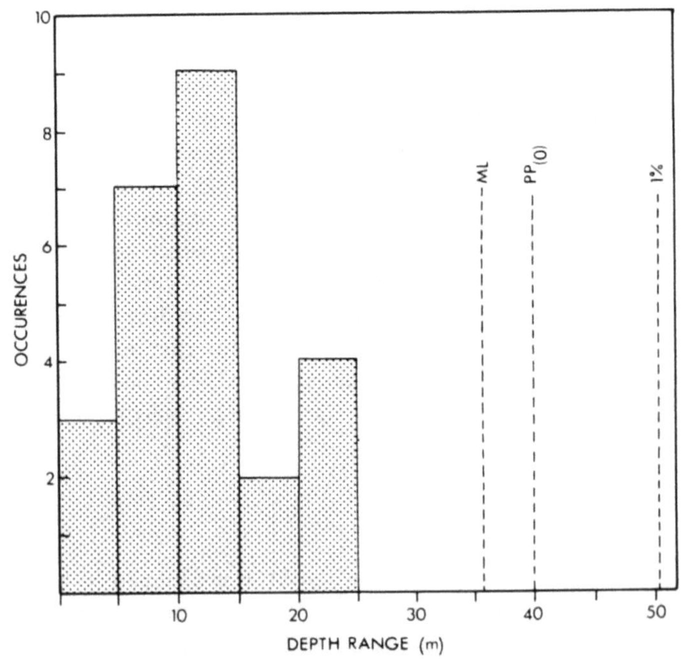

Fig. 10. Cumulative frequency of primary production maximum with depth (*ML* Mean depth of near-surface water layer; $PP_{(0)}$ Primary production "cut-off" depth; *1%* 1% light absorption depth)

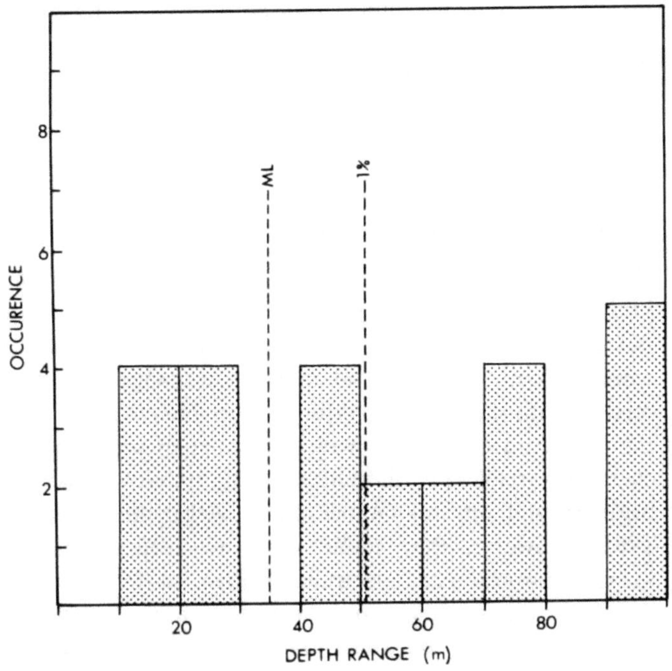

Fig. 11. Cumulative frequency of chlorophyll-a maximum with depth (*ML* Mean depth of near-surface water layer, *1%* 1% light absorption depth)

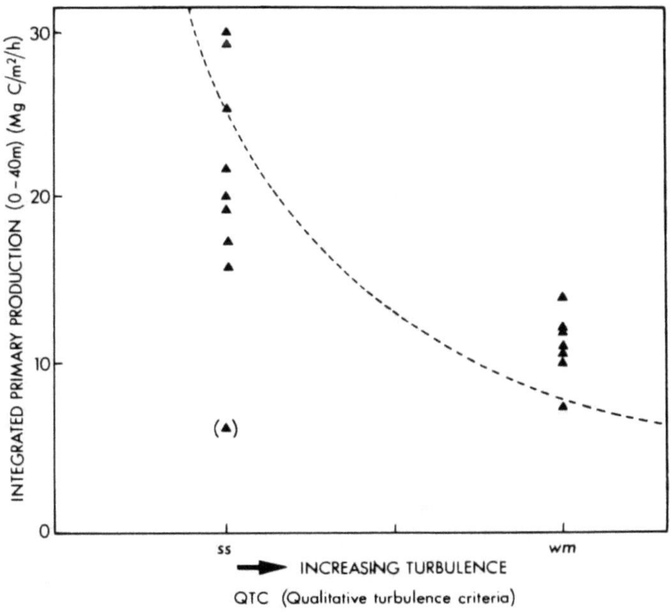

Fig. 12. Integrated primary production (mg C/m^2/h) versus QTC (Qualitative Turbulence Criteria) (*SS* stabely stratified; *WM* well mixed). The *curve* illustrates the expected relationship between primary productivity and turbulence if the latter could have been measured at each station (cf. Marra 1980). Integrated primary production at only one station (in *brackets*) fell well below the range expected for a stratified surface layer

grid (Stations 12, 16, 17 in the south, and 36, 37, 38 in the north). In these areas, increased stratification (i.e., minimum turbulence, see above) could be considered to result in conditions conducive to elevated phytoplankton activity. The deep chlorophyll was therefore probably detrital, as has been found in the Prydz Bay region (Monteiro et al. 1986).

Figure 12 further illustrates the close relationship between turbulence and primary production and shows an inverse relationship between these dynamic parameters. It was not possible to categorize the remaining stations for which primary productivity data were available, as turbulence lay between the two extremes represented by the SS and WM conditions respectively.

3.2.2 Krill Distribution

Mean krill density (grouped at 2-hourly intervals) over the whole survey area is shown in Fig. 13. A significant difference (P = 0.01) was found between mean krill density by day compared with night for both paired (day following night) and pooled data (cf. Hampton 1985). The respective overall means were 1.45 and 0.59 g/m^2 with the range of densities lying between 0–26.65 g/m^2 (day) and 0–15.78 g/m^2 (night). The coefficient of variation of the daytime density estimate for the whole survey area was 0.175. Sources of this variance are discussed at length by Hampton (1985).

Maximum krill densities were found to be concentrated in two regions of the survey grid – at the northwestern boundary (i.e., at the western edges of Legs 9 and 10) and along the most southerly leg (Leg 2) (Fig. 13).

A weak, negative correlation (r = –0.37) was found between mean krill density and latitude, although this was not significant. A similar trend was observed along Leg 1 (i.e., north-south leg along 30°E) (r = –0.19). Absence of any latitudinal trend in integrated krill density in the same area has also been reported by Weber (1984) and Weber and El-Sayed (1985), although the data were treated differently.

Both pooled interval data and the sum of individual transect correlations (derived by a Fisher transformation – Snedecor and Cochran 1972) showed a significant negative correlation between krill density and longitude (r = –0.1387, P = 0.01 for the pooled data and r = –0.01, P = 0.05 for the transformed data). Further consideration of mean density distributions west of 20°E and east of 25°E indicated significantly greater (P = 0.01) abundances at the western extreme of most of the transects (Fig. 14). The distribution of densities also showed generally higher krill abundance in the north than the south, with lower abundances at intermediate latitudes. This trend was most pronounced on the western survey boundary and may account for the lack of any significant correlation between density and latitude. In this connection, the density-latitude correlation in the west (r = 0.05) was reduced when compared with the overall correlation (r = –0.37) between these two para-

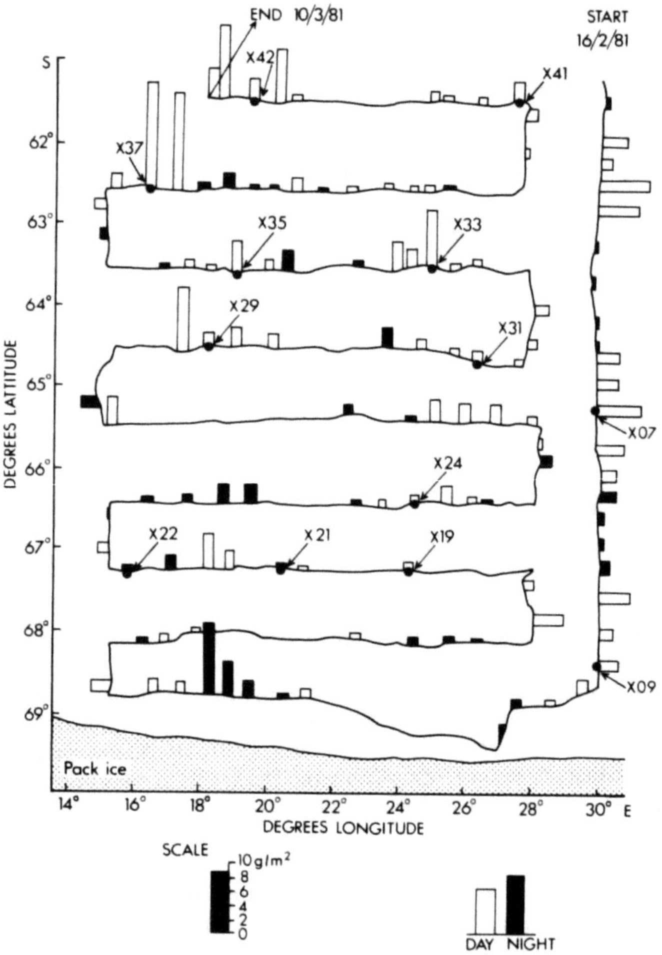

Fig. 13. Horizontal distribution of krill (g/m²) in the survey area. Thirty-minute echo-integration readings are grouped 2-hourly

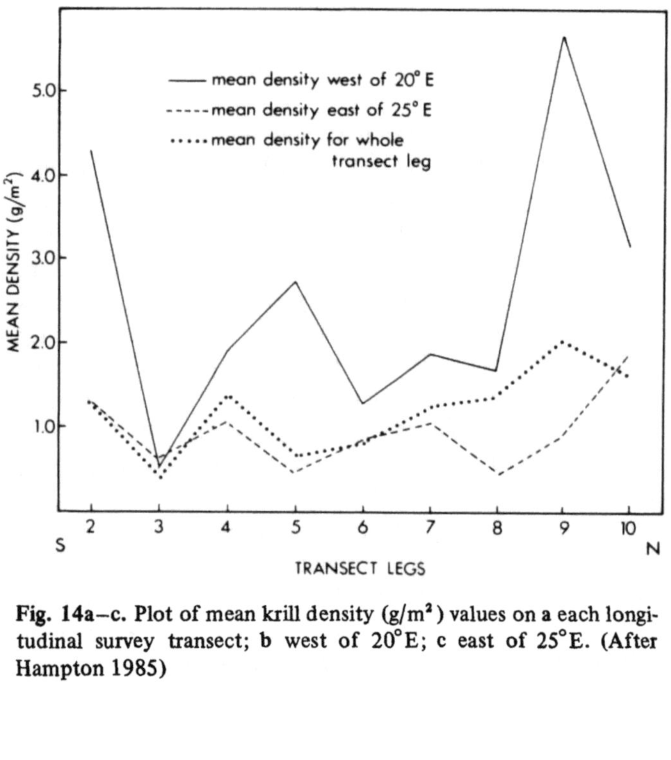

Fig. 14a–c. Plot of mean krill density (g/m²) values on a each longitudinal survey transect; **b** west of 20°E; **c** east of 25°E. (After Hampton 1985)

meters. It has to be emphasized, however, that the above correlations were derived without consideration of possible co-variance between successive echo-integration intervals. The effect of interval co-variance on observed distributional trends is thus undetermined.

3.3 Krill and the Environment

Using the same basic data, Hampton (1985) found that krill abundance, chlorophyll a and dissolved oxygen were not significantly correlated. Similarly, Weber and his co-workers found that krill abundance was negatively correlated with integrated phytoplankton biomass but that there was no correlation with a number of physical and chemical variables (i.e., latitude, density, stability, dissolved organics, and nitrate) (Weber 1984; Weber and El-Sayed 1985). However, they found that the steepness of the in vivo fluorescence spectrum relative to temperature, together with consistent coherence between phytoplankton and krill profiles, suggested that krill grazing may affect phytoplankton distribution locally (Weber et al. 1986). Krill also exhibited extensive patchiness and at a resolution of 2–20 km the variance spectrum was similar to white noise (Weber 1984; Weber et al. 1986).

Thus, the only bivariate correlation between krill abundance and any physical parameter that could be categorized was a significant association ($P = 0.01$) between peaks in krill abundance and temperature at the upper levels of the near-surface pycnocline (cf. Hampton 1985). Furthermore, some coherence appeared, although not statistically significant, between high krill densities and primary production maxima in the south and northwest (Figs. 9 and 13).

4 Discussion

The survey lay within the Antarctic open-ocean zone, although some influence of the continental water mass was evident in the south. Limited hydrographic data suggest that southward surface transport closes the Weddell Gyre east of about 25–30°E (Deacon 1979; Gordon et al. 1981). According to Marr (1962), increased krill abundances in both the East and West Wind Drift would be expected in the region. The survey can thus be considered to have been situated close to the Weddell Gyre retroflection zone and the observed meridional distribution of krill (i.e., at the "concentration" scale, cf. Murphy et al. this Vol.) would be in keeping with that proposed by Marr. On the other hand, the longitudinal distribution exhibited higher krill abundances at the northern and western extremities of the grid (± 60 to 62°S; 15°E). This may be an effect of thermohaline instability (Gordon 1978) associated with the Maud Rise (approx. 64–66°S; 1–5°E) to the west, inducing some local concentration of krill in the West Wind Drift.

The frontal discontinuity at 65°S was weak and its proximal position was close to that normally associated

Table 1. Important parameter means and ranges of variability

Parameter	Mean	Range	Variability (std. dev.)	Source
Fluorescence	–	–	3–5-fold	Weber et al. (1986)
Krill abundance	1.46 g/m²	0–26.65 g/m²	3–6 times as much as 18 times	Hampton (1985) This chapter
Water temp SL[a]	0.88°C	–1.2–1.69°C	(± 0.69°C)	This chapter
Salinity SL[a]	33.82‰	33.06–34.07‰	(± 0.21‰)	This chapter
Primary production	15.20 mg C/m²/h	5.79–30.58 mg C/m²/h	6 times	This chapter
Chlorophyll	3.65 mg Chl a/m²	1.47–17.14 Chl a/m²	11 times (± 3.36)	This chapter
Surface layer depth	34.63 m	12–50 m	(± 8.78 m)	This chapter

[a] SL = Near-surface layer (see text for definition).

with the Antarctic Divergence (cf. Lutjeharms 1985; Lutjeharms and Valentine 1984), although the Divergence's expression in this region is ephemeral (Lutjeharms and McQuaid 1986). Weber et al. (1986), however, make the point that both frontal mixing and bottom topography are likely to play a limited role in determining phytoplankton and krill distribution in the survey area. It is also interesting to note that unlike open-ocean areas farther to the east (Nasu 1983; Miller 1986) and krill-rich areas close to the Antarctic Peninsula (e.g., Marr 1962; Everson 1977; Amos 1984 and many others), no evidence could be found for any effect of the front on krill distribution per se. In contrast to similar surveys south of Australia (Inagake et al. 1985; Shirakihara et al. 1986), our acoustic results did not indicate any significant trend in krill abundance close to, and south of, the Divergence.

Since the area 20 to 30°E has been little studied, our results should offer an interesting basis for future comparisons of regional variation in the distribution of krill concentrations with respect to both latitude and prevailing environmental features.

As shown in Table 1, the vertical (i.e., fine-scale) variability of temperature in the SL (i.e., above the pycnocline) was not great (± 0.19°C) and variations in near-surface salinity were also small (± 0.12‰). Both integrated chlorophyll a and primary productivity varied as much as 11 and 6 times, respectively, while krill densities (over some 700 echo-integration intervals) were commonly 3–6 (as much as 18) times greater than the mean areal density estimate of 1.46 g/m². Similarly, Weber et al. (1986) found a three- to fivefold variation in the in vivo fluorescence which suggests intense phytoplankton patchiness. This contrasted well with the patchiness of krill aggregations over small spatial scales (Weber 1984) and suggests that, in addition to physical mixing processes, grazing plays an important role in determining the phytoplankton variance spectrum (see below) (Weber et al. 1986).

Both the present results and those of Weber and his co-workers indicate a general lack of empirical information of adequate resolution. This constitutes a major shortcoming when attempting to reconcile the wide range of temporal/spatial effects likely to influence interactions between krill and the physical environment in the open ocean. In particular, little that is constructive can be said about mechanistic links between the two.

Taking that we are attempting to relate coarse effects (with noticeable distributional variances and low signal amplitudes) to variables on which crucial parameter linkages are active, weak relationships between important variables can be expected. At a more global scale, such effects would tend to limit the level of statistical rigor given the low degree of freedom normally associated with broad areal surveys and the limited nature of available biological time-series data (Bakun 1986). Such shortcomings are generally applicable to any areal survey not having interactive linkage studies as an objective of the survey design. As emphasized by Murphy et al. (this Vol.), some account has to be taken of the temporal and spatial scales at which the most important interactions are thought to occur.

Therefore, it remains to be shown how demonstrably weak relationships between environmental parameters and important biotic components can best be used, taking into account that such relationships are representative of realistic levels in crucial mechanistic linkages in the system(s) concerned. Based on the data reported here, a conceptual model has been developed in an attempt to unify available knowledge with actual survey results. The model attempts to construct a practical and coherent foundation on which to link physical variability in open-ocean regions with the distribution of krill abundance during the austral summer.

In accordance with principles outlined by Longhurst (1981) and developed by Murphy et al. (this Vol.), adequate definition of krill's environmental matrix necessitates a description of physical and biological parameters at a level

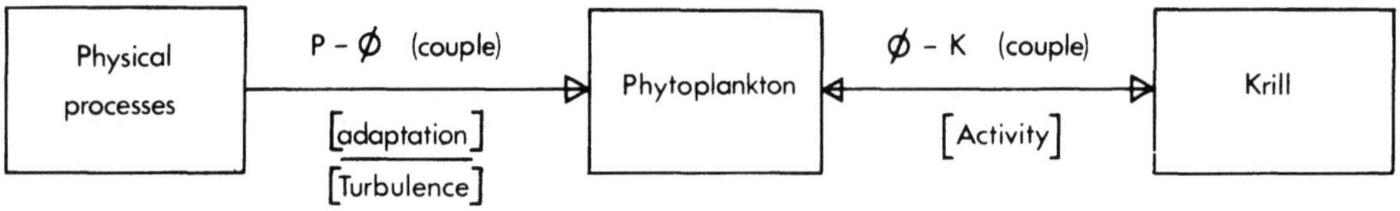

Fig. 15. Various components/linkages of the linear model. Refer to text for description of component elements

of spatial/temporal resolution one scale greater than that of the principle component interactions being considered. For this reason, a step-down approach is necessary to analyze the important component elements. Having considered the distribution/variability of both physical and biological parameters in the whole survey area, an in-depth appraisal of important mechanistic interactions can then be undertaken, taking into account the critical scales for such interactions.

The three principal components considered most important for our model are the physical environment, phytoplankton activity (i.e., primary production) and krill (represented by density distribution) (Fig. 15). The following coupling mechanisms are considered to link component elements sequentially within the temporal-spatial and interactive framework proposed by Haury et al. (1978) and modified by Murphy et al. (this Vol.) for krill individuals/ "swarms" (i.e., at a scale of hours to days and meters to tens of kilometers).

Physicobiological (P-Φ) Couple. The couple is predominantly regulated by turbulence in the SL and its relationship to the phytoplankton light adaptation rate (i.e., above the primary production cut-off depth). Thus the most important function is how turbulence affects the ratio:

$$\frac{[\text{Phytoplankton adaptation rate}]}{[\text{Mixing rate}]}$$

and how this ratio subsequently relates to primary production (Lewis et al. 1984a).

Biological (Φ-K) Couple. This refers directly to the regulation of the rate of production of a given phytoplankton population and in essence constitutes a two-way interaction. It is assumed that krill feed preferentially on large concentrations of growing (i.e., photosynthetically active) phytoplankton (cf. Holm-Hansen and Huntley 1984) and swarms are confined to feeding in near surface waters ($Z = < \pm 30$ m) as a result. By inference, individual krill are assumed to obtain food from a given volume of water and variations in food concentration (i.e., growing phytoplankton or that of a suitable size) would thus affect feeding rates (cf. Steele 1978). The krill-phytoplankton interaction can be considered "swathe" in nature (von Bodungen 1986) thereby accounting for small-scale patchiness in phytoplankton biomass (cf. Weber et al. 1986).

Component couples are dynamic and as a function of relatively short-term changes (i.e., light adaptation/turbulence and grazing) are reconcilable in terms of their respective scales of variability. This means that field observations must define areal variability in each principal component element. For this purpose, ***turbulence*** (represented by water column stability which can be measured directly) and *integrated primary productivity* set the scales of spatial variability for the physical and phytoplankton components respectively. These can then be examined with respect to variability in the third component (*krill abundance*).

Interactions between turbulence, irradiance, phytoplankton light adaptation, and primary production have recently been addressed by a number of workers (Platt and Gallegos 1980; Gallegos and Platt 1982; Lewis et al. 1984a,b; Falkowski 1984). As already hypothesized, the ratio of vertical mixing to phytoplankton light adaptation rates is critical (cf. Lewis et al. 1984a). If turbulence exceeds the adaptation rate, an overall decrease in primary production is observed. In these terms, elevated primary production in the present survey was associated with conditions of high irradiance (i.e., near the surface at between 5 and 15 m depth) and minimal turbulence. This association has since been further substantiated by results obtained during SIBEX (Second International BIOMASS Experiment) in the Prydz Bay region. Here the surface mixed layer was deeper (50–80 m) (Brundrit 1985) than observed during the present study and consequently primary productivity was generally lower (2.20–10.82 mg $C/m^2/h$). Only two stations near the ice edge (i.e., in more stable waters) exhibited productivity values in excess of 20 mg $C/m^2/h$ (Allanson 1985).

On the other hand, both results from this and Weber et al.'s analyses indicate some positive interaction between primary production and krill. The present results support the assumption that krill abundance and actively photosynthesizing phytoplankton form the basis of this interaction. Since enhanced primary production is also related to minimal turbulence, an inverse interaction between krill and turbulence would be expected. Although this could not be demonstrated from the present results and would not necessarily be expected from other published work (e.g., Witek et al. 1982), a positive correlation between krill abundance and thermal discontinuity at the upper levels of the pycnocline (Hampton 1985) suggests a direct linkage between these two components (i.e., via the "physicobiological" couple, Fig. 15). The link could also represent a second-order expression of a functional relationship between krill and phytoplankton. In these terms the pycnocline may act as a boundary to primary production (Angel 1968) or may provide a suitable physical milieu in which krill can feed (cf. discussion in Hampton 1985). Further definition of such interactive complexity would be expected as data and knowledge improve. For this reason, our basic model is sufficiently robust to allow future refinement and/or modification.

The present results therefore provide some indication of the range and variability of physicobiological parameters routinely measured in the estimation of krill abundance in the open ocean. They highlight one of the very real dangers of studies of this nature, where the search for interactive relationships between parameters is unlikely to be enlightening if simple statistical significance criteria alone are used (Bakun 1986), and/or no attempt is made to account for varying scales of variability in the interactive components being considered (von Bodungen 1986). Conceptual insights into important interactions are thus essential, and simple models, such as the one proposed, need to be tested in various Antarctic areas (e.g., Prydz Bay) to improve understanding of mechanistic interactions between abiotic and biotic components of the system. Furthermore, the accrual of suitable time-series data is essential for improving definition of the long-term variability of these components. Despite obvious practical difficulties, given the wide range of crucial scales, more direct experimentation in the field rather than additional areal surveys is required. Fulfilment of the statistical requirements referred to by Watkins et al. (1986) would entail increased time-series and spot sampling of all three model components (i.e., turbulence, primary production, and krill). As far as both temporal and spatial effects are concerned, Bakun (1986) has stressed that the greatest empirical return would be obtained from limited observations by sampling integrated results of important couple linkages on a small scale (i.e., readily sampled), in areas where the signal-to-noise ratios are comparatively high. Frontal discontinuities (e.g., the Weddell-Scotia Confluence and Antarctic Divergence) offer gradients of various component signatures and have already been implicated in studies of physical processes affecting krill aggregation behavior (e.g, Witek et al. 1982). For this reason, we suggest that frontal studies should assume priority in future investigations of dynamic and mechanistic interactions between Antarctic Ocean variability and krill.

In contrast, studies in the open ocean could provide inherently and spatially less variable baseline data. The acquisition of such data would be comparatively more costly, although more realistic comparisons of low-signal, component interactions should be possible in the long term. Such data collected over large temporal and spatial scales (e.g., using satellite remote sensing) may also provide useful information at the scalar dimensions which Murphy et al. (this Vol.) associate with biological "populations" (i.e., thousands of kilometers and years).

5 Conclusions

From the present analyses, it can be concluded that:

a) In terms of understanding important mechanistic interactions, there is considerable danger in trying to associate biotic and physical variability using statistical significance criteria alone. As stressed by Murphy et al. (this Vol.) some attempt must also be made to account for scale effects in important interactive linkages.
b) There is a need to formulate suitable models whereby interactive associations between biological and physical variability in the Southern Ocean can be tested. This will require more direct and specific experimental work in the field.
c) Studies in the open ocean are inherently difficult given the large spatial and temporal scales involved. Greater empirical returns would probably be achieved by concentrating research in areas of inherently high physical/biological variability (e.g., at fronts).
d) Comparative regional studies and various other techniques (e.g., spectral analysis of transect data, cf. Murphy et al.) should be employed such that some attempt is made to ordinate temporal/spatial structures in physical and biological data from a number of areas in the Southern Ocean.

Acknowledgements. We would like to thank L. Weber, S.Z. El-Sayed, and I. Hampton for access to a wide range of data and for helpful discussions. Prof. G. Brundrit, Drs. E. Murphy and J. Priddle, and an anonymous referee all contributed ideas at various stages. Financial support was provided by the South African Scientific Committee for Antarctic Research and Dept. of Transport, the Sea Fisheries Research Institute, and the University of Cape Town.

References

Allanson BR (1985) The South African SIBEX-I Cruise to the Prydz Bay region, 1984: VII. Light, chlorophyll a and primary production in the survey area. S Afr J Antarct Res 15:24–27

Allanson BR, Hart RC, Lutjeharms JRE (1981) Observations on the nutrients, chlorophyll and primary production of the Southern Ocean south of Africa. S Afr J Antarct Res 10/11:3–14

Amos AF (1984) Distribution of krill (*Euphausia superba*) and the hydrography of the Southern Ocean: large-scale processes. J Crust Biol 4 (Spec No 1):306–329

Angel MV (1968) The thermocline as an ecological boundary. Sarsia 34:299–312

Bakun A (1986) Definitions of environmental variability affecting biological processes in large marine ecosystems. In: Sherman K, Alexander LM (eds) Variability and management of large marine ecosystems. Select Symp Ser No 99. Westview Press, Boulder, AAS, pp 89–108

Bidigare RR, Frank TJ, Zastro C, Brooks JM (1986) The distribution of algal chlorophyll and their degradation products in the Southern Ocean. Deep-Sea Res 33A:923–938

Brundrit GB (1985) The South African SIBEX I Cruise to the Prydz Bay Region, 1984: II. Temperature, salinity and density overview. S Afr J Antarct Res 15:8–11

Carmack EC (1977) Water characteristics of the Southern Ocean south of the Polar Front. In: Angel M (ed) A voyage of Discovery: George Deacon 70th anniversary volume. (Deep-Sea Res Suppl). Pergamon, Oxford, pp 15–41

Deacon GER (1937) The hydrology of the Southern Ocean. Discovery Rep 15:1–124

Deacon GER (1979) The Weddell gyre. Deep-Sea Res 26:981–985

Deacon GER (1982) Physical and biological zonation in the Southern Ocean. Deep-Sea Res 29:1–15

El-Sayed SZ (1984) Productivity of Antarctic waters: A reappraisal. In: Holm-Hansen O, Bolis L, Gilles R (eds) Marine phytoplankton and productivity. Springer, Berlin Heidelberg New York Tokyo, pp 19–34

El-Sayed SZ, Taguchi S (1981) Primary production and standing crop of phytoplankton along the ice-edge in the Weddell Sea. Deep-Sea Res 28:1017–1032

El-Sayed SZ, Tees C, Warner R, Weber L, Meyer M (1980) Seasonal variability in biological productivity of the Scotia Sea and southwest Atlantic. Antarct J US 14:161–162

Evans CA, O'Reilly SE (1983) A manual for measurement of chlorophyll a, net phytoplankton and nanoplankton. BIOMASS Handb No 9:44 pp

Everson I (1977) The living resources of the Southern Ocean. FAO, Rome, (South Oceans Fish Surv Programme) GLO/SO/77/1, 156 pp

Falkowski PG (1984) Physiological responses of phytoplankton to natural light regimes. J Plankt Res 6:295–307

Fogg GE (1977) Aquatic primary production in the Antarctic. Philos Trans R Soc London Ser B 279:27–38

Foxton P (1964) Seasonal variations in the plankton of Antarctic waters. In: Carrick R, Holdgate MW, Prevost J (eds) Biologie antarcticque. Hermann, Paris, pp 311–318

Gallegos EL, Platt T (1982) Phytoplankton production and water motion in surface mixed layers. Deep-Sea Res 29A:65–76

Gordon AL (1978) Deep Antarctic convection west of Maud Rise. J Phys Oceanogr 8:600–612

Gordon AL, Martinson DG, Taylor HW (1981) The wind-driven circulation in the Weddell-Enderby Basin. Deep-Sea Res 28A: 151–163

Hampton I (1985) Abundance, distribution and behaviour of *Euphausia superba* in the Southern Ocean between 15° and 30°E during FIBEX. In: Siegfried WR, Condy PR, Laws RM (eds) Antarctic nutrient cycles and food webs. Springer, Berlin Heidelberg New York Tokyo, pp 294–303

Haury LR, McGowan JA, Wiebe PH (1978) Pattern and processes in the time-space scales of plankton distribution. In: Steele JH (ed) Spatial patterns in plankton communities. Plenum, London, pp 277–328

Hellmer HH, Bersch M, Augstein E, Grabeman I (1985) The Southern Ocean. A survey of oceanographic and marine meteorological research work. Ber Polarforsch 26:1–115

Heywood RB, Whitaker TM (1984) The marine flora. In: Laws RM (ed) Antarctic ecology, vol 2. Academic Press, London New York, pp 373–419

Holm-Hansen O, Huntley M (1984) Feeding requirements of krill in relation to food sources. J Crust Biol 4 (Spec No 1):156–173

Holm-Hansen O, El-Sayed SZ, Franceschini GA, Cuhel RL (1977) Primary production and factors controlling phytoplankton growth in the Southern Ocean. In: Llano GA (ed) Adaptations within Antarctic ecosystems. Smithsonian Inst, Washington, pp 11–50

Inagake D, Matsura N, Kurita Y (1985) Stock and quantitative distribution of the Antarctic krill (*Euphausia superba* Dana) in the Antarctic Ocean south of Australia in January and February 1984. Trans Tokyo Univ Fish 6:139–147

Jacques G, Minas M (1981) Production primaire dans le secteur indien de l'océan Antarctique en fin d'été. Oceanol Acta 4:33–41

Knox GA (1984) The key role of krill in the ecosystem of the Southern Ocean with special reference to the Convention on the Conservation of Antarctic Marine Living Resources. Ocean Manag 9:113–156

Lewis MR, Horne EPW, Cullen JJ, Oakey NS, Platt T (1984a) Turbulent motions may control phytoplankton photosynthesis in the upper ocean. Nature (London) 311:49–50

Lewis MR, Cullen JJ, Platt T (1984b) Relationships between vertical mixing and photoadaptation of phytoplankton: Similarity criteria. Mar Ecol Prog Ser 15:141–149

Longhurst AR (ed) (1981) Significance of spatial variability. In: Analysis of marine ecosystems. Academic Press, London New York: pp 415–441

Lutjeharms JRE (1985) Location of frontal systems between Africa and Antarctica: some preliminary results. Deep-Sea Res 32:1499–1509

Lutjeharms JRE, Valentine HR (1984) Southern Ocean thermal fronts south of Africa. Deep-Sea Res 31:1461–1475

Lutjeharms JRE, McQuaid LH (1986) Changes in structure of thermal ocean fronts south of Africa over a three-month period. S Afr J Sci 82:470–476

Lutjeharms JRE, Walters NM, Allanson BR (1985) Oceanic frontal systems and biological enhancement. In: Siegfried WR, Condy PR, Laws RM (eds) Antarctic nutrient cycles and food webs. Springer, Berlin Heidelberg New York Tokyo, pp 11–21

Marr JWS (1962) The natural history and geography of the Antarctic krill (*Euphausia superba* Dana). Discovery Rep 32:33–464

Marra J (1980) Vertical mixing and primary production. In: Falkowski PG (ed) Primary productivity in the sea. Environ Sci Res, vol 18. Plenum, New York, pp 121–137

Martinson DG, Killworth PD, Gordon AL (1981) A convective model for the Weddell polynya. J Phys Oceanogr 11:466–488

Miller DGM (1986) Results from biological investigations of krill (*Euphausia superba*) in the southern Indian Ocean during SIBEX I. Mem Natl Inst Polar Res Spec Issue 40:117–139

Miller DGM, Hampton I, Henry J, Abrams RW, Cooper J (1985) The relationship between krill food requirements and phytoplankton production in a sector of the southern Indian Ocean. In: Siegfried WR, Condy PR, Laws RM (eds) Antarctic nutrient cycles and food webs. Springer, Berlin Heidelberg New York Tokyo, pp 362–371

Monteiro PMS, Bacon EJ, Orren MJ (1986) The South African SIBEX I Cruise to Prydz Bay, 1984: Determination of photosynthetic pigments and their breakdown products by High Performance Liquid Gas Chromatography. S Afr J Antact Res 16:21–24

Morris DJ, Everson I, Ricketts C, Ward P (1984) Feeding of krill around South Georgia. II. Relation between feeding activity, environment and vertical distribution. Mar Ecol Prog Ser 20: 203–206

Nasu K (1983) On the geographic boundary of Antarctic krill distribution. Ber Polarforsch Sonderh 4:216–222

Oakey NS, Elliott JA (1982) Dissipation within the surface mixed layer. J Phys Oceanogr 12:171–185

O'Reilly JE, Thomas JP (1983) A handbook for the measurement of total daily primary production using ^{14}C simulated in situ sunlight incubation. BIOMASS Handb No 10:49 pp

Parsons TR, Takahashi M (1973) Biological oceanographic processes. Pergamon, Oxford, 186 pp

Plancke J (1977) Phytoplankton biomass and productivity in the Subtropical Convergence area and shelves of the western Indian Subantarctic islands. In: Llano GA (ed) Adaptations within Antarctic ecosystems. Smithsonian Inst, Washington, pp 51–82

Platt T, Gallegos CL (1980) Modelling primary production. In: Falkowski PG (ed) Primary productivity in the sea. Environ Sci Res, vol 18. Plenum, New York, pp 339–362

Pollard RT, Rhines PB, Thompson R (1973) The deepening of the wind-mixed layer. Geophys Fluid Dyn 4:381–404

Ruddick BR (1983) A practical indication of the stability of the water column to double diffusive activity. Deep-Sea Res 30A: 1105–1107

Sakshaug E, Holm-Hansen O (1986) Photoadaptation in Antarctic phytoplankton: Variations in growth rate, chemical composition and P versus I curves. J Plankt Res 8:459–473

Shirakihara K, Nakayama K, Komaki Y (1986) Acoustic estimation of krill biomass in R.V. Kaiyo Maru SIBEX I survey area (Indian Sector of the Southern Ocean). Mem Natl Inst Polar Res Spec Issue 40:140–152

Snedecor GW, Cochran WG (1972) Statistical methods, 6th edn. Iowa State Univ Press, Ames, 593 pp

Steele JH (ed) (1978) Some comments on plankton patches. In: Spatial patterns in plankton communities. Plenum, New York, pp 1–20

Strickland JDH, Parsons TR (1972) A practical handbook of seawater analysis. Bull Fish Res Board Can 167:201–203

Sverdrup HV (1953) On conditions for the vernal blooming of phytoplankton. J Cons Perm Int Explor Mer 18:287–295

Tchernia P (1980) Descriptive regional oceanography. Pergamon Mar Ser, Vol 3. Pergamon, Oxford, 253 pp

Tilzer MM, Dubinsky Z (1987) Effects of temperature and day length on the mass balance of Antarctic phytoplankton. Polar Biol 7:35–42

Tranter DJ (1982) Interlinking of physical and biological processes in the Antarctic Ocean. Oceanogr Mar Biol Ann Rev 20:11–35

von Bodungen B (1986) Phytoplankton growth and krill grazing during spring in the Bransfield Strait, Antarctica – Implications from sediment trap collections. Polar Biol 6:153–160

Watkins JL, Morris DJ, Ricketts C, Priddle J (1986) Differences between swarms of Antarctic krill and some implications for sampling krill populations. Mar Biol 93:137–146

Weber LH (1984) Spatial variability of phytoplankton in relation to the distributional patterns of krill (*Euphausia superba*). PhD Diss, Texas A&M Univ, College Stn, 82 pp

Weber LH, El-Sayed SZ (1985) Spatial variability of phytoplankton and the distribution and abundance of krill in the Indian Sector of the Southern Ocean. In: Siegfried WR, Condy PR, Laws RM (eds) Antarctic nutrient cycles and food webs. Springer, Berlin Heidelberg New York Tokyo, pp 284–293

Weber LH, El-Sayed SZ, Hampton I (1986) The variance spectra of phytoplankton, krill and water temperature in the Antarctic Ocean south of Africa. Deep-Sea Res 33:1327–1343

Witek Z, Grelowski A, Kalinowski J (1982) Formation of Antarctic krill concentrations in relation to hydrodynamic processes and social behaviour. ICES Biol Comm Memo CM 1982/L:59, 10 pp

Part 4
Krill Variability Detected from Predator Studies

Reproductive Performance of Seabirds and Seals at South Georgia and Signy Island, South Orkney Islands, 1976–1987: Implications for Southern Ocean Monitoring Studies

J.P. CROXALL, T.S. MCCANN, P.A. PRINCE, and P. ROTHERY[1]

Summary. Aspects of the reproductive performance over the last decade of Black-Browed, Grey-Headed and Wandering Albatrosses, Gentoo and Macaroni Penguins and Antarctic Fur Seals, at Bird Island, South Georgia and for Adélie and Chinstrap Penguins at Signy Island, South Orkney Islands, are summarized and reviewed. Breeding success of the Wandering Albatross, which breeds in winter and eats fish and squid, has remained constant, while population size has declined gradually but significantly. The other species at South Georgia, which breed in summer and feed extensively on krill, have shown major fluctuations in some or all of: breeding population size, breeding success, foraging trip duration and offspring growth rate. 1977–78 and 1983–84 were summers of particularly poor reproductive performance by almost all species; circumstantial evidence relating this to reduced availability of krill is discussed. The fluctuations in reproductive performance of the krill-eating, summer-breeding penguins at Signy Island are not synchronized with those at South Georgia; they correlate best (especially for Chinstraps, which suffered badly in 1980–81 and 1982–83) with the date of ice break-out in late spring. Numerous parameters of albatross, penguin and fur seal biology are reviewed in terms of their sensitivity and suitability for detecting changes in the marine environment.

1 Introduction

Accurate measurement of interannual variation in population parameters is difficult even in readily accessible terrestrial systems. Trying to account for these changes, i.e. to understand the processes which influence, or regulate, animal populations, has attracted a vast, and sometimes controversial literature (e.g. Lack 1954; Andrewartha and Birch 1955; Wynne-Edwards 1962) , which often fails to identify key factors and is usually unable to ascribe causal relationships.

We should, therefore, approach similar topics relating to the relatively inaccessible and highly dynamic marine system with appropriately realistic expectations. Among the most accessible components of marine ecosystems are those species which, while they depend on marine organisms for their food, have to leave the water to give birth and to rear their offspring. Seabirds and seals are familiar examples and it has been suggested that measuring aspects of their reproductive performance may furnish good indices of the state of the local marine system (Everson 1977; Croxall and Prince 1979; Gaston and Nettleship 1981; Wanless et al. 1982; Ricklefs et al. 1984). Furthermore, doing this may be more economical and/or accurate than trying to obtain regular measures of the prey stocks themselves. This logic was responsible for the gradual development of simple monitoring studies of seabirds and seals in Antarctic and Sub-Antarctic regions (Croxall and Prince 1979). This was developed by the Scientific Committee for Antarctic Research (SCAR 1979), incorporated as part of the international Biological Investigation of Marine Antarctic Systems and Stocks programme (BIOMASS 1980, 1983a,b) and has attained a prominent position in planning connected with the implementation of the Convention for the Conservation of Antarctic Marine Living Resources (CCAMLR 1985, 1986).

There are, however, few relevant published data and while Croxall (1987) provides an overview in the context of Southern Ocean conservation priorities, no critical appraisal has yet been attempted. Here we undertake such a task, treating most of the parameters which have been suggested as currently suitable for seabird and seal monitoring programmes (CCAMLR 1986), using data that have accumulated during seal and seabird research programmes at South Georgia, supplemented by data from Signy Island, South Orkney Islands. It should be stressed that most of these data were obtained as part of research programmes with specific aims, usually quite different from those of estimating interannual variability or providing time series of data.

The main aims of this paper, therefore, are first to describe and try to interpret variation in certain features (chiefly population size, breeding success, duration of foraging trips and offspring growth rates) of seabird and seal populations during the last decade at South Georgia. Second, to evaluate the utility of measuring these and other parameters for monitoring purposes.

2 Methods

The counting and sampling techniques used for estimating population size are described, or referenced, in Croxall and

1 British Antarctic Survey, Natural Environment Research Council, High Cross, Madingley Road, Cambridge, CB3 OET, UK

Antarctic Ocean and Resources Variability
(ed. by D. Sahrhage)

Prince (1979). Field counts of penguins at Signy Island were converted to numbers of pairs laying, chicks hatching and fledging using date-specific mortality data from Lishman (1983). Breeding success is expressed as chicks reared per egg laid, unless otherwise indicated. Techniques used for weighing penguins (adults and chicks) are summarized in Lishman (1985a, after Lishman 1983); procedures used in weighing Antarctic Fur Seals *Arctocephalus gazella* are detailed in Croxall and Prince (1979) and Doidge et al. (1984b). The fur seal data (from Bird Island only) are used to estimate growth rates over the period for which we have data in all years (age 55–85 days) and are not extrapolated to birth and weaning mass as in Doidge et al. (1984b). Determination of incubation (birds) and perinatal and foraging attendance (fur seals) routines were by daily or twice-daily observations of individually marked animals. The methods used to analyze these data for Antarctic Fur Seals are described more fully in Doidge et al. (1986); basicallly only the first six cycles are used. Foraging trip durations in penguins (and some data for fur seals) were also obtained using radio transmitters and automatic recording apparatus. These are described in Trivelpiece et al. (1986) and Croxall et al. (1988). Data for albatrosses came from automatic chick-weighing systems (Prince and Walton 1984). Activity budgets at sea for albatrosses were acquired using mechanical (Prince and Francis 1984) and electronic (P.A. Prince unpublished) recorders. Meal sizes for penguins were initially obtained by destructive sampling (Croxall and Prince 1980a; Lishman 1985a), later by water offloading techniques (Wilson 1984). Full details of dietary analyses for Macaroni and Gentoo Penguins *Eudyptes chrysolophus* and *Pygoscelis papua* are given in Croxall and Prince (1980a), for Grey-Headed and Black-Browed Albatrosses *Diomedea chrysostoma* and *D. melanophrys* in Prince (1980), for Antarctic Fur Seals in Doidge and Croxall (1985) and for Wandering Albatrosses *D. exulans* by Croxall and Prince (unpublished data). For albatrosses, data on meal size come from direct measurements (Prince 1980), from automatic weighing equipment (Prince and Walton 1984) and from regurgitated samples (corrected as in Prince 1980). The sources of other data are indicated as appropriate. Austral summer seasons are referred to by the year in which they end.

The main seabird and seal research programmes at Bird Island, including extensive compilations of data from which much of the information used in this paper derive, were reviewed by Croxall and Prince (1980b), Croxall et al. (1984, 1985a), Doidge and Croxall (1985) and Croxall and Prince (1987).

3 Background

To understand and interpet the patterns of variation within and between seasons in parameters of potential suitability for seabird and seal monitoring, we must consider the general environmental background and, in particular, the nature (and possible causes) of recent and current changes in seabird and seal populations.

Most important recent perturbations to the Southern Ocean ecosystem have been caused by human commercial activities. These commenced with the exploitation of the Antarctic Fur Seal in the 18th–19th century, reducing it to the verge of extinction, and continued with the development of the whaling industry (with associated exploitation of Southern Elephant Seals *Mirounga leonina*) and more recently of the fisheries for fin-fish (mainly *Notothenia* and *Champsocephalus* species) and Antarctic krill *Euphausia superba*. The changes in magnitude of the catches of whales, fish and krill are shown in Fig. 1. During this period, where adequate data exist, there have been significant increases in seabird and seal populations (Fig. 1, Table 1). The three penguin species increased at similar rates, whereas the Antarctic Fur Seal, which was recovering from intense exploitation, increased much more rapidly.

Thus, at least in the South Georgia and South Orkney Island areas (see also Poncet and Poncet 1985), there is good evidence for increases in seabird and seal species that are major consumers of krill (Antarctic Fur Seals, Adélie and Chinstrap Penguins *Pygoscelis adeliae* and *P. antarctica*). Similar increases have occurred in King Penguins *Aptenodytes patagonicus,* which do not eat krill but take fish (mainly lantern fish Myctophidae) and squid (Croxall and Lishman 1987; Adams and Klages 1987; Hindell 1988). There is some evidence from sites in the South Shetland Islands and Antarctic Peninsula for increases in Adélie and Chinstrap Penguins (see Sladen 1964; Croxall and Kirkwood 1979). There are also suggestions that Macaroni Penguin and Cape Pigeon *Daption capense* breeding populations at Bird Island may have increased, at least in the 1960's (Croxall and Prince 1979; Prince and Croxall 1983).

Sladen (1964) suggested that population increases in krill-eating species were a response to increased food availability following reduction of the baleen whale populations to one-fifth of their pre-exploitation levels. Implicit in this argument is that either summer food supply was previously limiting breeding population numbers, or that winter food in Antarctic and Sub-Antarctic regions has now increased (or is more readily available) commensurately to support the higher population levels. We cannot tell when the population increases started, but it appears that they are still continuing unabated, some 20 years after whaling effectively ceased in the area and include at least one species not directly dependent on krill. Consequently, although it is very likely that the massive reduction in stocks of krill-eating fur seals and whales promoted changes in other species dependent on the same basic resource, it is less certain to what extent this change is still fuelling the situation today. It is likely, anyway, that new predator-prey levels will be reached [and fur seals may well now have exceeded pre-exploitation levels at South Georgia (Bonner

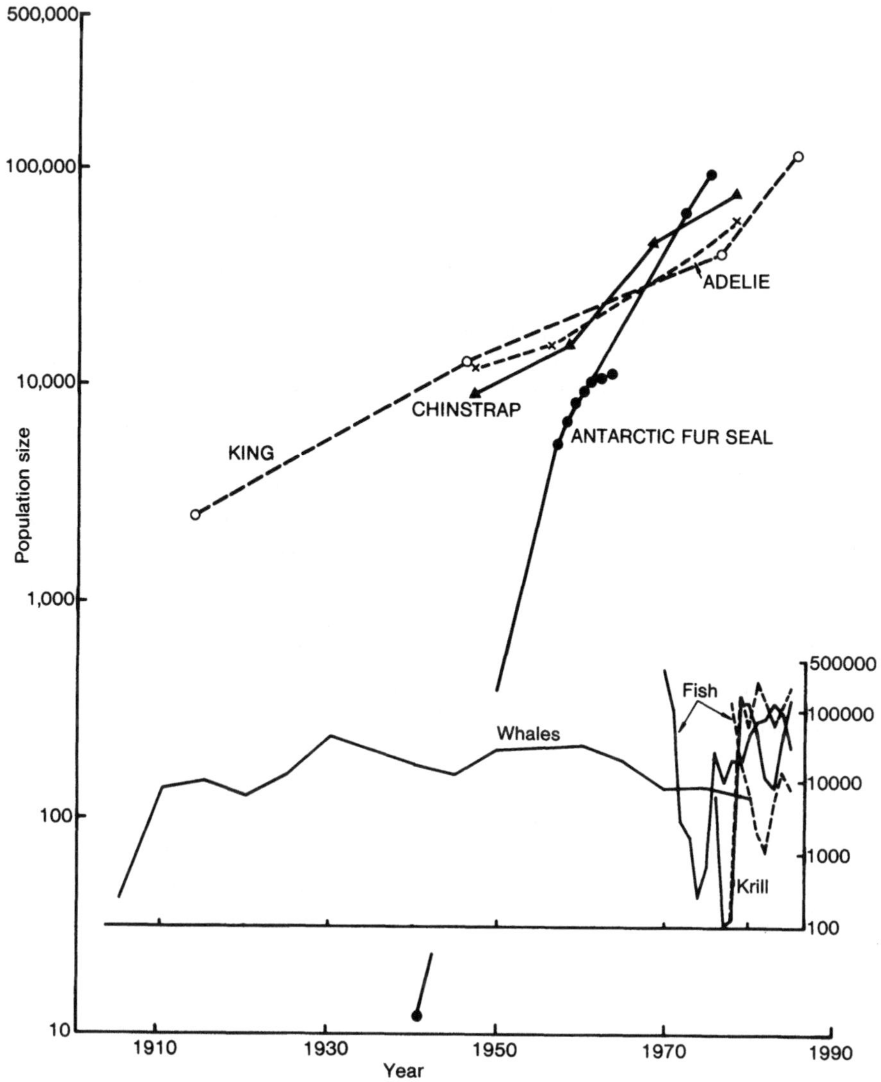

Fig. 1. Changes in population size of Adélie and Chinstrap Penguins (breeding pairs: Signy Island, South Orkney Islands), King Penguins (adults: South Georgia) and Antarctic Fur Seals (pups: South Georgia), in relation (*inset*) to commercial harvests of Antarctic whales (individuals; from Bonner 1984), fish and krill (tonnes; CCAMLR 1986) in the South Georgia (o—o) and South Orkney Islands (x- - -x) areas

Table 1. Average rates of increase of penguin and fur seal breeding populations at South Georgia and at Signy Island, South Orkney Islands

Species	Site	Period	Annual increase (%; mean ± SE)	Type of count	Data sources
Antarctic Fur Seal	South Georgia	1937–1976	22.6 ± 1.8	Pups	Bonner (1968), Payne (1977)
	South Georgia	1958–1976	15.5 ± 0.7	Pups	
King Penguin	South Georgia	1914–1986	5.0 ± 0.5	Adults	Smith and Tallowin (1980)
	South Georgia	1976–1986	12.3	Adults	BAS unpublished data
Adélie Penguin	Signy Island	1948–1979	3.6 ± 0.3	Breeding pairs	Croxall et al. (1981)
	Signy Island	1979–1986	4.0 ± 1.2	Breeding pairs	This chapter
Chinstrap Penguin	Signy Island	1948–1979	7.3 ± 0.8	Breeding pairs	Croxall et al. (1981)

1985)] and we may expect some stabilization of predator populations as natural limits to their population growth are reached.

Krill harvests have so far been very small in relation to seabird and seal food requirements (see Croxall 1987; Harwood and Croxall 1988). Thus even if harvests were to increase markedly and the rate of increase of seabird and seal populations to diminish, it would be difficult to exclude the possibility that these populations were, independently of krill harvesting, attaining the natural carrying capacity of the environment. As populations reach this level we may also expect them to fluctuate appreciably and so we are likely to have particular problems both in detecting and interpreting even statistically significant population trends.

Fish harvesting in the region has been much more intensive, especially in relation to estimates of fish stocks (Kock

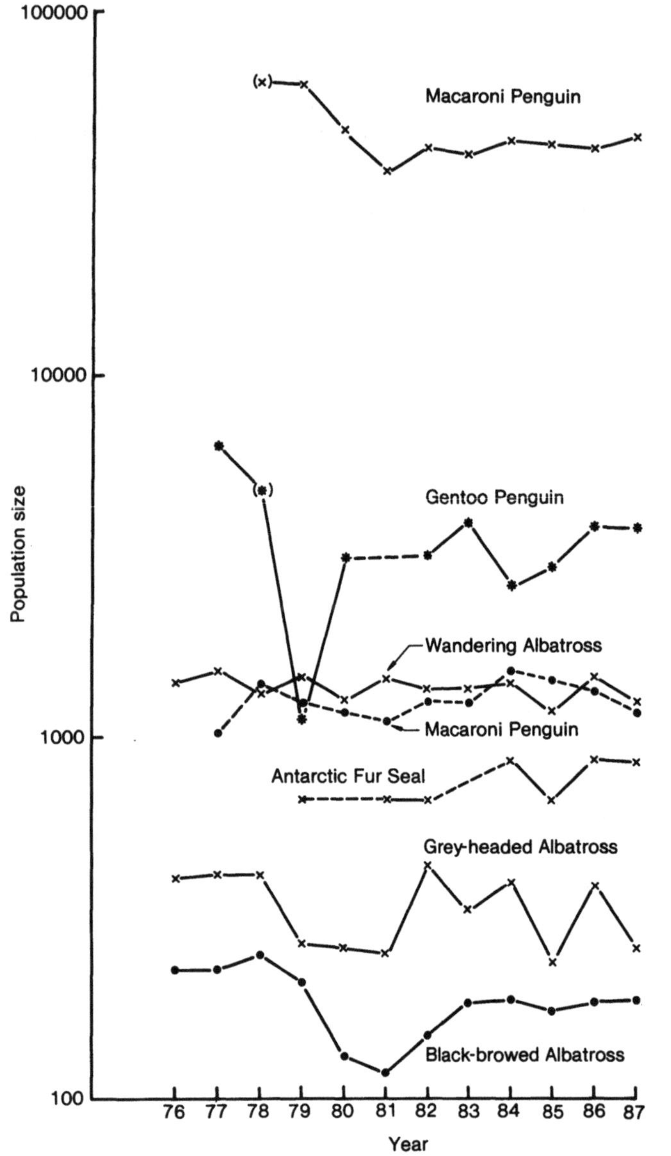

Fig. 2. Changes in population size of seabirds (breeding pairs) and Antarctic Fur Seals (pups born) at Bird Island, South Georgia, 1976–1987. Values in *parentheses* are estimates from incomplete data sets. Note log scale

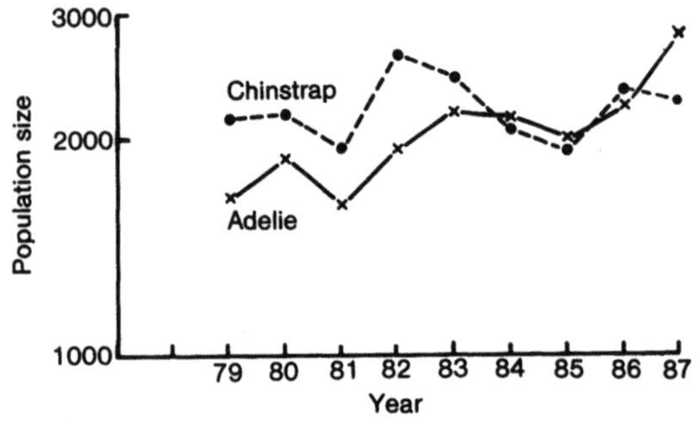

Fig. 3. Changes in population size (breeding pairs) of Adélie and Chinstrap Penguins at all study colonies combined on Signy Island, South Orkney Islands 1979–1987. Note log scale

1985; CCAMLR 1986). Any effects on birds and seals, however, are impossible to detect because no species, except perhaps Southern Elephant Seals (the nature and composition of whose fish diet is unknown) and male Antarctic Fur Seals in winter (North et al. 1983; Doidge and Croxall 1985), take significant quantities of the fish which are caught commercially. Even Blue-Eyed Shags *Phalacrocorax atriceps* and Gentoo Penguins, which do take *Champsocephalus* and *Notothenia* species, prey almost exclusively on immature age-classes. It is unlikely, therefore, that monitoring aspects of the reproductive performance of these species will furnish sensitive indices that can be used in relation to changes in fish stocks.

Not all seabird populations are currently increasing. Indeed the pattern at South Georgia over the last decade has been one of substantial fluctuations (Fig. 2). A common element was an abrupt decline in breeding numbers after 1978, which persisted for several years and was succeeded either by a gradual increase or by further fluctuations. Another notable decrease occurred after 1984, although that of Gentoo Penguins actually started in 1984. Fluctuations at Signy Island (Fig. 3) have not followed a similar pattern. Adélie Penguins have increased, averaging 4.0% per annum (0%, 3%, 4%, and 7% at the four study colonies), described by the equation $\log y = 4.35 + 0.00396x$ ($F_{1,6} = 11.29$, $P < 0.025$) whereas, overall, Chinstrap Penguin numbers did not change significantly over the same period (although one of the 11 colonies did increase by 21%). Breeding numbers of both species decreased abruptly in 1981 and, after a rapid recovery, Chinstraps decreased again in 1983–1985.

The only species whose breeding population has been decreasing is the Wandering Albatross (Fig. 4), the overall trend [at a rate of 1.1 ± 0.91 (s.e.) % per annum] being described by the equation $\log y = 29.5 - 0.0112x$ ($F_{1,16} = 33.27$; $P < 0.001$). A similar situation has been reported for the populations of the Crozet Islands (Jouventin et al. 1984). Neither can be ascribed to reduced breeding success (see later and Croxall 1979) and both seem to relate to reductions in adult survival and possibly recruitment (i.e. juvenile survival) rates. The factors influencing this may include changes in food availability in wintering areas but the main cause is probably mortality due to birds being caught in long-line fishing operations taking place outside the Southern Ocean (Croxall et al. 1984; Jouventin et al. 1984).

Interpretation of some of these patterns is complicated by differences in breeding frequency (e.g. Grey-Headed and Wandering Albatrosses breed biennially when successful in rearing chicks). Furthermore, even amongst annual breeders there are significant interannual differences in the proportion of birds returning to breed in successive years (see Prince 1985), only part of which is due to previous breeding success.

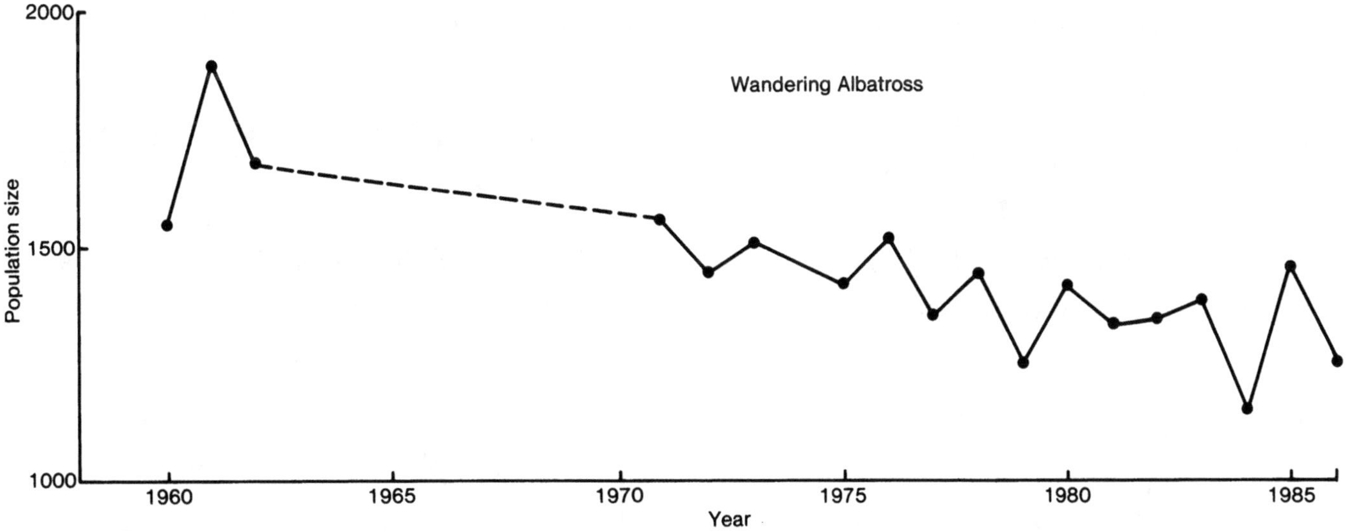

Fig. 4. Changes in population size (breeding pairs) of Wandering Albatrosses at Bird Island, South Georgia 1961–1986

4 Review of Data on Parameters Potentially Suitable for Monitoring

We divide parameters into two groups, those for which we have at least 5 years' data and those for which we have less than this. We have some data on most seabird and fur seal parameters so far suggested (CCAMLR 1986) as appropriate for monitoring studies starting now.

4.1 Parameters with 5 Years or more Data

4.1.1 Breeding Success

South Georgia. Fluctuations in the breeding success of three albatrosses, two penguins, and Antarctic Fur Seals are illustrated in Fig. 5. There are two main features to note. First, breeding success of Wandering Albatrosses, which rear their chicks on fish and squid throughout the winter, has varied very little over the last decade. The lowest value, after the 1977 winter, was a result of chick-rearing success 10% less than any subsequent value, but no cause for this is known. Extreme weather conditions (heavy snow, gales) are the most obvious causes of egg and chick mortality. Once brooding ceases, chicks receive food only every 3–4 days and can readily sustain fasts for 20–30 days; they are, therefore, well buffered against short-term changes in food availability. The long foraging trips allow adults to range far afield (records from 1000–1500 km distance) in search of food. Second, in some years all (or most) of the remaining species monitored (all of which breed in summer and most of which predominantly eat krill) showed poor breeding success. This is evident in 1978 (all species except Grey-Headed Albatross) and 1984 (all species, but with variations between species). If a common factor is the degree of dependence on krill, then the Grey-Headed Albatross, which shows the smallest fluctuations, is the only species for which krill is not the main food item. However, Macaroni Penguin, the species most dependent on krill, was less affected than the Black-Browed Albatross, which showed virtually complete reproductive failure in 3 years (including 1979, when other species were unaffected), even though krill forms only some 40% of its diet. Dependence on krill, however, is likely to reflect a complex of factors, including not just its importance in the diet but also its vertical and horizontal distribution and behaviour in relation to predator foraging abilities (see Croxall et al. 1985b, 1988) and these are likely to differ considerably between species and years.

Signy Island. Breeding success for Adélie and Chinstrap Penguins is summarized in Figs. 6 and 7. In general, most of the four Adélie colonies showed a rise in breeding success from 1980 to 1982, poor success in 1983 and 1984 and some increase thereafter. For Chinstrap Penguins, because of errors in the field in the timing of fledgling censuses, we must rely on data on hatching success. Fortunately this seems to show a close relationship with overall breeding success when both are available. Figure 7 indicates a very different pattern from Adélies, with 1981 and 1983 being seasons of exceptionally poor breeding performance.

4.1.2 Attendance Patterns of Antarctic Fur Seals at South Georgia

These patterns comprise three elements:

a) The period between the birth of the pup and the mother's departure to sea (perinatal attendance).
b) The foraging trips to sea.
c) The interval ashore between each of these trips when the pup is suckled.

All these are summarized in Fig. 8 (data in Table 2). There are three principal results. First, the duration of shore visits varies little between years and, although the

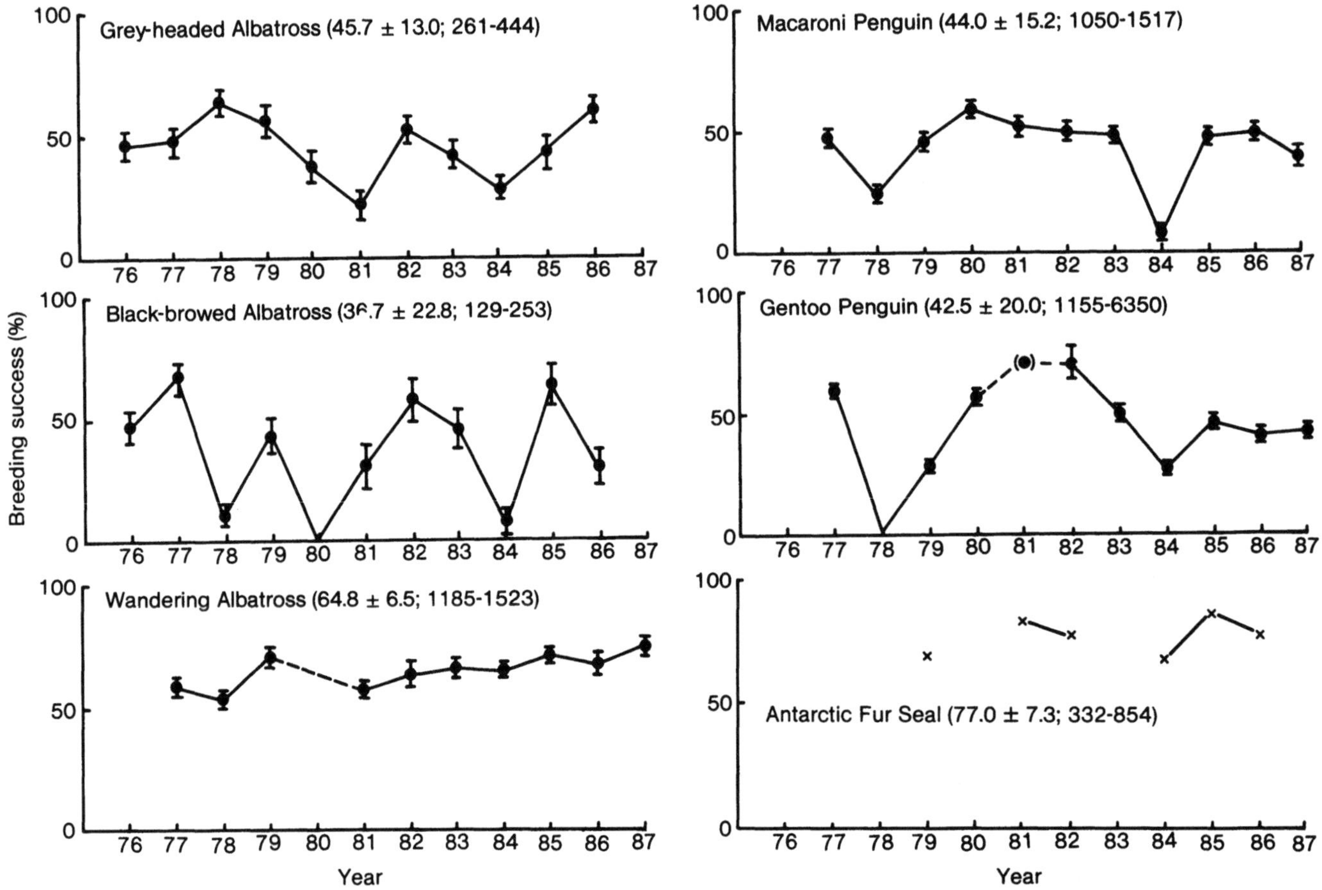

Fig. 5. Breeding success of seabirds [proportion (%) chicks reared per egg laid] and Antarctic Fur Seals (pups surviving) at Bird Island, South Georgia. Values are means (those in *parentheses* estimates for incomplete data sets), *vertical bars* two standard errors each side. Overall average, standard deviation and sample size range are also indicated

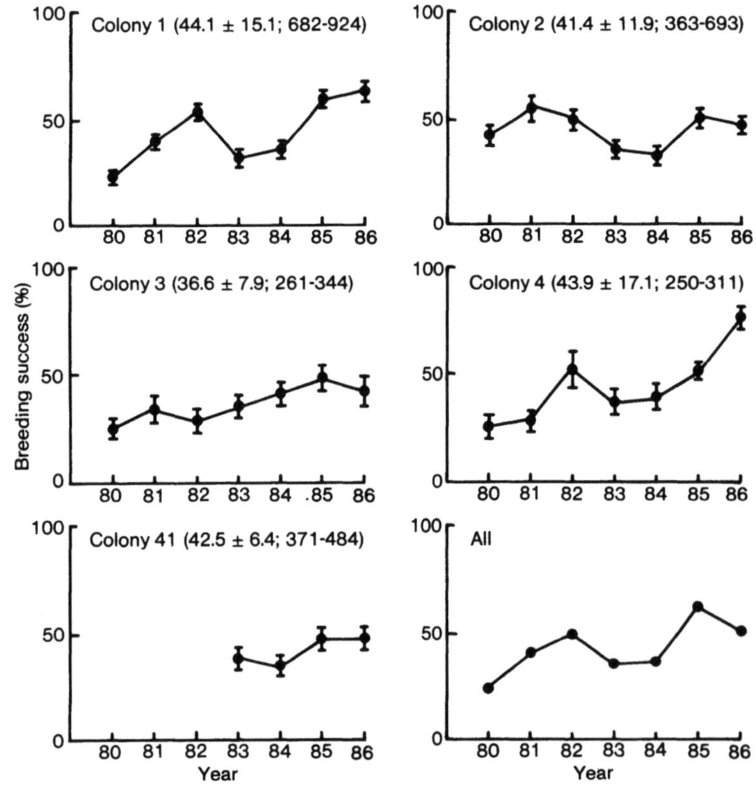

Fig. 6. Breeding success [proportion (%) of chicks reared per egg laid] of Adélie Penguins at five colonies (and all combined) on Signy Island, South Orkney Islands. Conventions as in Fig. 5

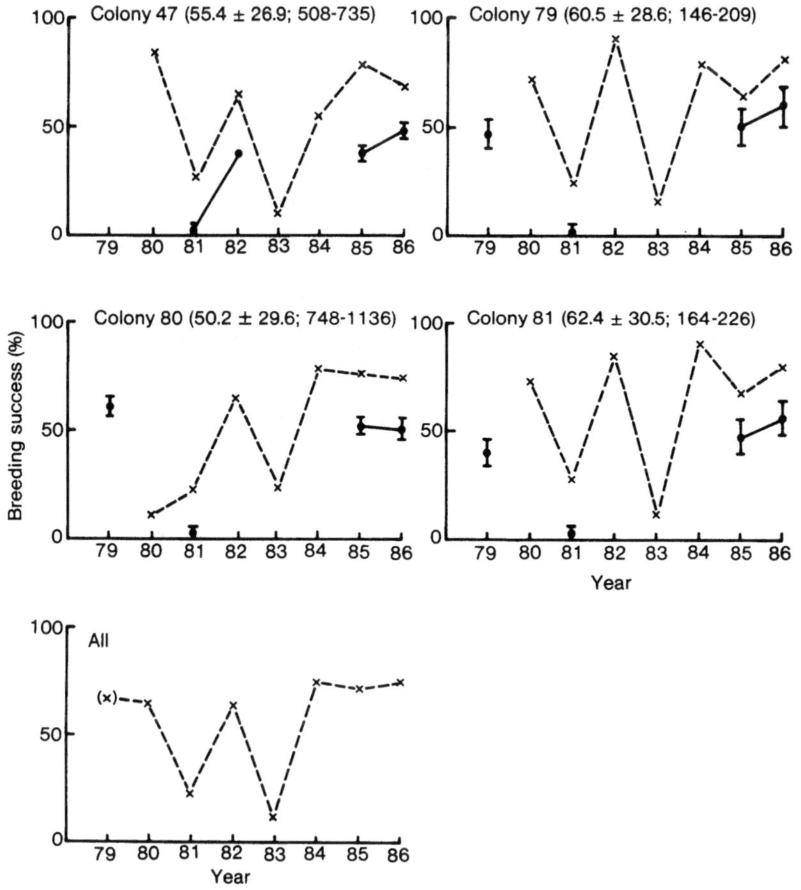

Fig. 7. Breeding success (●—●: chicks reared per egg laid; x- - -x: chicks hatched per egg laid) of Chinstrap Penguins at four colonies (and all combined) on Signy Island, South Orkney Islands. Conventions as in Fig. 5

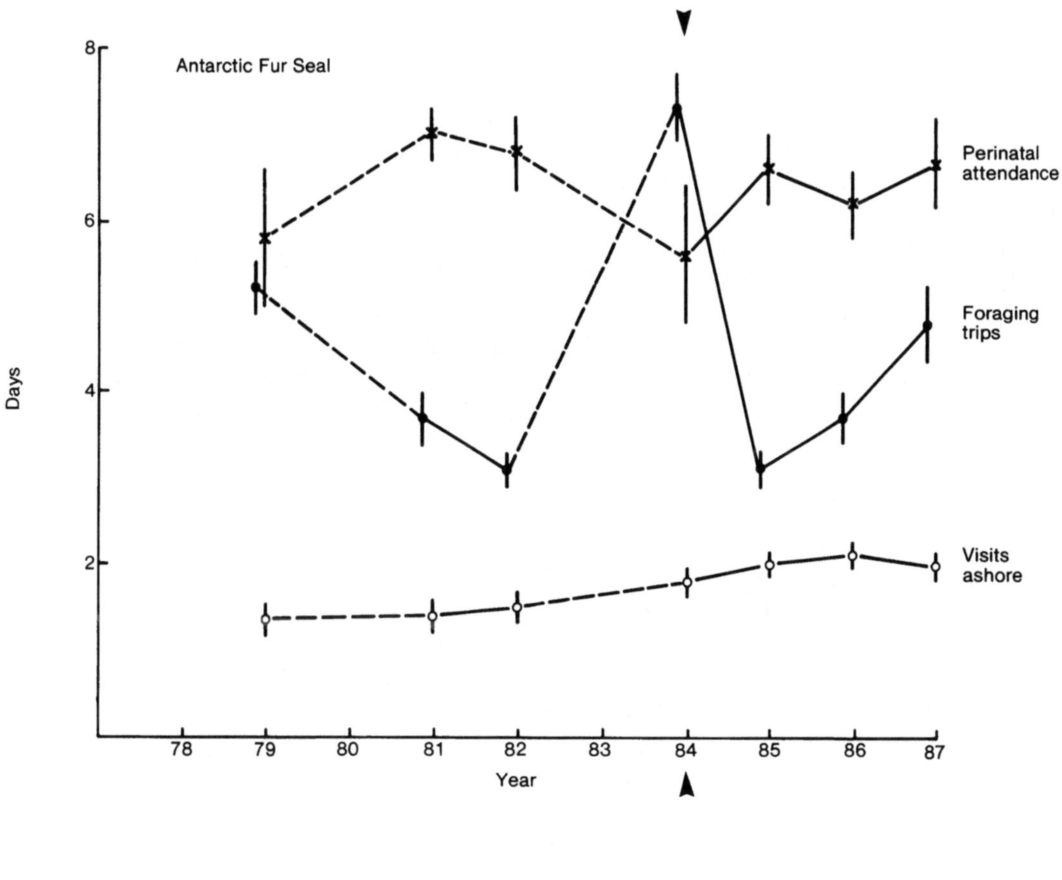

Fig. 8. Duration of attendance periods ashore and foraging trips to sea in Antarctic Fur Seals at Bird Island, South Georgia, 1979–1987. Perinatal attendance: x; subsequent periods ashore: 0; foraging trips: ●. Values are means ± 2 s.e. *Arrowheads* indicate anomalous years

Table 2. Attendance patterns, pup growth and survival of Antarctic Fur Seals at Bird Island, South Georgia. Values are means with standard error and/or sample size in parentheses

Year	Perinatal attendance (d)	Foraging trips (d)	Attendance ashore[a] (d)	Growth rate (g/d) Male[b]	Female[b]	Weight (kg) at day 70 Male[b]	Female[b]	Pup survival (%)	Starvation mortality (%)[c]
1973	–	–	–	101 (8)	77 (5)	13.6 (0.11)	11.5 (0.07)	–	–
1974	–	–	–	95 (9)	91 (9)	13.4 (0.12)	11.3 (0.11)	–	–
1978	–	–	–	69 (14)	86 (8)	11.8 (0.14)	10.2 (0.08)	–	–
1979	5.8 (0.39,29)	5.2 (0.14,252)	1.4 (0.04,252)	99 (16)	80 (12)	12.1 (0.16)	10.5 (0.12)	69 (672)	–
1980	–	–	–	94 (8)	105 (13)	12.6 (0.09)	10.9 (0.16)	–	–
1981	7.0 (0.29,27)	3.7 (0.16,180)	1.4 (0.04,180)	95 (8)	108 (12)	12.9 (0.29)	10.4 (0.11)	83 (669)	40 (75)
1982	6.8 (0.23,28)	3.1 (0.09,504)	1.5 (0.03,504)	–	–	–	–	78 (661)	–
1984	5.6 (0.40,14)	6.8 (0.17,180)	1.8 (0.05,180)	47 (10)	47 (10)	9.9 (0.18)	8.7 (0.16)	68 (850)	68 (185)
1985	6.6 (0.17,49)	3.1 (0.11,186)	2.0 (0.07,186)	72 (15)	59 (9)	13.1 (0.22)	11.1 (0.17)	86 (664)	39 (62)
1986	6.2 (0.22,42	3.7 (0.11,168)	1.6 (0.06,168)	97 (8)	105 (11)	13.3 (0.11)	11.0 (0.16)	78 (854)	22 (142)
1987	6.6 (0.24,20)	4.8 (0.19,120)	1.5 (0.05,120)	154 (11)	109 (11)	12.3 (0.16)	10.7 (0.15)	(838)	

[a] Sample sizes are of total number of trips (six for each animal).
[b] Sample of 50 pups of each sex.
[c] Of pups whose cause of death was determined (Doidge et al. 1984a; T.S. McCann unpublished data).

visits were significantly longer in 1984 and 1985 than in other years, this is not matched by changes in other parameters. Second, perinatal attendance was significantly shorter, and much more variable, in 1979 and 1984 than in other years, which show no obvious pattern (although there are some statistically significant differences, e.g. between 1981 and 1986). Third, foraging trips to sea were very long in 1979 and 1984 and exceptionally so in the latter. Trips in 1981, 1982, 1985 and 1986 are broadly similar to each other (though the first and last are significantly longer than the other two) and those in 1987 are intermediate between this group and the two anomalous years (1979 and 1984).

4.1.3 Pup Growth in Antarctic Fur Seals at South Georgia

Pup growth rates (Fig. 9) and a growth index (mass at age 70 days; Fig. 10) which serves as an approximation to pup mass at weaning (age ca. 110 days) showed broadly similar patterns (data in Table 2). There are considerable interannual variations in growth rates and in some years male pups grow fastest, while in others females do. There is an increase in growth rate from low values in 1978 and 1979 to values in the next two seasons which are comparable to those recorded in 1973 and 1974. Growth rates were exceptionally low in 1984 and have increased to normal levels since. The growth index shows clearly the exceptionally low mass of pups in 1984; 1978 and 1979 were also seasons in which pups were significantly lighter than in the other 6 years.

4.2 Parameters with Fewer than 5 Years' Data

Data for these analyses come from research programmes on the small *Diomedea* albatrosses (mollymauks) and penguins at Bird Island, South Georgia and on penguins at Signy Island. For convenience they will be treated in these three separate groups.

4.2.1 Grey-Headed and Black-Browed Albatrosses at South Georgia

Although demographic studies of both species have been maintained since 1975 (Prince 1985) and comparisons made of diet and growth (Prince 1980; Prince and Ricketts 1981; Ricketts and Prince 1981), most recent experimental work has been conducted on the biennially breeding Grey-Headed Albatross (Prince and Francis 1984; Prince and

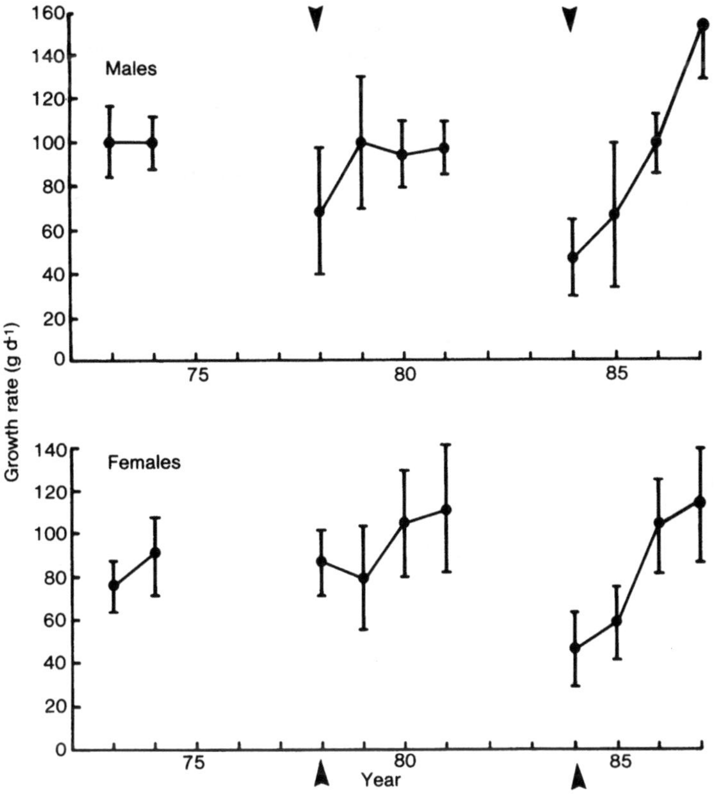

Fig. 9. Growth rates (g/d) of male and female Antarctic Fur Seal pups (n = 50 of each sex) at Bird Island, South Georgia 1978–1987. Conventions as in Fig. 8

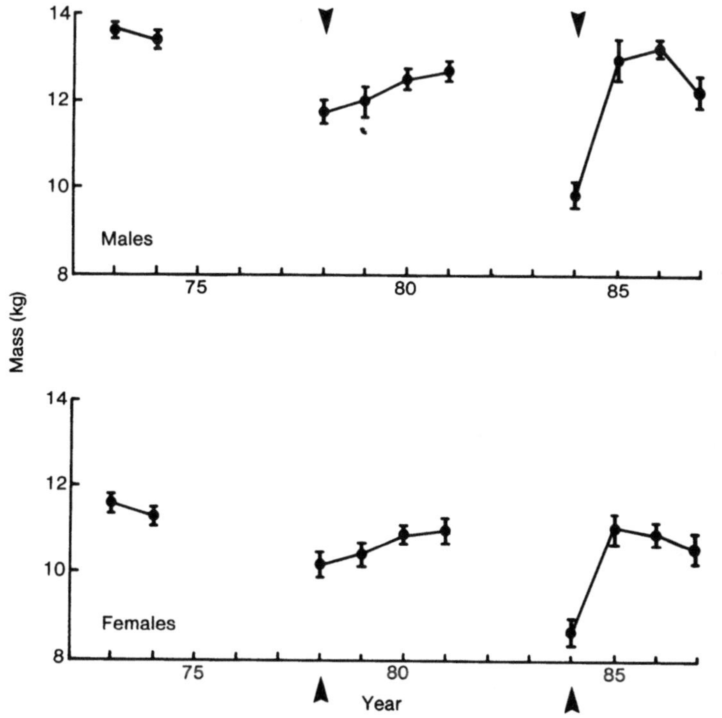

Fig. 10. Estimated average mass at age 70 days of male and female Antarctic Fur Seal pups (n = 50 of each sex) at Bird Island, South Georgia 1973–1987. Conventions as in Fig. 8

Walton 1984; Astheimer et al. 1985; Costa and Prince 1987). Despite numerous differences between the species, however, inter- and intra-annual variation in most parameters is similar. Relevant data are summarized in Table 3; they are presented separately for the abnormal (defined in terms of breeding success; see Fig. 5) years of 1980 (Black-Browed Albatrosses only) and 1984 (both species; but few data available for Black-Browed Albatrosses).

Breeding Timetable. We have no evidence of any significant interannual variation in laying date. However, Grey-Headed Albatross hatch date was significantly later in 1984 than in other years, but there were also significant differences within the latter. A similar situation prevails with fledging date and the duration of the fledging period. Black-Browed Albatrosses generally show less interannual variation in breeding timetable.

Foraging Trips. We have few data from direct measurements of foraging trips in albatrosses. The data used here are based on the intervals between meal delivery as recorded by the automatic weighing system and show a significant difference in the abnormal year of 1984.

Activity Budgets at Sea. The use of activity recorders (Prince and Francis 1984) enables at-sea activities to be partitioned. We can regard nighttime on water as an index of time potentially spent feeding in these predominantly nocturnal species, daytime on water as an index of time available for preening, sleeping, etc. (though it may also include time feeding on carrion, especially in Black-Browed Albatrosses) and time in flight as an index of time spent searching for food (and/or distance travelled). Comparing Grey-Headed Albatross data for 1983 and 1984 (abnormal year) we see that in 1984 less time was spent in flight, much more time was spent in potential feeding and significantly less time in other on-water activities.

Meal Mass. There is considerable intraseasonal variation in the meal mass delivered to chicks, though once brooding has ceased there is no trend for increasing meal size as the chicks grow, and no evidence of smaller meals in the abnormal year of 1984. For Black-Browed Albatrosses, however, meals were lighter in 1980 (a year of complete breeding failure) compared with two other apparently normal years.

Chick Growth. Various indices all indicate significant differences between 1984 and the normal year of 1986, but we have few data with which to assess interannual variation.

4.2.2 Macaroni and Gentoo Penguins at South Georgia

Research on these species chiefly involves studies of diet (Croxall and Prince 1980a), energy budgets (Davis et al. 1983) and diving (Croxall et al. 1988), though other data on general breeding biology, seasonal mass changes and

Table 3. Breeding season parameters in Grey-Headed and Black-Browed Albatrosses at Bird Island, South Georgia. Values are means, with standard deviation and sample size in parentheses. Differences between normal and abnormal years significant at $P < 0.05$ = *, at $P < 0.01$ = **, at $P < 0.001$ = ***. Significant differences within normal seasons are indicated similarly

Parameter	Grey-Headed Albatross		Black-Browed Albatross	
	Normal	Abnormal[a]	Normal	Abnormal
Laying date	19.4 Oct[b] (3.4,278)	18.8 Oct (3.1,54)	27.0 Oct[b] (4.0,302)	
	19.3 Oct[h] (3.8,232)		28.3 Oct[h] (3.8,60)	
Hatch date	30.0 Dec[b] (3.1,105) ***		3.6 Jan[k] (3.7,143)	
	28.0 Dec[i] *** (3.4,102) ***	30.3 Dec (3.2,118)	3.0 Jan[b] (3.4,111)	
	29.1 Dec[k] (2.9,312) ***		4.6 Jan[h] (3.9,52)	
			5.0 Jan[i] (3.6,148)	
			3.6 Jan[k] (3.7,143)	
Fledge date	18.0 May[i] *** (5.3,102) ***	26.3 May (9.3,36)	29.8 April[h] (4.5,40)	
	22.6 May[k] ** (5.0,231)		2.0 May[i] (3.7,110)	
			1.4 May[k] (3.8,57)	
Fledging period (d)	141[i] *** (5.1,102) ***	149.1 (8.4,36)	117[i] (3.4,110)	
	144.4[k] *** (5.1,265)		117.5[k] (3.7,57)	
Foraging trip duration (h)	52.2[h] *** (36.8,41)	94.4 (75.4,79)		
Activity budget at sea: % day on water	15.1[g] ** (9.1,13)	6.2 (3.3,13)	27.6[h] (8.6,2)	
% night on water	49.9[g] *** (14.6,13)	76.0 (6.6,13)	85.0[h] (12.8,2)	
% time in flight	71.8[g] * (9.0,13)	59.8 (17.1,13)	50.7[h] (7.5,2)	
Meal mass (g)	566[c] * (203,117)	687 (180,180)	425[c] * (149,95)	340[f] (140,20)
	621[d] (203,93)		561[d] (197,71)	
	593[e] (170,91)		523[e] (204,135)	
	560[f] (110,20)		520[k] (228,42)	

Table 3 (continued)

Parameter		Grey-Headed Albatross				Black-Browed Albatross			
		Normal		Abnormal[a]		Normal		Abnormal	
		670[h] (259,41)							
		601[k] (95,53)							
		1976	1977	1980	1986	1976	1977	1980	1986
Meal composition:	krill	11.2	32.3	10.5	20.6	41.6	34.2	27.1	37.8
(% by mass)	fish	42.3	8.7	29.5	13.8	40.5	28.4	39.5	29.2
	squid	44.1	58.7	59.5	65.5	13.2	37.3	30.0	32.9
	other	2.4	0.3	0.5	0.1	4.7	0.1	3.4	0.1
	n	92	39	20	53	94	42	20	42
Chick mass (g):									
day 80		4281[c] (649,45)							
		4063[k] (538,22)	***	3023 (553,28)		4850[c] (711,25)			
day 100		4642[k] (727,33)	***	2985 (581,24)					
day 140		3561[k] (428,22)	***	2454 (382,14)					
Chick growth rate (g/d):									
days 20–80		51.4[c] 48.6[k]	***	33.7		58.5[c]			
20–100		43.7[c]	***	24.8					

[a] 1984 (Prince and Morgan 1987; P.A. Prince and R. Lidstone-Scott unpublished data).
[b] 1964 (Tickell and Pinder 1975).
[c] 1976 (Prince 1980; Ricketts and Prince 1981).
[d] 1977 (Prince 1980; Prince and Ricketts 1981 and unpublished data).
[e] 1978 (P.A. Prince unpublished data).
[f] 1980 (P.A. Prince and J.P. Croxall unpublished data).
[g] 1982 (Prince and Francis 1984).
[h] 1983 (Prince and Morgan 1987; P.A. Prince and P.G. Copestake unpublished data).
[i] 1985 (P.A. Prince and M.J. O'Connell unpublished data).
[k] 1986 (P.A. Prince and S. Delany unpublished data).

chick growth are available (J.P. Croxall unpublished data). Data are summarized in Table 4; there are none for the abnormal seasons of 1978 or 1984. Comments are confined here to parameters for which several years' data are available or where there are special points of interest.

Foraging Trip Duration. The distribution of Macaroni Penguin foraging trips is discontinuous, distinct modes corresponding to whether trips cover one night or one or more days. This complicates the use of this parameter in monitoring studies with this species, and perhaps also with other penguins.

Meal Mass and Composition. There is considerable intra- and interyear variation in meal mass in both species. Meal composition has been very consistent in 4 years' sampling of Macaroni Penguins; for Gentoos the 1984 diet was anomalous, with a preponderance of fish. For both species krill was significantly smaller and more variable in 1985 than in other years, in which krill size has been very consistent.

Chick Growth. Intrayear variations in chick growth rates are rather large, being least when measured over longer time spans, and none of the differences is statistically

Table 4. Breeding season parameters in Macaroni and Gentoo Penguins at Bird Island, South Georgia. Conventions as in Table 3

Parameter		Macaroni				Gentoo			
Arrival mass (g)[a]:									
males		4685 (350,14)							
females		4800 (380,14)							
Incubation shift (d)[a]:									
first by male		10.1 (1.8,13)				}			
first by female		23.0 (1.8,12)				} 1.4 (0.8,150)			
second by male		11.0 (1.4,8)				}			
Foraging trip duration (d)[d]		11.8 (2.1,130)				9.8 (2.7,128)			
		28.2 (5.5,52)							
Stomach content mass (g)		692[a] (227,40)				857[a] (227,40)			
		508[b] (235,10)				790[b] (230,10)			
		520[c] (241,14)				830[c] (256,15)			
		448[d] (259,40)				878[d] (262,40)			
		1977	1980	1985	1986	1977	1980	1985	1986
Meal composition:	krill	98.0	97.0	96.5	93.5	68	75	38.5	80
(% by mass)	fish	2.8	2.5	2.9	5.0	32	25	61.3	20
	other	–	0.5	0.6	1.5	–	–	0.2	–
	n	40	10	14	40	43	10	15	40
Krill standard length (mm)		53.2 (3.7,626)	52.7 (35,200)	35.4 (12.9,50)	52.4 (3.3,1376)	54.0 (3.4,597)	53.5 (3.2,200)	39.1 (13.7,50)	c53
Chick growth rate (g/d)[a]:									
to creche	(5–23)	63.2 (14.4,30)				(7–31)	84.3 (13.2,36)		
creche-fledge	(23–52)	57.5 (18.1,26)				(31–50)	71.3 (11.2,34)		
to fledge	(5–52)	59.4 (10.2,25)				(7–50)	82.5 (8.8,34)		
						(7–70)	76.2 (9.3,31)		
Chick mass at fledging (g):									
day 50[d]		3002 (354,30)							
day 52[a]		3048 (456,26)							
day 60[a]		3106 (394,14)							
Adult mass at chick fledging (g)[a]:									
male		4200 (305,18)							
female		3600 (280,16)							
Adult mass at return to moult (g)[a]:									
male		6780 (640,13)							
female		6100 (505,16)							

[a] 1977 (Croxall and Prince 1980a; Croxall 1984 and unpublished data). Chick growth rates with first and last days in parentheses.
[b] 1980 (J.P. Croxall unpublished data).
[c] 1985 (J.P. Croxall unpublished data).
[d] 1986 (Croxall et al. 1988).

significant. Comparison of weights at age with weights on equivalent calendar dates (i.e. assuming chicks weighed are of the same age, which would greatly reduce the fieldwork required) results in errors for Macaroni Penguins of –5% to +6% and for Gentoo Penguins of –4.5% to +3%. Such are the variances involved that significant errors would only be introduced when the actual and assumed dates of hatching differ by more than 6 days (Macaroni) or –4 to +2 days (Gentoo). This is a realistic situation for Macaroni Penguins but not for Gentoos, whose laying date may vary by 2–3 weeks between years.

4.2.3 Adélie and Chinstrap Penguins at Signy Island

After pioneer studies by Sladen (1958), little research was carried out on these species until the brief but intensive study of breeding biology and diet by Lishman (1983, 1985a,b). This study included one apparently normal year

Table 5. Breeding season parameters in Adélie and Chinstrap Penguins at Signy Island, South Orkney Islands. Conventions as in Table 3. Data from Lishman (1983, 1985a,b), unless otherwise indicated

Parameter	Adélie 1981	Adélie 1982	Chinstrap 1981	Chinstrap 1982
Ice break-out (1/10 cover)	5 Feb	10 Nov	5 Feb	10 Nov
Breeding success (per pair)				
a) eggs hatched	1.60 (394)	1.30 (456)	0.50 (294)	1.30 (348)
b) eggs to fledged chicks	0.77	1.27	0.04	0.76
Clutch size	1.90 (0.4,394)	1.65 (0.5,456)	1.85 (0.4,294)	1.87 (0.4,343)
First egg (date)	3 Nov (3;60)	3 Nov (3;74)	8 Dec *** (3;48)	4 Dec (3;73)
Laying interval (d)	3.1 (0.7;130)	3.1 (0.7;72)	3.4 (0.8;91)	3.2 (1.1;134)
Duration first incubation shift (d)	12.6[a] *** (1.7,123)	13.7[a] (1.7;77)	6.1[b] (2.0;67)	6.0[b] (2.4;128)
Duration second incubation shift (d)	14.0[b] *** (2.0;112)	12.8[b] (2.1;73)	11.2[a] (2.8;45)	9.8[a] (2.9;103)
Foraging trip duration (h)				
a) guard period	39.3 (25.1;316)	40.7 (22.1;111)	61.5 *** (36.3;113)	42.9 (29.8;149)
b) creche period	54.5 (31.0;215)			30.4 (16.5;53)
c) S. Shetland Islands, 1981 and 1983[c]	22.0 *** (0.9;19)	26.7 (1.2;18)	16.2 * (0.8;22)	17.1 (1.6;25)
Stomach content mass (g)				
a) end creche	374 (72;3)	441 (161;5)	227 ** (66;5)	457 (67;5)
b) late-rear	398 (189;5)	540 (151;5)	492 (51;4)	563 (203;5)
c) S. Shetland Islands, throughout (1978–1980)[d]	610 (139;97)			475 (156;84)
Chick hatch mass (g)	85.5 (130;53)	85.4 (9.8;52)	66.0 (9.8;52)	77.4 (8.1;40)
Chick mass: day 20	933 * (294;39)	1207 (343;12)		1089 (317;11)
day ca. 48	2032 * (355;28)	2463 (382;9)		3086 (499;7)
Chick age at creche (d)	19.1 ** (1.7;22)	21.0 (2.3;43)		28.7 (4.0;19)
Fledging period (d)		60.3 (2.4;21)		
Diet (% krill by mass)	98.3 (3.1;15)	99.1 (1.9;13)	97.0 (1.9;21)	99.8 (0.6;14)

[a] By male.
[b] By female.
[c] Volkman et al. (1987).
[d] Jablonski (1985).

(1982) and another (1981) in which the date of ice break-out was the latest ever recorded (early Feb. 1981), which had a very marked effect on Chinstrap Penguin breeding success. Data are summarized in Table 5 and selected points commented on below.

Clutch Size. In a variety of studies at several sites (reviewed in Lishman 1985b and Ainley et al. 1983) this has varied only between 1.8 and 1.9, so the value of 1.65 for Adélies in 1982 is very anomalous. Single-egg clutches are mainly laid by inexperienced birds (Ainley et al. 1983) and variations may, therefore, mainly reflect changes in recruitment rate rather than conditions just prior to arrival ashore.

Breeding Timetable. Laying dates are somewhat variable both at Signy [Adélie mean 27 October, range 22 October–2 November, n = 12; Chinstrap mean 25 November, range 19 November–3 December, n = 8 (Lishman 1985b)] and at Cape Crozier, Ross Island, Antarctica [18 November ± 4.9 days (SD) (Ainley et al. 1983)]; it is likely that this mainly reflects physical environmental conditions, especially the nature and extent of ice cover (Ainley and LeResche 1973).

Incubation Shift Duration. Adélies and Chinstraps differ, at least at Signy, in the identity of the sex which takes the first shift. Apart from the short first shift in Chinstraps, the other shift durations are significantly different between years for both species (note, though, that sample sizes were very large).

Foraging Trip Duration. As estimated by the intervals between feeds to the chick, these were long and variable for both species in both seasons, and significantly longer for Chinstrap Penguins in 1981. In contrast, data for penguins foraging chiefly in Admiralty Bay, South Shetland Islands (Volkman et al. 1987), recorded by radiotelemetry, showed small intra-annual variation and the much smaller differences between 1982 and 1983 were statistically significant for both species.

Chick Growth. Due to large variances and relatively small sample sizes, interannual differences in Adélie Penguin chick mass of 30% and 21%, at days 20 and 48 respectively, are only just statistically significant, but clearly reflect the underlying environmental differences between the two seasons. Very few Chinstrap Penguin chicks survived in 1981.

Meal Size. Few samples of adult stomach content mass were taken each year; they were variable in size but generally smaller than Jablonski (1985) found at the South Shetland Islands, particularly for Adélies. Nevertheless, Chinstrap Penguin samples from early chick rearing were significantly larger in 1982 than 1981, but there was no difference later on, at least for the few 1981 chicks that were still surviving. Lishman (1985a) showed that there is a significant trend for meal mass to increase with chick age, even after creches have formed.

General. The physical environment at Signy differed very greatly between 1981 and 1982. The persistence of ice was a major factor in Chinstrap Penguin breeding failure (Lishman 1985a,b), forcing adults to spend much longer away from their chicks, which died of starvation. For Adélies, although conditions were extreme by Signy standards, they were little worse than populations breeding on the Antarctic Continent must regularly encounter, and both adults and chicks seem physiologically better adapted to cope than do Chinstraps. Nevertheless, we could still detect significant interannual differences in incubation shifts and chick growth, though foraging trip duration remained the same.

5 Variation in Reproductive Performance

Here we highlight certain aspects of the variation in reproductive performance of seabirds and seals that we have reviewed and suggest possible reasons for such patterns as emerge. In addition we summarize the evaluation of potential monitoring parameters, based on the numerical data presented, particularly from the standpoint of their implications for sample size (and therefore field methodology) and for detection of changes which might be induced by commercial harvests.

5.1 Variation at South Georgia

The main anomalies are summarized in Fig. 11. This analysis ignores much small-scale interannual variation, some of which is statistically significant, but we wish to focus here on major differences which we believe to have additional biological significance, rather than just reflecting the typical pattern of variation.

By these criteria, there is no doubt that 1984 was a highly anomalous year; its effects may even have continued into 1985 for Antarctic Fur Seals. The significant decrease in krill length-frequency between 1984 and 1985 may also reflect events in 1984. Breeding populations of albatrosses, fur seals and Macaroni Penguins were depressed the following year but generally seemed to have recovered by 1986 or 1987. Gentoo Penguin breeding populations decreased in 1984 and recovered by 1986.

Because of the paucity and unevenness of data, interpretation of events in the early years of the study is more uncertain. In 1978, seabird breeding success (apart from Grey-Headed Albatrosses) was very poor and Antarctic Fur Seal growth rates (the only data available for this species in that year) were low. Albatross and penguin breeding populations (no data for fur seals) had decreased substantially in 1979 and took several years to recover.

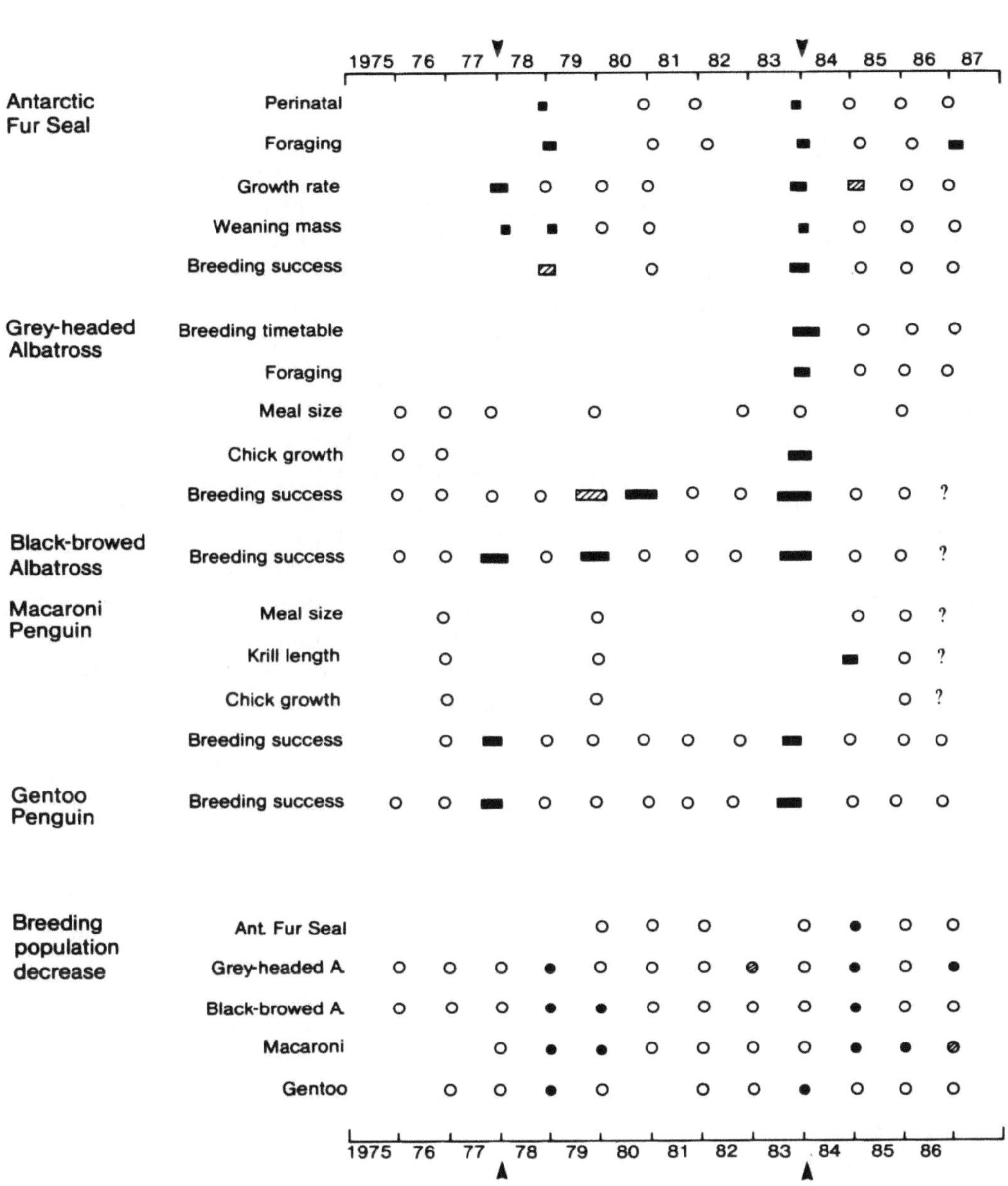

Fig. 11. Normal (o) and anomalous (•, ▪) values of parameters of seabird and fur seals breeding biology at South Georgia, 1975–1987. *Gaps* indicate no data; *hatched symbols* indicate weak effects

Antarctic Fur Seals, however, showed several significant anomalies in apparently sensitive parameters in 1979, the season in which the few data for other species were normal. Black-Browed Albatrosses had complete reproductive failure in 1980 and Grey-Headed Albatrosses sustained a very poor season in 1981. Again, the limited data for other species do not indicate any more extensive effects in either of these seasons.

It is not surprising to find years in which one species or another experiences problems in rearing offspring, nor that we have difficulty relating these to specific aspects of their feeding ecology. However, in seasons where several species of differing ecological adaptation are affected, we may reasonably look for a more general explanation.

We have some evidence of medium- to large-scale environmental anomalies in both 1978 and 1984 (and either no data or no such indication for the other years). In 1978 a variety of fishing and research vessels failed to locate krill concentrations around South Georgia (Bonner et al. 1978), the only abundant planktonic crustacean being the amphipod *Themisto gaudichaudii.* Krill was, however, present around the South Shetland and South Orkney Islands. The events leading up to the 1984 summer season have been extensively documented (Heywood et al. 1985; Priddle et al. this Vol.), suggesting the existence of substantial oceanographic anomalies and the absence of krill on a broader scale and possibly over a longer time span. In fact, krill was not entirely absent from the South Georgia area but it was clearly much harder for predators to find. The best strategy for long-lived vertebrates operating as long-distance central place foragers (Orians and Pearson 1979) is to stay away foraging until they have met their own requirements and collected enough extra food to make the return to their offspring energetically worthwhile (Charnov et al. 1976). Because meal sizes (Grey-Headed Albatross) and suckling periods ashore (Antarctic Fur Seal) were normal in 1984 and foraging trips in both species were the longest yet recorded, it is suggested that adults were adopting this strategy. Because the situation persisted for so long it had the almost inevitable result of major offspring mortality and, eventually, judging from the reduction in breeding population size the next year, significant mortality of breeding adults. Juveniles, unconstrained by the need to rear an offspring, and presumably able to leave the area altogether, may prove to have suffered less (we can only tell this by changes in the pattern of recruitment over the next few years). Similarly, adults which failed earlier in the year would be at an advantage compared with those whose offspring survived for longer. We already have indications that survival was higher in Black-Browed and Grey-Headed Albatrosses that suffered breeding failure early rather than later in the season (P.A. Prince and P. Rothery unpublished data).

Themisto did not appear to be unduly abundant in 1984 and this may explain some differences from 1978. Thus Grey-Headed Albatrosses eat more squid (and fish) than krill; but many of these prey themselves depend extensively on catching krill (Targett 1981; Nemoto et al. 1985; BAS unpublished data). In 1978 at least some of the squid had stomachs full of *Themisto* and this switch in squid diet may have enabled Grey-Headed Albatrosses to rear chicks successfully. The failure of Black-Browed Albatrosses in 1980 (when meals were significantly lighter than usual) might reflect the fact that few krill swarms were recorded at the surface during acoustic surveys at South Georgia (Everson 1983), which would explain why a surface-feeding albatross was affected but not penguins. We cannot explain the poor performance of Antarctic Fur Seals in 1979, unless this might have been delayed recovery from the 1978 event, or the parameters we measure with them are more sensitive than those of other species.

Only one further speculation seems warranted. The two years when there is evidence, from the predators, of some relatively widespread effects were each one year after strong (massive in 1983) El Niño-Southern Oscillation (ENSO) events. These events are now known to have profound effects on seabirds and seals, extending not only throughout the Pacific but also affecting Atlantic areas such as the Benguela Current (Duffy et al. 1984; Schreiber and Schreiber 1984; Trillmich and Limberger 1985; La Cock 1986). We have been aware of the possibility that our data could reflect oceanographic perturbations initiated by ENSO events, but in the absence of any oceanographic evidence or climatic anomalies we were unwilling to make this speculation. However, with the discovery of various lines of evidence indicating possible connections between Southern Ocean phenomena and ENSO events (Carleton 1986; Priddle et al. this Vol.), the predator data no longer stand entirely alone. It will be interesting in future to see if strong ENSO's are followed by similar atmospheric and oceanographic events and if predators respond in broadly similar fashion. If effects are consistently detectable one year after an ENSO, this will have profound effects on resources management policy.

5.2 Variation at Signy Island

Even from the limited long-term data (Figs. 2, 6, 7), it is clear that at Signy there is no pattern resembling that at South Georgia. Adélie Penguin breeding success was low in 1980, 1983 and 1984; population size decreased in 1981 but has increased otherwise. Chinstrap Penguin breeding populations decreased in 1981 and from 1983 to 1985; breeding success was very low indeed in 1981 and 1983. The poor breeding performance in 1981 was confidently attributed to the exceptional ice conditions (Lishman 1985b). Might the relationship be more general than this? In Fig. 12 there is indeed a significant relationship (described by $y = -0.5x + 31.7; F_{1,7} = 4.5, P < 0.01$) between the date of ice break-out at Signy and Chinstrap Penguin hatching success. The relationship would probably be stronger still if we had fledging success data for all years. It shows clearly the poor performance in 1981 and 1983 compared with the generally good success in the other years when ice break-out had occurred before late November. That there is even a weak relationship ($y = -0.17x + 31.8; F_{1,6} = 5.9, P = 0.05$) for Adélie Penguins is perhaps surprising in view of their ability to breed successfully at sites like Cape Crozier at 77°S, where they do not lay until mid-November.

It is, of course, possible that the extent and persistence of ice cover is correlated with changes in krill abundance. However, even in 1981, Chinstrap Penguins with surviving chicks were able to deliver meals of normal size and it appears that the only delay imposed was by having to spend so long walking across ice to find open water suitable for foraging. We have no evidence from the Signy data for

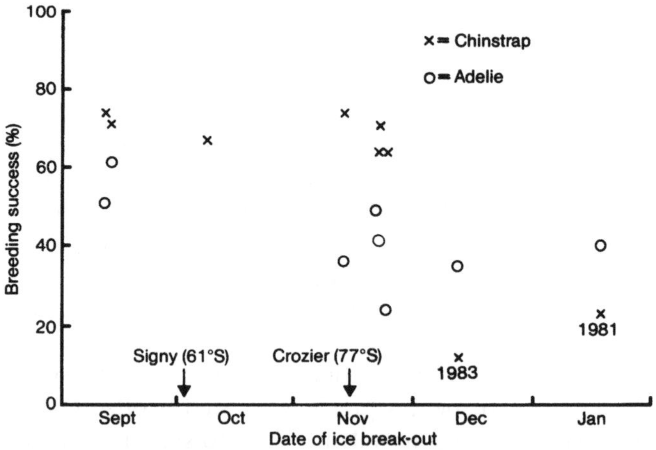

Fig. 12. Average breeding success of Adélie Penguins (x: % chicks reared from eggs laid) and Chinstrap Penguins (0: % chicks hatched from eggs laid) in relation to date of ice break-out at Signy Island, South Orkney Islands 1979–1987. Ice data from Antarctic Ice Charts, prepared by Naval Polar Oceanography Center, Washington DC, USA. *Arrows* indicate mean dates of first arrival at Signy Island (Lishman 1985b) and Cape Crozier (Ainley et al. 1983) colonies

any ENSO effect, which is not surprising because ice is *least* extensive at the time of ENSO events (Carleton 1986).

It would seem, therefore, that a rather different balance of proximate mechanisms influences reproductive performance at Signy and South Georgia. This is particularly significant in the context of monitoring with respect to detection of changes in response to food availability, where it is essential to be able to discount the effects of the physical environment. While these are obviously very important (especially to Chinstraps) at the South Orkney Islands, they might not be quite so important at, say, the South Shetland Islands, where ice break-out is usually considerably earlier than at Signy. However, there is an obvious need to monitor physical environmental parameters if sensible interpretation of reproductive performance data is to be achieved.

6 Evaluation of Potential Monitoring Operations

It is assumed that the object of this monitoring is to detect, as quickly as possible, changes that result from changes in food availability and, if possible, to distinguish between responses due to natural events and those induced by commercial exploitation.

6.1 Choice of Species

Criteria for this have been formulated on several occasions (SCAR 1979; BIOMASS 1980; CCAMLR 1985, 1986) and, in essence, they are that species should:

a) be important components (in terms of prey consumption) of the Southern Ocean system,
b) be specialist predators on harvestable prey, especially krill,
c) have broad geographical breeding ranges, including sites near to and far from areas likely to be subject to intensive fishing,
d) be readily accessible at breeding sites and tolerant of human presence and activity.

We saw earlier that there are really no suitable species of seabirds and seals for monitoring changes in fish stocks. Squid are important to predators only in Sub-Antarctic regions (especially some albatrosses and Southern Elephant Seal). Commercial exploitation of squid is likely to start soon, although it may be less rewarding, mainly because there are fewer suitable target species in the Southern Ocean system than in more temperate waters north of the Antarctic Polar Front (e.g. around the Falkland Islands). Existing data from Grey-Headed and Wandering Albatrosses probably provide an adequate monitoring baseline for the South Georgia area. There is an urgent need for data on the squid element of the diet of Elephant Seals, whose population at South Georgia appears, even in the face of the intensive fin-fish fishery, to be stable (Rothery and McCann 1988; McCann and Rothery 1988), in contrast to the declines in Indian Ocean populations.

Consequently, it is appropriate that the predator species should be selected particularly on the basis of their dependence on krill. The most important species are, therefore, Minke whale *Balaenoptera acutorostrata,* Crabeater Seal *Lobodon carcinophagus,* Adélie, Chinstrap and Macaroni (including Royal *E.c. schlegeli*) Penguins and Antarctic Fur Seals, with Gentoo Penguin and Black-Browed Albatross the most important additional species and several fulmarine petrels (e.g. Antarctic Petrel *Thalassoica antarctica* and Cape Pigeon) worthy of serious consideration. Two of these species (Chinstrap Penguin and Antarctic Fur Seal) are of relatively restricted longitudinal distribution.

We have dealt here fully with all but Minke Whales, Crabeater Seals and fulmarine petrels. Reviews of Crabeater seal parameters (BIOMASS 1983b; CCAMLR 1985, 1986) indicated that the only ones for which adequate data may be available now are reproductive rates, age at sexual maturity and cohort strength. As noted below, these are manifestly unsuitable for programmes requiring indices sensitive to relatively short-term changes in food availability. It is clearly important to investigate whether aspects of feeding ecology could provide suitable indices in this respect. Evaluation of Minke Whale parameters is being conducted by the International Whaling Commission (CCAMLR 1986), but similar conclusions seem inescapable. We have few relevant data for fulmarine petrels. Here we shall consider further only the species for which detailed results were provided earlier. We shall largely ignore Wandering Albatrosses (because they do not eat krill) and consider Grey-Headed Albatrosses mainly because they are the best substitute for Black-Browed Albatrosses, for which we have fewer data.

6.2 Choice of Parameters and Detection of Interannual Change

6.2.1 Criteria

The main criteria in assessing suitability are ease and accuracy of measurement, because the object of monitoring is to detect significant change as soon as possible, which is likely to require large samples; and sensitivity to changes in food availability.

Sensitivity has a number of more or less distinct components. It is likely to be influenced by:

a) Species' dependence on krill, a critical element of which relates to a predator's horizontal and vertical foraging abilities in relation to krill distribution and behaviour.
b) Time over which the change is being measured; short timespans are likely to provide better chances of identifying causal relationships.
c) Intra-annual variation in parameter; small variations (which are also more likely to be associated with shorter

Table 6. Timespan (days) over which the main parameters are measured, and the calendar time (months of the year below) when this occurs

Parameter	Albatrosses	Penguins	Fur seals
Breeding success (eggs laid – chicks fledged)	185–220 (–360[a]) Oct.–May (Dec.–Dec.)	90–100 Nov–Feb.	115 Dec.–Mar.
Arrival mass	120–150 (May) June–Sept.	ca. 250 May–Nov.	ca. 230 Apr.–Nov.
Clutch size	– –	20–25 Oct.–Nov.	– –
Incubation shift (first)	10–30	5–15	– –
Perinatal attendance	– –	– –	5–10 Dec.
Foraging trip duration	1–4 (–4) Jan.–Mar.	0.5–2 Dec.–Feb.	3–6 (–10) Dec.–Feb.
Activity budget at sea	1–2 (–4) Jan.–Mar.	?	?
Meal mass	1–2 (–4) Jan.–Mar.	0.5–2 Dec.–Feb.	– –
Offspring growth rate (also fledge/wean mass)	120–150 (–270[a]) Jan.–May (Apr.–Dec.)[a]	50–70 Dec.–Feb.	ca. 110 Dec.–Mar.
Adult mass at offspring departure	210–240 Sept.–May	100–110 Oct.–Feb.	ca. 130 Dec.–Mar.
Adult mass at start moult[b]	– –	10–15 Mar.	– –

[a] Wandering Albatross only.
[b] Macaroni Penguin only.

measurement durations) will help detect significant changes quickly.

d) Interannual variation in parameter; parameters which show little variation in response to natural interannual fluctuations in food availability are unlikely to respond to artificially induced changes. Large interannual variation in baseline data, however, makes detection of harvest-induced change very difficult.

6.2.2 Evaluation of Parameters

By these definitions and standards, certain parameters (e.g. survival and fecundity rates), of great importance in influencing the demography of populations, must be regarded as of low suitability for monitoring studies because of the long lag times (up to 5–15 years) involved and the very large samples required. Similarly, changes in population size take several years to detect. They also reflect the changes in some or all of the following factors: adult survival, breeding frequency, juvenile survival and emigration/immigration rates, and interpretation requires that all these factors are measured. This is only possible with detailed long-term studies. It must be stressed, however, that there is limited point starting extensive monitoring programmes unless there are ongoing complementary long-term studies of the fundamental demographic processes, because knowledge of these will be essential to the interpretation of monitoring results.

Several other potential parameters for seabird and seal monitoring studies have been suggested (summarized in CCAMLR 1986). Evaluating these according to the criteria above is difficult because we lack much of this information for most of them. We can, however, specify the measurement timespan fairly precisely for many parameters of the main predator groups (Table 6). This shows that only very few parameters (arrival mass, mean clutch size) reflect conditions prior to the start of breeding activities (i.e. in winter) and that parameters are broadly divisible into those with measurement timespans of days (foraging trips, activity budgets at sea, meal mass, perinatal attendance), weeks (clutch size, incubation shifts, adult mass at moult) and months (the rest).

To provide further material for this evaluation we have calculated the changes detected with high probability using samples of different sizes and compared these changes with the observed differences between years of normal and abnormal breeding success (Table 7). Such calculations necessarily involve the arbitrary choice of dectection level but they provide useful guidelines and a framework for evaluating the use of different parameters. It must be stressed also that we assume that the results reflect food availability, that many of our data derive from the highly abnormal year of 1984 and therefore that much larger sample sizes will be needed for some parameters to detect the more usual degrees of variation. Samples smaller than 20 are seldom likely to be effective.

Breeding Timetable. Evidence from penguins suggests that variation mainly relates to changes in the physical environment prior to laying. The Grey-Headed Albatross data suggest that duration of the fledging period may reflect differences in food supply. However, accurate determination of such dates requires large samples of individually marked birds, which is extremely labour-demanding. Coupled with uncertainty over the timespan integrated by such measurements, they would have a low priority at present.

Breeding Success. There are two main problems with this parameter. First, it integrates events over a long period. Second, it is susceptible to biases induced by birds of different ages, or position in the colony (e.g. centre, edge), or in different colonies, or in colonies of different size (Hunt et al. 1986), having different breeding success. The first can be improved (for birds) by recording hatching and fledging success separately. Even better would be an expression representing the stochastic pattern of mortality (e.g. Mayfield 1961) as done by Ainley et al. (1983) and Lishman (1985b). This may show differences not apparent from overall mortality values. Thus Antarctic Fur Seal pup mortality in 1984, when starvation was paramount (see Table 2), was spread relatively evenly over the first 30 days of life. In normal seasons (when pup losses may only be 10–15% less), with most deaths due to injuries caused by overcrowding, most mortality occurs in the first week of life (T.S. McCann unpublished data). The second effect can be reduced by selecting complete colonies, ideally well distanced from others, and if possible using several, each of different size. Notwithstanding the long integration period, accurate measures of breeding success should be fundamental elements of monitoring studies.

Foraging Trip Duration. This parameter has proved very useful with Antarctic Fur Seals and appears to have considerable potential with other species, providing the trip durations do not exhibit markedly discontinuous distributions. Fairly large sample sizes are needed and this could be greatly helped by using radio transmitters and automated recording units. For each species, however, it should be demonstrated that the performance of instrumented individuals does not differ from uninstrumented ones.

Activity Budgets at Sea. The few data so far available suggest some promise if many individuals can be instrumented and providing that the recorders used are sufficiently small, reliable and easy to deploy and decode. Except for albatrosses, all present devices are probably best regarded as still at the experimental stage. This is a vital field in which to promote development, particularly for seals and penguins.

Meal Mass and Diet Composition. The substantial intra-annual variation in these parameters, the difficulty of obtaining large sample sizes, the time needed for diet analysis and, especially, the tendency for birds to return with full stomachs even in bad years, makes them unsuitable for routine monitoring. It is very important, however, that some systematic attempt be made to determine the diet of target species at the time other parameters are being monitored.

Offspring Growth. There is substantial individual variation in growth rates (see Table 4), so large samples are needed even when the age of the offspring is accurately known. Such sampling requires either intensive fieldwork, marking many individuals of known birth date and recapturing them at intervals, or being able to estimate age from other measurements (e.g. wing length). For species which are highly synchronous in breeding timetable, like Antarctic Fur Seals where the mean dates of pupping have only differed by +2 days from the mean over 8 years, little, if any, extra variation is introduced by weighing large samples of pups at intervals. For this species, both growth rates and especially the growth index (mass near weaning) have proved very useful. Providing mean hatching date is established, similar data for albatrosses (at near peak mass and near fledging) and penguins (at creche and near fledging) may be the best compromise between accuracy and practicality.

There is need for a critical study, ideally using data from daily weighings of large numbers of known-age individuals, to determine optimum growth indices. A general problem with growth data is that measurements of older individuals are, by definition, those of survivors, whereas the part of the population that has been most affected by conditions and may have shown poorest growth, is already dead.

Adult Mass. This parameter, whether measured at arrival or departure, shows substantial variation, so large samples would be required. This will be aided if birds were to cross electronic balances, linked to automatic recording units. This approach would seem to be especially feasible for Macaroni Penguins which typically enter and leave colonies along traditional narrow pathways; doubtless other species could be induced to do the same. Arrival mass is also one of the few parameters that may reflect conditions at sea prior to the breeding season. The subsequent status (breeding or non-breeding) of birds arriving may need to be recorded. For interannual comparisons it may help to plot

Table 7. Magnitude of absolute changes (in parentheses as percentage of normal year) detectable with 85% chance, for various samples size. Values for an abnormal year (and the changes this represents) are also shown, together with the sample sizes required to detect this change at the 85% and 95% levels[a]

Species	Parameter		Normal year[b]			Detectable change[c]				Abnormal year[d]			
			n	SD	Mean	n = 30	n = 50	n = 100	n = 200	Mean	Change	n[8e]	n[9e]
Antarctic Fur Seal	Perinatal attendance (d)		42	1.43	6.2	0.78 (13)	0.61 (10)	0.43 (7)	0.30 (5)	5.7	0.5 (8)	75	100
	Foraging trip (d)		168	1.43	3.7	0.78 (21)	0.61 (16)	0.43 (12)	0.30 (8)	5.2	1.5 (40)	10	10
	Attendance ashore (d)		168	0.78	1.6	0.43 (27)	0.33 (21)	0.23 (15)	0.17 (10)	–			
	Growth rate[e] (g/d)	M	50	1.17	97	30 (31)	23 (24)	16 (16)	12 (12)	72	25 (26)	40	60
		F	50	1.56	91	40 (44)	31 (34)	22 (24)	16 (17)	59	32 (35)	50	60
	Mass at 70 d (kg)	M	50	1.17	12.9	0.45 (3)	0.35 (3)	0.25 (2)	0.18 (1)	11.8	1.1 (9)	5	10
		F	50	1.56	10.9	0.60 (6)	0.47 (4)	0.33 (3)	0.23 (2)	10.2	0.8 (7)	20	25
Grey-Headed Albatross	Laying date (d from 1 Oct.)		232	3.8	19.3	2.1	1.6	1.1	0.8	18.8	0.5	520	710
	Hatch date (d from 1 Dec.)		312	2.9	29.1	1.6	1.2	0.9	0.6	30.3	1.2	55	70
	Fledging period (d)		102	5.1	141	2.8 (2)	2.2 (2)	1.5 (1)	1.1 (1)	144	3	25	35
	Foraging trip (d)		41	36.8	52.2	20.1 (39)	15.6 (30)	11.0 (21)	7.8 (15)	94.4	42.2 (81)	10	25
	Day on water (%)		13	9.1	15.1	5.0 (33)	3.9 (26)	2.7 (18)	1.9 (13)	6.2	8.9 (60)	10	35
	Night on water (%)		13	14.6	49.9	8.0 (16)	6.2 (12)	4.4 (9)	3.1 (6)	76.0	26.1 (52)	5	10
	Time in flight (%)		13	9.0	71.8	4.9 (7)	3.8 (5)	2.7 (4)	1.9 (3)	50.7	21.1 (29)	5	10
	Meal mass (g)		95	206	621	111 (18)	86 (14)	61 (10)	43 (7)				
	Chick mass at 80 d (g)		22	538	4063	295 (7)	228 (6)	161 (4)	114 (3)	3023	1040 (26)	5	10
	Chick mass at 140 d (g)		22	727	3561	398 (11)	308 (9)	218 (6)	154 (4)	2454	1107 (31)	5	10
Adélie (A) and Chinstrap (C) Penguins	Clutch size	A	394	0.4	1.90	0.22 (12)	0.17 (9)	0.12 (6)	0.08 (4)	–			
		C	343	0.4	1.87	0.22 (12)	0.17 (9)	0.12 (6)	0.08 (4)	–			
	First egg date (d)	A	74	3	3 Nov.	1.6	1.3	0.9	0.4	–			
		C	73	3	4 Dec.	1.6	1.3	0.9	0.4	8 Dec.	4	5	10
	Duration of shift 1 (d)	A	77	1.7	13.7	0.93 (7)	0.72 (5)	0.51 (4)	0.36 (3)	12.6	1.1 (8)	20	30
		C	128	2.4	6.0	1.31 (22)	1.02 (17)	0.72 (12)	0.51 (8)				
	Duration of shift d (d)	A	73	2.1	12.8	1.15 (9)	0.89 (7)	0.63 (5)	0.45 (3)	14.0	1.2 (9)	30	35
		C	103	2.9	9.8	1.58 (16)	1.23 (13)	0.87 (9)	0.62 (6)	11.2	1.4 (14)	40	55
	Chick hatch mass (g)	A	52	9.8	85.4	5.4 (6)	4.2 (5)	2.9 (3)	2.1 (2)	–			
		C	40	8.1	77.4	4.4 (6)	3.4 (4)	2.4 (3)	1.7 (2)	66.0	11.4 (15)	5	10

Table 7 (continued)

Species	Parameter		Normal year[b]			Detectable change[c]				Abnormal year[d]			
			n	SD	Mean	n = 30	n = 50	n = 100	n = 200	Mean	Change	n[8e]	n[9e]
	Chick mass at 48 d (g)	A	9	382	2463	209 (8)	162 (7)	115 (5)	81 (3)	2032	431 (18)	10	10
		C	7	499	3086	273 (9)	212 (7)	150 (5)	106 (3)	–			
	Chick creche age (d)	A	43	2.3	21.0	1.3 (6)	1.0 (5)	0.7 (3)	0.5 (2)	19.1	1.9 (9)	15	20
		C	19	4.0	28.7	2.2 (8)	1.7 (6)	1.2 (4)	0.8 (3)	–			
	Foraging trip (d)	A	18	1.2	26.7	0.7 (2)	0.5 (2)	0.4 (1)	0.3 (1)	22.0	4.7 (18)	5	10
		C	25	1.6	17.1	0.9 (5)	0.7 (4)	0.5 (3)	0.3 (2)	16.2	0.9 (5)	30	40
	Stomach contents (g)	A	5	151	540	83 (15)	64 (12)	45 (8)	32 (6)	398	142 (26)	10	15
		C	5	203	563	111 (20)	86 (15)	61 (11)	43 (8)	492	71 (13)	75	100
Macaroni (M) and Gentoo (G) Penguins	Arrival mass (g)	M	14	350	4685	192 (4)	148 (3)	105 (2)	74 (2)				
	Foraging trip (d)	G	128	2.7	9.8	1.5 (15)	1.1 (12)	0.8 (8)	0.6 (6)				
	Stomach content (g)	M	14	241	520	132 (25)	102 (20)	72 (14)	51 (10)				
		G	40	227	857	124 (15)	96 (11)	68 (8)	48 (6)				
	Chick growth: to creche (g/d)	M	30	14.4	63.2	7.9 (12)	6.1 (90)	5.4 (9)	3.1 (5)				
		G	36	13.2	84.3	7.2 (9)	5.6 (7)	4.0 (5)	2.8 (3)				
	to fledge (g/d)	M	25	10.2	59.4	5.6 (9)	4.3 (7)	3.1 (5)	2.2 (4)				
		G	34	8.8	82.5	4.8 (6)	3.7 (5)	2.6 (3)	1.9 (2)				
	Adult mass at departure	M♂	18	305	4200	167 (4)	129 (3)	91 (2)	65 (2)				
	Adult mass at moult	M♂	13	640	6780	350 (5)	272 (4)	192 (3)	136 (2)				

[a] Calculations are based on the power of the t-test to detect changes (Pearson and Hartley 1972).
[b] Values used are those for the year closest to overall mean of all years, excluding 1978 (all species), 1979 (Antarctic Fur Seal) and 1984 (all species).
[c] Values used are for 1978 or 1984 (both all species) or 1979 (Antarctic Fur Seal), whichever is closest to the normal year value.
[d] Based on comparison with the data from the normal year. If the baseline is estimated from the same sample sizes, differences need to be increased by 40%.
[e] Calculations for Antarctic Fur Seal growth parameters are based on two samples of size n at 55 and 85 days respectively. Quoted values of SD refer to variations in weight of animals of given age.

mass data in relation to arrival date. One potential problem is that it may be mainly adults in good condition which bother to turn up at the breeding colony in time for breeding. Adult seals will be more difficult to weigh routinely; where this can be done, measurement of blubber thickness by ultrasonic techniques (Gales and Burton 1987) would seem likely to provide a particularly good index of condition.

6.3 Detection of Changes Following Commercial Harvesting

The data so far obtained relevant to monitoring studies on krill- (and squid-) eating seabirds and seals can reasonably be regarded as representing baseline data from the period before any significant commercial exploitation of these resources. We have seen in the previous section the size of

Table 8. Changes in parameters (expressed as absolute values and percentage of baseline in parentheses) that could be detected with 85% and 95% chance with a further 5 and 10 years monitoring[a]

Species	Parameter		Detectable change: With abnormal years[b] Baseline Mean	S.D.	85% detection 5 y	85% detection 10 y	95% detection 5 y	95% detection 10 y	Without abnormal years[c] Baseline Mean	S.D.	85% detection 5 y	85% detection 10 y	95% detection 5 y	95% detection 10 y
Wandering Albatross	Breeding success (%)		65	6.5	10 (16)	8 (13)	12 (19)	10 (16)	63	6.8	11 (17)	9 (14)	13 (20)	11 (17)
Grey-Headed Albatross	Breeding success (%)		46	13.0	18 (39)	16 (34)	20 (44)	18 (40)	46	11.5	16 (35)	14 (31)	18 (39)	17 (36)
Black-Browed Albatross	Breeding success (%)		37	22.8	26 (70)	24 (66)	29 (79)	25 (67)	43	20.6	25 (58)	23 (53)	28 (65)	27 (62)
Macaroni Penguin	Breeding success (%)		44	15.2	20 (45)	18 (41)	22 (51)	21 (47)	51	4.2	7 (13)	6 (11)	9 (18)	7 (14)
Antarctic Fur Seal	Foraging trip (d)		4.3	1.3	1.8 (41)	1.6 (37)	2.0 (47)	1.8 (43)	3.7	0.69	0.81 (22)	0.74 (20)	1.2 (32)	1.0 (28)
	Mass at 70 d	M	12.5	1.2	1.9 (15)	1.5 (12)	2.1 (17)	1.8 (14)	13.0	0.52	0.91 (7)	0.78 (6)	1.0 (8)	0.91 (7)
		F	10.6	0.8	1.4 (13)	1.2 (11)	1.7 (16)	1.4 (13)	11.0	0.40	0.66 (6)	0.44 (4)	0.77 (7)	0.66 (6)

[a] Harvesting is assumed to cause a proportionate decrease in both the parameter and its year to year standard deviation. The calculations are based on the power of a t-test to compare pre- and postharvesting periods.
[b] Ten years data, taken from Table 2 and Fig. 5.
[c] 1978 and 1984 for all species; also 1979 for Antarctic Fur Seals.

interannual differences that we might expect to detect in many of these parameters, given appropriate sample sizes. We have also seen the large variation in some parameters that has occurred in the absence of any harvest. For most species at South Georgia and at least for Chinstrap Penguins at Signy it is unlikely that even an intensive commercial harvest would, in a single year, depress reproductive performance by an amount greater than that recorded in our abnormal years. However, a persistent localized harvest might result in a small but consistent reduction in breeding performance, which, if it continued for long enough, might be detectable.

We examine this in a very simple fashion (Table 8) using data on parameters for which we have the longest series of data. The evaluation of monitoring schemes to detect this type of change requires more qualification than the earlier calculations of sample sizes to detect specified interannual changes. The essential difference is that an average change measured over several years must be assessed against a background of year-to-year variation (baseline); within-year variability is of secondary importance. To date, our baseline information on this is limited to 10 or so years. Furthermore, statistical tests based on assumptions of random samples of normally distributed data are less tenable. Conclusions drawn are necessarily tentative, but could serve to expose limitations of particular schemes. There are four main conclusions from this analysis. First, identification of abnormal years (where major environmental changes are affecting all or most species to some significant extent) results in clear improvements in detectability of change for all species (except the Wandering Albatross) and most strikingly for the Macaroni Penguin. Second, with or without abnormal years, the required reductions in breeding success are probably unrealistically large (for all species except the Wandering Albatross) to occur naturally. Third, the magnitude of changes in parameters with shorter measurement timespans (especially mass data) are more realistic, especially if anomalous years can be independently identified. Finally, there is little prospect of detecting statistically significant changes, even using the best parameters, with fewer than 5 years post-baseline data.

For fisheries management in the short term, e.g regulating annual catch limits, it may not be acceptable to wait until a deleterious change has become statistically significant. A more pragmatic approach might be to close the krill fishery, in appropriate CCAMLR subareas, in the year following an anomalous year and to reduce allowed catches when two successive normal years produce decreases in the values of, say, a majority of the parameters being measured.

7 Conclusions

We have shown here that certain aspects of the biology of Antarctic seabirds and seals could, if recorded accurately and with adequate sample sizes, act as indices of change in the local marine environment. At South Georgia, we suggest that observed major variations in these parameters have related to changes in food availability. At Signy Island, however, we can only relate them at present to differences in physical environmental conditions. This suggests that there may be a major difficulty at higher latitudes in separating effects due to changes in the physical conditions from those relating to food supply. Considerable research is needed, particularly on Adélie Penguins and Crabeater Seals, if progress is to be made.

Critical interpretation of smaller-scale interannual variation, even in Sub-Antarctic areas, will only be feasible if:

a) a network of monitoring sites is established, covering both regularly fished and control areas,
b) where possible, several of the selected species are monitored simultaneously and annually,
c) for each species, several parameters are monitored, including ones with measurement timespans of days, weeks and months,
d) appropriate physical environmental factors are monitored also,
e) ideally, some long-term demographic studies are running concurrently.

Detection and interpretation of changes following harvesting will depend critically on the existence of monitoring sites, both near to and far from commercial fishing areas. The possibility of conducting experiments involving intensive fishing close to selected breeding colonies should be seriously considered if cause-effect relationships are to be plausibly evaluated.

However, even under the most favourable circumstances, it will take several years to detect harvest-induced changes and may still be impossible to demonstrate the probability of causal links.

Acknowledgements. We are greatly indebted to everyone who has worked at Bird Island for their help in collecting the data on which this paper is based, and especially to P.G. Copestake, S. Delany, R. Lidstone-Scott and M.J. O'Connell. Similarly we are grateful to many people at Signy Island who have helped count penguins, especially P. Burren, R. Forster, G.S. Lishman, R.A. Price and D. Rootes. We also thank C.S. Harcourt for much assistance in the preparation of material for this paper, A. Sylvester for drawing the illustrations, G.D. Hurrell for typing the manuscript and I. Everson, G.L. Hunt Jr., C.M. Phillips, and J. Priddle for helpful comments on it.

References

Adams NJ, Klages NT (1987) Diet of the King Penguin *Aptenodytes patagonicus* at Sub-Antarctic Marion Island. J Zool 212:303–324

Ainley DG, LeResche RE (1973) The effects of weather and ice conditions on breeding in Adélie Penguins. Condor 75:235–239

Ainley DG, LeResche RE, Sladen WJL (1983) Breeding biology of the Adélie Penguin. Univ California Press, Berkeley Los Angeles

Andrewartha HG, Birch LC (1955) Distribution and abundance of animals. Univ Chicago Press, Chicago

Astheimer LB, Prince PA, Grau CR (1985) Egg formation and the pre-laying period of Black-Browed and Grey-Headed Albatrosses at Bird Island, South Georgia. Ibis 127:523–529

BIOMASS (1980) Antarctic bird biology. BIOMASS Rep Ser 8: 21 p

BIOMASS (1983a) Report of meeting of BIOMASS Working Party on Bird Ecology, Wilderness, South Africa, September 1983. BIOMASS Rep Ser 34:33 p

BIOMASS (1983b) Meeting of the SCAR Group of Specialists on Seals. BIOMASS Rep Ser 35:41 pp

Bonner WN (1968) The fur seal of South Georgia. Sci Rep Br Antarct Surv No 56

Bonner WN (1984) Conservation and the Antarctic. In: Laws RM (ed) Antarctic ecology, vol 2. Academic Press, London New York, pp 821–850

Bonner WN (1985) Impact of fur seals on the terrestrial environment at South Georgia. In: Siegfried WR, Condy PR, Laws RM (eds) Antarctic nutrient cycles and food webs. Springer, Berlin Heidelberg New York Tokyo, pp 641–646

Bonner WN, Everson I, Prince PA (1978) A shortage of krill *Euphausia superba,* around South Georgia. Int Counc Explor Sea Ser C, CM 1978 L/22, 4 p

Carleton AM (1986) Antarctic sea ice – atmosphere signal of the Southern Oscillation. In: 2nd Int Conf South Hemisphere Meteorol, Dec 1–5, 1986, Wellington, New Zealand. Am Meteorol Soc, Boston, pp 431–434

CCAMLR (1985) Report of the fourth meeting of the Scientific Committee. Annex 7. Rep Ad hoc Work Group Ecosyst Monitor. SC-CAMLR-IV, Hobart, Australia

CCAMLR (1986) Report of the fifth meeting of the Scientific Committee. Annex 6. Rep Work Group CCAMLR Ecosyst Monitor Program. SC-CAMLR-V. Hobart, Australia

Charnov EL, Orians GH, Hyatt K (1976) The ecological implications of resources depression. Am Nat 110:247–259

Costa DP, Prince PA (1987) Foraging energetics of Grey-Headed Albatrosses *Diomedea chrysostoma* at Bird Island, South Georgia. Ibis 129:149–158

Croxall JP (1979) Distribution and population changes in the Wandering Albatross *Diomedea exulans* L. at South Georgia. Ardea 67:15–21

Croxall JP (1984) Seabirds. In: Laws RM (ed) Antarctic ecology, vol 2. Academic Press, London New York, pp 533–619

Croxall JP (1987) The status and conservation of Antarctic seabirds and seals: a review. Environ Int 13:55–70

Croxall JP, Kirkwood ED (1979) The distribution of penguins on the Antarctic Peninsula and islands of the Scotia Sea. Br Antarct Surv, Cambridge

Croxall JF, Lishman GS (1987) The food and feeding ecology of penguins. In: Croxall JP (ed) Seabirds: feeding ecology and role in marine ecosystems. Cambridge Univ Press, Cambridge, pp 101–133

Croxall JP, Prince PA (1979) Antarctic seabird and seal monitoring studies. Polar Rec 19:573–595

Croxall JP, Prince PA (1980a) The food of Gentoo Penguins *Pygoscelis papua* and Macaroni Penguins *Eudyptes chrysolophus* at South Georgia. Ibis 122:245–253

Croxall JP, Prince PA (1980b) Food, feeding ecology and ecological segregation of seabirds at South Georgia. Biol J Linn Soc 14: 103–131

Croxall JP, Prince PA (1987) Seabirds as predators on marine resources, especially krill, at South Georgia. In: Croxall JP (ed) Seabirds: feeding ecology and role in marine ecosystems. Cambridge Univ Press, Cambridge, pp 347–368

Croxall JP, Rootes DM, Price R (1981) Increases in penguin populations at Signy Island, South Orkney Islands. Bull Br Antarct Surv 54:47–56

Croxall JP, Ricketts C, Prince PA (1984) The impact of seabirds on marine resources, especially krill, at South Georgia. In: Whittow GC, Rahn H (eds) Seabird energetics. Plenum, New York, pp 285–318

Croxall JP, Prince PA, Hunter I, McInnes SJ, Copestake PG (1984) The seabirds of the Antarctic Peninsula, islands of the Scotia Sea and Antarctic Continent between 80°W and 20°W: their status and conservation. In: Croxall JP, Evans PGH, Schreiber RW (eds) Status and conservation of the world's seabirds. ICBP, Cambridge, pp 635–664

Croxall JP, Prince PA, Ricketts C (1985a) Relationships between prey life-cycles and the extent, nature and timing of seal and seabird predation in the Scotia Sea. In: Siegfried WR, Condy PR, Laws RM (eds) Antarctic nutrient cycling and food webs. Springer, Berlin Heidelberg New York Tokyo, pp 516–533

Croxall JP, Everson I, Kooyman GL, Ricketts C, Davis RW (1985b) Fur seal diving behaviour in relation to vertical distribution of krill. J Anim Ecol 54:1–8

Croxall JP, Davis RW, O'Connell MJ (1988) Diving patterns in relation to diet in Gentoo and Macaroni Penguins, *Pygoscelis papua* and *Eudyptes chrysolophus*, at South Georgia. Condor (in press)

Davis RW, Kooyman GL, Croxall JP (1983) Water flux and estimated metabolism of free-ranging Gentoo and Macaroni Penguins at South Georgia. Polar Biol 2:41–46

Doidge DW, Croxall JP (1985) Diet and energy budget of the Antarctic Fur Seal *Arctocephalus gazella*. In: Siegfried WR, Condy PR, Laws RM (eds) Antarctic nutrient cycling and food webs. Springer, Berlin Heidelberg New York Tokyo, pp 543–550

Doidge DW, Croxall JP, Baker JR (1984a) Density-dependent pup mortality in the Antarctic Fur Seal *Arctocephalus gazella* at South Georgia. J Zool 202:449–460

Doidge DW, Croxall JP, Ricketts C (1984b) Growth rates of Antarctic Fur Seal *Arctocephalus gazella* pups at South Georgia. J Zool 203:87–93

Doidge DW, McCann TS, Croxall JP (1986) Attendance behaviour of Antarctic Fur Seals *Arctocephalus gazella*. In: Gentry RL, Kooyman GL (eds) Fur seals: maternal strategies on land and at sea. Princeton Univ Press, Princeton, pp 102–114

Duffy DC, Berruti A, Randall RM, Cooper J (1984) Effects of the 1982-3 Warm Water Event on the breeding of South African seabirds. S Afr J Sci 80:65–69

Everson I (1977) The living resources of the Southern Ocean. FAO GLO/SO/77/1, Rome, 156 pp

Everson I (1983) Variations in the distribution of krill swarms in the vicinity of South Georgia. Mem Natl Inst Polar Res No 27: 84–92

Gales NJ, Burton HR (1987) Ultrasonic measurement of blubber thickness of the Southern Elephant Seal *Mirounga leonina* (Linn.). Aust J Zool 35:207–217

Gaston AJ, Nettleship DN (1981) The Thick-Billed Murres of Prince Leopold Island. Can Wildl Serv Monogr Ser No 6:350 pp

Harwood J, Croxall JP (1988) The assessment of competition between seals and commercial fisheries in the North Atlantic and Antarctic. Mar Mammal Sci (in press)

Heywood RB, Everson I, Priddle J (1985) The absence of krill from the South Georgia zone, winter 1983. Deep-Sea Res 32: 369–378

Hindell MA (1988) The diet of penguins at Macquarie Island. 1. The King Penguin *Aptenodytes patagonicus*. Ibis (in press)

Hunt GL, Eppley ZA, Schneider DC (1986) Reproductive performance of seabirds: the importance of population and colony size. Auk 103:306–317

Jablonski B (1985) The diet of penguins on King George Island, South Shetland Islands. Acta Zool Cracov 29:117–186

Jouventin P, Stahl JC, Weimerskirch H, Mougin JL (1984) The seabirds of the French Sub-Antarctic islands and Adélie Land, their status and conservation. In: Croxall JP, Evans PGH, Schreiber RW (eds) Status and conservation of the world's seabirds. ICBP, Cambridge, pp 609–625

Kock KH (1985) The state of exploited Antarctic fish stocks around South Georgia. Arch Fischereiwiss 36:155–183

Lack D (1954) The natural regulation of animal numbers. Clarendon, Oxford

La Cock GD (1986) The Southern Oscillation, environmental anomalies, and mortality of two Southern African seabirds. Climat Change 8:173–184

Lishman GS (1983) The comparative breeding biology, feeding ecology and bioenergetics of Adélie and Chinstrap penguins. D Phil Thes Univ Oxford

Lishman GS (1985a) The food and feeding ecology of Adélie Penguins *Pygoscelis adelie* and Chinstrap Penguins *P. antarctica* at Signy Island, South Orkney Islands. J Zool 205:245–263

Lishman GS (1985b) The comparative breeding biology of Adélie and Chinstrap Penguins *Pygoscelis adeliae* and *P. antarctica* at Signy Island, South Orkney Islands. Ibis 127:84–99

Mayfield HF (1961) Nesting success calculated from exposure. Wilson Bull 73:255–261

McCann TS, Rothery P (1988) Population size and status of the Southern Elephant Seal *Mirounga leonina* at South Georgia, 1951–1985. Polar Biol (in press)

Nemoto T, Okiyama M, Takahashi M (1985) Aspects of the roles of squid in food chains of marine Antarctic ecosystems. In: Siegfried WR, Condy PR, Laws RM (eds) Antarctic nutrient cycles and food webs. Springer, Berlin Heidelberg New York Tokyo, pp 415–420

North AW, Croxall JP, Doidge DW (1983) Fish prey of the Antarctic Fur Seal *Arctocephalus gazella* at South Georgia. Bull Br Antarct Surv 61:27–38

Orians GH, Pearson WE (1979) On the theory of central place foraging. In: Horn DI, Mitchell RD, Stairs GR (eds) Analysis of ecological systems. Ohio State Univ Press, Ohio, pp 155–177

Payne MR (1977) Growth of a fur seal population. Philos Trans R Soc London Ser B 279:67–79

Pearson ES, Hartley HO (1972) Biometrika tables for statisticians, vol 2. Cambridge Univ Press, Cambridge

Poncet S, Poncet J (1985) A survey of penguin breeding populations at South Orkney Islands. Bull Br Antarct Surv 68:71–81

Prince PA (1980) The food and feeding ecology of Grey-Headed Albatross *Diomedea chrysostoma* and Black-Browed Albatross, *D. melanophris*. Ibis 122:476–488

Prince PA (1985) Population and energetic aspects of the relationships between Black-Browed and Grey-Headed Albatrosses and the Southern Ocean marine environment. In: Siegfried WR, Condy PR, Laws RM (eds) Antarctic nutrient cycling and food webs. Springer, Berlin Heidelberg New York Tokyo, pp 473–477

Prince PA, Croxall JP (1983) Birds of South Georgia: new records and re-evaluations of status. Bull Br Antarct Surv 59:15–27

Prince PA, Francis M (1984) Activity budgets of foraging Grey-Headed Albatrosses. Condor 86:297–300

Prince PA, Morgan RA (1987) Diet and feeding ecology of Procellariiformes. In: Croxall JP (ed) Seabirds: feeding ecology and role in marine ecosystems. Cambridge Univ Press, Cambridge, pp 135–171

Prince PA, Ricketts C (1981) Relationships between food supply and growth in albatrosses: an interspecies chick-fostering experiment. Ornis Scand 12:207–210

Prince PA, Walton DWH (1984) Automated measurement of feed size and feeding frequency in albatrosses. J Appl Ecol 21:789–794

Ricketts C, Prince PA (1981) Comparison of growth of albatrosses. Ornis Scand 12:120–124

Ricklefs RE, Duffy DC, Coulter M (1984) Weight gain of Blue-Footed Booby chicks: an indicator of marine resources. Ornis Scand 15:162–166

Rothery P, McCann TS (1988) Estimating pup production of Southern Elephant Seals *Mirounga leonina* at South Georgia. In: Harris S (ed) Mammal population studies. Symp Zool Soc, London (in press)

SCAR (1979) Fifteenth meeting of SCAR, Chamonix, 16–26 May 1978. Appendix A. Work Group Biol. Polar Rec 19:304–312

Schreiber RW, Schreiber EA (1984) Central Pacific seabirds and the El Niño-Southern Oscillation: 1982 to 1983 perspectives. Science 225:713–716

Sladen WJL (1958) The pygoscelid penguins. I Methods of study. II The Adélie Penguin *Pygoscelis adeliae* (Hombrom & Jacquinot). Sci Falkl Is Depend Surv No 17

Sladen WJL (1964) The distribution of the Adélie and Chinstrap Penguins. In: Carrick R, Holdgate MW, Prevost J (eds) Biologie antarctique. Hermann, Paris, pp 359–365

Smith RIL, Tallowin JRB (1980) The distribution and size of King Penguin rookeries on South Georgia. Bull Br Antarct Surv 49: 259–276

Targett TE (1981) Trophic ecology and structure of coastal Antarctic fish communities. Mar Ecol Progr Ser 4:243–263

Tickell WLN, Pinder R (1975) Breeding biology of the Black-Browed Albatross *Diomedea melanophris* and Grea-Headed Albatross *D. chrysostoma* at Bird Island, South Georgia. Ibis 117:433–450

Trillmich F, Limberger D (1985) Drastic effects of El Niño on Galapagos pinnipeds. Oecologia (Berlin) 67:19–22

Trivelpiece WZ, Bengtson JL, Trivelpiece SG, Volkman NJ (1986) Foraging behaviour of Gentoo and Chinstrap Penguins as determined by new radiotelemetry techniques. Auk 103:777–781

Volkman NJ, Jazdzewski K, Kittel W, Trivelpiece SG, Trivelpiece WZ (1987) Adélie, Chinstrap and Gentoo Penguin diets during chick-rearing. Condor

Wanless S, French DD, Harris MP, Langslow DR (1982) Detection of annual changes in the numbers of cliff-nesting seabirds in Orkney 1976–1980. J Anim Ecol 51:785–795

Wilson RP (1984) An improved stomach pump for penguins and other birds. J Fld Orn 55:109–112

Wynne-Edwards VC (1962) Animal dispersion in relation to social behaviour. Oliver and Boyd, Edinburgh London

Long-Term Trends in the Foraging Patterns of Female Antarctic Fur Seals at South Georgia

J.L. BENGTSON[1]

Summary. The number of feeding trips to sea made by female Antarctic Fur Seals during lactation may reflect the relative availability of local prey resources. Experimental work utilizing tetracycline-marked teeth confirmed that the feeding trip/suckling cycles of females are reflected as starving/suckling layers in the teeth of their pups. A collection of unmarked Antarctic Fur Seal teeth from Bird Island, South Georgia, was analyzed to estimate: (1) birth year of individuals, and (2) the number of feeding trips made by an individual's mother during lactation.

This analysis showed that between 1962 and 1981 the mean number of feeding trips made by female fur seals varied markedly. From 1962 to 1979 there were several significant increasing and decreasing trends in the mean number of feeding trips, with 1979 being the year with the fewest trips made during the entire 20-year period.

1 Introduction

Each year between November and April, Antarctic Fur Seals (*Arctocephalus gazella*) haul out on the beaches at South Georgia for their reproductive season. The patterns of pupping and postnatal behavior have been described well elsewhere (Bonner 1968, 1981; Payne 1977, 1979; Doidge et al. 1984a, 1986), but a brief summary is given here. Following the birth of pups, which reaches its peak in early December, females suckle their offspring until weaning at approximately 4 months. During this period, females undertake a series of feeding trips to sea to replenish their energy reserves. Female fur seals at South Georgia may make as many as 25 of these trips, during which time they prey principally on Antarctic krill (*Euphausia superba*) (Croxall and Pilcher 1984; Doidge and Croxall 1985).

Because natural selection favors the production of robust pups with enhanced chances for survival, females in all years are likely to attempt to provide their pups with as much milk as possible. To do this, females must forage efficiently, regain condition, return ashore to suckle pups, and depart again on the next feeding trip expeditiously. Hence, it is likely that interannual variability in the onshore attendance patterns of lactating females at South Georgia is strongly influenced by the local availability of prey and the rate at which females can successfully exploit such resources. The foraging behavior of lactating female fur seals is therefore likely to be a sensitive indicator of local changes in the relative availability of prey. Comparing the changes in the number of feeding trips made over a period of years could provide insight into the interannual variability in key components of the marine food web.

Examination of the fine structure of the dentine in fur seal teeth reveals abundant layers of varying widths and patterns (Fig. 1A). Some of these layers had been postulated to be caused by changes in nutrition during suckling/fasting periods in Northern Fur Seal (*Callorhinus ursinus*) pups (Scheffer and Peterson 1967) and suckling/feeding periods in Antarctic Fur Seal adult females (Payne 1978). The ability to accurately identify the layers laid down between birth and weaning as being caused by feeding trips would allow one to determine the number of feeding trips that a pup's mother had made in a given year. Such a technique would enable analyzing the variability in the number of feeding trips made over a series of years. In the case of Antarctic Fur Seals, it would be possible to examine patterns over the past few decades, at times when visual observations had not necessarily been made.

The study described here sought to: (1) determine whether it was possible to use tooth fine structure to estimate the number of feeding trips made in various years, and (2) evaluate the interannual variability of feeding trips made by female fur seals, and (3) assess indirectly the relative availability of local prey resources over a 20-year period.

2 Methods

To determine whether the fine layers in the dentine of teeth laid down shortly after birth were indeed related to pups' suckling/fasting events, an experimental group of Antarctic Fur Seals was studied during the 1983 austral summer at Bird Island, South Georgia (i.e., the 1982/83 season; in this chapter, summer seasons are denoted by the second year of the split season). The feeding/suckling/fasting behavior of 21 adult females and their pups was closely

1 National Marine Mammal Laboratory, Northwest and Alaska Fisheries Center, 7600 Sand Point Way N.E., Seattle, WA 98115-0070, USA

Antarctic Ocean and Resources Variability
(ed. by D. Sahrhage)

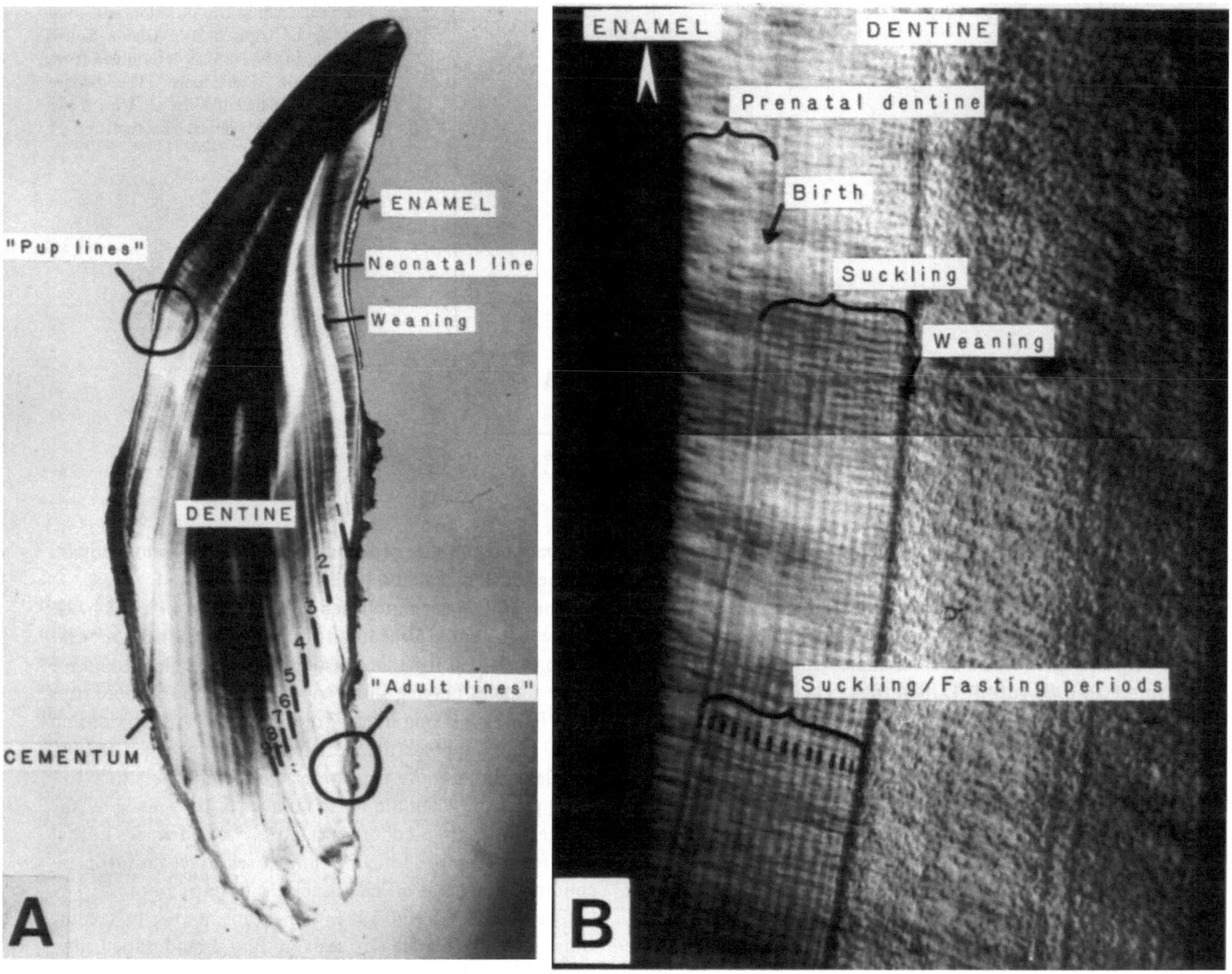

Fig. 1A,B. Longitudinal thin-section of the canine tooth from a known-aged 9-year-old female Antarctic Fur Seal. **A** General view showing the region of "pup lines" between neonatal (birth) line and weaning line, which reflect the number of feeding trips made by this individual's mother during lactation. Some dentine layers in the teeth of adult females ("adult lines") reflect the number of feeding trips, if any, made by the female during her lactation period in that year. **B** Magnified view of "pup lines", indicating the regular layers of dentine laid down between birth and weaning, in relation to the number of suckling/fasting periods experienced

monitored from the pups' birth (early December) until late March. Small radio frequency transmitters (164 MHz by Cedar Creek Bioelectronics Laboratory, Bethel, MN, USA) affixed to adult females were automatically monitored by radio receivers (multi-channel scanning receivers by Cedar Creek Bioelectronics Laboratory) that recorded the females' presence or absence ashore to feed their pups. This record documented the dates and times at which pups were being suckled and allowed comparing the relationship between the total number of feeding trips and the time spent foraging at sea during the lactation period.

Intramuscular injections (20 mg/kg body weight) of oxytetracycline (Terramycin/LA by Pfizer Limited, Sandwich, Kent, UK) were administered periodically to the pups of the 21 radio-tagged females to mark specific layers of dentine being laid down. Oxytetracycline, a broad-spectrum antibiotic, is incorporated into hard tissues being formed at that time (Gurevich et al. 1980). Teeth from monitored pups were collected, thin-sectioned on a high speed diamond saw (Tyslide by Tycet Limited, Hemel Hempstead, Herts., UK), and examined microscopically under polarized and ultraviolet light (M5A stereomicroscope with EPI-fluorescence by Wild Heerbrug Limited, Heerbrug, Switzerland). The position of fluorescent bands in the dentine (indicating a tetracycline injection) was compared to the total number of layers observed under polarized light (presumed to represent suckling/fasting events) to determine the timing of layer deposition in pup teeth prior to weaning.

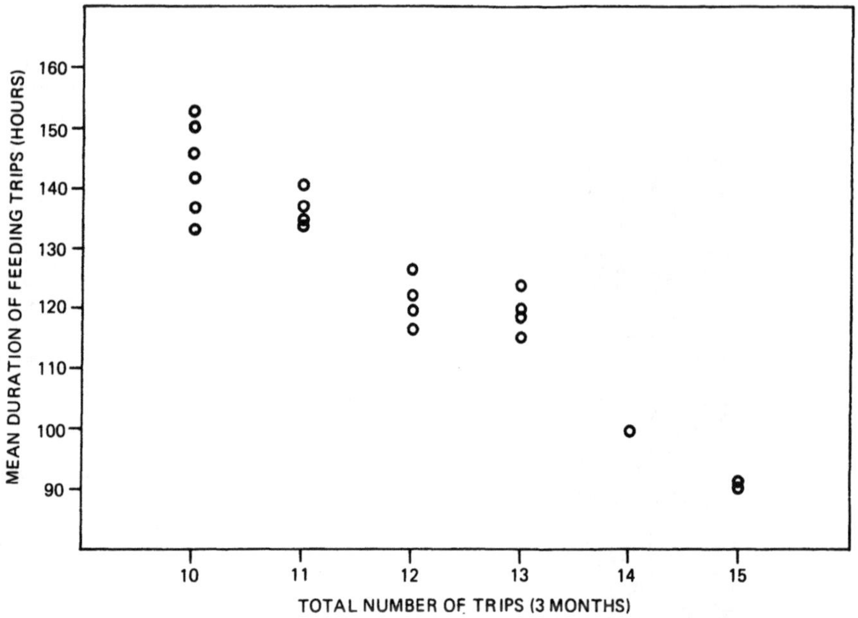

Fig. 2. Relationship of female Antarctic Fur Seal feeding trip duration and total number of trips at Bird Island, South Georgia, in 1982/83, as determined from radio-tagged individuals. The values shown are calculated for all trips made within the first 3 months postpartum

Teeth from a series of collections made at Bird Island were examined to provide information on the number of feeding trips made by the mothers of pups of both sexes in the seasons 1962 to 1981. The sample included 399 canine teeth collected from males and females in 1972 (n = 56), 1973 (n = 49), 1974 (n = 11), and 1983 (n = 283). After being thin-sectioned, each tooth was examined under polarized, transmitted light to estimate: (1) the number of suckling/fasting cycles (mother's feeding trips/suckling sessions) that occurred between birth and weaning, and (2) the individual's age (to determine cohort year). Teeth were read in blind replicates to obtain estimates for age (at least three replicates) and suckling/suckling cycles (at least two replicates).

3 Results

Analysis of the teeth from the closely-monitored experimental group of pups at Bird Island in 1983 confirmed that the characteristic layers laid down just after birth are related to pups' suckling/fasting events caused by their mothers' feeding trips to sea (Fig. 1B). Both a birth (neonatal) line and weaning line were apparent in the thin-sectioned teeth examined. Alternating opaque and translucent layers of dentine, corresponding to the number of feeding trips made by its mother, were clearly observed when examined under polarized light. Fluorescing bands in the dentine layers deposited during the study, indicating tetracycline injections, were obvious when viewed under ultraviolet light. The position of these tetracycline benchmarks in the series of fine dentine layers ("pup lines") confirmed that the timing of each injection matched the known record of the pup's mother's feeding trips to sea. By counting these layers in unmarked teeth from either males or females of any age, it is therefore possible to estimate the number of feeding trips made by an individual's mother during lactation in that year.

The 21 females monitored at Bird Island in 1983 made between 10 and 15 feeding trips during the first 3 months postpartum. During that season, there was a clear trend for those females making short trips to sea to make more trips than females making longer trips (Fig. 2). This relationship from 1983 supports the notion that in any given season, females making more trips to sea are spending less time foraging per trip. When the mean number of trips made by the mothers of a given cohort is relatively low, females in that year required more foraging time at sea to regain sufficient condition to enable efficient suckling of pups ashore.

Between 1962 and 1981, the mean number of feeding trips made by female fur seals at Bird Island varied markedly (Fig. 3). In order to determine whether there were significant trends over this period, values calculated for the mean number of feeding trips in each year were analyzed using a weighted linear regression (Seber 1982) (Table 1). Groups of years were chosen for this analysis if visual examination indicated the presence of apparent trends. From 1962 to 1979, there were several significant increasing and decreasing trends in the mean number of feeding trips. From 1976 to 1979, there was a particularly sharp reduction in the mean number of trips, with 1979 being the year with the fewest trips made during the entire 20-year period. Although there were insufficient data for the years 1979 to 1981 to indicate a trend using weighted regression analysis, an analysis of variance demonstrated that these years differed significantly from each other. The variance around calculated mean values was relatively constant for all years, with higher variation at the beginning and end of the sample, probably related to the small sample sizes (Table 2).

Because the teeth used in this analysis were obtained from essentially two collections separated by about 10 years

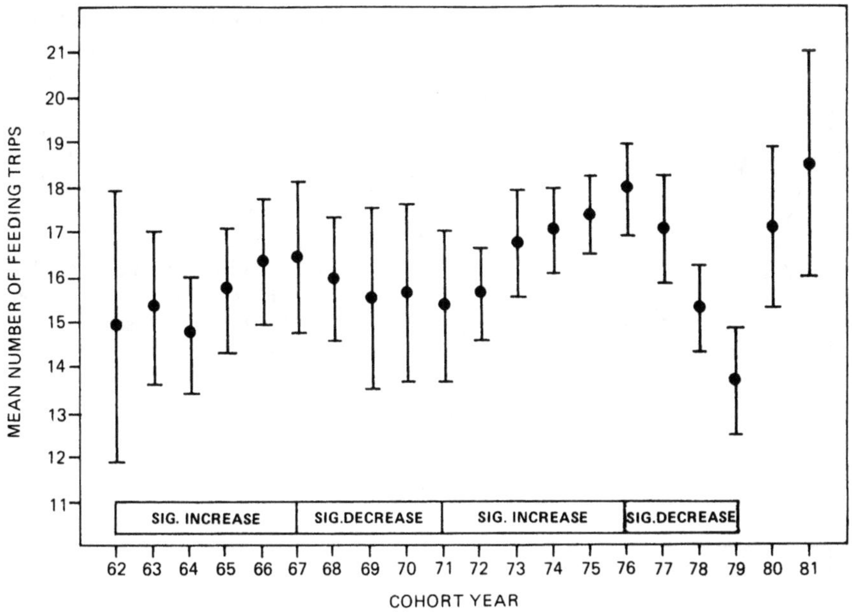

Fig. 3. Trends in the mean number of feeding trips made by female Antarctic Fur Seals at Bird Island, South Georgia, as backcalculated from fine structure of teeth. *Vertical bars* indicate 95% confidence limits around mean

Table 1. Variation in numbers of feeding trips by female Antarctic Fur Seals made in different seasons at Bird Island, South Georgia, as determined from fine structure of teeth

Cohort year	Sample size	Range	Mean	Standard deviation	Mean std error
1962	6	12–22	15.0	3.58	1.46
1963	14	10–22	15.4	2.98	0.796
1964	22	11–21	14.8	2.84	0.605
1965	18	10–21	15.8	2.90	0.684
1966	19	12–21	16.4	2.81	0.646
1967	15	12–24	16.5	3.18	0.822
1968	20	11–22	15.9	3.01	0.672
1969	12	11–21	15.6	3.37	0.973
1970	10	10–19	15.7	3.02	0.955
1971	10	11–19	15.4	2.63	0.833
1972	21	11–20	15.7	2.37	0.518
1973	31	12–23	16.8	3.05	0.548
1974	37	11–22	17.0	2.69	0.442
1975	42	11–22	17.4	2.81	0.434
1976	38	13–25	18.0	2.74	0.445
1977	24	10–22	17.1	2.81	0.574
1978	27	11–20	15.3	2.45	0.471
1979	19	10–18	13.7	2.40	0.552
1980	10	11–21	17.1	2.77	0.875
1981	4	16–21	18.5	2.38	1.19

Table 2. Significance of regression coefficients for mean number of feeding trips on cohort

Cohort years compared	Slope	F-statistic	df_1	df_2	P-value	Significance
1962–1967	0.35	10.660	1	4	0.031	+
1967–1971	−0.24	13.312	1	3	0.036	+
1971–1976	0.54	103.435	1	4	0.001	+
1976–1979	−1.48	138.569	1	2	0.007	+
1979–1981	2.41	20.923	1	1	0.137	–
1962–1981	0.08	2.828	1	18	0.110	–

(1972–74 and 1983), the possibility of a sampling bias associated with tooth back-calculation was examined. Previous studies involving back-calculating reproductive patterns from Crabeater Seal teeth had found a rather significant sampling bias of this nature (Bengtson and Laws 1985). To test the comparability of the two samples, mean feeding trip values were compared for the 6 years in which the two data sets overlapped. Mean number of feeding trip values calculated independently from the 1970's and 1983 samples were compared using a two-way analysis of variance. There was no significant difference in the estimated mean number of feeding trips calculated from either the 1970's or the 1983 samples (P = 0.842).

4 Discussion

Trends in the mean number of feeding trips made by lactating females at Bird Island provides insight into the presumed local availability of prey resources, especially Antarctic krill, over a 20-year period. Throughout this period, the number of feeding trips required by females to find sufficient food exhibited significant fluctuations. The drop in the number of feeding trips made in 1979 is intriguing not only because it is so sharp, but also because it was the only event of that magnitude to occur during the 20 years investigated in this study.

The sharp dip in the number of fur seal feeding trips made in 1979 is corroborated by behavioral data from Bird Island at that time (Fig. 4). Doidge et al. (1986) described a significant increase in the duration of feeding trips made by females in the 1979 season (compared to previous observations at Bird Island and sites nearby on South Georgia). An increased length in feeding trips to sea is consistent with a reduced total number of trips for that year as estimated from tooth analysis. This relationship

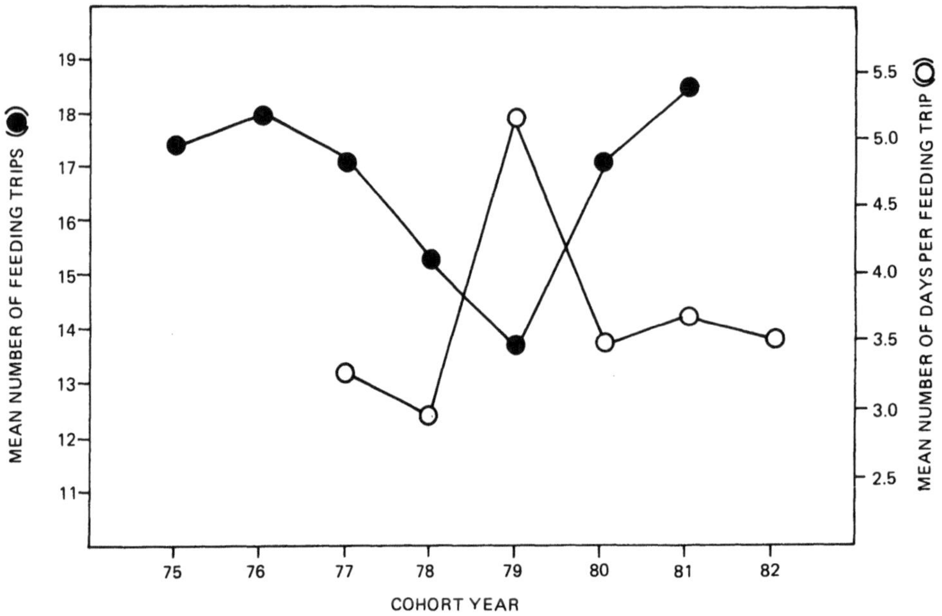

Fig. 4. Comparison of the mean number of feeding trips made (calculated from fine tooth structure) and mean number of days per feeding trip (calculated from visual observations) made by female Antarctic Fur Seals at South Georgia. Values for feeding trip duration are from Doidge et al. (1986), and were calculated for the first six trips postpartum at Bird Island (1978, 1980, 1981), Schlieper Bay, South Georgia (1977, 1979), and Elsehul, South Georgia (1976)

for change between seasons is similar to the pattern within a season, as demonstrated by the 1983 data on trip duration (Fig. 2): the mean duration of trips increases as the total number of trips decreases. Additional parameters cited by Doidge et al. (1986) indicate that 1979 was a year in which local prey resources were sparse: short perinatal attendance by female fur seals, short time spent ashore suckling pups, low pup weights (Croxall and Prince 1979; Doidge et al. 1984b), high pup mortality (Doidge et al. 1984a), and research and commercial fishing vessels reported low concentrations of krill offshore South Georgia.

Although the behavioral and tooth data both indicate that 1979 was a poor year for fur seals (Fig. 4), only the tooth data show a decrease in the 2 years preceding and the 1 year following 1979. The behavioral data show little change except for the major difference in 1979. It is not immediately apparent if this difference in the two data sets is caused by reality (e.g., one set of data is more sensitive to prey availability than the other) or by artifacts (e.g., sampling bias). Two potential sources of error in the tooth data that might affect estimates are the readings of total age and the counts of the number of suckling/fasting periods. As in all analyses of layers in teeth, there is error associated with the readers' counts and interpretation of layers. In this case, errors in estimating the total age of individuals (to identify their cohort year) would tend to smooth the curve shown in Figs. 3 and 5. For example, if individuals from the 1979 cohort (whose mothers made relatively few feeding trips) were assigned in error to the 1977, 1978, or 1980 cohorts, the estimates of mean number of feeding trips made in these years would be lowered.

The results from krill surveys over the past 10 years have drawn attention to the high annual variability in the estimated abundance of krill at various localities in the Scotia Arch (Hempel 1985; Heywood et al. 1985; Brinton et al. 1986; Macaulay et al. 1988). In particular, results from the late 1970's and early 1980's indicate major annual fluctuations in the estimated local abundance and distribution of krill. The tooth back-calculation estimates also reveal significant fluctuations in presumed prey abundance over two decades. Investigators will be challenged to determine what factors are involved in the interannual variability in local prey availability to fur seals and its consequences on upper trophic levels. Shifts in the local abundance and availability of prey resources could be affected by biological as well as physical features in the ecosystem. A further challenge to be addressed is determining to what extent such changes are caused by either natural events (e.g., density-dependent effects, sea temperature, sea ice) and/or human activities (e.g., fisheries, pollution).

Long-term research and monitoring of the behavior and ecology of key components of the Antarctic marine ecosystem are currently being planned and implemented (Ecosystem Monitoring Program of the Scientific Committee for the Conservation of Antarctic Marine Living Resources). These studies will focus on ecological links between different trophic levels over time. For fur seals, there are several useful parameters that should be monitored in addition to the number of feeding trips made by lactating female fur seals. For example, detailed data on the at-sea diving and hunting behavior may prove to be especially sensitive to changes in the abundance and distribution of local prey resources. Such data, coupled with information on onshore attendance patterns, number of feeding trips, gup growth rates, and the characteristics of prey items taken will provide essential information needed to better understand the ecological interactions among Antarctic Fur Seals, their prey, and environment.

Integrating data on upper trophic level predators, estimates of prey abundance, and background information on physical parameters will bring the monitoring program together toward a systems view of ecological relationships, and will assist scientists in their efforts to understand the complex interactions occurring among various components of the ecosystem. This understanding is fundamental if the CCAMLR Ecosystem Monitoring Program is to succeed in providing guidance toward enlightened management and conservation of Antarctic marine living resources.

Acknowledgments. The author wishes to thank the many persons who contributed to this study, including: D.J. Schneider, who assisted in most of the field work; M.R. Payne, for the 1970's tooth samples; B.D. Ebberts and M.M. Muto, who assisted in the tooth analysis; and J. Breiwick, M.E. Goebel, and A.E. York, who helped with data analysis. Appreciation is also extended to W.N. Bonner, J.P. Croxall, and R.M. Laws for support and helpful discussions during the author's tenure at the British Antarctic Survey, when the project was conceived. This work was jointly supported by the British Antarctic Survey, the US National Science Foundation's Division of Polar Programs, and the US National Marine Fisheries Service's Antarctic Marine Living Resources Program.

References

Bengtson JL, Laws RM (1985) Trends in crabeater seal age at maturity: an insight into Antarctic marine interactions. In: Siegfried WR, Condy PR, Laws RM (eds) Antarctic nutrient cycles and food webs. Springer, Berlin Heidelberg New York Tokyo, pp 669–675

Bonner WN (1968) The fur seal of South Georgia. Sci Rep Br Antarct Surv 56:1–81

Bonner WN (1981) Southern fur seals, *Arctocephalus* (Geoffroy Saint-Hilaire and Cuvier 1826). In: Ridgway SH, Harrison RJ (eds) Handbook of marine mammals, vol 1. Academic Press, London New York, pp 161–208

Brinton E, Huntley M, Townsend AW (1986) Larvae of *Euphausia superba* in the Scotia Sea and Bransfield Strait in March 1984 – development and abundance compared with 1981 larvae. Polar Biol 5:221–234

Croxall JP, Pilcher MN (1984) Characteristics of krill, *Euphausia superba,* eaten by Antarctic Fur Seals, *Arctocephalus gazella,* at South Georgia. Br Antarct Surv Bull 63:117–125

Croxall JP, Prince PA (1979) Antarctic seabird and seal monitoring studies. Polar Rec 19:573–595

Doidge DW, Croxall JP (1985) Diet and energy budget of the Antarctic Fur Seal *Arctocephalus gazella* at South Georgia. In: Siegfried WR, Condy PR, Laws RM (eds) Antarctic nutrient cycles and food webs. Springer, Berlin Heidelberg New York Tokyo, pp 543–550

Doidge DW, Croxall JP, Baker JR (1984a) Density-dependent pup mortality in the Antarctic Fur Seal *Arctocephalus gazella* at South Georgia. J Zool 202:449–460

Doidge DW, Croxall JP, Ricketts C (1984b) Growth rates of Antarctic Fur Seal *Arctocephalus gazella* pups at South Georgia. J Zool 203:87–93

Doidge DW, McCann TS, Croxall JP (1986) Attendance behavior of Antarctic Fur Seals. In: Gentry RL, Kooyman GI (eds) Fur seals, maternal strategies on land and at sea. Princeton Univ Press, Princeton, pp 102–114

Gurevich VS, Stewart BS, Cornell LH (1980) The use of tetracycline in age determination of common dolphins, *Delphinus delphis.* In: Perrin WF, Myrick AC Jr (eds) Age determination of toothed whales and sirenians. Rep Int Whal Comm Spec Issue 3 Cambridge, pp 165–169

Hempel I (1985) Variation in geographical distribution and abundance of larvae of Antarctic krill, *Euphausia superba,* in the southern Atlantic Ocean. In: Siegfried WR, Condy PR, Laws RM (eds) Antarctic nutrient cycles and food webs. Springer, Berlin Heidelberg New York Tokyo, pp 305–307

Heywood RB, Everson I, Priddle J (1985) The absence of krill from the South Georgia zone, winter 1983. Deep-Sea Res 32:369–378

Macaulay M, Daly KL, Mathisen OA (1988) Interyear variability in the abundance and distribution of Antarctic krill near Elephant Island and the South Orkney Islands. (in prep)

Payne MR (1977) Growth of a fur seal population. Philos Trans R Soc London Ser B 279:67–79

Payne MR (1978) Population size and age determination in the Antarctic Fur Seal *Arctocephalus gazella.* Mammal Rev 8: 67–73

Payne MR (1979) Fur seals *Arctocephalus tropicalis* and *Arctocephalus gazella* crossing the Antarctic Convergence at South Georgia, South Atlantic Ocean. Mammalia 43:93–98

Scheffer VB, Peterson RS (1967) Growth layers in teeth of suckling fur seals. Growth 31:35–38

Seber GAF (1982) The estimation of animal abundance and related parameters, 2nd edn. Griffin, London, 654 p

Squid as Predators on Krill (*Euphausia superba*) and Prey for Sperm Whales in the Southern Ocean

T. NEMOTO, M. OKIYAMA, N. IWASAKI, and T. KIKUCHI[1]

Summary. This chapter deals with information on the role of squid, and on feeding by sperm whales on squid, in the Antarctic. Seasonal variation in the vertical distribution of krill swarms may possibly affect the distribution of smaller squid which feed on krill. Larger squid are distributed in deeper water layers where they do not feed on krill alone. Sperm whales eat larger squid in the deeper waters of the Southern Ocean. The diet of these larger squid is unknown. It is suggested that the food chain from krill to sperm whale has several links, including fish and squid. The importance of studies on deep-sea systems in the Southern Ocean is emphasized.

1 Introduction

Investigations during the BIOMASS Program between 1977 and 1986 have revealed many food chain pathways relating to krill in the Antarctic. These include the krill-squid-sperm whale complex. Nemoto et al. (1985) reported on research on squid ecology and predator-prey relationships in the Antarctic. It has long been known that squid play an important role in the Southern Ocean ecosystem. However, squid are very difficult to catch with conventional plankton nets and commercial trawls, and so there is still considerable scope for research.

2 Material and Methods

Al specimens of squid (n = 324) were collected during the Japanese BIOMASS project in the Antarctic summer seasons from 1980/81 to 1982/83.

The squid were caught in the 10–200 m layer by trawls with a large opening of 615.4 m^2, a length of 62 m, and 26 m cod end. The inner mesh of the anterior part of the net was 20 mm, and that of the posterior part 13 mm. The net was designed to catch krill.

Specimens of squid were deep frozen immediately and later thawed in the laboratory, and fixed in formalin.

Data on squid in the stomach of sperm whales and biological parameters of sperm whales were obtained during Japanese whaling expeditions to the Antarctic in the 1951/52 season.

1 Ocean Research Institute, University of Tokyo, Tokyo, Japan

Table 1. Species and number of specimens of cephalopods collected in the Antarctic by trawling operations during the summers of 1980/81, 1981/82, 1982/83

Species	1980/81	1981/82	1982/83	Total
Alluroteuthis antarcticus	–	3	5	8
Moroteuthis knipovitchi	5	8	12	25
Moroteuthis robsoni	–	–	5	5
Moroteuthis ingens	–	–	1	1
Kondakovia longimana	20	31	81	132
Pholidoteuthis boschmai	–	1	1	2
Brachioteuthis picta	–	3	74	77
Galiteuthis glacialis	–	4	18	22
Gonatus antarcticus	–	–	48	48
Unknown	–	1	3	4
Total	25	51	248	324

3 Results

Ten species of squid (one unknown) were identified (Table 1). Nemoto et al. (1985) have reported on six of these species. Four further species, including the unidentified specimens, were found in the catches during the 1982/83 season. These are all common species in the Antarctic (Filippova 1972; McSweeney 1978).

The large species *Kondakovia longimana* accounts for 41% of all squid sampled in the 10–200 m layer. The other common species caught were: *Brachioteuthis picta, Gonatus antarcticus, Moroteuthis knipovitchi* and *Galiteuthis glacialis.* Catches of squid were highest in the 1982/83 season.

The size frequency distributions of seven of the species of squid in Table 1 are shown in Fig. 1. The frequency distributions for *B. picta* and *G. antarcticus* indicate that these squid are generally of less than 150 mm mantle length. It is likely, however, that the net simply did not catch larger specimens. Kubodera and Okutani (1986) described three specimens of *G. antarcticus* of 218–235 mm mantle length. The size range for *K. longimana* and *M. knipovitchi* was from a few centimeters to 400 mm mantle length. On the basis of the relationship between lower rostral length and mantle length, *K. longimana* is estimated to grow to 800 mm mantle length (Clarke 1980). As shown

Antarctic Ocean and Resources Variability
(ed. by D. Sahrhage)

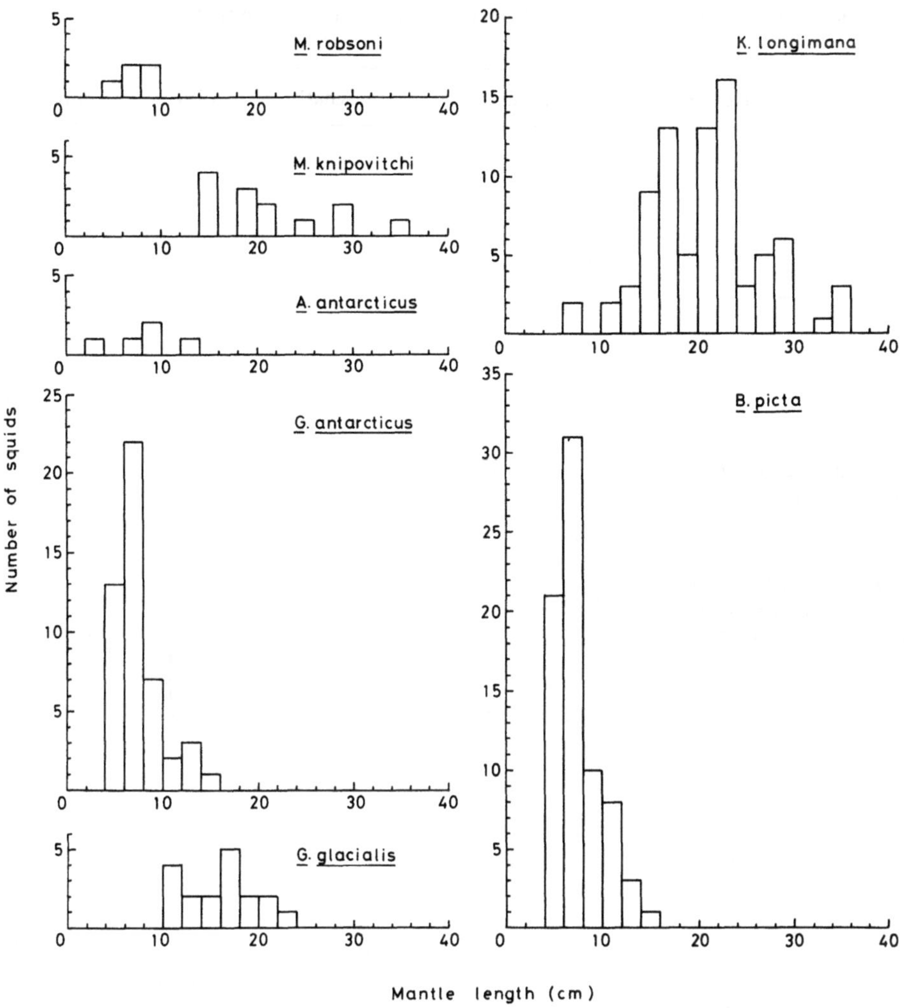

Fig. 1. Size frequency distribution of squid collected by trawling operations during the summers of 1980/81, 1981/82 and 1982/83

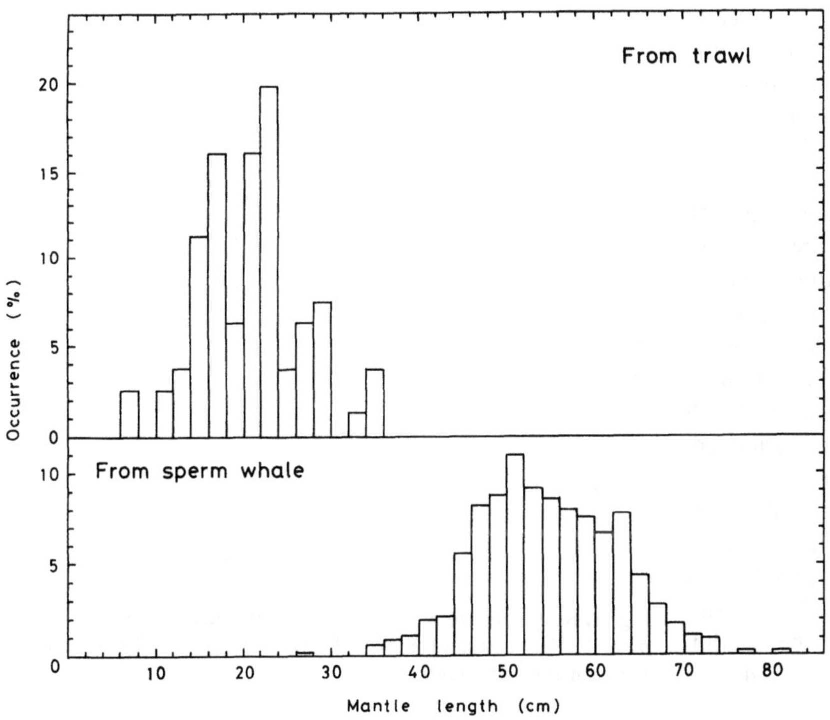

Fig. 2. Size frequency distribution of *Kondakovia longimana* collected by trawling operations and found in the stomach contents of sperm whales

Table 2. Food items of 300 specimens of squid collected in the Antarctic by trawling operations during the summers of 1980/81, 1981/82, and 1982/83

Squid species	n	*E. superba*	*T. macrura*	Amphipods	Chaetognath	Fish	Squid
K. longimana	121	69	3	10	3	3	3
M. knipovitchi	23	4	–	–	–	8	–
M. ingens	1	1	–	–	–	–	–
M. robsoni	5	3	–	–	–	–	–
A. antarcticus	7	2	–	1	–	1	1
G. glacialis	19	7	–	1	5	–	–
G. antarcticus	48	29	–	–	–	1	–
B. picta	75	29	–	–	–	–	1
Unknown	1	–	–	–	–	1	–

Table 3. Percentage of squid found to have been feeding on *E. superba* in catches from trawling operations during summer 1982/83 in the Antarctic

Species	Number of specimens examined	No. of specimens which had fed on *E. superba*	Percent of specimens which had fed on *E. superba*
A. antarcticus	2	1	50.0
B. picta	71	22	31.0
K. longimana	31	28	90.3
G. glacialis	13	6	46.2
G. antarcticus	49	24	49.0
M. ingens	1	1	100.0
M. robsoni	5	2	40.0

in Fig. 2, specimens of *K. longimana* larger than 350 mm mantle length and up to about 800 mm are found in the stomachs of sperm whales, while the trawl catches only smaller specimens up to 350 mm mantle length (Nemoto et al. 1985). Photographs taken during the Antarctic whaling expeditions (Maeda and Teraoka 1952) clearly show that larger *Kondakovia* of about 800 mm mantle length were found in the stomachs of sperm whales. A similar difference in body size between specimens in trawl catches and sperm whale stomachs has been observed for *Moroteuthis knipovitchi* (Nemoto et al. 1985).

The food of trawl-caught squid was investigated (Table 2). Krill (*Euphausia superba*) was identified as the predominant prey item (see also Nemoto et al. 1985). *K. longimana, G. antarcticus* and *B. picta* feed mainly on plankton, especially krill. *M. knipovitchi* feeds on fish and on krill. The percentages of specimens from each of seven species found to have been feeding on *E. superba* are shown in Table 3. Although for some species only a few stomachs could be examined, the role of krill as a major prey species is evident.

Investigations on the size composition of krill in the stomachs of *Kondakovia longimana* and *Brachioteuthis picta* reveal that larger squids tend to feed upon krill of larger sizes especially in *B. picta* (Figs. 3 and 4). The size of krill, *Euphausia superba,* is estimated from the relation between eye size and body length (Nemoto et al. 1984).

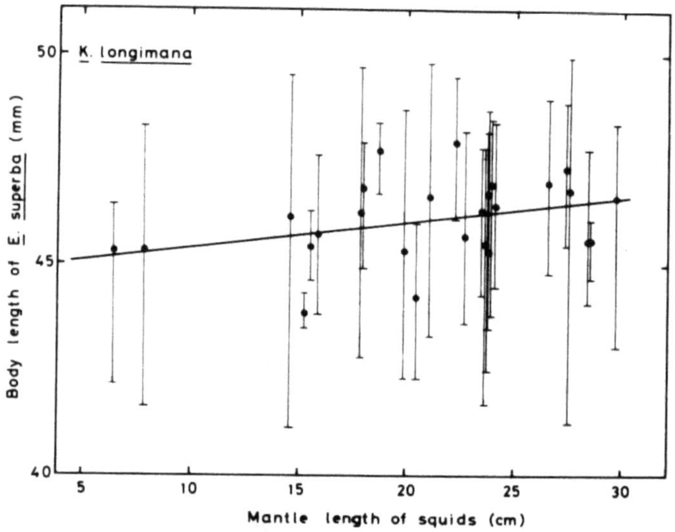

Fig. 3. Size of *E. superba* found in the stomach of *Kondakovia longimana* in the Antarctic. Size of *E. superba* calculated from relation between eye and body sizes

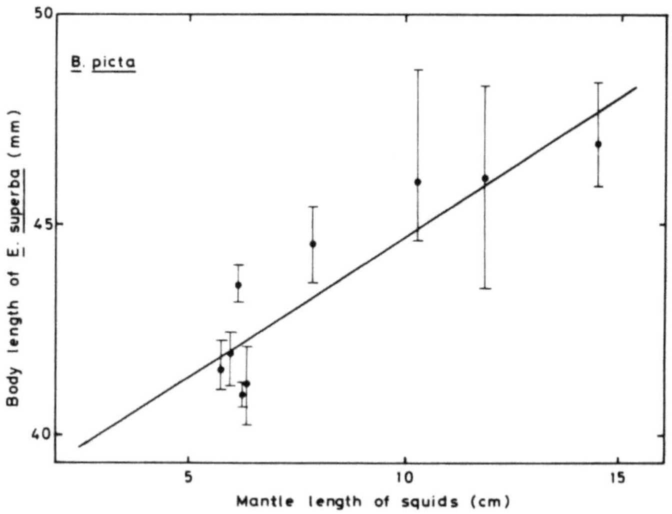

Fig. 4. Size of *E. superba* found in the stomachs of *Brachioteuthis picta* in the Antarctic. Size of *E. superba* calculated from relation between eye and body sizes

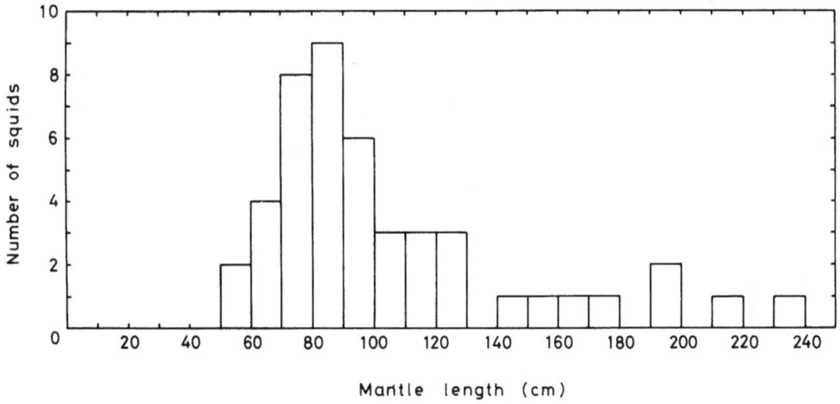

Fig. 5. Size of squid consumed by sperm whales caught in 1951/52 season in the Antarctic (*Baikal Maru* expedition)

Table 4. Quantity and freshness of squid in stomachs of sperm whales caught during the 1951/52 season in the Antarctic (*Baikal Maru* expedition)

Quantity	Freshness				Total number of stomachs investigated
	Very fresh	Fresh	Half digested	Mostly digested	
Full	2	2	3	0	7
Abundant	3	8	7	0	18
Moderate	1	6	14	9	30
Few	0	0	9	44	53
None	–	–	–	–	112
					220

The size distribution of squid consumed by sperm whales was examined during the 1951/52 summer season in the Antarctic (Fig. 5). The main fraction in the sperm whale diet consists of squid of 500–1200 mm mantle length, mostly of larger *Kondakovia longimana* or a related species. Larger squid of 2 m or more (up to 2.4 m) mantle length from sperm whale stomachs were also described from these expeditions and were identified as *Mesonychoteuthis hamiltoni*. However, their numbers were rather small.

Data on the quantity and freshness of squid found in sperm whale stomachs from the *Baikal Maru* expedition in 1951/52 are presented in Table 4. Of the 220 stomachs examined only 108 contained food; sperm whales with full stomachs were rare, and there were few fresh remains of squid. The low percentage of sperm whales with full stomachs (much lower than for baleen whales in the Antarctic) suggests that the feeding grounds for the sperm whales are rather widespread throughout the deep pelagic ocean and so also is the distribution of the target squid. A low percentage of sperm whales with full stomachs has also been noted in catches from waters off Australia and South Africa (Mathews 1938).

Seasonal variation in blubber thickness is an important parameter associated with the feeding status of sperm whales in the Antarctic. There is a clear increase in the thickness of blubber of baleen whales such as Blue, Fin and Sei Whales (Mackintosh and Wheeler 1929) in the Antarctic summer season, but there is no such seasonal variation in the sperm whale (Fig. 6). The reason for this difference may be that the availability of squid to the sperm whales is more regular than for the baleen whales feeding on krill. Squid in the deeper layers of the ocean, which are the sperm whales prey, are perhaps more stable in abundance

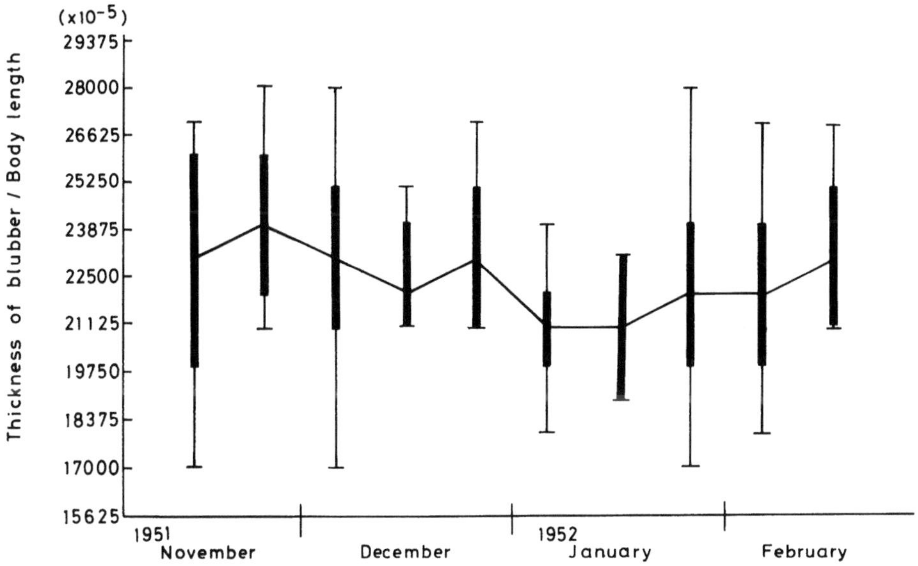

Fig. 6. Seasonal variation of thickness of blubber of sperm whales caught in 1951/52 season in the Antarctic (*Baikal Maru* expedition)

than the biomass in the upper layers of the Antarctic which have large seasonal fluctuations in productivity (Hart 1942).

The depth distribution of younger, small squid, feeding on krill and other zooplankton, showed seasonal variation (Nemoto et al. 1984). Specimens of *Kondakovia longimana* were caught at depths between 60 and 80 m in the early summer. In January/February this squid was caught in the upper 40 m. This trend could possibly be associated with a seasonal variability in the depth distribution of krill. It is possible that krill and squid form different vertical distribution patterns at different seasons.

4 Discussion

Squid are rarely taken by small plankton nets but large krill trawls can catch squid up to 400 mm mantle length. Squid caught in this way are apparently feeding on krill and other zooplankton in the Antarctic (Nemoto et al. 1985). Feeding on *E. superba* may be affected by the distribution and swarming behavior of krill. Seasonal variation in krill abundance possibly affects squid distribution. Young *Kondakovia longimana* may follow the krill swarms seasonally and move up in the water column to the krill layers or swarms when they feed. Male sperm whales probably do not depend on small *K. longimana* and other small squid, since they appear to prey on larger squid from greater depths. In the North Pacific, particularly larger squid is eaten by big male sperm whales (Tarasevich 1968).

The percentage of sperm whales actively feeding in the Antarctic is apparently similar to that from other oceanic regions, and the relative thickness of the blubber of sperm whales does not show a clear increase in the Antarctic, as it does in the baleen whales. This suggests that the food chain from krill to sperm whales has more than one link. Clarke (1980) described already fish and squid as diet of *K. longimana* along with crustaceans. When we can analyze food of larger *K. longimana*, which may live in deeper waters, this link in the deep sea system in the Antarctic may be clarified.

Further studies are required to investigate the food of fish and squid in the Antarctic Ocean, and the links in the food chain from secondary production near the surface, mainly of krill, to these fish and squid.

References

Clarke MR (1977) Beaks, nets and numbers. Symp Zool 38:89–126

Clarke MR (1980) Cephalopods in the diet of sperm whales of the Southern Hemisphere and their bearing on sperm whale biology. Discovery Rep 27:1–324

Filippova JA (1972) New data on the squids (Cephalopoda; Oegopsida) from the Scotia Sea (Antarctica). Malacologia 11 (2): 391–406

Hart TJ (1942) Phytoplankton periodicity in Antarctic surface waters. Discovery Rep 21:261–356

Kubodera T, Okutani T (1986) New and rare cephalopods from the Antarctic waters. Mem Natl Inst Polar Res Spec Issue 44:129–143

Mackintosh NA, Wheeler JFG (1929) Southern blue and fin whales. Discovery Rep 1:257–540

Maeda K, Teraoka Y (1952) Hogei (Whaling). Jpn Whal Assoc, Tokyo, 450 pp

Mathews LH (1938) The sperm whale, *Physeter catodon*. Discovery Rep 17:93–168

McSweeney ES (1978) Systematics and morphology of the Antarctic cranchiid squid *Galiteuthis glacialis* (Chum). Antarct Res Ser 27:1–39

Nemoto T, Okiyama M, Takahashi M (1984) Aspects of the roles of squid in food chains of the Antarctic marine ecosystems (Extended abstract). Mem Natl Inst Polar Res Spec Issue 32: 89–92

Nemoto T, Okiyama M, Takahashi M (1985) Aspects of the roles of squid in food chains of marine Antarctic ecosystems. In: Siegfried WR, Condy PR, Laws RM (eds) Antarctic nutrient cycles. Springer, Berlin Heidelberg New York Tokyo, pp 415–420

Tarasevich MN (1968) Dependence of distribution of the sperm whale males upon the character of feeding. Zool Z Ukrayiny 47 (11):1685–1688

Summary and Conclusions

D. SAHRHAGE[1]

The Seminar provided not only a forum for the presentation of scientific contributions (most of them contained in this book) but also for extensive discussions. Needs for future research were identified and recommendations made accordingly to promote further practical steps through CCAMLR, IOC and possibly other bodies. In the following an overview is given on the main aspects discussed, structured according to the sections of the Seminar.

1 Meso/Large-Scale Variability in the Environment

On the first two questions of the Seminar: What is known on the variability in the Antarctic Ocean circulation system? and: What are the possible causes for such variations? reviews were provided on the atmospheric elements at the ocean's surface forming the driving forces for the water circulation, with information on long-term, interannual and seasonal variability in the atmospheric circulation. Overviews on the spatial and temporal variability within the Southern Ocean were supplemented by presentations of results in special areas.

From the contributions it is evident that there is large variability in the environmental conditions between seasons, from year to year, even with long-term trends, and also with differences from one region to another. The importance of wind-induced variations in the Antarctic water circulation was recognized. It was noted that the Southern Ocean is not radially symmetric but many circulation and water mass features vary markedly with longitude and are also dependent on topographic influences (ridges, islands). Based on data from ships of opportunity, variation in the position of oceanic fronts was demonstrated. Relatively good knowledge exists already on variations in the Weddell-Scotia Confluence in the Atlantic sector and related meanders and eddies which seem to be of considerable influence on krill distribution.

Generally, however, results presented showed mainly the mean distribution of atmospheric circulation of water characteristics and flow, together with average seasonal and interannual fluctuations. Much less is known so far about environmental conditions in specific seasons or years and in defined areas. The reason for this is that meteorologists and oceanographers mostly tend to use average values of parameters in their investigations. However, it is the interannual change and seasonal abnormality in the weather, ice, and hydrographic conditions which is of greatest influence on the biology and distribution of the living marine organisms. Therefore, the biologists are mainly interested in the environmental conditions during a specific year or season in defined areas. The Seminar may have contributed to improving the joint understanding between meteorologists, oceanographers, and biologists on these aspects and related research needs.

There appears to be a wealth of environmental data, published and unpublished, that have not yet been analyzed in this respect, and this should be done during the coming years through interdisciplinary collaboration. Biologists need better access to available publications in the field of meteorology, therefore a bibliography is given in the first contribution to this book.

More detailed information for specific periods and areas is available on the variations in the ice cover of the Southern Ocean through continuous satellite observations since the early 1970's. Continuation of these observations is essential in view of the great influence of ice cover on the Antarctic marine ecosystem.

A special aspect is the irregular appearance of the El Niño/Southern Oscillation (ENSO) events which are likely to influence the Southern Ocean. This influence is, however, not yet well understood, and further investigations are encouraged.

Observations showed that variations in the environmental conditions may differ from one area of the Southern Ocean to another. Developments in one region have direct consequences for the other sectors. This system and the driving forces behind such processes are so far little known. It is dangerous to use data from few stations in one area to derive conclusions concerning another sector of the Southern Ocean. For certain areas, particularly the Atlantic sector, there are more stations and data available than for other regions. It is important to develop a well-designed system

1 Institut für Seefischerei, Bundesforschungsanstalt für Fischerei, Palmaille 9, 2000 Hamburg 50, FRG

Antarctic Ocean and Resources Variability
(ed. by D. Sahrhage)

for environmental monitoring around Antarctica complementary to the biological monitoring being implemented by CCAMLR, with compatible standardized series of long time observations. The value of such series was highly appreciated. New stations need to be established in hitherto unsampled or poorly covered regions and, if possible, in places that are representative of conditions of the real oceanic environment.

However, such long-term monitoring alone will not be sufficient to elucidate the influence of the environmental factors on the living organisms. Moreover, interdisciplinary experiments will be required which, besides macro- and mesoscale investigations, should concentrate particularly on micro scale processes in selected areas. Upper layer dynamics (vertical and horizontal mixing processes) should be studied more closely.

Another aspect deserving special attention is the further investigation of the mechanisms, involving interaction between meteorological and oceanographic factors, which lead to substantial variations in the circulation of Bellingshausen and Weddell Sea waters in the Antarctic Peninsula region, Scotia Sea, and around South Georgia, and which can have tremendous effects on the living resources, particularly krill distribution.

2 Meso/Large-Scale Variability in the Biota (Related to the Environment)

An overview of the spatial and temporal variabilities in the distribution of Antarctic phytoplankton biomass showed that the fluctuations in this biomass and in primary production are very large. As with other parts of the world oceans, timing of peak phytoplankton abundance varies from year to year. However, the seasonal variability is much more pronounced than any interannual variability, and furthermore large spatial differences at the same time overshadow the seasonal fluctuations. Within the marginal ice zone of the Southern Ocean the variability of primary production is closely linked with the variations in the advance and retreat of the seasonal pack ice, and such fluctuations in ice cover play an important role in regulating the flux and transformation of biogenic material.

An extensive discussion developed on the results of biological studies under and in the pack ice. Recent investigations in this zone in winter have shown that a rich flora and fauna exist. The ice cover provides shelter and food and appears to be a most important area in relation to the survival strategies of the organisms, and particularly krill, in winter. Further studies were much encouraged. However, it was also stressed that caution should be exercised during such investigations in extrapolating from one season or one region to the entire Southern Ocean, in view of the size of the ocean and the present limited data from few locations.

It was stated that despite many years of intensive field work in the Southern Ocean, we are still unable to answer some of the most fundamental questions in Antarctic marine ecology, for example, why are parts of the Southern Ocean so much more productive than the rest? Major progress in our knowledge may be expected with the increasing application of advanced technology, especially research satellites and moored instruments for obtaining long-term records of environmental and biological data. Again, the importance of proper monitoring and of long time-series of observations was stressed.

A conceptual framework for the analysis of the Southern Ocean ecosystem in terms of time and space was presented. Such a framework allows an overview of ecosystem structure and function. The temporal and spatial scale relationships of biotic and abiotic processes were found to be important in the investigation of ecosystem structure. Any survey design should maintain a scale view of the ecosystem. An important implication of the scale concept is that localized predator population processes will not provide an overall index of krill abundance unless the predators forage over sufficiently large scales. The central role of krill within the Southern Ocean food web is not only the result of their large number but is also related to their availability to predators over a wide range of temporal and spatial scales. This scale view of the ecosystem has important consequences for the management of the resources and related monitoring activities.

3 Krill Variability in Relation to the Environment

Fluctuations in the distribution and abundance of krill in the Southern Ocean and a range of possible causes of this variability in space and time were illustrated in a considerable number of presentations. Most contributions were concerned with the krill in the southwest Atlantic sector but there were also a few papers covering other regions.

A distinction can be made between short-term changes, seasonal, and interannual variation. Large-scale fluctuations can be observed in certain years (like the drastic decrease of krill aggregations around South Georgia in 1977/78 and 1983/84). Within-year variation was identified as a highly significant component. If one wants to quantify the extent of such variation, the question arises what the range of fluctuation is in the average during "normal" years or seasons. It is necessary to establish such a baseline through long-term monitoring investigations under standardized conditions, and a first attempt was made to describe the "usual" seasonal variation in the krill distribution in the Antarctic Peninsula area. A continuation of such studies as part of the CCAMLR monitoring program should be much encouraged.

Besides the seasonal variations there are interannual fluctuations which can be very large in certain areas and certain years/seasons. If year-to-year oscillations are modest, they are overshadowed by the observed seasonal variations. It appears that the variability is generally the result of changes in the krill distribution rather than changes in abundance and biomass, although fluctuations in year-classes of krill can also be observed.

South Georgia was identified as an area of high variability with the irregular occurrence of catastrophic events due to unusual conditions, possibly two to three times in a decade. In contrast, the waters around Elephant Island and north of Coronation Island (South Orkneys) were considered as areas with less dramatic changes. The latter two areas are known to be major and persistent krill fishing grounds.

It was shown that the commercial fishery provides a large data source, not only in the Atlantic sector but also in other parts of the Southern Ocean, which can provide information on krill distribution and abundance. However, there is evidence that catch per unit effort (CPUE) estimated from data available to CCAMLR does not provide a realistic index of krill abundance. Further studies are required to relate CPUE to broader-scale distribution of krill.

It was evident that considerable difficulties exist with the comparability of data as a result of different sampling methods. This holds true particularly for CPUE, net catches and acoustic data. There is an urgent need to standardize, as far as possible, equipment and methods in order to allow more accurate comparisons.

Several contributions are concerned with the influence of environmental factors causing variability in the distribution and density of krill accumulations. The best information on this comes from the waters around Elephant Island, where the correlations are such that a prognosis of krill concentrations from environmental data might be feasible. There seems also to be a good relationship, on a coarse spatial scale, between the high abundance of krill to the northwest and north of Elephant Island and the relatively high biomass of fish in this area.

As an explanation for the substantial interannual variation of krill distribution in the Scotia Sea and Bransfield Strait region, an ocean-atmosphere mechanism which initiates such disturbances was suggested. This may be related to changes in the cyclone tracks resulting in southward airflow which forces surface water to the south in the Scotia and northern Weddell Seas and disrupts eddy activity. Testing of this hypothesis will be dependent on long series of synoptic mesoscale observations and on the development of fine-resolution Southern Ocean circulation models. In this connection studies on the mechanisms which seem to regulate the inflow of Bellingshausen Sea water and Weddell Sea water into the Bransfield Strait, or block this inflow at other times, are also encouraged.

There is a need to quantify the sources of krill reaching these areas, including South Georgia. Despite a large amount of biological and physical data, it is still very difficult to understand the mechanisms involved. Furthermore there is still a considerable lack of knowledge on krill population parameters, on the genetic variability of krill as an indicator of geographic differences, on the ability of krill to migrate actively, and on the proportion of krill dispersed and aggregated in swarms.

It was suggested that the sectors of the Southern Ocean should not be treated in isolation, and that also atmospheric and ocean processes outside the Antarctic water circulation system, including the ENSO events, are of considerable influence. It was suggested that a comparison between sectors of the Southern Ocean should be made to test whether any circumpolar decreases or increases in krill abundance can be observed.

Again, the special importance of the pack-ice zone for the survival of krill in winter and the likely effects of the advance and retreat of the ice on the variability of krill were underlined. Further intensive studies on these problems during the winter months are very important.

Biological investigations in the Southern Ocean during the last decade have led to the conclusion that the mixed-layer development in ice-covered and open-water areas, the turbulent mixing processes in the oceanic surface layer (including sea-ice development), and the convection penetrating into the deeper layers of the water column seem significantly to affect the development of phytoplankton and zooplankton, including krill. It was agreed that well-planned interdisciplinary studies in selected places to study these processes and their variability would be of particular value. Related recommendations from the Seminar were further considered at the Fifth Session of the IOC Regional Committee for the Southern Ocean, held immediately after the Seminar. The result was a recommendation that IOC, in consultation with other international organizations, should organize in 1987/88 an expert consultation for the preparation of an interdisciplinary research plan for field and model investigations of the above-mentioned processes. At the same time two other expert consultations were recommended: one for the development of a monitoring program to study the extent of the inflow of water masses in the Bransfield Strait, and their outflow, and to study the influence of topographic features in the Bransfield Strait on the movement and exchange of the water masses with particular reference to the distribution and abundance of krill larval stages. The other expert consultation is expected to develop research plans to study the physical factors contributing to the formation and dispersal of cold-core rings, eddies, and their influence on the distribution of the planktonic components, and to investigate the abiotic factors contributing to the enhancement of the phytoplankton and zooplankton biomass in frontal zones.

4 Krill Variability Detected from Predator Studies

A review of the reproductive performance of seabirds and seals at South Georgia and Signy Island (South Orkneys) also revealed major interannual variations. Around South Georgia almost all species feeding extensively on krill suffered particularly poor reproductive success during the summers 1977/78 and 1983/84. It appears that this was the result of changes in food availability, i.e., scarcity of krill in swarms. At Signy Island observed fluctuations in reproductive performance of krill-eating summer-breeding penguins are not synchronized with those at South Georgia, but can be related to differences in physical environment conditions, especially to unfavorably late dates of ice break-out in late spring around the island. There appears to be no obvious link to changes in food supply.

Another way to detect variations in krill density is to investigate the number of feeding trips to sea made by female Antarctic fur seals during lactation. It was shown that the feeding trip/suckling cycles of females are reflected as fasting/suckling layers in the teeth of their pups. From investigations on fur seals from Bird Island (South Georgia) it appeared that throughout the 1960's foraging patterns fluctuated slightly but were not significantly different. In contrast, the number of feeding trips required by females to find sufficient food (mainly krill) in the 1970's did show significant fluctuations. In 1978 there was a sharp drop in the number of feeding trips, and the duration of these trips increased significantly. This was the same time when research and commercial fishing vessels reported low concentrations of krill offshore South Georgia. Thus, the foraging behavior of Antarctic fur seals represents a useful monitoring parameter that can provide important insights into potential ecosystem changes.

Another contribution outlined some aspects of trophic interactions between krill, squid, and sperm whales. It is suggested that the food chain from krill to sperm whale has several links, including fish and squid. These seems to be a relationship between the distribution of smaller squid which feed on krill and variations in the vertical distribution of krill. Blubber thickness of baleen whales could possibly be used as an indicator of variations in prey (krill) availability, as suggested for minke whales during the Seminar, but there is apparently no seasonal variation of blubber thickness in sperm whales.

In connection with the present development of an ecosystem monitoring program by CCAMLR, the question of the extent to which predator studies can provide useful information on the variability of physical and biological, temporal and spatial components of ecosystem interactions is important. Related aspects have been discussed extensively by the Working Group for the CCAMLR Ecosystem Monitoring Program; details are available in their reports. During the Seminar it was cautioned that the results of studies in localized areas may not be suitable for any extrapolation to larger regions or the whole system. The issue of temporal and spatial scale must be carefully considered when evaluating future research activities related to monitoring. The application of modern technology may open new possibilities also for studies in this field.

Acknowledgments. In preparing this chapter the author used partly the summaries of Seminar discussions, prepared by the rapporteurs of the various sections, and published in the Report of the CCAMLR-IOC Scientific Seminar on Antarctic Ocean Variability and its Influence on Marine Living Resources, Particularly Krill (IOC Workshop Report No. 50). The author wishes to thank Drs. J. Croxall, J. Priddle (BAS, Cambridge), V. Siegel and M. Stein (ISH, Hamburg) for reviewing the manuscript and making useful suggestions for amendments.

Subject Index

MIX
Papier aus verantwortungsvollen Quellen
Paper from responsible sources
FSC® C105338

If you have any concerns about our products,
you can contact us on
ProductSafety@springernature.com

In case Publisher is established outside the EU,
the EU authorized representative is:
Springer Nature Customer Service Center GmbH
Europaplatz 3, 69115 Heidelberg, Germany

Printed by Libri Plureos GmbH
in Hamburg, Germany